SUSTAINABLE BUILDINGS AND INFRASTRUCTURE

The second edition of *Sustainable Buildings and Infrastructure* continues to provide students with an introduction to the principles and practices of sustainability as they apply to the construction sector, including both buildings and infrastructure systems. As a textbook, it is aimed at students taking courses in construction management and the built environment, but it is also designed to be a useful reference for practitioners involved in implementing sustainability in their projects or firms. Case studies, best practices and highlights of cutting edge research are included throughout, making the book both a core reference and a practical guide.

Annie R. Pearce is an Associate Professor in the Building Construction Department of the Myers-Lawson School of Construction at Virginia Tech, Blacksburg, VA, USA.

Yong Han Ahn is an Assistant Professor in the School of Architecture and Architectural Engineering at Hanyang University, ERICA, Ansan-si, South Korea.

SUSTAINABLE BUILDINGS AND INFRASTRUCTURE

Paths to the Future

Second Edition

Annie R. Pearce, Yong Han Ahn and HanmiGlobal

LONDON AND NEW YORK

Second edition published 2018
by Routledge
2 Park Square, Milton Park, Abingdon, Oxon OX14 4RN

and by Routledge
711 Third Avenue, New York, NY 10017

Routledge is an imprint of the Taylor & Francis Group, an informa business

First edition published by Routledge 2012

British Library Cataloguing-in-Publication Data
A catalogue record for this book is available from the British Library

Library of Congress Cataloging-in-Publication Data
Names: Pearce, Annie R., author. | Ahn, Yong Han, author. |
Hanmi Kullobol Chusik Hoesa, author.
Title: Sustainable buildings and infrastructure : paths to the future /
Annie R. Pearce, Yong Han Ahn and HanmiGlobal.
Description: 2nd edition. | New York, NY : Routledge, 2018.
| Includes bibliographical references and index.
Identifiers: LCCN 2017007169 | ISBN 9781138672239
(hardback : alk. paper) | ISBN 9781138672253 (pbk. : alk. paper) |
ISBN 9781315562643 (ebook: alk. paper)
Subjects: LCSH: Sustainable construction.
Classification: LCC th880 .P43 2018 | DDC 720/.47 – dc23
LC record available at https://lccn.loc.gov/2017007169

ISBN: 978-1-138-67223-9 (hbk)
ISBN: 978-1-138-67225-3 (pbk)
ISBN: 978-1-315-56264-3 (ebk)

Typeset in Palatino and Frutiger
by Florence Production Ltd, Stoodleigh, Devon, UK

This book is dedicated to a sustainable future for the built environment and new ways of thinking for the people who create and live in it.

Contents

Case studies and features

Note: cases and features not specifically attributed to anyone were written by Pearce or Ahn

Figures

Tables

Foreword by Jong Hoon Kim

In the 5 years since the first edition of this book was published, we have been contacted numerous times on contents that have evolved over time and suggestions on how it could be improved. We are happy to announce the publication of the second edition with more up-to-date contents, including a number of global case studies. We hope this publication will continue to be useful to those who pursue to accomplish sustainability goals for the construction industry.

HanmiGlobal has established environmental standards for high performance buildings and eco-friendly infrastructures. These environmental standards conform to the standards of sustainability we have established. It is our goal to create future construction management practices in our projects that will employ project/sustainable techniques through continuous research and development of new technologies.

Our effort is not only a passive approach to reducing the corporate carbon footprint, but also to proactively participate in the global sustainability business. In moving forward with these issues, our company is interested in preparing a sustainability business which includes: sustainable design management, sustainability knowledge support and sustainability due diligence. We will also undertake research and consulting on Green Buildings and new renewable energy resources.

In order to initiate our vision of sustainable management practices, we have enacted sustainability policies with the joint application of ISO 14001, LEED (Leadership in Energy and Environmental Design), BEMS (Building Energy Management System) and other global environmental policies. By providing a total solution concept using optimized sustainable guidelines, HanmiGlobal plans to develop into a sustainable PM/CM consultancy, from the design process to construction. However, is it possible to provide sustainable construction using only pre-established environmental standards and systems?

It is more important to establish a workable tool for construction professionals to understand both the administrative and the technical requirements of a sustainable construction project. This book draws attention to the current way of thinking and offers practical approaches to sustainability. In addition, the book will call for contributions concerning sustainability concepts, theories and case studies to combine better decision making with practical relevance to sustainable developments.

We are pleased to collaborate with Professor Annie Pearce at Virginia Tech and Professor Yong Han Ahn at Hanyang University in order to conduct research about sustainability opportunities throughout the construction lifecycle to establish the best practices in industry standards. We are very satisfied with the outcome of this research and look forward in seeing the results implemented as a guideline for sustainable project management.

One of the most critical issues in the current construction industry concerns not only green building issues and high efficiency structures, but also the need to form an ecological balance with the environment. We hope this publication provides readers with an overall understanding of sustainable building and practical business cases that will encourage the government, construction experts and students who are interested in sustainability objectives to present approaches to future projects.

Jong Hoon Kim, CEO & Chairman of
HanmiGlobal Co., Ltd.

Foreword by Norbert Lechner

In July 2016 the temperatures in both Iraq and Kuwait reached 54 degrees Celsius (129 degrees Fahrenheit) with people dying from the heat in many countries. Worldwide, 2014 was the hottest year until 2015 was significantly hotter and 2016 was hotter still. In 2013, CO_2 levels surpassed 400 ppm which was for a long time considered the upper limit that the world should not exceed. Climate change is not a future problem – it is a problem now. Consequently, the major subset of sustainability issues is the set of issues restricting the growth of CO_2 levels in the atmosphere. It is time to act at every level and in every sector of the economy to severely restrict the growth of atmospheric CO_2.

The building sector is especially important in fighting climate change, because it is a major consumer of both materials and energy, with buildings consuming about 50 per cent of the energy in the United States, thereby causing about 50 per cent of climate change. The process of getting raw materials from the ground into the fabric of a building being constructed requires huge amounts of energy that then becomes embodied in the finished building. And the endless depletion of our material and energy resources is certainly also not sustainable. Both the operational and embodied energy required by buildings comes mainly from the burning of fossil fuels which is the main source of the CO_2 in the atmosphere. There is no sector of the economy that is causing climate change as much as the construction sector.

In the process of creating a building, the early decisions have the greatest impact. Although the actual construction comes late in the process, constructors have critically important knowledge at all points in the design process. Before the first line is drawn, the architect needs the knowledge of both the engineers and constructors in order to design a sustainable building. Thus, from the very first step in the building process, it is necessary to use the team approach in order to create high-performance sustainable buildings. Low-performing buildings are not acceptable because they will not help create a sustainable future.

This very comprehensive and scholarly book presents the knowledge needed by the constructor to help create sustainable buildings. The numerous case studies anchor this book in reality. The book not only presents the history and drivers of sustainable construction, but also presents the required policies and programs that promote sustainable construction. The book gives an in-depth description of the available green rating systems as well as the opportunities in project delivery and pre-design. It also provides information on sustainable design strategies to help the constructor be a more effective member of the design team.

The many opportunities for saving energy and materials during the actual construction process are covered in the book. Since constructors are in an excellent position to help in post-occupancy activities and strategies, the book also explains how they can help to maximize energy savings for the life of the building.

Fortunately, besides controlling climate change, there are many immediate benefits to sustainable design. The book makes it clear that many sustainable practices are good for business. The last chapter lays out the trends for the future of sustainable design and construction. Since the world's population and affluence will both keep increasing, there will be a continuing need for not only more buildings but ever more complicated sustainable buildings. Building professionals well informed about sustainable design will be in great demand, and this book is a great resource for creating sustainable constructors.

Norbert Lechner, Professor Emeritus,
Auburn University

Foreword by Jorge A. Vanegas

Over the last three decades, the call in 1987 from the United Nation's World Commission on Environment and Development for an approach to development that '. . . *meets the needs of the present without compromising the ability of future generations to meet their own needs'*, has been discussed among, and debated by, thought leaders across all sectors of society, industry, business, government and academia. From attempting to ensure that development explicitly addresses the Triple Bottom Line of social, ecological and economic performance, to promoting the development of environmentally friendly products, goods, processes, services, business models, and lifestyles, the quest for sustainability is no longer a fringe activity, and instead has become a global quest that has galvanized a wide and diverse range of constituencies to action. However, three challenges still remain.

The first challenge is the need for explicit collaboration and alignment among this range of constituencies in this global quest for sustainability: (1) from private and public sector organizations ranging from federal, state and local government agencies, corporations, the military services, non-governmental organizations (NGOs), and civic activist groups; (2) through practitioners, researchers and educators in the physical, life and social sciences, in the transportation, energy, healthcare, education, financial, agricultural, manufacturing and other industries, and in planning, architecture, the many disciplines of engineering and construction, among so many more; and (3) to individuals, families and communities, in urban, suburban and rural settings. Bringing together such a diverse range of constituencies requires identifying common imminent threats, common compelling opportunities, and common values and beliefs, all from the perspective of the full scope and context of sustainability.

The second challenge is the need to clearly understand that the intellectual foundation of sustainability cuts across existing disciplines and cultural practices, each tending to have its own specialized vocabulary, disciplinary composition, and selective focus. In addition, within the current context of existing and extremely large data sets, which are associated with the multiple dimensions of sustainability, some of the data on patterns, trends, and associations, especially those relating to human behavior and interactions, are aligned with each other, and some are in direct conflict with each other. For example, there are sustainability principles, concepts, heuristics, strategies, guidelines, specifications, standards, best practices, lessons learned, processes, and tools. Knowing how to access and select what is relevant and applicable for a given constituency from this extensive body of knowledge and experience is not an easy task.

Finally, the third challenge is establishing what sustainability is from a specific Architecture, Engineering and Construction (AEC) industry point of view, and more specifically, establishing what Built Environment Sustainability (BES) is, particularly given that the AEC industry is both a provider and a custodian of industrial, residential and non-residential facilities, as well as the civil infrastructure systems that provide the lifelines of any society including water, energy, transportation, communication, solid waste, sewage, among others. As documented extensively over the last few years, what the AEC industry does (its projects), how the industry does it (the processes it follows in doing them), with what the industry does what it does and how (the resource base it uses in their delivery and use), and where and when (their specific contextual envelope), results in a significant yearly volume of economic activity, and also, in major direct and indirect impacts, such as (1) natural resource consumption, degradation and depletion, (2) excessive energy and water consumption, (3) waste generation and accumulation, and (4) environmental impact and degradation. First, and foremost, BES is about *People*; it is about continuously enabling, maintaining and enhancing the quality of life of people within families, communities, organizations and society. Second, to achieve this goal, BES requires continuously enabling, maintaining and enhancing the quality and the integrity of four other elements: (1) the performance of the *Built Environment*, i.e. facilities and civil infrastructure systems; (2) the performance of the *Industrial Base* of production systems for products, goods and services; (3) the health of the *Natural Environment*, i.e. air, water, soil and biota; and (4) the availability of the *Resource Base*, i.e. social, built, industrial, natural and economic capital. In addition, BES requires managing effectively the influences that affect it, which stem from *Social, Cultural, Political, and Regulatory Systems*, from *Economic and Financial Systems*, and from *Environmental and Ecological Systems*. These systems define the contextual envelope, within which people, the built environment, the industrial base, the natural environment and the resource base co-exist, and furthermore, they also establish the nature and the complexity of the interrelationships and interdependencies among them. Finally, BES requires framing these elements and their contextual envelope within a *spatial scale* (e.g. from a site, local, state and regional footprints, to a national and global footprints), and a *temporal scale*, from today (the present) to tomorrow (the future).

This ambitious, comprehensive and timely textbook provides an excellent response to these challenges, and offers the AEC industry both a solid reference book and a practical guide for transitioning towards BES. With the drivers and history of sustainability within the built environment from a global perspective as a point of departure, followed by a general overview of sustainability policies, programs and systems for rating green buildings, the authors have done an outstanding job in identifying specific opportunities and best practices for embedding sustainability across the main stages of execution of building and infrastructure capital projects: from project planning and pre-design, through design and construction; to post-occupancy. They conclude by making the business case for BES, and by identifying future trends for sustainable design and construction. One of the strengths of the book is the presentation and discussion of a wide range of relevant case studies that illustrate the specific content of each chapter. While the book is primarily targeted at students of construction management and the built environment, the wealth, breadth and depth of information provided and topics covered should be of great value and interest for anyone interested in the practical implementation of BES.

This book is an excellent reminder that the journey to BES may be long and complex, and full of inhibitors, obstacles and barriers; but by improving the sustainability of facilities and civil infrastructure systems, the destination of a potentially sustainable AEC industry can be reached: whether one decision, one choice, or one action at a time; one paradigm, one product or one process at a time; one capital program or project at a time; one AEC enterprise or industry sector at a time; all as part of a gradual shift to a sustainable future.

Dr. Jorge A. Vanegas
Dean, College of Architecture, and
Professor, Department of Architecture,
Texas A&M University,

and

Research Professor, Texas Engineering Experiment Station
Texas A&M University System
College Station, Texas

Preface to the second edition

Imagine this: you are given a very important task to complete, one that is too complex to do alone. You must spend millions of dollars to get it done, work closely with people you have never met before, and finish the work with limited time and resources. Every decision will be monitored closely, and you are likely to have less money to complete the project than you need. Now imagine that everyone on your team speaks a different language, has different and sometimes conflicting interests, and uses different tools and techniques to do their part. Imagine further that you may not even be allowed to talk to each other some of the time. Your solution has to remain valid and useful for 30 years, 50 years or even 100 years or more, although the environment in which it is used and the uses to which it will be put will change dramatically over that time in ways you cannot imagine. Imagine that if you fail to complete your task as assigned, very bad things could happen – people could become ill, be injured or even die. If you make a mistake, the penalties could be staggering.

Feel the pressure? This is the world of modern construction, in all its technical and legal complexity. Now add a whole range of additional challenges: new codes, standards and environmental requirements, a very competitive market with extremely low profit margins, well-informed clients who have new ideas that they aggressively want to implement on their projects, and thousands of components with different attributes and sources in the marketplace. Imagine further that your clients want you to keep track of each of those components, where they came from, and the problems that were created in their manufacture or will be created during and after their service life. You also have to understand how they should work together, connect them all in the right ways, and verify that those connections were made correctly so that the project meets your predictions for performance. Finally, put this all in the context of a world where our actions as a species appear to be leading to runaway changes on a global scale. We don't know what those changes will be, exactly, but we know they will affect the kinds of resources available in the future, the ecological and socio-political conditions in which our buildings must perform, and the expectations of the people who use and rely upon our creations during their lifecycle.

We wrote the first edition of this book because we recognize the challenges facing the people who create the built environment, and these challenges are even more relevant and urgent today. As educators and practitioners, we are acutely aware of the complexity inherent in the construction industry, both technical and otherwise. Even as new technologies emerge to improve the performance of constructed systems, there is evidence to suggest that humans do not use existing technologies as effectively as we could. We have dedicated our lives to try to make things better, and we believe this book will play a role in achieving that outcome.

We want the next generation to inherit a world that has fewer, not more, problems than our generation has had to face. Many of those problems have been caused in large part by the built environment and the legacies of how it has been constructed over time. We see enormous challenges in figuring out how to fix those problems, but we also see incredible potential for change across a range of opportunities. Solving the problems that result from our industry will take the contributions of many people working together, but we believe it can be done.

This book is a tool for those who seek to change the architecture, engineering, and construction (AEC) industry. It was designed to offer a new way to think about how we create the built environment, along with a sampling of promising technologies and practices, and a look ahead to what the future may hold. The first two chapters provide an introduction to why sustainability is important for stakeholders of the built environment and an overview of the theory that underlies our efforts to make the built environment more sustainable. These chapters lay the foundation for understanding both the human and organizational dimensions of the problem, discussed in Chapter 3, and the technological and process dimensions over the building lifecycle, discussed in Chapters 4 to 8. In Chapter 9, we use a detailed case study to illuminate how these dimensions come together in a real project, and we show how decisions can be analysed according to the economic criteria that shape how most organizations operate. Finally, in Chapter 10, we offer an updated projection of what the future holds for our industry. Each chapter also has a section on upcoming trends, questions for reflection and discussion, and an updated and expanded list of resources and references for more information. The second edition builds significantly on the content in the first edition, including many new case studies of different types of projects, new features from cutting edge researchers in the field, and updated information on the constantly evolving population of policies and rating systems in the field. We hope these resources will help our readers consider what is to come as they develop a strategy for sustainability for their own organizations.

We welcome your comments about what this book does well, and how it could be better in future editions. We hope you find it useful both for learning more about built environment sustainability and as a tool for day-to-day practice. Please contact us with your comments and good ideas, and we wish you well in your quest for sustainability.

Dr. Annie Pearce, urbangenesis@mac.com
Dr. Yong Han Ahn, yonghan77@gmail.com

Acknowledgements

We would like to thank the many people who contributed to this book. Particular thanks go to the following:

HanmiGlobal Co., Ltd. for support to write this book

Soojung Kim for producing figures and images

Youoh Choi for collecting case data and translating portions of the book into Korean

Hyuksoo Kwon, Jeehee Lee and Kyungwon Lee for translating the book into Korean

C. Jacob Brown for providing photographs of the City Center project

Rossana Merida for providing data and photographs of the Lodging Complex in Panama City, Panama

Ron Rademacher and Jason Wirick of the Phipps Conservatory and Botanical Gardens for hosting multiple tours and providing data on the Center for Sustainable Landscapes at Phipps

Susan A. Frieson for providing Bardessono Hotel project data

Eric Stackhouse of the Antigonish Public Library and Archibald and Frasier Architects for sharing their information about the Antigonish Public Library

Brent Zern of Emory University for providing insights and photographs of the WaterHub at Emory

Mark File for sharing photos of the Proximity Hotel

Lee Cravey of USAF Air Combat Command and Kenneth Hahn of Kenneth Hahn Architects for providing background on the Air Force Weather Agency Headquarters Building

POSCO A&C and Dr. Min Jae Suh for their support in developing information featured in Chapter 7 on green construction methods

Roberto Meza for information and photographs on the Bosques de Escazú project in Costa Rica

Staff from Aldridge Electric for providing information and photographs on the Chicago Red Line Retrofit Project

Stacey-Ann Hosang for providing Empire State Building project information and photos

Bill Cadle of USAF Reserve Command HQ for providing information and background about the Homestead ARB Fire Station

Baewon Koh for providing data for the Reedy Fork Elementary project

Golnar Ahmadi for providing photos of the Sharifi-Ha House

Dr. John Mogge of CH2MHill for providing information and insight on multiple sustainable infrastructure projects including Masdar City, the London Olympics and Freedom Park

Dr. Rodolfo Valdes-Vasquez, Dr. Leidy Klotz, Dr. Tripp Shealy, Dr. Freddy Paige, Dr. Sheila Bosch, Dr. Craig Shillaber, Dr. Ken Sands, Dr. William Rhoads, Mr. Ben Chambers, Dr. Min Jae Suh, Dr. Tolga Ozbakan and Dr. Sandeep Langar for contributing features on their cutting edge sustainability research.

We also thank our universities and academic programs – the Department of Building Construction and Myers-Lawson School of Construction at Virginia Tech, and the Architectural Engineering Department at Hanyang University ERICA – for their inspiration and support. Thanks also goes to the Virginia Tech International Faculty Development Program under the leadership of Dr. John Dooley and the Virginia Tech College of Architecture and Urban Studies under the leadership of Dean Jack Davis for the opportunity to travel internationally to develop case studies for this book.

We would like to extend our gratitude to the following organizations for providing project data and images:

Wells Fargo Bank, N.A.
tvsdesign
Empire State Building
Innovative Design, Inc.
Cook + Fox Architects
WATG
Cello & Maudru Construction Co, Inc.
Bardessono Hotel
Trees Atlanta
Smith Dalia Architects, LLC
Eberly & Associates, Inc.
US Air Force Reserve Command
US Air Force Air Combat Command
CH2MHill
Proximity Hotel
Olive Architecture
Weaver Cooke Construction LLC
Superior Mechanical, Inc.
Hyundai Engineering and Construction Co, Ltd.

POSCO Engineering & Construction Co, Ltd.
POSCO Architects & Consultants Co, Ltd.
HanmiGlobal Co., Ltd.
Pelli Clarke Pelli Architects
Swire Properties, Ltd.
Kenneth Hahn Architects, Inc.
Archibald and Frasier Architects, Inc.
United Nations Environment Programme (UNEP)
Collier County Growth Management Division, Collier, FL
Aldridge Electric
Emory University

In addition, many thanks go to the reviewers, who provided guidance and suggestions throughout the development of this book:

Professor Norbert M. Lechner
    Auburn University
    Alabama, USA
Min Jae Suh
    Sam Houston State University
    Texas, USA
Alice Aldous
    Earthscan/Taylor & Francis
    Oxford, UK
Chairman Jong Hoon Kim
    HanmiGlobal Co, Ltd.
    Seoul, South Korea
Dr Sang Hyuk Park
    HanmiGlobal Co, Ltd.
    Seoul, South Korea

This book was based in part upon work supported by the U.S. National Science Foundation (NSF) under Grants No. 0828779, 1336650, and 0935102, and by an exploratory grant from the National Institute of Occupational Safety and Health. Any opinions, findings and conclusions or recommendations expressed in this work represent those of the authors and do not necessarily reflect the views of the NSF or the National Institute of Occupational Safety and Health (NIOSH).

This book was based in part upon work supported by the Basic Science Research Program of the National Research Foundation of Korea (NRF) under Grant No. 2015R1A5A1037548. Any opinions, findings and conclusions or recommendations expressed in this work represent those of the authors and do not necessarily reflect the views of the NRF.

Finally, we graciously appreciate our family including Susan Pearce, Larry Pearce, Hee Jung Kim, Jinna Ahn and Jae Hyuk Ahn. Without their love and strong support, we would not have completed this book.

Annie R. Pearce and Yong Han Ahn

1

Drivers and definitions of sustainability in the built environment

The concept of sustainability has gained popular momentum over the last 30 years. The goals of sustainability are to enable all people to meet their basic needs and improve their quality of life, while ensuring that the natural systems, resources and diversity upon which they depend are maintained and enhanced, for both their benefit and that of future generations. The construction industry is beginning to adopt the concept of sustainability in all construction activities and has significant opportunity to mitigate environmental problems associated with construction activities while contributing to a high quality of life for its clients. This chapter presents drivers and definitions of sustainability in the built environment by describing how construction activities affect the natural environment, the economy and society.

Construction and its impacts

Construction is one of the most significant industries in the world. It provides critical civil infrastructure including bridges, roads, rail, water and wastewater treatment, plants for the production and transmission of energy, and facility assets such as office buildings and the houses in which we live, work and play (Russell et al. 2007). The economic activities of the global construction industry are worth around $8.5 trillion U.S. dollars annually, or 13 per cent of gross domestic product (GDP) worldwide (CIC 2015).

In the United States, the construction industry is a major player in the nation's economy, contributing over $1.1 trillion including $851 billion of private construction and $282.5 billion in the public sector as of June 2016 (US Census 2016). In addition, the construction industry accounts for around 4 per cent of the US national GDP (BEA 2015). The U.S. construction industry employs about 6.5 million people annually. Within the U.S. industry, approximately 1.4 million people worked for building construction contractors, over 950,000 worked in heavy and civil engineering construction or highway construction and nearly 4.2 million worked for speciality trade contractors (DOL 2016).

Due to their cumulative magnitude, construction activities have a major impact on physical development, government policies, community activities and welfare programmes. Construction projects can improve social welfare, well-being and quality of life. Construction activities including the construction, operation, maintenance and demolition of built facilities are also connected with the broader problems and issues affecting the environment, including climate change, ozone depletion, soil erosion, desertification, deforestation, eutrophication, acidification, loss of bio diversity, land pollution, water pollution, air pollution, depletion of fisheries and consumption of valuable resources such as fossil fuels, minerals and aggregates. In addition to these ecological and resource impacts, built facilities also significantly impact human health, comfort and productivity.

Energy use

A great amount of energy is consumed by construction activities, mainly the operation of buildings. The building sector, including both residential and commercial buildings, accounts for 20.1 per cent of total delivered energy consumed worldwide (EIA 2016a). In the United States, about 43 per cent of all energy is consumed by buildings (EIA 2016b). Trends in building energy use vary greatly among developed vs. developing countries as a function of economic growth, and are influenced not only by the growth of populations needing buildings, but also increasing expectations of building functionality to achieve higher standards of living (EIA 2016a). In the residential sector, for instance, building energy consumption is affected by factors including income levels, energy prices and policies, available energy sources, building and household characteristics, location and weather, and others (ibid.). In the commercial sector, energy use is affected by economic and population growth trends and the kinds of economic activity driving that growth. It is also affected by factors including climate, availability of resources and technology, and efficiency of buildings and equipment (ibid).

Heating, cooling and lighting in buildings are a major share of energy consumption in the United States (Figure 1.1), and the share of energy used for appliances and electronics has grown substantially over time (EIA 2016c). By 2025, electricity is expected to surpass natural gas as the largest source of energy in residential buildings in developed countries due to increased demand for household electronics, with even greater demand growth for electricity in developing countries (EIA 2016a).

Air quality and atmosphere

Around 18,000 people die each day worldwide as a result of air pollution, more than deaths from HIV/AIDS, tuberculosis and road injuries combined (IEA 2016). Concurrently, climate change due to greenhouse gas emissions poses significant short- and long-term threats to human health and prosperity. Energy production and use is the primary cause of air pollution as well as greenhouse gas emissions, and so efforts that can

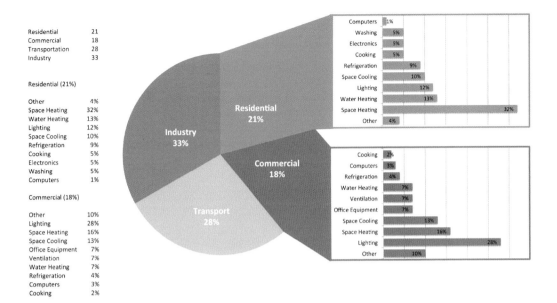

Residential	21
Commercial	18
Transportation	28
Industry	33

Residential (21%)

Other	4%
Space Heating	32%
Water Heating	13%
Lighting	12%
Space Cooling	10%
Refrigeration	9%
Cooking	5%
Electronics	5%
Washing	5%
Computers	1%

Commercial (18%)

Other	10%
Lighting	28%
Space Heating	16%
Space Cooling	13%
Office Equipment	7%
Ventilation	7%
Water Heating	7%
Refrigeration	4%
Computers	3%
Cooking	2%

improve energy performance have great relevance for improving air quality.

It has been estimated that as much as one-third of global greenhouse gas emissions can be attributed to buildings (UNEP 2009). In the United States, buildings contribute approximately 34 per cent of the nation's total carbon dioxide (CO_2) emissions, including emissions from fossil fuel combustion and electricity production, refrigerant leaks, and emissions from waste decomposition in landfills and wastewater treatment plants (EPA 2016a). Drinking water and wastewater systems account for 3 to 4 per cent of total energy use in the United States, making water efficiency a significant opportunity to reduce greenhouse gas emissions as well (ibid.). Worldwide, energy use in buildings is responsible for 55 per cent of all fine particulate matter emissions from human activity, as well as 5 per cent of all nitrogen oxide emissions and 7 per cent of sulphur dioxide emissions (IEA 2016). Overall, the main source of air pollutants from buildings is energy use, consumed during:

- Harvesting and manufacturing of building materials ('embedded' or 'embodied' energy).
- Transport of these materials from production plants to building sites ('grey' energy).
- Construction of the building ('induced' energy).
- Operation of the building ('operational' energy).
- Demolition of the building and recovery or disposal of its parts.

Scientific consensus has established the link between human activity and climate change, including rising sea levels, increased occurrence of severe weather events and resulting critical resource shortages (IPCC 2014). The construction sector has significant potential to help reduce

Figure 1.1
US energy consumption by end-use (LBL 2009)

greenhouse gas emissions as a means of addressing these problems, not only through improvements in energy and water use but also by reducing other non-CO_2 greenhouse emissions such as halocarbons, chlorofluorocarbons (CFCs) and hydrochlorofluorocarbons (HCFCs) used for cooling, refrigeration, fire suppression and insulation materials.

The construction sector also has the potential to influence climate change through changes to land use and development patterns. According to U.S. EPA, ecological land use and forestry reduced gross annual greenhouse emissions in the United States by acting as a carbon sink for 11 per cent of total emissions in 2014 (EPA 2016a). By choosing more dense types of developments, incorporating vegetative and living systems as part of the built environment, and preserving or restoring natural ecosystems on development sites, the construction industry can directly offset emission of greenhouse gasses from other activities.

Development choices also have significant influence on the micro-climate in and around the built environment (see *Case study: CityCenter, Las Vegas*). For example, the annual mean air temperature of a city with 1 million people or more can be 1–3°C (1.8–5.4°F) warmer than its surroundings. In the evening, the temperature difference can be as high as 12°C (22°F) (EPA 2016b). This phenomenon, known as the urban heat island effect, can also increase summertime peak energy demand, air conditioning costs, air pollution and greenhouse gas emissions, heat-related illness and mortality, and water quality.

Water use

Clean water is critical for human life. In many contexts, water has become a limiting factor for future development and prosperity. Developed countries such as the United States struggle with aging infrastructure and reduced public investment in infrastructure repair and upgrades and consequent problems that arise in providing safe potable water (see *Case study: the 2016 Flint Water Crisis*). At the same time, developing nations face their own water shortages. The World Health Organization estimates that by 2025, half of the world's population will be linving in water-stressed areas (WHO 2016a).

Case study: the 2016 Flint Water Crisis

A series of decisions made by public officials in Flint, Michigan in the United States left over 8,000 children in the city with neurological damage and other health problems after city officials made a decision to change the city's water source without also adjusting its treatment process to account for changes in water chemistry. The city of Flint has experienced decades of economic decline with the decline in the U.S. automotive and steel industries. Beginning in 2011, Flint officials found themselves in a serious financial crisis and were faced with the challenge of continuing to provide essential infrastructure services with diminishing resources due to the city's declining tax base.

continued

The decision to change water sources was not necessarily a problem in and of itself. Rather than continuing to procure its water supply from the neighbouring city of Detroit, in April 2014 officials decided to change its water source to the nearby Flint River and provide its own treatment in an effort to save costs. However, while offering considerable savings in operational costs to Flint, the water provided from the Flint River was significantly more corrosive than the Detroit water supply, leading to erosion of joints and fittings in water supply piping and releasing poisonous lead into the city's drinking water supply. Many critical components of the city's water distribution system had been installed between 1901 and 1920, during which period it was common to use lead piping to connect cast iron water mains to buildings due to low cost and workability.

Before the switch, water treatment for the Detroit water supply was sufficient to prevent leaching of lead from these pipes at dangerous levels. However, more acidic water accelerates the leaching of lead from lead pipes, solder and brass fittings, releasing lead into the water supply where it can be absorbed through consuming the water. Relatively inexpensive corrosion-inhibiting chemicals could have been added to the water to prevent these problems. However, city officials decided not to add the inhibitor chemicals to the water in hopes of saving additional costs. Both city and state officials were aware of the risks this decision posed to water quality, but did not properly disclose the risk and in some cases even falsified test data to suggest that the water quality in the city was safe.

In addition to the health impacts and decreased quality of life caused to Flint citizens by these decisions, Flint is now faced with the challenge of even worse deterioration in its water distribution system due to accelerated corrosion. Aging pipes and fittings that already cost too much to replace under the city's limited budget now will have to be replaced even sooner, posing an even greater burden on city decision makers. Estimates of the cost to repair or replace water infrastructure in Flint range from millions of U.S. dollars up to $1.5 billion, and loans and funding have been allocated at the local, state and federal levels to begin fixing the problem.

The Flint water crisis is even more tragic due to its disproportionate effects on the city's poorest citizens. During the first year after the switch to Flint River water, a medical study revealed that the number of children with elevated blood-lead levels doubled on average, and increased by more than 400 per cent in certain hot spot levels. Increases in deaths from Legionnaires' Disease have also been correlated with the switch, leading some to conclude that contaminated water in Flint may also be responsible for the growth of opportunistic pathogens in the city's water supply. However, the long-term costs to citizens suffering from lead poisoning are difficult to fathom. Childhood exposure to lead is known to cause central nervous system problems, which affect intellectual function and cause other problems such as attention deficit disorder that reduce learning ability. Children with elevated blood-lead levels have also been found to have a higher likelihood as adults to commit crimes, be imprisoned, be unemployed or under-employed or depend on government services for their livelihood. Although the city ultimately switched its water supply back to Detroit in October 2015, the long-term social and economic impacts of the Flint crisis are only just beginning.

Additional information about the Flint Water crisis is available at http://flintwaterstudy.org and https://en.wikipedia.org/wiki/Flint_water_crisis. A Journalist's Resource on *Lead poisoning: Sources of exposure, health effects, and policy implications* is also available online at http://journalistsresource. org/studies/society/public-health/lead-poisoning-exposure-health-policy.

Case study: CityCenter, Las Vegas

The changing face of green building

Completed in 2009, the CityCenter project was developed on 67 acres of the famous Las Vegas Strip (Figure 1.2). Employing 12,000 people and providing over 16,797,000 sq ft (1,560,500 m²) of developed space, CityCenter includes:

- ARIA, a 61-storey, 4,004 room gaming resort
- luxury hotels including Mandarin Oriental and Vdara Hotel & Spa
- 2,400 residential units, including Veer Towers
- Crystals, a 500,000 sq ft retail and entertainment district.

Figure 1.2 CityCenter under construction
Source: Courtesy of C. Jacob Brown

Six individual projects in the CityCenter project have been awarded Leadership Energy and Environmental Design (LEED) Gold certification, which set a new bar for green development in Las Vegas. Las Vegas is located in the hot, arid, drought-stricken desert climate of Nevada in the United States (Salvaggio and Futrell 2013). Marketed as an 'oasis in the desert', the Las Vegas Strip and its many resort and casino developments have treated water features as a status symbol, with energy and water needs that place significant demands on their ecological context (The Project on Vegas 2015). One developer has even recently proposed creating Paradise Park, a new major water park on the site of the Wynn Resorts in the city. The developer has said that this project, which will operate on solar power and include a 38-acre man-made lake, is 'very green' because it will use less water than a golf course (Ho 2016).

continued

CityCenter was such a large project that it was able to achieve economies of scale in purchasing and procurement that influenced multiple industry segments. This market transformation has led to greater opportunities for other developers in Las Vegas to follow in CityCenter's footsteps and develop more sustainably. Multiple innovative tactics were employed in the project, including:

- Waste heat from the on-site natural gas cogeneration plant is used to provide all domestic hot water throughout the project.
- Slot machine bases were designed to function as displacement ventilation units, cooling casino spaces more efficiently and increasing the comfort of guests using those machines.
- A large-scale waste reuse/recycling program during site preparation and construction resulted in recovery or reuse of more than 80 per cent of demolition waste from the Boardwalk Hotel that previously occupied the site, including bathroom fixtures shipped to other countries for reuse after being wrapped in Boardwalk curtains and carpets as recycled packing materials.
- Reclaimed water from a neighbouring casino along with crushed blocks from demolition were used for dust control and abatement during construction, saving an estimated 2.4 million gallons of water during construction.
- All concrete for the project was produced using an on-site batch plant, avoiding both waste and transportation impacts.
- Custom-designed water fixtures throughout the development use a third less water than conventional fixtures while delivering satisfying shower experiences to guests and residents.
- Building exteriors are designed to reduce heat gain in this extremely hot environment, including reflective roofs, high performance windows, spectrally selective coatings and light shelves or 'air-brows' (Figure 1.3).

A variety of other operational features, including alternative fuel stretch limos, a location good for walkability and transit use, green cleaning practices, and farm-to-table food also improve the guest experience while reducing environmental impacts.

Project team:

- **Developers**: MGM Resorts International and Infinity World Development Corp. (Joint venture)
- **Design Team**: Pelli Clarke Pelli, Kohn Pedersen Fox, Helmut Jahn, RV Architecture LLC, Foster + Partners, Studio Daniel Libeskind, David Rockwell and Rockwell Group, and Gensler
- **Construction Team**: Perini Building Company

Figure 1.3 'Air-brows' are incorporated on the exterior of tower buildings to deflect solar radiation from the building's interior
Source: Courtesy of C. Jacob Brown

continued

Despite its sustainable design and construction goals, the CityCenter project became a major turning point for public opinion about construction safety in Las Vegas and on green projects. Six workers were killed during the construction of CityCenter, a much higher death rate than experienced even 20 years earlier during the 1990s' building boom in the city. Safety problems were attributed in part to a rush to complete the project before New Year's Eve in 2009, to take advantage of one of the busiest gambling days of the year.

Subsequent investigation revealed a pattern of dangerous safety problems on construction sites throughout the city. Identified problems included inadequate training, crowded sites, faulty equipment, job speed-ups and worker fatigue from excessive overtime (Berzon 2008). Workers on the CityCenter project went on strike to protest unsafe working conditions in June of 2008. The strike was quickly ended after an agreement was reached to improve safety conditions, but many believed that insufficient actions were taken. Although the project was ultimately completed without additional loss of life, the complexity of local relationships between labour unions, safety regulators, contractors and casino companies provided incentives to preserve the status quo rather than aggressively provide needed training and other safety interventions (*Las Vegas Sun* 2016).

The CityCenter project has had a significant impact on social sustainability issues considered in green certification systems like LEED. Some argue that no project with significant injuries or fatalities ought to be eligible for certification. Others question the sustainability of resource-intensive developments in locations where development places extreme loads on the natural environment. Green building rating systems must address these issues in deciding what is green as part of their role in influencing market transformation. Driven in part by experiences on the CityCenter project, the U.S. National Institutes of Occupational Safety and Health worked with the U.S. Green Building Council to develop a new LEED Pilot Credit (Prevention through Design) focused on identifying and mitigating hazards early in the planning and design process. Released in 2015, this credit is part of the LEED v.4 rating system and is being used by projects in the United States and other countries.

More information on City Center's sustainability initiatives is available at http://citycenter.com/. Information on the LEED Prevention through Design Pilot Credit is available at http://usgbc.org/credits/preventionthroughdesign.

References

Berzon, A. (2008). "Workers walk off CityCenter site in protest." *Las Vegas Sun*, (June 3, 2008). http://lasvegassun.com/news/2008/jun/03/workers-walk-citycenter-site-protest/.

Ho, S. (2016). "Casino magnate Wynn envisions water paradise in dry Nevada." *Associated Press*, (May 25, 2016). http://bigstory.ap.org/journalist/sally-ho.

Las Vegas Sun. (2016). "Construction Deaths." Archive of articles related to construction safety in Las Vegas, *Las Vegas Sun*. http://lasvegassun.com/news/topics/construction-deaths/.

Michaels, J. (2010). "Las Vegas' CityCenter one of the world's largest green developments." *Press Release* (June 18, 2010). http://citycenter.com/press_pdf/CityCenter%20Sustainability%20Release.pdf.

Salvaggio, M., and Futrell, R. (2013). "Environment and sustainability in Nevada," *The Social Health of Nevada: Leading Indicators and Quality of Life in the Silver State*, D.N. Shalin, ed. UNLV Center for Democratic Culture, Las Vegas, NV. http://cdclv.unlv.edu/healthnv_2012/environment.pdf.

The Project on Vegas. (2015). *Strip Cultures: Finding America in Las Vegas*. Duke University Press Books, Durham, NC.

Figure 1.4
Sources of drinking water

Sources of
Drinking Water

Buildings are one of the largest water consumers. Buildings in the United States consume about 10 per cent of total water withdrawals from ground and surface sources, or over 39 billion gallons per day (DOE 2012). This water comes from a variety of sources (Figure 1.4) and is generally treated to drinking water standards with the use of significant energy resources expended for treatment and conveyance, even though such a high level of water quality is not required for most building uses. Much of the water used in the developed world comes from aquifers, which are typically harvested at a much higher rate than they are re-charged naturally. Even surface water sources may become increasingly unreliable in the future, with growing demand and significant changes in patterns of precipitation due to climate change. Improving the effi-ciency with which buildings use water to meet our needs reduces the demands placed by water extraction and treatment on the natural environment and maintains water capacity for future generations.

Indoor environmental impacts

The indoor environment is very important for human health, comfort and productivity. However, pollutant levels indoors are often higher than those outside, typically 2.5 times and occasionally more than 100 times higher than outdoor levels. These high levels of indoor air pollutants are of particular concern because it is estimated that most people spend as much as 90 percent of their time indoors (EPA 2010).

Health problems resulting from indoor air pollution have become one of the most acute environmental problems related to the built environ-ment. High levels of pollutants are generated by building materials and components including furnishings, carpets, finishes, paints and backing materials; household cleaning, maintenance, personal care or hobby products; central heating and cooling systems and humidification devices; and outdoor sources such as radon, pesticides and outdoor air pollution (Figure 1.5). These pollutants can lead to a variety of health problems, such as irritation of the eyes, nose and throat, headaches, dizziness and even long-term irreversible health impacts. They are also thought to contribute to thousands of cancer deaths and hundreds of thousands of respiratory health problems each year.

For example, an estimated 235 million people worldwide suffer from asthma, which is strongly correlated with building-related indoor air pollutants. Urbanisation has also been associated with increase in asthma, although the specific relationship has not been determined (WHO 2013). In the United States, asthma accounted for over 3,600 deaths in 2014 and is responsible for an estimated 10.5 million physician visits, 1.8 million emergency department visits, and nearly 440,000 instances of in-patient hospitalization each year, with an average length of stay of 3.6 days (CDC 2016). Other substances in the built environment also cause significant problems. For instance, exposure of children to lead in paints and plumbing components is estimated to contribute to 600,000 new cases of children with intellectual disabilities annually, leading the World Health Organization to call it a 'major public health concern' (WHO 2016b).

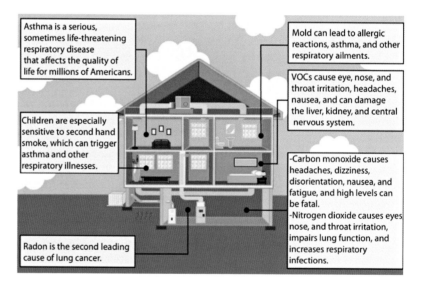

Figure 1.5 Sources of air pollutants

The indoor environment also offers significant opportunities for realizing benefits. Studies have noted positive effects on human performance when properly adjusting ventilation, temperature, indoor air quality, daylight and views, and lighting levels. Controlling levels of indoor pollutants have also been noted to have positive effects on performance (EPA 2016b).

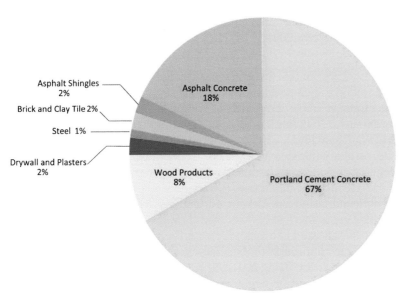

Figure 1.6
Construction waste in the USA
Source: EPA 2015

Materials and waste impacts

In addition to disseminating air- and water-borne pollutants, construction activities inevitably produce a significant amount of solid waste from the production, transportation, use and end-of-lifecycle of materials (Figure 1.6). In 2013, 530 million tons of construction and demolition (C&D) debris were generated in the United States, which was more than twice the amount of municipal waste generated that year (EPA 2015). While nearly half of the municipal waste generated was managed through recycling, composting or combustion with energy recovery, recycling rates for C&D waste are not presently tracked. With the growth in green building rating systems and increasing costs of transportation and disposal, however, recycling is definitely on the rise in the United States.

Upstream in the building lifecycle, the harvest and extraction of raw materials also has significant impacts. Global extraction of material resources continues to grow, and is expected to increase from 62 billion metric tonnes (Gt) per year in 2008 to 100 Gt by 2030. An additional 44 Gt of materials were extracted in 2008 as byproducts of harvesting, but did not actually enter the economy as useful materials (OECD 2015). This unused domestic extraction (UDE) contributes significantly to the impacts associated with material use, particularly for fossil fuels and metals. Recent European studies have estimated that the construction of buildings in the EU consumes between 1,200 and 1,800 million tonnes of materials annually for new building and refurbishment. Specifically, buildings consume approximately 25 per cent of all aluminum consumed in total, 70 per cent of bricks and clay, 75 per cent of concrete, 65 per cent of aggregate materials, 35 per cent of copper, 65 per cent of glass, 35 per cent of stone, 21 per cent of steel and 38 per cent of wood (Ecorys 2014). The specific distribution varies greatly across EU member countries. Given that the density of a material significantly affects the energy required for transportation and handling, the large share of dense materials devoted to buildings also has climate change implications.

Land use impacts

In addition to generating waste, construction activities also irreversibly transform valuable land such as farmland and forests into physical assets such as buildings, roads, dams or other civil infrastructure (Spence and Mulligan 1995). Construction activities also disturb the soil on a site, resulting in damage to the ecological viability of the land. About 7 per cent of the world's farmland was lost between 1980 and 1990 mainly through construction activities (Langford et al. 1999), and the conversion of agricultural land to developed land has continued (USDA 2012). Arable land is also lost or destroyed through quarrying and mining the raw materials used in construction, and construction contributes to the loss of forests through timber harvest for construction materials and energy production (Ding 2004). By reducing and destroying forests and farmland, construction negatively affects biodiversity, crop production, photosynthesis, air purification and other ecosystem services. It also diminishes the ability of natural ecosystems to sequester CO_2, thereby affecting global climate change.

Based on growing awareness of environmental and health issues and problems associated with construction activities and built facilities, the construction industry has begun to embrace sustainable design and construction as a means to reduce or mitigate these impacts.

Sustainable design and construction

The concept of sustainability in the Architecture/Engineering/Construction (AEC) industry is becoming broadly accepted as part of efforts to reduce negative environmental impacts and natural resource depletion due to human activity. In particular, the construction industry has begun to emphasize the use of green building products and technologies and incorporate green goals and objectives as part of comprehensive green project delivery because of the growing public awareness of the industry's significant negative impact on both the natural environment and human health. Other important drivers include economic benefits of energy efficiency, increasing resiliency with regard to uncertain future energy prices and global conditions, and a variety of incentives have emerged to encourage energy efficiency in particular for these projects.

In the context of construction, many definitions of *sustainable* and *green* have been proposed that address issues of environmental benefit and human health, examples of which are shown in Table 1.1. Common themes across definitions include environmental responsibility, resource efficiency and reduction of negative impacts on human health.

The term 'green' is often used interchangeably with 'sustainable' or 'high performance' to describe projects, technologies and strategies used for sustainable design and construction. However, there are fundamental differences between these terms despite some overlapping scope (Figure 1.7). Among these terms, green is the most inclusive since it

Table 1.1 Existing Definitions of Green and Sustainable Construction (Pearce and Suh 2013)

Source	Definition	Components
U.S. Environmental Protection Agency (EPA 2013)	"Green building is the practice of creating structures and using processes that are environmentally responsible and resource-efficient throughout a building's lifecycle from siting to design, construction, operation, maintenance, renovation and deconstruction. This practice expands and complements the classical building design concerns of economy, utility, durability, and comfort. Green buildings are also known as sustainable or high performance buildings."	• Environmental responsibility • Resource efficiency • Economy • Utility • Durability • Comfort
U.S. Office of the Federal Environmental Executive (OFEE 2003)	"The practice of (1) increasing the efficiency with which buildings and their sites use energy, water, and materials, and (2) reducing building impacts on human health and the environment, through better siting, design, construction, operation, maintenance, and removal: the complete building lifecycle."	• Energy efficiency • Water efficiency • Material efficiency • Reduced impacts on human health • Reduced environmental impacts
American Institute of Architects (AIA 2013)	"A green building incorporates design, construction, and maintenance practices that significantly reduce or eliminate the negative impact of the building on occupants and the environment."	• Reduced or eliminated negative impact on occupants • Reduced or eliminated impact on environment
Green Construction Market Outlook (McGraw-Hill 2008)	"[A building] built to LEED standards, an equivalent green building certification program, or one that incorporates numerous green building elements across five category areas: energy efficiency, water efficiency, resource efficiency, responsible site management and improved indoor air quality."	• Energy efficiency • Water efficiency • Resource efficiency • Responsible site management • Improved indoor air quality
The Green Revolution (Yudelson 2008)	"A green building is a high-performance property that considers and reduces its impact on the environment and human health."	• High-performance property (for users) • Reduced impact on the environment • Reduced impact on human health

requires only a net benefit to the environment, resource bases or project stakeholders to qualify. Specifically, a project or technology is considered 'green' if it has *any* of the following outcomes compared to a conventional project or technology (Pearce and Suh 2013):

- reduced negative impacts on or enhancements to natural ecosystems or their function;
- reduced depletion or enhanced recovery/generation/reuse of resources, including materials, energy or water;
- reduced negative impacts on or enhancements to project stakeholders both current and future.

The more stringent term 'sustainable' has a specific definable meaning with respect to the built environment (Pearce 1999; Pearce and Vanegas 2002). Overall, sustainability includes a broader range of considerations and requirements needed to ensure system stability over time

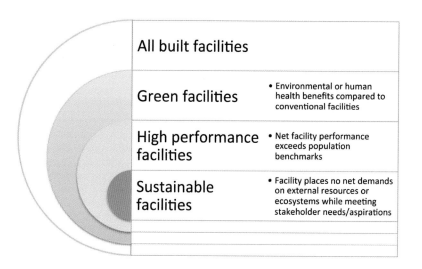

All built facilities	
Green facilities	• Environmental or human health benefits compared to conventional facilities
High performance facilities	• Net facility performance exceeds population benchmarks
Sustainable facilities	• Facility places no net demands on external resources or ecosystems while meeting stakeholder needs/aspirations

without depleting or damaging resource bases and ecosystems to ensure their ongoing viability (Table 1.2), whereas green focuses primarily on environmental impacts alone. Specifically, a built facility's sustainability depends on how well it meets the needs and aspirations of its stakeholders without compromising the ability of non-stakeholders to meet their own needs and aspirations. Non-stakeholders include both future generations as well as other humans on the planet today beyond those who directly influence or are influenced by the facility.

The primary ways in which a building or infrastructure project influences non-stakeholders is through competing with them for available resources and damaging or destroying natural systems that are the primary means of sustainably regenerating those resources (Pearce and Vanegas 2002). Therefore, a project that can meet the needs and aspirations of its own stakeholders while remaining net neutral or positive with regard to resource use and damage to ecosystems will fall within the octant of sustainability (Figure 1.8). Together, the three axes define a set of constraints that provide guidance for making choices in managing facilities that can improve the sustainability of those facilities.

More concisely, a 'sustainable facility' is one for which the current and probable future states of the facility cause no net negative impacts to resource bases or ecosystems (the two means by which humanity now and in the future will meet its needs), while satisfying the needs and aspirations of its stakeholders (Pearce 1999; Pearce and Fischer 2001). While these three constraints do not explicitly incorporate the economic pillar of sustainability as part of the framework, economics is bound within the stakeholder satisfaction constraint, since stakeholders will not be satisfied with facility options that do not meet their feasibility or investment criteria. The underlying assumption is that stakeholders will behave rationally and choose to remain in a situation where their needs and aspirations are met, as long as that solution performs at least as well as other options in terms of all three variables.

Social sustainability is an emerging focus of the construction industry to consider the impacts of projects on stakeholders throughout the facility lifecycle as well as indirect stakeholders in surrounding communities and the world at large (see *What does social sustainability mean for construction projects?*). Key concepts addressed under the rubric of social sustainability include safety through design, social design, corporate social responsibility and community involvement (Valdes-Vasquez and Klotz 2010).

Table 1.2 Aspects of facility sustainability

Environmental Aspects	Social Aspects	Economic Aspects
• Protecting air, water, land ecosystems • Conserving natural resources (fossil fuels) • Preserving animal species and genetic diversity • Protecting the biosphere • Using renewable natural resources • Minimizing waste production or disposal • Minimizing CO_2 emissions and other pollutants • Maintaining essential ecological processes and life support systems • Pursuing active recycling • Maintaining integrity of environment • Preventing global warming	• Improving quality of life for individuals, and society as a whole • Alleviating poverty • Satisfying human needs • Incorporating cultural data into development • Optimizing social benefits • Improving health, comfort and well-being • Having concern for inter-generational equity • Minimizing cultural disruption • Providing education services • Promoting harmony among human beings and between humanity and nature • Understanding the importance of social and cultural capital • Understanding multidisciplinary communities	• Improving economic growth • Reducing energy consumption and costs • Raising real income • Improving productivity • Lowering infrastructure costs • Decreasing environmental damage costs • Reducing water consumption and costs • Decreasing health costs • Decreasing absenteeism in organizations • Improving Return on Investments (ROI)

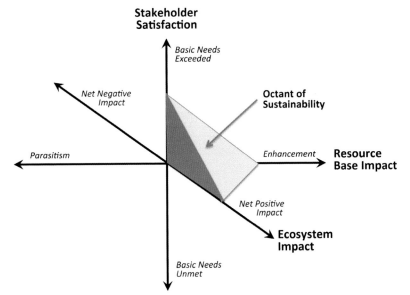

Figure 1.8
Sustainability criteria for built facilities
Source: Pearce and Vanegas 2002

What does social sustainability mean for construction projects?

Contributors: Drs. Rodolfo Valdes-Vasquez, Leidy Klotz and Eugenio Pellicer

Sustainable construction requires improvements not only in the environmental and economic pillars of sustainability but also the social one. Social sustainability is fundamentally about people, both current and future. In the construction industry, social sustainability is defined in different ways, depending upon the stakeholder's perspective and where it is applied during the project lifecycle. For instance, during the planning and design phase, one focus involves estimating the impact of construction projects in relation to where users live, work, play and engage in cultural activities (Burdge 2004). These estimations are normally embedded in the environmental impact assessments required by government agencies. In addition, in these early phases, community involvement approaches such as public hearings are used by external stakeholders and government agencies to influence design decisions (Solitare 2005). Social sustainability also relates to the aspects required to improve the decisions during the design phase such as transparency (Kaatz et al. 2006).

During construction, a focus on social sustainability from the perspective of construction firms relates to the application of corporate social responsibility practices (Lamprinidi and Ringland 2008), which consider how the organization can meet the needs of stakeholders affected by its operations (Kolk 2003). Construction companies and designers advocate for worker safety by eliminating potential safety hazards from the work site during the design phase (Gambatese et al. 2008). Other considerations include engagement among employees, local communities, clients and the supply chain to ensure meeting the needs of current and future populations and communities (Herd-Smith and Fewings 2008). A concept map summarizing these social sustainability concepts is shown in Figure 1.9. Also, Table 1.3 defines the four main categories of considerations seen in this figure.

Additionally, the operation phase is critical to guarantee the service life of the facility or infrastructure for as much time as possible. Generally, conducting post-occupancy evaluations, performing timing inspections and carrying out adequate maintenance work is relevant (Pellicer et al. 2014). For this phase, social sustainability can be perceived as having follow-up plans that improve those goals to be achieved by the facility or infrastructure in the long term (Morrissey et al. 2011). These goals are generally related to the population living in the area (Sierra et al. 2016),

Table 1.3 Social sustainability considerations in construction projects

Conceptual areas	General description
Corporate social responsibility	Considers the accountability of an organization to care for all of the stakeholders affected by its operations.
Community involvement	Emphasizes the influence of public constituencies on private and governmental proposed projects.
Safety through design	Ensures worker safety by eliminating potential construction/operation safety hazards from the work site during the design phase.
Social design	Focuses on enhancing the safety, health, productivity and inclusion of the end users and on improving the decision-making process of the design team.

continued

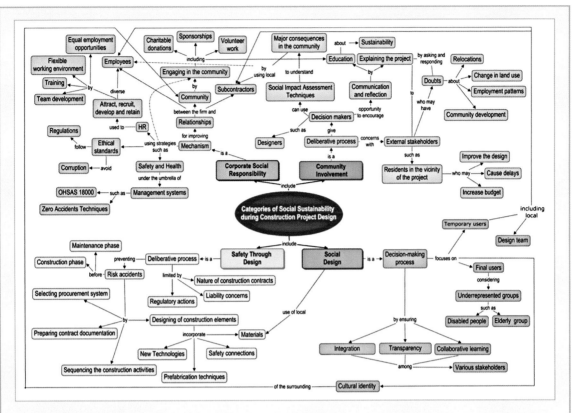

Figure 1.9 Concepts related to social sustainability for the construction industry
Source: Valdez-Vasquez and Klotz (2010)

for instance: services (such as sanitary, electrical or telecommunications); education, training and career development; communal cohesion and identity; business enterprises and others. Ultimately, the improvement generated by building facilities and infrastructure implies social benefits for the people living in the area as well as opportunities for the industrial and commercial network at the regional and national level, which will also bring benefit back to those people (Sierra et al. 2016).

Figure 1.10 shows practical considerations for social sustainability in construction. Although it focuses on individual categories, the reality is this application operates as an integrated combination, representing a system perspective. For this reason, feedback loops have been included in the diagram to represent the influence of one region on the others, allowing for adjustment in implementation and self-monitoring. In addition, according to the type of project (e.g. commercial buildings or highways), some considerations may or may not be pertinent. For instance, when considering infrastructure projects such as highways or bridges, the range of stakeholders affected may be more extensive than for a commercial building; thus, community participation will be more significant in the former than in the latter. Even when considering the same type of projects and locations, different stakeholders will have various levels of understanding of the concept of sustainability and their needs, affecting the dynamics of the processes that should be applied at any given phase.

continued

Stakeholder Engagement
Engage the final and
temporary users
Involve the community
Use partnering strategies
Establish a stakeholder
engagement plan
Document lessons learned

Health Impact Assessment
Use previous post-
occupancy evaluation
studies
Evaluate the security and
connectivity of the final
and temporary users
Include a social life cycle
of products/materials

Sustainable Outcomes
Safety and Health
Security
Pride in ownership
Productivity
Transparency
Equity

Project Considerations
Use and integrated delivery
method
Form a diverse and
integrated design team
Create a project team with
a sustainability focus
Establish zero harm and
zero accidents targets

Social Impact Assessment of
Infrastructure needs
Culture
New social classes
Population changes
Zoning decisions
Access to public
transportation
Others

Follow-up plans for
Safety and health
Transparency
Productivity
Security
Education
Infrastructure needs
Post-occupancy
evaluation

Figure 1.10 Practical considerations for social sustainability in construction
Source: Valdes-Vasquez and Klotz (2012)

In summary, a truly sustainable construction project needs to include social considerations such as its effect on the community and the safety, health and education of the workers, as well as the benefits to the users and affected population. Failure to include these considerations will impact both people and performance of projects in the long run. Furthermore, the decision-makers, such as designers and project managers, need to analyse the social dimension of sustainability from a lifecycle perspective (Pellicer et al. 2016). The first step in this direction is to educate future professionals across all three dimensions of sustainability: social, environmental and economic (Valdes-Vasquez and Klotz 2011).

Sources

Burdge, R.J. (2004). *The concepts, process, and methods of social impact assessment*. Social Ecology Press, USA.

Gambatese, J., Behm, M., and Rajendran, S. (2008). "Design's role in construction accident causality and prevention: Perspectives from an expert panel." *Safety Science* 46(4), 675–691.

Herd-Smith, A., and Fewings, P. (2008). "The implementation of social sustainability in regeneration projects: Myth or reality?" RICS, COBRA 2008, *Building and Real Estate Research Conference*, Dublin, Ireland, January, 2008.

Kaatz, E., Root, D., and Bowen, P. (2005). "Broadening project participation through a modified building sustainability assessment." *Building Research & Information, 33*(5), 441–454.

Kolk, A., (2003). "Trends in sustainability reporting by the Fortune Global 250." *Business Strategy and the Environment.* 12(5), 279–291.

continued

Lamprinidi, S., and Ringland, L. (2008). "A snapshot of sustainability reporting in the construction and real estate sector." Global Reporting Initiative (GRI). https://globalreporting.org/resourcelibrary/A-Snapshot-of-sustainability-reporting-in-the-Construction-Real-Estate-Sector.pdf (accessed 10 October 2011).

Morrissey, J., Iyer-Raniga, U., McLaughlin, P., and Mills, A. (2012). "A strategic project appraisal framework for ecologically sustainable urban infrastructure." *Environ. Impact Assess*, 33(1), 55–65.

Pellicer, E., Yepes, V., Teixeira, J.C., Moura, H., and Catalá, J. (2014). *Construction management*. Wiley-Blackwell, Oxford, UK.

Pellicer, E., Sierra, L.A., and Yepes, V. (2016). "Appraisal of infrastructure sustainability by graduate students using an active-learning method." *Journal of Cleaner Production*, 113, 884–896.

Sierra, L.A., Pellicer, E., and Yepes, V. (2016). "Social sustainability in the lifecycle of Chilean public infrastructure." *Journal of Construction Engineering and Management*, 142(5), 05015020.

Solitare, L. (2005). "Prerequisite conditions for meaningful participation in brownfields redevelopment." *Journal of Environmental Planning and Management*, 48(6), 917–935.

Valdes-Vasquez, R., and Klotz, L. (2010). "Considering social dimensions of sustainability during construction project planning and design." *International Journal of Environmental, Cultural, Economic, & Social Sustainability*, 6.

Valdes-Vasquez, R., and Klotz, L. (2011). "Incorporating the social dimension of sustainability into civil engineering education." *J. Prof. Issues Eng. Educ. Pract*, 137(4), 189–197.

Valdes-Vasquez, R., and Klotz, L.E. (2012). "Social sustainability considerations during planning and design: Framework of processes for construction projects." *Journal of Construction Engineering and Management*, 139(1), 80–89.

Benefits of implementing sustainable design and construction

Three types of benefits arise from implementing sustainable design and construction strategies on capital projects, including both new and existing facilities. First, sustainable construction offers significant *environmental benefits* compared with conventional construction. By reducing the amount of energy and resources needed to construct and operate built facilities, the loads imposed by the built environment on the natural environment can be reduced, and the Earth's finite resources can be conserved for use by future generations. Not only do sustainable facilities and infrastructure systems reduce demands placed on the natural environment, they can also contribute to and enhance the carrying capacity and environmental quality of the natural environment, which is diminished by conventional construction practices. For example, green roofs, rain gardens and other similar structures treat stormwater arising from impervious surfaces as well as providing green space as part of a development that can serve as habitat for local species of insects, birds and other animals and plants. These and other building features help to reduce the negative impacts of the built environment on natural ecosystems and buffer the effects of human development. The increased efficiency in use of energy and resources also helps to reduce the impacts of humans on global climate change by reducing the amount of energy that is needed from fossil fuels. Incorporating renewable energy generation as part of sustainable construction projects further reduces energy demand from non-renewable sources, reducing the amount of carbon produced during energy production.

Selling points for sustainable design and construction

Environmental benefits

- Enhance and protect biodiversity and ecosystems
- Improve air and water quality
- Reduce waste streams
- Conserve and restore natural resources
- Reduce global warming

Economic benefits

- Reduce operating and maintenance costs
- Create, expand and shape markets for green product and services
- Improve occupant productivity
- Minimize occupant absenteeism
- Optimize lifecycle economic performance
- Improve the image of building
- Reduce the civil infrastructure costs

Social benefits

- Enhance occupant comfort and health
- Heighten aesthetic qualities
- Create new and enhanced employment and business opportunities
- Minimize strain on local infrastructure
- Improve overall quality of life

These environmental benefits also have direct implications for the economics of capital projects. Accordingly, sustainable design and construction can offer significant *economic benefits* in addition to their benefits for the natural environment. Sustainable facilities and infrastructure systems require fewer resources to meet the needs of their users, thus reducing the costs over time to operate and maintain these facilities. The array of new technologies that is evolving to improve the sustainability of the built environment creates jobs and establishes new markets. For example, new technologies for renewable energy have resulted in the growth of new manufacturing enterprises and supply chains to produce these technologies. New jobs are also created for people who install and maintain these systems, and new educational programmes and degrees are needed to train these new workers in the green economy. New business models such as long-term product leases have also been developed as an alternative to conventional purchase of building technologies, resulting in the need for new businesses who manage the leasing and upgrade process. All of these changes in construction technology help to stimulate the economy and provide new opportunities to replace and enhance the inefficient and unsustainable industries of the past.

Finally, sustainable design and construction tactics can result in significant *social benefits* to both facility stakeholders and society at large in addition to their environmental and economic benefits. In the building sector, sustainable facilities have been observed to enhance occupant comfort, health and productivity. For example, daylighting in green buildings has been correlated with reduced absenteeism among employees in general, increased productivity among factory workers and improved test scores, growth rates and dental health in schoolchildren (EPA 2016b). The use of sustainable technologies and practices creates new employment opportunities that can stimulate local economies and help to stabilize communities, especially those that are suffering from loss of jobs as a result of industry downsizing and outmoded technologies. Sustainable buildings, with their reduced needs for energy, water and other resources, can reduce the loads imposed on centralized infrastructure systems, thereby reducing the need to expand already strained capacity. New approaches to development that reduce dependence on automobile transportation, such as mixed-use development and alternative transportation systems, also reduce infrastructure demand while improving quality of life and reducing threats to environmental quality.

The world of the future: why sustainability is essential

Together, the benefits of sustainable design and construction offer the potential to change the way in which we as humans face the challenges of the next decades. These challenges are not insignificant. Both at the level of the built environment and for global development overall, the coming years will bring major changes that must be addressed in how we create and maintain the built environment.

Overall global trends for the next decade

From a global standpoint, both sociopolitical and environmental trends will shape the next 10 years. In 2009, author Thomas Friedman, in his bestselling book *Hot, Flat, and Crowded,* characterized the current state of global development in terms of three primary influences: climate change exacerbated by human activities (*hot*), reduction of geopolitical and social stratification due to ubiquitous and cheap telecommunications (*flat*), and having a growing global population that is increasingly concentrated in large urban centres (*crowded*). Together with major shifts in the balance of geopolitical and economic power worldwide, these three characteristics provide the context throughout the coming years for a new way of thinking about the built environment. In a world where energy and water are the lifeblood of civilization, competition for scarce resources is already becoming fierce as nations dam their rivers,

"[E]xpect the dragon ascendant, the eagle descending, the South rising, and the planet possibly trumping all of these."
(Dr. Michael Klare, author of *Rising Powers, Shrinking Planet: The new geopolitics of energy*)

mine their aquifers, and strive to establish long-term contracts with oil-producing nations to ensure future energy security.

We have been aware since at least the 1970s of the reinforcing relationship between world population, the standard of living/level of affluence expected and sought by that population, and the technology used to achieve the desired standard of living. Together, these factors define the influence of human beings on the world's finite resource bases and the ecosystems that renew and restore them. Critical to our management of those impacts is the evolution of three complementary strategies (Ehrlich and Holdren 1971):

- Control of population growth to reduce consumption of Earth's finite resources.
- Evolution of our expectations for quality of life to favour increases in standards of living that are not dependent upon increasing resource consumption.
- Development of new technologies that contribute to increased standards of living with dramatically reduced impacts on the planet's resources and natural ecosystems.

Although birth rates have become more moderate over time due to the influence of education, availability of birth control, improved access to medical care and other factors, increased life expectancy worldwide has led to continued population growth. The increasing global population has led to ever-increasing demand for resources so that people in both developed and developing countries can meet their aspirations for the improved standards of living modelled by the richest countries in the developing world.

At the same time that demand for resources is increasing, the unwanted side-effects of using those resources are continuing to grow and interact in unpredictable ways. Current predictions about the exact impacts of climate change vary widely, but there is widespread consensus that the world in the coming years will be subject to more severe weather patterns, rising sea levels, melting ice-caps and glaciers, and changes in precipitation levels in key agricultural areas (IPCC 2014). These changes will have impacts on food production and security, water resources, production of biofuels to replace dwindling fossil fuels, and the vulnerability of human settlements to flooding, droughts, and other severe weather patterns.

Consensus is also widespread that we are very close to or past the peak of global crude oil production (Figure 1.11). Similar peaks are also approaching for a number of other key resources critical to the prosperity of modern society, including coal, platinum, uranium and aquifer-level groundwater (as described by Heinberg 2007 and others).

In summary, the next decades are likely to see growing pressure from increased demand for resources by a population hungry for a better life, coupled with reduced supply and a reduced basis for resource regeneration as a result of the ecological impacts of human activities.

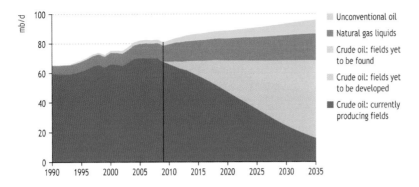

Figure 1.11 Global oil and gas production forecast
Source: OECD/IEA (2010)

Changes to climate on a global scale are beyond our current ability to predict with great certainty, but direct observations in recent history suggest that we will see changes in weather, especially rainfall and severe storm events, that will negatively impact our ability to meet human needs. The concentration of human populations in coastal areas will increase the vulnerability of development to the negative effects of climate change such as sea level rise and increasingly violent weather patterns (IPCC 2014). The challenge is to find new ways of developing the built environment that are robust in the face of these coming changes at a global scale.

Significant trends related to the built environment

All of the challenges described in the previous section have implications for the built environment. Overall reduction in supply of nonrenewable resources, coupled with an awareness of the negative climate impacts of fossil fuel consumption, will significantly increase the cost of energy and the volatility of its supply, leading to a demand for buildings that have reduced reliance on external energy supply. New technologies and approaches to extract fossil fuels such as hydraulic fracturing ('fracking') to recover natural gas from shale and new techniques for extraction of oil from bituminous oil sands have resulted in a glut of new fuel sources on the market in countries like the United States and Canada, with significant additional known reserves in Venezuela, the Congo and Russia. However, the environmental damage and risks of these techniques to human health and safety are becoming increasingly apparent. Thus, the future of fossil fuel supplies remains an uncertain tradeoff between demand for the benefits it affords, and the negative risks and impacts its extraction requires.

Energy cost and raw material shortages will also impact the cost and availability of other types of materials used in construction, leading to a need for low- or net zero energy alternatives. These trends will also

Top ten green trends shaping the future

1. **Renewable fuels for electric power generation**
   - Increased demand, especially in China and India
   - Awareness of greenhouse gas problems
   - Development of new fuels for power production

2. **Water resource management, including reuse and recycling of water**
   - Better technologies for conservation, treatment, reuse, and desalination

3. **Carbon regulations and policy**
   - Limitations on carbon production
   - Increase in renewable energy generation
   - Market-based solutions

4. **'Green' as good business**
   - Increased industrial efficiency
   - Balanced scorecard reporting
   - Green labelling

5. **Greening of transportation**
   - Renewable and sustainable fuels
   - New hybrid, plug-in and fuel cell vehicles
   - Carbon capture technologies

6. **Increased availability of green products/services**
   - Better consumer education and information availability
   - Product design for end-of-lifecycle
   - Reduction in waste by design

7. **Systems approach to environmental analysis**
   - Increased computational abilities
   - Evaluation of products and systems in terms of their context at a macro scale

8. **Increased resource needs of growing urban population**
   - Increased stress on urban infrastructure
   - Increased life spans

9. **Information and communication technologies use instead of travelling**
   - Use of telecommuting to reduce impacts and increase productivity
   - Increased use of the internet for shopping, recreation, and socialization

10. **Green building**
    - Increased focus on whole lifecycle
    - Integrated systems optimization
    - Planned communities and eco-cities
    - Integrated energy production

Source: *Battelle (www.battelle.org).*

enhance the attractiveness of resource recovery from existing buildings and infrastructure, as will a reduction in landfill capacity for waste products and significant increase in the costs of transportation.

Demographic changes, in both developed and developing countries, will lead to a need to build facilities and infrastructure that can accommodate the needs of the changing world population. Due to increased life expectancy and improved medical care, an overall increase in the population of aging people in developed countries has led to new types of functional requirements for the built environment and infrastructure to provide for their needs that can allow these people to remain in their communities and 'age in place'. This demographic segment of the population is expected to continue to grow over the coming years, and with it will grow the market for accessible, affordable communities and 'lifespan friendly' housing that can accommodate the aging in place for the current generation.

The movement of population to urban centres will create a need for new types of development that can accommodate higher human density without compromising quality of life. These new developments must provide for human needs such as food, water and energy which in the past were imported from non-urban areas, leading to net zero energy and net zero water buildings. Vertical farming, which involves using urban infrastructure and buildings as a base for agriculture and food growing, will increase as the cost of transportation increases and city dwellers seek to become less dependent on external supplies of resources. New solutions to address the challenges of urban farming, such as limitations of solar geometry and energy requirements for pumping and lighting, will need to be developed. Ubiquitous communication systems via the internet will also affect transportation patterns of people within urban areas as an increasing number of people telecommute and socialize online instead of physically travelling from place to place.

While the built environment of tomorrow is certain to look and behave differently from today's facilities and infrastructure, no one knows exactly what form it will take or what functions it will provide. What *is* certain is that it must be dramatically more resource efficient than the built environment of today to accommodate the ever-increasing needs of a changing and growing population. Sustainable practices for the planning, design, construction, operations and end-of-lifecycle of facility and infrastructure systems offer an approach to achieve this end.

Overview of the book

With all of the potential benefits associated with sustainable design and construction and the challenges to be faced in the coming years, many organizations are seeking ways to improve the sustainability of their buildings. The purpose of this book is to provide an introduction on how to improve the sustainability of the built environment.

Chapter 2, Evolution of sustainability: theory and applications describes the evolution of sustainability as a concept that has led to the current state of practice in sustainable construction. It covers the history of sustainability thinking that has shaped contemporary policy and practices, and describes the ideas involved in the ongoing evolution of sustainability theory and its application to the built environment. It concludes with an overview of trends in sustainability thinking for the built environment.

Chapter 3, Sustainability policies and programmes presents an overview of the elements of sustainability policies in the public and private sector. It covers the social, economic and environmental considerations of a policy, the issues associated with implementability of a policy, and the pros and cons of major policy options. It presents the major components of a green building programme, and provides examples from existing programs in the United States and elsewhere. It concludes with a look at leading corporate sustainability policies and a look ahead to sustainability-related policy in the future.

Chapter 4, Green rating systems introduces major rating systems available around the world to guide decision making for sustainability at the raw material, product, building and infrastructure scales. It discusses the ways in which stakeholders use rating systems, the two major types of tools (threshold and profile) that are presently in use, and the labelling and logo systems that can be found on products for sustainable construction. It also describes considerations and information sources that are important in product selection and decision making, and provides an overview of the systems available for corporate rating and sustainability reporting. It concludes with a look at trends in rating systems that are likely to be prevalent in the coming years.

Chapter 5, Project delivery and pre-design sustainability opportunities describes the overall process for planning, programming and delivering a capital facility and the players who are involved at various stages, with a focus on integrative processes and practices. It identifies best practices and opportunities that can be employed prior to the facility design process, and introduces a case study of the Trees Atlanta Kendeda Center that is used in this and the following three chapters to provide specific examples of best practices in use. It also provides an overview of the charrette process that is often employed in the earliest phases of a project to obtain stakeholder input, and describes likely changes to the pre-design process that may occur in the coming years.

Chapter 6, Sustainable design opportunities and best practices introduces the integrative design process employed on green projects. It provides an overview of best practices for the design of green buildings in the areas of sustainable sites, energy optimization, water and wastewater performance, materials optimization and indoor environmental quality. It also reviews overarching best practices that affect multiple systems and have multiple benefits at once. The second half of the chapter presents design-phase sustainability strategies and best practices employed at the Trees Atlanta Kendeda along with other cases. It concludes with an overview of how the design process is likely to change in the future.

Chapter 7, Sustainable construction opportunities and best practices covers the major practices and technologies that can be employed during preconstruction and construction to improve and maintain project sustainability. It presents key components of major sustainability implementation plans that are used during the construction phase to ensure that sustainability goals are met. The second half of the chapter presents construction-phase sustainability strategies and best practices employed at the Trees Atlanta Kendeda Center along with other cases. It concludes with a look ahead to how construction is likely to change in the coming years.

Chapter 8, Post-occupancy sustainability opportunities and best practices describes the array of strategies that can be employed during the operations, maintenance and end-of-lifecycle phases of a facility's lifecycle to enhance sustainability and maintain ongoing high performance.

The second half of the chapter introduces sustainability strategies and best practices for sustainability used after occupancy began at the Trees Atlanta Kendeda Center. It also covers end-of-lifecycle tactics used for deconstruction, demolition and adaptive reuse of the old warehouse that occupied the site where the Trees Atlanta Kendeda Center was later built. A case study of major renovations to the Empire State Building in New York City is also included to illustrate the process of prioritizing energy-related decisions in high-rise building rehabilitation. The chapter concludes with a look at how post-occupancy practices are likely to change in the coming years.

Chapter 9, The business case for sustainability presents a detailed case study of the Reedy Fork Elementary School in North Carolina as a means of illustrating the decision process used to select strategies that make good business sense for a sustainable project. It also provides an over-view of the evidence to support the business case for project sustain-ability, including a holistic perspective on project costs. This holistic perspective includes both tangible and intangible project costs along with externalities that should be considered in developing sustainable projects.

Chapter 10, The future of sustainable buildings and infrastructure concludes the book by presenting an overview of upcoming research, trends and philosophies that will shape the future of sustainable design and construction for both buildings and infrastructure systems. It introduces emerging technologies that are likely to significantly change the marketplace, significant process improvements that will influence practice and broader trends that should be taken into account in planning for a future in this field.

Discussion questions and exercises

1.1   The built environment is directly or indirectly responsible for many types of impacts on resource bases and natural ecosystems. How does your home, office or classroom building contribute to those impacts? Conduct an inventory of your building and identify all of the resources necessary to operate it over the course of a year, including energy, water, materials and others. From what sources do those resources come? What impacts do these flows of matter and energy have on the world at large?

1.2   How do buildings contribute to global climate change? Choose a construction project in your community and identify the ways in which that project will contribute emissions of greenhouse gases throughout its lifecycle. Then inventory the ways in which the project will affect the ability of natural systems to absorb and mitigate greenhouse gases. What are some strategies for reducing these impacts?

1.3   What are the possible threats to indoor environmental quality in your building? Consider possible sources of air pollutants as well as sources of unwanted noise, light and visual impacts. Are there areas where thermal comfort is a problem?

1.4   What waste streams – solid, liquid or other – leave your building, and what is their destination? What impacts do these flows of matter and energy have on the world at large?

1.5   How has land use in your community changed over time? Using a source for archival aerial photographs such as www.historicaerials.com, examine how land use has changed over time. Visit your local government office to determine the history of development on your lot or site. What was the site originally before it was developed?

1.6   Sustainable design and construction of the built environment offers the potential for numerous benefits to those who implement it. Which of the major selling points for sustainable design and construction applies to your situation? Who makes decisions about operating practices and remodels/retrofits for your facility? How would you formulate an argument to convince decision makers to invest in sustainable technologies and practices in your home, school or workplace?

1.7   What is the population growth rate in your region or country? Plot the historical trends as well as future projections. How does it compare with the projected global population growth through 2050 shown in Chapter 1?

1.8   Which of the top ten green trends listed in the chapter will be most important to you, your organization or your community? Why?

References

AIA – American Institute of Architects. (2013). "Toolkit: Sustainability 2030." https://info.aia.org/toolkit2030/design/what-building-green.html (accessed 2013).

Battelle. (2010). "Top ten green trends for the next decade." http://reliableplant.com/Read/11575/battelle-top-10-list-forecasts-emerging-green-trends (accessed 8 August 2016).

BEA – U.S. Bureau of Economic Affairs. (2015). "Gross Domestic Product by Industry: First Quarter 2016." (21 July, 2016, Table 5a) https://bea.gov/newsreleases/industry/gdpindustry/2016/pdf/gdpind116.pdf (accessed 9 August 2016).

BLS – Bureau of Labor Statistics. (2014). "CPS Annual Report 2014–2015", United States Bureau of Census for the Bureau of Labor Statistics.

CDC U.S. Centers for Disease Control and Prevention. (2016). "Asthma." FastStats, National Center for Health Statistics, CDC. https://cdc.gov/nchs/fastats/asthma.htm (accessed 9 August 2016).

CIC – Construction Intelligence Center. (2015). "Outlook for global construction industry is looking brighter." Press Release Summary, Global Construction Outlook to 2020, Construction Intelligence Center. (7 April, 2016) http://construction-ic.com/pressrelease/outlook-for-global-construction-industry-is-getting-brighter-4859732 (accessed 9 August 2016).

Ding, G.K. (2004). "The development of a multi-criteria approach for the measurement of sustainable performance for built projects and facilities," PhD dissertation, University of Technology, Sydney, Australia.

DOE – U.S. Department of Energy. (2012). "8.1 Buildings Sector Water Consumption," *Buildings Energy Data Book*: 2011. U.S. Department of Energy, Washington, DC. http://buildingsdatabook.eren.doe.gov/default.aspx (accessed 9 August 2016).

DOE – U.S. Department of Energy. (2015). "1.1: Buildings Sector Energy Consumption." *Buildings Energy Data Book*, http://buildingsdatabook.eren.doe.gov/TableView.aspx?table = 1.1.5 (accessed 17 September 2016).

DOL – U.S. Department of Labor. (2016). "The Employment Situation-December 2015." U.S. Department of Labor (January 8). http://bls.gov/news.release/archives/empsit_01082016.pdf (accessed 9 August 2016).

Ecorys. (2014). Resource efficiency in the building sector: Final report. DG Environment, Rotterdam, The Netherlands. (23 May). http://ec.europa.eu/environment/eussd/pdf/Resource%20efficiency%20in%20the%20building%20sector.pdf. (accessed 9 August 2016).

Ehrlich, P.R., and Holdren, J.P. (1971). "Impact of population growth." *Science*, 171, 1212–17.

EIA – U.S. Energy Information Administration. (2016a). "International Energy Outlook 2016." (June 30). http://eia.gov/forecasts/ieo/pdf/0484(2016).pdf (accessed 9 August 2016).

EIA – U.S. Energy Information Administration. (2016b). "How much energy is consumed in residential and commercial buildings in the United States?" (6 April). http://eia.gov/tools/faqs/faq.cfm?id = 86&t = 1 (accessed 9 August 2016).

EIA – U.S. Energy Information Administration. (2016c). "2012 Commercial Buildings Energy Consumption Survey (CBECS): Energy Usage Summary." *U.S. Energy Information Administration*, Washington, DC. (Release date 18 March, 2016). http://eia.gov/consumption/commercial/reports/2012/energyusage/index.cfm (accessed 9 August 2016).

EPA – U.S. Environmental Protection Agency. (2008). *Reducing urban heat islands: Compendium of strategies*. U.S. EPA, Washington, DC. https://epa.gov/heat-islands/heat-island-compendium (accessed 9 August 2016).

EPA – U.S. Environmental Protection Agency. (2010). "The inside story: a guide to indoor air quality." https://epa.gov/indoor-air-quality-iaq/inside-story-guide-indoor-air-quality (accessed 9 August 2016).

EPA – U.S. Environmental Protection Agency. (2013). "Definition of green building." *EPA green building basic information*. https://archive.epa.gov/greenbuilding/web/html/about.html (accessed 2013).

EPA – U.S. Environmental Protection Agency. (2015). "Advancing sustainable materials management: 2013 fact sheet." U.S. Environmental Protection Agency. (June). https://epa.gov/sites/production/files/2015–09/documents/2013_advncng_smm_fs.pdf (accessed 9 August 2016).

EPA – U.S. Environmental Protection Agency. (2016a). "Inventory of U.S. Greenhouse Gas Emissions and Sinks: 1990–2014." U.S. Environmental Protection Agency. https://epa.gov/ghgemissions/inventory-us-greenhouse-gas-emissions-and-sinks-1990–2014 (accessed 9 August 2016).

EPA – U.S. Environmental Protection Agency. (2016b). "Indoor Air Quality Scientific Findings Resource Bank. Lawrence Berkeley National Laboratory Indoor Environment Group." https://iaqscience.lbl.gov (accessed 9 August 2016).

Friedman, T.L. (2009). *Hot, Flat, and Crowded*. Picador Press, New York, NY.

Heinberg, R. (2007). *Peak everything: Waking up to the century of declines*. New Society Publishers, Gabriola Island, BC, Canada.

IEA – International Energy Agency. (2016). *Energy and Air Pollution: World Energy Outlook Special Report*. Directorate of Sustainability, Technology, and Outlooks, International Energy Agency, Paris, France. http://iea.org/publications/freepublications/publication/WorldEnergyOutlookSpecialReport2016EnergyandAirPollution.pdf (accessed 9 August 2016).

IPCC – Intergovernmental Panel on Climate Change. (2014). IPCC 2014: Climate Change 2014: Synthesis Report. Contribution of Working Groups I, II and III to the Fifth Assessment Report of the Intergovernmental Panel on Climate Change [Core Writing Team, R.K. Pachauri and L.A. Meyer (eds.)]. IPCC, Geneva, Switzerland, p. 151.

Klare, M. (2008). *Rising powers, shrinking planet: The new geopolitics of energy*. Metropolitan Books, New York, NY.

Langford, D.A., Zhang, X.Q., Maver, T., MacLeod, I., and Dimitrijeic, B. (1999). "Design and managing for sustainable buildings in the UK." Profitable partnering in construction procurement CIB W92 and CIB 23 Joint Symposium, London, 373–82.

LBL – Lawrence Berkeley Laboratory. (2009). "Working toward the very low energy consumption building of the future." http://newscenter.lbl.gov/2009/06/02/working-toward-the-very-low-energy-consumption-building-of-the-future/ (accessed 13 January 2017).

McGraw Hill Construction. (2008). *2009 Green outlook: Trends driving change report*. McGraw Hill Construction, New York, NY.

OECD/EIA. (2010). "World energy outlook 2010." www.worldenergyoutlook.org/ (accessed 10 January 2011).

OECD – Organisation for Economic Co-operation and Development. (2015). *Material resources, productivity, and the environment*. OECD Publishing, Paris, France. http://oecd.org/env/waste/material-resources-productivity-and-the-environment-9789264190504-en.htm (accessed 9 August 2016).

OFEE. (2003). *The federal commitment to green building: Experiences and expectations*. Office of the Federal Environmental Executive, Washington, DC.

Pearce, A.R. (1999). *Sustainability and the built environment: A metric and process for prioritizing improvement opportunities*. Ph.D. Dissertation, School of Civil & Environmental Engineering, Georgia Institute of Technology, Atlanta, GA.

Pearce, A.R., and Fischer, C.L.J. (2001). *Systems-based sustainability analysis of building 170, Ft. McPherson, Atlanta, GA*. U.S. Army Environmental Policy Institute, Atlanta, GA, May.

Pearce, A.R., and Kleiner, B.M. (2013). The Safety and health of construction workers on "green" projects: A systematic review of the literature and green construction rating system analysis. Technical report to National Iinstitute of Occupational Safety and Health. Washington, DC. http://mlsoc.vt.edu/sites/mlsoc/files/attachments/constructionworkersgreenprojects.pdf.

Pearce, A.R., and Suh, M.J. (2013). *Green construction technologies and strategies in the U.S.* POSCO A&C, Seoul, South Korea.

Pearce, A.R., and Vanegas, J.A. (2002). "Defining sustainability for built environment systems." *International Journal of Environmental Technology and Management*, 2(1), 94–113.

Russell, J.S., Hanna, A., Bank, L.C. and Shapira, A. (2007). "Education in construction engineering and management built on tradition: blueprint or tomorrow." *Journal of Construction Engineering and Management*, 133(9), 661–8.

Spence, R., and Mulligan, H. (1995). "Sustainable development and the construction industry." *Habitat International*, 19(3), 279–92.

UNEP – United Nations Environmental Programme. (2009). *Building and Climate Change*. UNEP, Paris, France.

US Census. (2016). "Construction Spending." U.S. Census Bureau. (June). http://census.gov/construction/c30/c30index.html (accessed 9 August 2016).

USDA – U.S. Department of Agriculture. (2012). "National Resources Inventory." U.S. Department of Agriculture Natural Resources Conservation Service. http://nrcs.usda.gov/wps/portal/nrcs/main/national/technical/nra/nri/ (accessed 9 August 2016).

Valdes-Vasquez, R., and Klotz, L. (2010). "Considering social dimensions of sustainability during construction project planning and design." *International Journal of Environmental, Cultural, Economic, & Social Sustainability*, 6.

WHO – World Health Organization. (2013). "Asthma." Fact sheet No. 307, World Health Organization. (November). http://who.int/mediacentre/factsheets/fs307/en/ (accessed 9 August 2016).

WHO - World Health Organization. (2016a). "Drinking water." Media Centre Fact Sheet, World Health Organization. www.who.int/mediacentre/factsheets/fs391/en/ (accessed 9 July 2017).

WHO – World Health Organization. (2016b). "Lead." International Programme on Chemical Safety, World Health Organization. http://who.int/ipcs/assessment/public_health/lead/en/ (accessed 9 August 2016).

Yudelson, J. (2008). *The Green Revolution*. Island Press, Washington, DC.

Chapter 2
Evolution of sustainability: theory and applications

To understand the motivation and context for sustainability in the built environment today, it is important to review the evolution of the sustainability movement and the major ideas and concepts that emerged along the way. At the most fundamental level, sustainability emerged as a theoretical concept that has guided thinking in many different fields including economics, social and political science, biological and ecological systems, geology and others. These basic theories have also been translated into ideas that can be applied to how we make choices, including how we design solutions to our problems, how we determine what is and is not sustainable, and how we measure progress toward our goals. Even more specifically, sustainability ideas and concepts have been applied to how we make choices and interact with technologies as humans. This chapter provides an overview of the major streams of thought that have evolved in each of these areas.

> Sustainability is based on a simple and long-recognized factual premise: Everything that humans require for their survival and well-being depends, directly or indirectly, on the natural environment . . . The environment provides the air we breathe, the water we drink, and the food we eat. It defines in fundamental ways the communities in which we live and is the source for renewable and non-renewable resources on which civilization depends. Our health and well-being, our economy, and our security all require a high quality environment. When we act on that understanding, we tend to prosper; when we do not, we suffer.
>
> (National Research Council's Committee on Incorporating Sustainability in the U.S. Environmental Protection Agency, 2011)

Evolution of sustainability as a theoretical concept

Some of the key factors leading to the sustainability movement in the 1970s and 1980s included (1) rapid worldwide population growth; (2) a resultant increase in economic growth as members of the growing population sought higher standards of living and (3) extraction and consumption of natural resources as a means of achieving higher living standards (NRC 2011). All of these factors emerged as a natural result of humans both individually and collectively seeking to better their own lives and the lives of their families. However, the unintended consequences emerging from all three occurring together on an enormous and escalating scale led some to realize that this trajectory could not continue indefinitely (Figure 2.1).

As shown in Figure 2.1, sustainable development has emerged as potential solution to the problems posed by human development patterns as the expectations of a growing population increase over time.

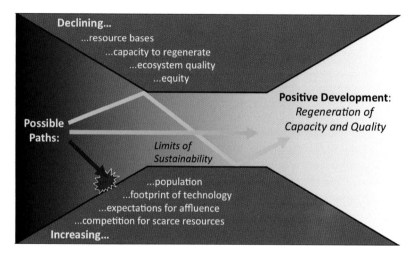

Figure 2.1 Future trajectories of human development (adapted from Nattrass and Altomare 1999)

Drivers of sustainable development

For much of human history, the demands placed on the natural environment to support human development were negligibly small compared to the carrying capacity of Earth's ecosystems to support human activity. Some cultures evolved to live within the limits of these systems, and a few of these cultures continue to exist within the limits of their local ecosystems and available resource even today. In other cultures, as concentrations of populations in cities grew to displace and overwhelm local ecosystems and stocks of natural resources, explorers established new modes of mobility to enable exploration beyond the boundaries of established territory, discovering new lands that could be claimed for expansion and exploited for new resources.

The social impact of colonial expansion on indigenous cultures is beyond the scope of this book, although it forms a very important baseline for the attitudes of contemporary society about the relationship between and among human cultures and the natural environment. Environmentally, the actions taken by humans as they attempt to sustain and improve their standards of living have had increasingly undeniable consequences as human population grows. Improved standards of living and the technologies used to provide that standard of living have also become increasingly impactful.

The conservation movement

Two key realizations became progressively evident as early as the late 1800s. First, human activities were fundamentally changing the landscape of the planet in irreversible ways. During early colonization of new territories, explorers unfamiliar with the ecological and social fabric of those territories often operated under the paradigm of 'taming the wilderness', converting large areas into agricultural production to support their communities, extensively harvesting resources to send home to their countries of origin, and displacing native people and ecosystems from the landscape to reduce potential threats to their ongoing survival.

Initially, these activities may have seemed to hardly make a difference. Over time, however, the significance and scale of impacts became increasingly apparent. For instance, by the late 1800s, the clearing of virgin forests to harvest timber and convert land for agriculture in the United States resulted in permanent biodiversity loss through the extinction of species (NRC 2011). The ability of lands to support agriculture subsequently diminished as poor practices led to wide-scale erosion and destruction of fragile topsoils, which, when coupled with droughts, led to further soil loss and degradation of the land's biological integrity and productive capability. Conservation policies have been subsequently adopted in the United States with the aim of abating or even reversing some of the damage caused by these kinds of actions.

However, even today, economic conditions in other parts of the world lead people to take the same actions, sometimes just to meet basic survival needs. For instance, colonization of rainforests in the Amazon was originally promoted by land laws starting in the 1960s that encouraged both large- and small-scale deforestation to convert land into cattle ranches and other agricultural enterprises. Farmers able to demonstrate agricultural productivity for 1 year were given ownership of the land they farmed, creating an economic incentive to convert further land into agricultural use. Increasing worldwide demand for beef further increased the economic attractiveness of this activity, leading to investments in infrastructure such as the Trans Amazonian highway that made exploitation even easier. Methods used to convert land to agricultural purposes include burning existing vegetation, which not only eliminates the ability of that land to capture carbon and help stabilize climate, but also contributes to additional air pollutants that must be mitigated by remaining forested areas. Cattle on the converted land also produce significant amounts of methane, which is a potent contributor to climate change. These self-reinforcing feedback loops highlight the ways in which policy decisions without considering systems effects can have serious unanticipated consequences.

The environmental movement

The second realization emerged with the recognition of increasingly destructive impacts of chemical and physical agents released into the biosphere by human industrial activities. While the negative human and ecological health impacts of many of these agents is well known, it was widely believed that the ambient environment had sufficient carrying capacity to absorb and diffuse hazardous discharges to a level at which they were no longer hazardous. 'The solution to pollution is dilution' became a common philosophy in managing industrial wastes, leading to discharges that gradually came to overwhelm any ecological capability to handle them. The combination and interactions of hazardous substances after discharge to the environment also poses hazards that are still not well-understood even today, leading to extensive ecological damage as well as negative impacts on human health.

Many of the worst health impacts have been borne disproportionately by poor and minority populations, with landfills, incinerators and other potentially toxic facilities located disproportionately in poor or minority neighbourhoods. Known as environmental discrimination, the location of such facilities in these areas is often influenced by the lobbying power of majority populations to locate undesirable activities 'Not In My Back Yard' (NIMBY). Poor and minority populations may have neither the economic wherewithal to move away from these areas, nor the voice and political power to keep toxic developments out of their neigh-bourhoods. Environmental justice is a major element of the ecological preservation movement that seeks to address these injustices (see box), including equitable distribution of environmental risks and benefits, fair and meaningful participation in environmental decision-making, and recognition of the value of local context in finding solutions (Schlosberg 2007).

Limits to growth

Following the first Earth Day, public awareness of the need for concern and action on sustainability-related issues continued to grow. In the early 1970s, a hearty debate among thought leaders Barry Commoner, Paul Ehrlich, John Holdren and others resulted in what is known as the I = PAT (pronounced 'eye-pat') equation (Figure 2.2), which defines the total impact of human activity on the planet as a function of population (P), the per capita level of affluence of that population (A) and the unit impact of the technologies used to achieve that level of affluence (T). In his book *The Closing Circle* (1971), Commoner asserted that post-World War II changes in the type and scale of technologies used in industry were the leading cause of changes observed in the natural environment. Ehrlich, author of the influential book *The Population*

Environmental justice is the fair treatment and meaningful involvement of all people regardless of race, color, national origin, or income with respect to the development, implementation, and enforcement of environmental laws, regulations, and policies.

(U.S. Environmental Protection Agency)

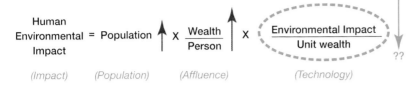

Figure 2.2 Human impact on the Earth as a function of population, affluence and technology (I = PAT)

Bomb (1968), and Holdren countered that population growth was critical in establishing the scale of impact, no matter what technologies might be employed by the population (Ehrlich and Holdren 1972, further countered by Commoner 1972).

Debate about the role of overpopulation as a threat to the global environment continued with the Club of Rome's report *The Limits to Growth* in 1972. This book described the 'World3 model', which used five main variables (world population, industrialization, pollution, food production and resource depletion) to simulate the consequences of interactions between the Earth and human systems (Meadows et al. 1972). This detailed simulation model was based on the same basic ideas captured in the I = PAT equation. In particular, the model assumed that each of the five variables would grow exponentially, while at the same time the ability of technology to increase available resources would grow only linearly. The outcome indicated potentially dire environmental consequences of continuing the current development path, with several scenarios ultimately leading to societal overshoot of available resources and subsequent collapse.

Many readers were skeptical of these predictions, and their reactions sparked intense debate and accusations that the modellers were 'crying wolf' about potential global disasters. More recent updates of the model have been developed (see Meadows et al. 1992, 2012) that continue to support the overall trends identified in the first study. Although the most dire scenarios predicted in the original model have not yet materialized, the authors maintain that key aspects of the model such as ecological overshoot have in fact come to pass, citing evidence of climate change and other significant ecological trends. A 2008 external study comparing the model data with actual data showed a close match for nearly all of the model outputs between 1970 and 2000, supporting its validity (Turner 2008). A more recent review by the same author (Turner 2014) replicated these findings.

One important outcome from the debate on factors important for sustainability has been clarity that there is a significant role for disciplines involved in the design and development of technological solutions to meet human needs. These early models set the stage to better understand the ideas of sustainable development and understand its application to the challenge of developing more sustainable technological solutions.

Sustainability limits and resource peaking

A fundamental assumption of sustainability is the notion of constrained resources, based on the idea that the Earth is essentially a closed system (Figure 2.3). In systems theory, systems are defined as related and organized groups of parts whose behaviour together is more than the sum of its parts (Von Bertalanffy 1969). Two types of systems exist: open and closed. Open systems exchange both matter and energy with their surroundings, whereas closed systems can exchange energy but not matter. The Earth does exchange very limited amounts of matter with its context, including meteors that enter the atmosphere and human-created vehicles for space travel that leave the boundaries of the planet. However, the amount of matter coming into the Earth system is miniscule and unpredictable, and at present, the amount of matter leaving without also returning is extremely small and also very energy-intensive.

For all practical purposes, the Earth may be considered a closed system. It regularly exchanges energy across the boundary of the outer atmosphere, including solar radiation coming in, and waste heat and light radiating back out to space. We also rely on the ability of the Earth to disseminate heat resulting from internal metabolic processes back out to space, preventing global temperature from rising. Climate change is a result of this process happening less efficiently, when greenhouse gases accumulate in the atmosphere and trap waste heat inside the system.

Given that Earth is a closed system, we have limited resources, for all practical purposes at this time, including the matter already existing within our planet's boundaries, plus the energy entering those boundaries from our sun. The sun's incoming energy ultimately drives all

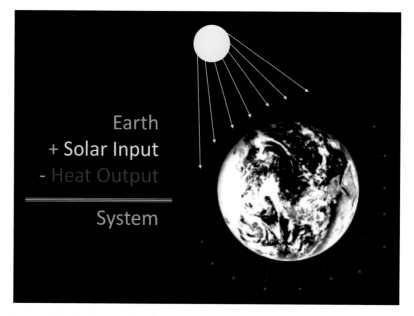

Figure 2.3 Earth as a closed system

processes we use to convert existing matter into useful products to meet human needs. Some of the sun's energy has been stored in the form of fossil fuels over very long periods of time, and it is through the recovery and use of this stored energy that we have been able to support extremely rapid human development and growth. Incoming energy can be tapped using certain human-developed technologies such as photo-voltaics, through the temperature and pressure differentials it creates (e.g. through wind energy), or through passive measures such as heat sinks.

The primary means of transforming incoming solar energy into useful products and services occurs through natural ecosystems, which is one reason that environmental protection and conservation of these systems is so critical. Solar radiation coming into the Earth system is the primary source of energy for all life on Earth, since it drives photosynthesis at the very foundation of the food chain. We also depend on photosynthesis to support plant processes that produce bio-based resources for our use, such as food (fruits, vegetables) and other materials (wood, plant fibre), and to support the biological functions that provide essential ecosystem services such as purifying air and water, and decomposing waste.

Resource peaking happens when we reach the maximum point of production for a finite resource, at which point additional production becomes increasingly more difficult and requires additional energy and resources to accomplish. Resource peaking was originally proposed by geophysicist M. King Hubbert, who created a method to model the expected production curve for oil, given an assumed ultimate recovery volume (Hubbert 1956). The point on the curve where production is highest is known as Hubbert's Peak.

For stored energy such as fossil fuels, it is possible to reach a point where extracting the energy source actually requires more energy than is stored in that which is extracted. At this point, the Energy Return on Investment (EROI), which is the ratio of energy extracted to energy consumed by its extraction, drops below 1.0, and the energy source does not make sense to exploit. Future technological advancements may increase extraction efficiency, at which point the EROI may become great enough to justify extraction at that time. Additional reserves may also be discovered, and the future may bring significant changes in consumption patterns due to factors such as population growth or the development of new technologies requiring a particular limited resource. Such uncertainties make developing reliable resource curves challenging.

Hubbert's peak theory has been applied to finite resources of many types, including fossil fuels such as oil, natural gas and coal; fissionable materials such as uranium; metals such as copper, lithium, platinum and gold; phosphorous and others. It has also been used to analyse overexploitation of water resources, particularly aquifer-stored water resources that are recharged very slowly over long periods of time, but which can quickly be depleted if extraction rates exceed recharge rates. Others have applied the theory to renewable resources such as fisheries and oxygen production by phytoplankton.

Peak resource theory has been criticized by some for the inherent uncertainty involved in predicting key factors including the development of new technologies for resource production, discovery of new resources, the impact of geopolitics on production, and economic demand for resources (Jackson 2006). While it is difficult to produce precise predictions given these uncertainties, overall the patterns predicted by Hubbert's peak theory have correlated well with the overall shape of production curves.

Scenarios for energy descent and climate change

Peak resource theory has provided a basis for the construction of energy descent scenarios that pose possible futures for society subject to post-peak resource conditions, coupled with anticipated environmental stressors such as climate change. David Holmgren, known for his work in the permaculture field, proposed a set of four possible energy descent scenarios based on the speed with which fossil fuels decline and the level of stress imposed by climate change (Holmgren 2009). He calls these scenarios (Figure 2.4):

- Brown Tech, characterized by slow oil decline and fast climate change.
- Green Tech, characterized by slow oil decline and slow climate change.
- Earth Steward, characterized by fast oil decline and slow climate change.
- Lifeboats, characterized by fast oil decline and fast climate change.

The most benign scenario, Green Tech, allows time for new technologies to develop as replacements for old technologies dependent on fossil fuels. The level of chaos and crisis resulting from climate change is also slow, encouraging social developments such as empowerment of women to lower birth rates and redirection of efforts away

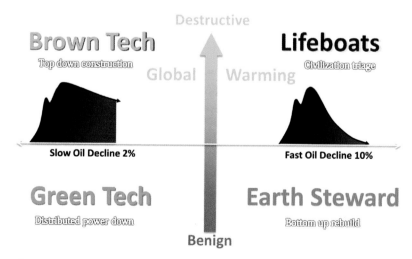

Figure 2.4 Global climate change and energy descent scenarios (adapted from Holmgren 2009)

from defense and resource capture. Instead, these efforts would be directed toward resource conservation and technological innovation.

The most pessimistic scenario, Lifeboats, involves rapid decline in high quality fossil fuels coupled with escalation in the number of situations where those resources are critical. For instance, rapidly rising temperatures are more survivable with mechanical air conditioning technologies, but since these technologies require significant energy to operate, energy shortages might render them unusable. The speed with which change occurs would prevent enough time for innovation, instead requiring energy to be devoted to defense and dealing with societal chaos.

Between these two extremes, the Brown Tech scenario provides a longer period for society to adjust to scarce oil resources, but a shorter time until the threats of climate change cause serious societal disruption such as severe weather patterns, sea level rise, droughts and other problems that require extensive investment and attention to solve. The Earth Steward scenario is opposite in many ways, with rapid decline in oil forcing abandonment of many of the technologies that are presently contributing to climate change such as motorized transport. These changes could potentially slow the rate at which climate change occurs, allowing more time for society to retool and develop alternative approaches.

The four scenarios provide an interesting view of possible futures along with the kinds of factors that may influence human development. They can serve as a useful starting point for the task of defining sustainability, the next step on the path to making more sustainable choices in solving societal problems.

Defining sustainability

The actual term 'sustainable development' was first expressed at the World Conservation Strategy meeting in 1980. The aim of the World Conservation Strategy, developed as a result of that meeting, was to help advance the achievement of sustainable development through the conservation of living resources (IUCN 1980). Following that initial Strategy meeting, definitions of sustainability and sustainable development have continued to evolve as people seek to understand what it means and how this important goal can be achieved in practice.

Our Common Future

Many of the most well-known definitions of sustainability have evolved from the definition of sustainable development coined in *Our Common Future*, the report describing the outcomes of the United Nations World Commission on Environment and Development in 1987 (WCED 1987). This report is also frequently called the *Brundtland Report* in honour of the commission chair Gro Harlem Brundtland, the Prime Minister of Norway. The definition of sustainable development coined in this report (see *The Brundtland Definition*) became widely used in many circles as a baseline for thinking about sustainability ideas.

The Brundtland Definition of Sustainable Development:

A sustainable society is one that meets the needs of the present without compromising the ability of future generations to meet their own needs.

(Our Common Future WCED 1987)

Actions necessary to achieve sustainable development

- Eliminate poverty and deprivation
- Conserve and enhance natural resources
- Encapsulate the concepts of economic growth and social and cultural variations into development
- Incorporate economic growth and ecological decision-making

(The Brundtland Report 1987)

The Brundtland Report emphasized a variety of actions needed to achieve the goals of sustainable development (see *Actions necessary to achieve sustainable development*). To achieve its stated goals, the report emphasized three fundamental components for sustainable development: environmental protection, social equity and economic growth. In the context of these goals, sustainable development must not only minimize environmental problems including global climate change and depletion of fossil fuel resources, but also achieve social components including poverty reduction, equity and well-being, along with economic components such as economic growth and prosperity.

World Bank report on biogeophysical sustainability

In 1995, led by senior managers Mohan Munasinghe and Walter Shearer, the World Bank produced a comprehensive report *Defining and Measuring Sustainability: The Biogeophysical Foundations* that made great strides in clarifying the meaning of sustainability and its applications across many disciplines. In particular, they focused on biogeophysical sustainability, defining it as 'the maintenance and/or improvement of the integrity of the life-support system on Earth'. Driving the group's discussion was the goal of maximizing future options for current and future generations of humans, using the Earth's biogeophysical life-support systems. Inherent in this definition is the idea that sustainability is ultimately an anthropocentric or 'people-centred' concept, going beyond the notions of deep ecology that natural systems comprising the Earth's life-support system should be preserved for their own sake.

One of the most well-known ideas incorporated in the report is Munasinghe's Triangle of approaches to sustainable development (Figure 2.5). The triangle relates the perspectives of three distinct but complementary disciplines working toward societal sustainability:

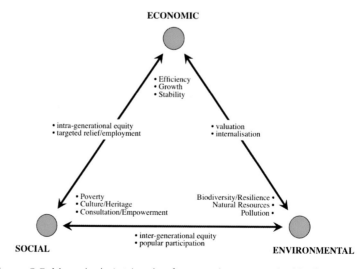

Figure 2.5 Munasinghe's triangle of approaches to sustainable development (adapted from Munasinghe 1993)

economics, physical science and social science. All three perspectives are important to achieve sustainability.

The economist's perspective focuses primarily on preserving production capabilities of capital stock for meeting human needs and aspirations, including human-created capital such as factories, natural capital such as the ecosystems that provide raw materials for those factories, or both together as a system. The physical sciences represent the environmental perspective, focusing primarily on the integrity of biological and physical systems that comprise natural capital, which supply the human capital of interest to economists. The third perspective, social science, considers the demand side of the equation and how diverse social and cultural systems can adapt to changing conditions to meet human needs. Connecting these different perspectives are considerations such as equity and justice, stakeholder participation, and valuation of ecological and social systems as part of transactions between them.

The World Bank Report compiled multiple perspectives on sustainability across key disciplines and brought together thought leaders to discuss important considerations necessary to achieve it, including representatives of both social and natural sciences. One interesting outcome was an inventory of the social ills that human development must address (see *Ills that development must address*).

Mohan Munasinghe went on to serve as Vice-chair of the Intergovernmental Panel on Climate Change (IPCC), author of a series of reports establishing the credibility of and emphasizing the serious implications of climate change, both for Earth systems and for human prosperity and survival as Earth's inhabitants. In 2007, the IPCC shared the Nobel Peace Prize with former Vice-President of the United States Al Gore for its role in bringing climate change to the forefront of attention.

Pillars of sustainability and the Triple Bottom Line (TBL)

Munasinghe's triangle and the ideas identified in the Brundtland Report set the stage for what would become a widely accepted way of conceptualizing sustainability at a societal level: the three pillars of sustainability (Figure 2.6). This idea has been employed as a way to organize indicators when evaluating or measuring the sustainability of systems. It has also been the source of fierce debate regarding what should be the relationship among the three aspects of sustainability and how tradeoffs should be resolved among them.

One argument within the debate pertains to 'Weak' vs. 'Strong' sustainability (Figure 2.7). The question in this debate is whether it is reasonable or possible to think of human capital and natural capital as interchangeable or *fungible* in meeting human needs. In the weak sustainability paradigm (based on the work of Robert Solow among others), human needs are met in various ways using resources both from the natural environment as well as developed by humans. For example, suppose a forest was depleted by logging to produce timber for house construction. Under weak sustainability, depleting the forest would not

Ills that development must address

- Perverse conditions
- Poverty
- Impoverishment of environment
- Possibility of war
- Oppression of human rights
- Wastage of human potential
- Driving forces
- Excessive population growth
- Maldistribution of consumption and investment
- Misuse of technology
- Corruption and mismanagement
- Powerlessness of the victims
- Underlying human frailties
- Greed, selfishness, intolerance and shortsightedness
- Ignorance, stupidity, apathy and denial

(Holdren et al. 1995)

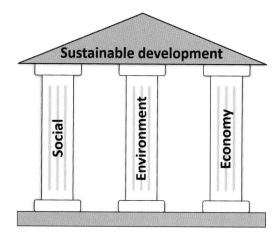

Figure 2.6 Pillars of sustainability

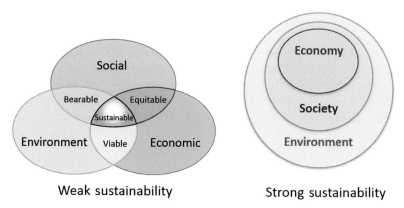

Weak sustainability   Strong sustainability

Figure 2.7 Weak vs. strong sustainability

be unsustainable if new technologies for manufacturing homes were concurrently being developed to replace the forest, perhaps by workers living in the timber homes who worked for a factory making bricks. In other words, as long as the overall capacity for meeting human needs remains the same through growth of human-created capital, ecological capital may be depleted. Natural capital stocks are allowed to diminish over time as long as human capital is increased.

The other school of thought, strong sustainability, asserts that not all resources are interchangeable, although they may be complementary (based on the work of Herman Daly and others). This perspective asserts that some functions of natural ecosystems cannot be duplicated by humans or human-made capital, and often human-made capital is dependent upon complex interactions with natural capital to serve its purpose (see *Ecosystem services*). One commonly cited example of natural capital that cannot be easily replaced by human technology is the ozone layer, which provides a barrier to harmful radiation for plants and

Ecosystem services

Provisioning services for human activities

- Food and conditions to grow it
- Raw materials for human enterprise
- Fresh water
- Medicinal resources
- Heat and light energy

Regulating services for human needs

- Local climate stability
- Global climate stability
- Regulation of atmospheric chemistry and air quality
- Protection against UV radiation
- Carbon sequestration and storage
- Moderation of extreme weather events
- Treatment of wastewater

- Decomposition of solid waste
- Pollination for food production
- Soil fertility and stabilization
- Biological control of pests

Cultural services for humans

- Recreation and mental/physical health
- Tourism
- Aesthetic appreciation
- Inspiration for culture, art and design
- Spiritual experiences
- Sense of place

Habitat/supporting services for other species

- Habitats for other species
- Genetic diversity of other species

animals on the Earth's surface. As a way of illustrating the essential relationships in the strong sustainability perspective, Daly and Cobb (1989) ask the question, 'What use is a sawmill without a forest?'

John Elkington is commonly credited with transforming the three pillars concept into a business-focused framework called the Triple Bottom Line in his 1997 book *Cannibals with Forks: Triple Bottom Line of 21st Century Business.* He begins the book by asking, in the words of Polish poet Stanislaw Lec, 'Is it progress if a cannibal uses a fork?' The cannibal's fork, in his book, is a metaphor for a new way of doing business in what has historically been a climate focused only on cutthroat profit. The fork has three prongs: economic prosperity, environmental quality and social justice. Other authors have proposed that additional concepts such as future generations should be included as a separate category, resulting in a 'quadruple bottom line' (e.g. Waite 2013).

Heinberg's axioms of sustainability

In 2007, Richard Heinberg, known for his extensive work in resource peaking implications for society, proposed five axioms based on a review of the literature on sustainability in multiple domains (Heinberg 2008). Heinberg's axioms draw upon existing work such as Bartlett's 17 Laws of Sustainability (Bartlett 1998), Joseph Tainter's work *The Collapse of Complex Societies,* and other work as part of his synthesis of the state of knowledge. The result is a total of five axioms or self-evident truths that define sustainability. Together, the axioms had to meet multiple criteria to be considered valid, including capability of being tested using scientific methods, mutual exclusivity, collective exhaustiveness and

Heinberg's five axioms of sustainability (2008)

1 **Tainter's Axiom:** Any society that continues to use critical resources unsustainably will collapse.
Exception: A society can avoid collapse by finding replacement resources.
Limit to the exception: In a finite world, the number of possible replacements is also finite.

2 **Bartlett's Axiom:** Population growth and/or growth in the rates of consumption of resources cannot be sustained.

3 To be sustainable, the use of renewable resources must proceed at a rate that is less than or equal to the rate of natural replenishment.

4 To be sustainable, the use of non-renewable resources must proceed at a rate that is declining, and the rate of decline must be greater than or equal to the rate of depletion.
The rate of depletion is defined as the amount being extracted and used during a specified time interval (usually a year) as a percentage of the amount left to extract.

5 Sustainability requires that substances introduced into the environment from human activities be minimized and rendered harmless to biosphere functions.
In cases where pollution from the extraction and consumption of non-renewable resources that has proceeded at expanding rates for some time threatens the viability of ecosystems, reduction in the rates of extraction and consumption of those resources may need to occur at a rate greater than the rate of depletion.

being worded in terms understandable to laypeople. The resulting axioms (see *Heinberg's five axioms of sustainability*) concisely summarize the broad and sometimes conflicting notions of sustainability in the literature. Heinberg explicitly opted to not include an axiom pertaining to social equity, since he believes that adherence to the five axioms as stated will 'tend to lead to relatively greater levels of economic and political equality'. Together, these axioms serve as a concise framework for thinking about the minimum requirements for sustainability at the societal scale.

Achieving sustainable development

Both individuals and organizations have continued to debate what sustainability actually means, although there is broad consensus that our current path is not sustainable. However, in the early 1990s, a group convened to address the challenge of what would be necessary to actually achieve sustainability. The 1992 Earth Summit was the first significant international initiative following the Brundtland Commission to address this task, followed by other summits and collective plans including the United Nations Millennium Development Goals and Sustainable Development Goals.

Agenda 21

The United Nations Conference on Environment and Development (UNCED) was held in Rio de Janeiro in 1992 and was known popularly

Agenda 21 principles (1992)

Principle 1: Human beings are at the centre of concerns for sustainable development. They are entitled to a healthy and productive life in harmony with nature.

Principle 2: States have, in accordance with the Charter of the United Nations and the principles of international law, the sovereign right to exploit their own resources pursuant to their own environmental and developmental policies, and the responsibility to ensure that activities within their jurisdiction or control do not cause damage to the environment of other States or of areas beyond the limits of national jurisdiction.

Principle 3: The right to development must be fulfilled so as to equitably meet developmental and environmental needs of present and future generations.

Principle 4: In order to achieve sustainable development, environmental protection shall constitute an integral part of the development process and cannot be considered in isolation from it.

Principle 5: All States and all people shall cooperate in the essential task of eradicating poverty as an indispensable requirement for sustainable development, in order to decrease the disparities in standards of living and better meet the needs of the majority of the people of the world.

Principle 6: The special situation and needs of developing countries, particularly the least developed and those most environmentally vulnerable, shall be given special priority. International actions in the field of environment and development should also address the interests and needs of all countries.

Principle 7: States shall cooperate in a spirit of global partnership to conserve, protect, and restore the health and integrity of the Earth's ecosystem. In view of the different contributions to global environmental degradation, States have common but differentiated responsibilities. The developed countries acknowledge the responsibility that they bear in the international pursuit of sustainable development in view of the pressures their societies place on the global environment and of the technologies and financial resources they command.

Principle 8: To achieve sustainable development and a higher quality of life for all people, States should reduce and eliminate unsustainable patterns of production and consumption and promote appropriate demographic policies.

Principle 9: States should cooperate to strengthen endogenous capacity-building for sustainable development by improving scientific understanding through exchanges of scientific and technological knowledge, and by enhancing the development, adaptation, diffusion and transfer of technologies, including new and innovative technologies.

Principle 10: Environmental issues are best handled with the participation of all concerned citizens, at the relevant level. At the national level, each individual shall have appropriate access to information concerning the environment that is held by public authorities, including information on hazardous materials and activities in their communities, and the opportunity to participate in decision-making processes. States shall facilitate and encourage public awareness and participation by making information widely available. Effective access to judicial and administrative proceedings, including redress/remedy, shall be provided.

Principle 11: States shall enact effective environmental legislation. Environmental standards, management objectives and priorities should reflect the environmental and developmental context to which they apply. Standards applied by some countries may be inappropriate and of unwarranted economic and social cost to other countries, in particular developing countries.

continued

Principle 12: States should cooperate to promote a supportive and open international economic system that would lead to economic growth and sustainable development in all countries, to better address the problems of environmental degradation. Trade policy measures for environmental purposes should not constitute a means of arbitrary or unjustifiable discrimination or a disguised restriction on international trade. Unilateral actions to deal with environmental challenges outside the jurisdiction of the importing country should be avoided. Environmental measures addressing transboundary or global environmental problems should, as far as possible, be based on an international consensus.

Principle 13: States shall develop national law regarding liability and compensation for the victims of pollution and other environmental damage. States shall also cooperate in an expeditious and more determined manner to develop further international law regarding liability/compensation for adverse effects of environmental damage caused by activities within their jurisdiction or control to areas beyond their jurisdiction.

Principle 14: States should effectively cooperate to discourage or prevent the relocation and transfer to other States of any activities and substances that cause severe environmental degradation or are found to be harmful to human health.

Principle 15: In order to protect the environment, the precautionary approach shall be widely applied by States according to their capabilities. Where there are threats of serious or irreversible damage, lack of full scientific certainty shall not be used as a reason for postponing cost-effective measures to prevent environmental degradation.

Principle 16: National authorities should endeavour to promote the internalization of environmental costs and the use of economic instruments, taking into account the approach that the polluter should, in principle, bear the cost of pollution, with due regard to the public interest and without distorting international trade/investment.

Principle 17: Environmental impact assessment, as a national instrument, shall be undertaken for proposed activities that are likely to have a significant adverse impact on the environment and are subject to a decision of a competent national authority.

Principle 18: States shall immediately notify other States of any natural disasters or other emergencies that are likely to produce sudden harmful effects on the environment of those States. Every effort shall be made by the international community to help States so afflicted.

Principle 19: States shall provide prior and timely notification and relevant information to potentially affected States on activities that may have a significant adverse transboundary environmental effect and shall consult with those States at an early stage and in good faith.

Principle 20: Women have a vital role in environmental management and development. Their full participation is therefore essential to achieve sustainable development.

Principle 21: The creativity, ideals and courage of the youth of the world should be mobilized to forge a global partnership in order to achieve sustainable development and ensure a better future for all.

Principle 22: Indigenous people and their communities and other local communities have a vital role in environmental management and development because of their knowledge and traditional practices. States should recognize and duly support their identity, culture and interests, and enable their effective participation in the achievement of sustainable development.

continued

Principle 23: The environment and natural resources of people under oppression, domination and occupation shall be protected.

Principle 24: Warfare is inherently destructive of sustainable development. States shall therefore respect international law providing protection for the environment in times of armed conflict and cooperate in its further development, as necessary.

Principle 25: Peace, development and environmental protection are interdependent and indivisible.

Principle 26: States shall resolve all their environmental disputes peacefully and by appropriate means in accordance with the Charter of the United Nations.

Principle 27: States and people shall cooperate in good faith and in a spirit of partnership in the fulfillment of the principles embodied in this Declaration and in the further development of international law in the field of sustainable development.

as the Earth Summit. At this Summit, sustainable development was discussed with the primary goal of 'com[ing] to an understanding of "development" that would support socio-economic development and prevent the continued deterioration of the environment, and to lay a foundation for a global partnership between developing and the more industrialized countries, based on mutual needs and common interests, that would ensure a healthy future for the planet' (UNCED 1992). At the summit, 178 countries' governments adopted three major agreements aimed at changing the traditional approach to development:

Agenda 21 – a comprehensive programme for global action in all areas of sustainable development (see *Agenda 21 principles*).

The Rio Declaration on Environment and Development – a series of principles defining the rights and responsibilities of States.

The Statement of Forest Principles – a series of principles to underlie the sustainable management of forests worldwide.

Implementing Agenda 21 was a key task given to the United Nations to help governments integrate the concept of sustainable development into all relevant policies and areas. The other two documents were legally binding international treaties.

In 2012, a 20-year follow-up to the original Earth Summit was held in the same city (Rio+20) to revisit the economic and environmental goals of the global community. The ten-day summit included participation from 192 UN member states, in many cases high-level elected heads of state, with the aim of developing a focused political document to shape global environmental policy. The primary result of the conference was a 49-page working paper titled 'The Future We Want'. While the document was non-binding, it contained political commitments from the participating heads of state to promoting a sustainable future. It also set the stage for the development of new Sustainable Development Goals to comprise the 2030 Agenda for Sustainable Development reflected in the report *Transforming Our World*. These goals build on the Millennium Development Goals created in 2000.

Case study: United Nations Office Complex, Nairobi, Kenya

A significant organization for sustainability today is the United Nations Environment Programme (UNEP) that was established after the UN Conference on the Human Environment in 1972. The mission of UNEP is to provide leadership and encourage partnership in caring for the environment by inspiring, informing and enabling nations and peoples to improve their quality of life without compromising that of future generations. UNEP currently undertakes a variety of actions to promote sustainability, including (UNEP 2016):

- Assessing global, regional and national environmental conditions and trends.
- Developing international agreements and national environmental instruments.
- Strengthening institutions for the wise management of the environment.
- Facilitating the transfer of knowledge and technology for sustainable development.
- Encouraging new partnerships and mind-sets within civil society and the private sector.

One of the key tasks UNEP has undertaken to achieve the goals of sustainability was to make UNEP itself climate-neutral through employing sustainable design and construction features in its new headquarters building in Nairobi, Kenya.

Opened in spring 2011, the 215,000 sq ft United Nations Office Complex located in Nairobi, Kenya has received acclaim for its innovative features designed to reduce the carbon footprint of the building. Part of the UN 'Greening the Blue' initiative, the project's four buildings house the headquarters of the United Nations Environment Programme (UNEP) and the UN Human Settlements Programme (UN-HABITAT). Approximately 1200 staff are housed in the complex, which is net zero energy or energy neutral due to its 6000 sq m of solar panels and energy-efficient design features including extensive daylighting, innovative IT design and elimination of mechanical heating, ventilation and air conditioning (HVAC). While the facility is not completely energy independent, over the course of a given year, the grid-tied complex is expected to produce as much or more power than it consumes overall. Additional features of the complex include:

- A smart location in Nairobi, where the mild climate means that no mechanical ventilation, heating, or cooling is required, and within the city itself, as part of a 'green lung' area on the city's outskirts that serves as a habitat for indigenous trees, birds and small mammals.
- Rainwater collection from roof areas that feeds fountains and ponds at the four entrances to the complex, eliminating the use of potable water for this purpose.
- Water reuse for landscaping from sewage treated on site through a state-of-the-art aeration system, and use of indigenous plants to minimize irrigation needs.
- Daylighting design using central atria and light wells of toughened glass located at floor level, stacked to allow light penetration to ground level and supplemented with low-energy automated lighting systems.
- Central atrium using a thermal chimney effect to naturally ventilate the buildings.
- Four distinct landscaped areas along the atria in the centre of the buildings (representing Kenya's four major climatic zones of coastal, desert, savanna and high-altitude forest) that are designed to require minimal water, encourage biodiversity, and provide cool and beautiful interior gardens.
- Innovative data centre design using information technology pre-assembled component (ITPAC) external server rooms, which employ negative air pressure to draw air through cool water to manage server temperatures, thus removing the need for air conditioning.

continued

- Use of notebook computers instead of desktop PCs to reduce plug loads.
- Solar hot water heating for kitchen areas.
- Operable windows of high quality solar glass and open office plans.
- North–south orientation to achieve maximum daylight with minimum heat intake.
- High recycled content carpet that is also recyclable, and environmentally friendly paint.
- Dual-flush toilets, expected to reduce water use in bathrooms by up to 60 per cent.
- Sustainability guidelines for occupants to explain the building's function and how to use it correctly.

In keeping with its commitment to environmental sustainability, UNEP together with the United Nations Office at Nairobi (UNON) and the UN Human Settlements Program (UN-Habitat) signed a new Environmental Commitment Charter on World Environment Day on June 2015 as an affirmation of their ongoing efforts toward greater sustainability (UNEP 2015a). Efforts undertaken at the complex in Nairobi included reduction of meat consumption by replacing meat with vegetarian options in the cafeteria one day per week, and encouraging staff to make individual commitments to reduce their own environmental footprints through commitments on social media, their website and a writing-wall. UNEP is also undertaking efforts to offset carbon emissions associated with its meetings and air travel of staff (UNEP 2015b), as well as implementing an Environmental Management System, maintaining climate neutrality in operations via the use of offsets, and developing an online environmental sustainability tutorial for its staff (UNEP 2016).

Figure 2.8 The central atrium in each building contains a landscaped area featuring one of the four major climatic zones of Kenya. All atria use minimal water and highlight biodiversity.

Figure 2.9 Translucent glass blocks provide a path for light from rooftop skylights to pass through to lower floors.

continued

Figure 2.10 Local contractors were used during construction to help build capacity for green building.

Sources

UNEP. (2011). 'Building for the future: a United Nations showcase in Nairobi.' www.unep.org/gc/gc26/Building-for-the-Future.pdf (accessed September 17, 2016).

UNEP. (2015a). 'UNON commits to greening UN Nairobi compound.' www.greeningtheblue.org/news/unon-commits-greening-un-nairobi-compound June 15. (accessed September 17, 2016).

UNEP. (2015b). *Moving Towards a Climate Neutral UN: The UN system's footprint and efforts to reduce it, 2015.* United Nations Environment Programme (UNEP). www.greeningtheblue.org/sites/default/files/MTCNUN-24.11.15-sequential.pdf. (accessed September 17, 2016).

UNEP. (2016a). 'About UNEP.' www.unep.org/newyork/AboutUNEP/tabid/52259/Default.aspx (accessed September 17, 2016).

UNEP. (2016b). 'UNEP sustainability: Our Achievements.' web.unep.org/sustainability/who-we-are/our-achievements (accessed September 17, 2016).

Source: All photos courtesy of United Nations Environment Programme

Millennium Development Goals (MDGs)

The Millennium Development Goals are a set of eight international development goals established following the Millennium Summit of the United Nations in 2000. All 189 participating member states committed to help achieve these eight goals by the year 2015, and funding was allocated by the G8 group of nations in 2005 to support achieving these goals in heavily indebted poor countries.

Most of the MDGs address issues of great relevance for social and economic sustainability. Goal 7 focused specifically on environmental sustainability, including four specific targets:

Target 7A: Integrate the principles of sustainable development into country policies and programs; reverse loss of environmental resources.

Target 7B: Reduce biodiversity loss, achieving, by 2010, a significant reduction in the rate of loss.

Target 7C: Halve, by 2015, the proportion of the population without sustainable access to safe drinking water and basic sanitation.

Target 7D: By 2020, achieve a significant improvement in the lives of at least 100 million slum-dwellers.

Targets 7C and 7D have direct implications for stakeholders of the built environment, given the role of buildings and infrastructure in achieving them. Other MDGs such as Goal 8 (Target 8F: In cooperation with the private sector, make available the benefits of new technologies, especially information and communications) also have implications for the built environment.

Criticisms were levelled against the MDGs and the process used to develop them, claiming that the process lacks legitimacy due to its failure to include the voices of the people they were meant to affect. Specific criticism was levelled against Goal 7: Environmental Sustainability, arguing that insufficient emphasis was placed on this topic and key elements were missing. In particular, critics point out that agriculture was not included in this or any other MGD, despite the fact that many of the people living in poverty who would benefit from the goals are farmers.

As of 2013, progress toward the goals was uneven across countries, with some nations achieving most of their targets and other nations making little to no progress. In 2014, The Sustainable Development Goals were developed to set the stage for the next 15 years of development, superseding the MDGs.

Sustainable Development Goals (SDGs)

Developed in 2014 as a response to the Rio+20 report, the Sustainable Development Goals build on the Millennium Development Goals to continue progress toward sustainable development. Documented in the report 'Transforming our world: The 2030 agenda for sustainable development', there are 17 total goals included (see *United Nations Sustainable Development Grab for 2030*).

United Nations Millennium Development Goals for 2015

1 To eradicate extreme poverty and hunger.
2 To achieve universal primary education.
3 To promote gender equality and empower women.
4 To reduce child mortality.
5 To improve maternal health.
6 To combat HIV/AIDS, malaria and other diseases.
7 To ensure environmental sustainability.
8 To develop a global partnership for development.

United Nations Sustainable Development Goals for 2030

1 **No poverty** – End poverty in all its forms everywhere.
2 **Zero hunger** – End hunger, achieve food security and improved nutrition, and promote sustainable agriculture.
3 **Good health and well-being** – Ensure healthy lives and promote well-being for all at all ages.
4 **Quality education** – Ensure inclusive and equitable quality education and promote lifelong learning opportunities for all.
5 **Gender equality** – Achieve gender equality and empower all women and girls.
6 **Clean water and sanitation** – Ensure availability and sustainable management of water and sanitation for all.
7 **Affordable and clean energy** – Ensure access to affordable, reliable, sustainable and clean energy for all.
8 **Decent work and economic growth** – Promote sustained, inclusive and sustainable economic growth, full and productive employment, and decent work for all.
9 **Industry, innovation and infrastructure** – Build resilient infrastructure, promote inclusive and sustainable industrialization, and foster innovation.
10 **Reduced inequalities** – Reduce inequality within and among countries.
11 **Sustainable cities and communities** – Make cities and human settlements inclusive, safe, resilient and sustainable.
12 **Responsible consumption and production** – Ensure sustainable consumption and production patterns.
13 **Climate action** – Take urgent action to combat climate change and its impacts.
14 **Life below water** – Conserve and sustainably use the oceans, seas and marine resources for sustainable development.
15 **Life on land** – Protect, restore and promote sustainable use of terrestrial ecosystems, sustainably manage forests, combat desertification, halt and reverse land degradation, and halt biodiversity loss.
16 **Peace, justice and strong institutions** – Promote peaceful and inclusive societies for sustainable development, provide access to justice for all, and build effective, accountable and inclusive institutions at all levels.
17 **Partnerships for the goals** – Strengthen the means of implementation and revitalize the global partnership for sustainable development.

As with previous sets of goals, criticism has been levelled against these goals for not being ambitious enough, and some have said that the goals may ultimately work against one another. For instance, there is concern that achieving goals for economic development may make achieving ecological goals difficult. While some critics believe the goals are too many, the SDGs cover a far more comprehensive set of issues than their predecessors, the MDGs. Too many targets may be overwhelming to some, but to others, having more specific goals could help them see how to engage in finding specific solutions.

Links have been identified between the SDGs and subsequent policy initiatives including the 2015 Paris Climate Deal. Ultimately, all of the issues identified in the implementation frameworks in this section are part of the larger ecological and societal system of which we are a part, and they cannot be considered in isolation. Understanding these complex issues require a larger perspective that considers consequences across multiple dimensions, from a whole systems perspective.

Five core principles of sustainability

In 2005, Michael Ben-Eli of non-profit organization The Sustainability Laboratory proposed five core principles of sustainability as a comprehensive and holistic road map to develop model sustainability practices (Ben-Eli 2006). These principles address all of the sustainability considerations presented so far in this chapter in five key domains: the Material Domain, the Economic Domain, the Domain of Life, the Social Domain and the Spiritual Domain. Each domain has a defining principle that offers guidance for action to move toward sustainability, at the individual and societal levels (see *Five core principles of sustainability*). In addition to the basic science that is the focus of many who try to operationalize sustainability, Ben-Eli's principles extend into the realm of ethics and values, which are critical for prosperity and happiness. The fifth principle helps shift decision makers into an attitude that values sustainability not just as a necessity for basic survival, but also as a stepping stone for our transformation as humans to thrive in partnership with others on Earth and the systems of which it is a part.

The five core principles of sustainability

1 **The Material Domain**: Contain entropy and ensure that the flow of resources, through and within the economy, is as nearly non-declining as is permitted by physical laws.

2 **The Economic Domain**: Adopt an appropriate accounting system to guide the economy, fully aligned with the planet's ecological processes and reflecting true, comprehensive biospheric pricing.

3 **The Domain of Life**: Ensure that the essential diversity of all forms of life in the biosphere is maintained.

4 **The Social Domain**: Maximize degrees of freedom and potential self-realization of all humans without any individual or group adversely affecting others.

5 **The Spiritual Domain**: Recognize the seamless, dynamic continuum of mystery, wisdom, love, energy and matter that links the outer reaches of the cosmos with our solar system, our planet and its biosphere, including all humans with our internal metabolic systems, and their externalized extensions. Embody this recognition in a universal ethics for guiding human actions.

(Ben-Eli 2006)

Evolution of applied sustainability concepts and ideas

In parallel with the evolution of sustainability theory, many important advances have been made to apply this theory to the solution of societal problems. This section of the chapter presents some of the most well-known examples of applied sustainability frameworks that have been used to guide design and decision making in multiple fields.

The Natural Step system conditions

The Natural Step is a non-profit, non-governmental organization originally founded by Dr. Karl-Henrik Robèrt in Sweden in 1989. Dr. Robèrt is an oncologist who was originally inspired to act when he noticed an increase in cancer rates among pediatric patients. The fact that this increase was observed in children rather than adults, who might have influenced disease through their lifestyle choices, led Robèrt to the notion of 'molecular garbage' that accumulates in the environment and can be found at the molecular level (e.g. in the breast milk of nursing mothers). In 1989, he authored a working paper describing basic system conditions for sustainability at a societal scale, and sent it to 50 scientists for review and comment (see Robèrt et al.1994). After repeated cycling of the ideas with participating scientists, what are now known as the Natural Step system conditions (see *Natural Step system conditions*) resulted after 22 different cycles. Two additional iterations were later undertaken to further refine the conditions.

At first glance, the Natural Step system conditions may appear to be similar to previous sets of axioms and principles expressed in the first part of the chapter. However, they are expressed in a way that makes their application to design much easier to understand, at the society, organization or product level. DuBose and Pearce (1997), for instance, applied the system conditions to develop a framework for evaluating the sustainability of built facilities.

The system conditions focus on problems which threaten overall health and stability of the natural and human systems on which society

---

Natural Step system conditions

In a sustainable society, nature is not subject to systematically increasing . . .

1   . . . concentrations of substances from the earth's crust (such as fossil CO_2 and heavy metals),
2   . . . concentrations of substances produces by society (such as antibiotics and endocrine disruptors),
3   . . . degradation by physical means (such as deforestation and draining of groundwater tables),
    . . . and in that society . . .
4   . . . there are no structural obstacles to people's health, influence, competence, impartiality and meaning.

(thenaturalstep.org)

depends for its existence. Specific problems underlying the system conditions include the negative impacts on nature of substances being extracted and used from beneath the Earth's crust, such as fossil fuel combustion and heavy metals. Problems such as bioaccumulation of antibiotics and endocrine disrupters in nature are noted in the second system condition. The third condition focuses on avoiding damage to natural ecosystems through physical displacement, over-harvesting or other manipulation. The final system condition focuses on removing barriers to human health, well-being and happiness, without which humans would tend to seek solutions that may be potentially damaging according to the other three conditions.

The negative wording of the statements (i.e. the use of the word 'not' to express what makes something unsustainable) has been both criticized and lauded. Those who favour the wording argue that it leaves to the user of the conditions the opportunity to be creative in finding solutions. In other words, the conditions are not prescriptive but rather act as screening criteria to help evaluate what is *not* sustainable from among a set of possible ideas.

The Natural Step is also known for its use of 'backcasting' as a way to promote sustainable design. Backcasting is a technique that involves working backward from a vision of future success to the present, determining what actions are necessary to end up in the desired state (Figure 2.11). This technique has been applied successfully in many contexts to help stakeholders reach a consensus about how to proceed with changes to complex systems to improve their sustainability. The shared language provided by the system conditions helps overcome disagreement by being clear about what is important to everyone, helping people to focus on what they have in common rather than their differences.

The Hannover Principles

Another systematic application of sustainability to built environment design was begun in 1991 when Hannover, Germany was selected as the site for the EXPO 2000 World's Fair. Officials in Hannover made

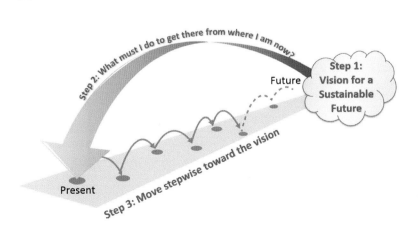

Figure 2.11
Backcasting as a way to move toward sustainability
(The Natural Step)

The Hannover Principles

1 Insist on rights of humanity and nature to co-exist.
2 Recognize interdependence.
3 Respect relationships between spirit and matter.
4 Accept responsibility for the consequences of design.
5 Create safe objects of long-term value.
6 Eliminate the concept of waste.
7 Rely on natural energy flows.
8 Understand the limitations of design.
9 Seek constant improvement by the sharing of knowledge.
(McDonough and Braungart 1992)

the decision to develop the city and facilities for the event to be environmentally sustainable, and sought the input of architect William McDonough and scientist Michael Braungart to develop a set of principles that would guide development in a more sustainable direction. The result was the Hannover Principles (see *The Hannover Principles*), a set of nine guidelines to inspire design solutions to be more sustainable, not only for EXPO 2000 but for other design problems as well (McDonough and Braungart 1992).

Places to intervene in a system

Concurrent with these other activities, efforts were underway to develop a better understanding of the behaviour of large, complex systems to influence their sustainability. In 1997, Donella Meadows, a scientist and systems analyst, proposed a set of nine leverage points representing places to intervene in a system (see *Places to intervene in a system*) in order to change its behaviour. Her work was inspired by her involvement in the North American Free Trade Agreement (NAFTA) in the early 1990s, as she looked for better ways to understand the behaviour and possible effects of this influential policy on global economic, social and environmental systems. In 1999, she revised the list to include twelve specific leverage points.

Meadows arranged her list in ascending order of importance, with the most influential and effective leverage points at the end of the list. She believed that these leverage points apply to complex systems in general, from a societal scale to individual product design. Her aim was to identify specific leverage points, where a 'small shift in one thing can produce big changes in everything'. These leverage points can be used to change the state of the system from the status quo to some desired or goal state.

Places to intervene in a system

1   Constants, parameters and numbers (such as subsidies, taxes and standards).
2   The size of buffers and other stabilizing stocks, relative to their flows.
3   Structure of material stocks and flows (such as transport networks and population age structures)
4   Length of delays, relative to the rate of system changes.
5   Strength of negative feedback loops, relative to the effect they are trying to correct against.
6   Gain around driving positive feedback loops.
7   Structure of information flow (who *does* and *does not* have access to what kinds of information).
8   Rules of the system (such as incentives, punishment and constraints).
9   Power to add, change, evolve or self-organize system structure.
10 Goal of the system.
11 Mindset or paradigm from which the system arises (goals, structure, rules, delays, parameters).
12 Power to transcend paradigms.

(Meadows 1999)

Coupled with a backcasting approach such as used in The Natural Step, these leverage points offer powerful ways to direct change to present conditions to achieve a more sustainable future. Understanding how complex systems function is critical to successfully apply the leverage points. Meadows uses an example of a lake used for water supply and recreation in her paper to show how each of the leverage points can be applied to improve the behaviour of the lake system with respect to citizens who desire an unpolluted lake with water levels adequate to meet their needs. The original work is widely referenced by engineers and designers of human systems and is available free online at http://donellameadows.org.

Eco-efficiency, eco-effectiveness and regenerative design

One of the major ongoing debates in applied sustainability is between eco-efficiency and eco-effectiveness. These two perspectives differ in the philosophy they believe should govern society's approach to achieving sustainability. The eco-efficiency approach focuses on maximizing the efficiency with which resources are used by society to meet human needs, seeking to do 'more with less'. In contrast, the eco-effectiveness approach looks for ways to establish regenerative metabolisms that are supported and enhanced by the flow of materials. In the eco-effectiveness paradigm, material flows are not problems to be minimized but rather essential assets for the ongoing health of the overall system.

The eco-efficiency perspective was articulated by the World Business Council for Sustainable Development in its 1992 publication 'Changing Course' (WBCSD 1992), and it was later endorsed at the 1992 Earth Summit as an approach for achieving Agenda 21 goals in the private sector. In essence, this approach entails continuing to produce goods and services that meet human needs and achieve desired quality of life, while 'progressively reducing environmental impacts of goods and resource intensity throughout the entire lifecycle to a level at least in line with the Earth's estimated carrying capacity' (WBCSD 2000). Eco-efficiency involves the pursuit by business of seven objectives using a technological innovation approach (see *Critical objectives of eco-efficiency*). Eco-efficiency advocates believe that technological innovation, developed by the business sector while pursuing economic growth under free market principles, will ultimately be sufficient to overcome biogeophysical constraints to sustainability.

Eco-efficiency advocates suggest various ways to achieve these key objectives. The Factor Ten approach has been suggested as a guide for the magnitude of improvement that must be achieved to avoid overshooting critical ecological limits. Factor Ten was developed by Friedrich Schmidt-Bleek from the Wuppertal Institute for Climate, Environment, and Energy in the early 1990s. In order to achieve sustainability, Schmidt-Bleek proposed that humanity must reduce resource turnover by 90 per cent on a global scale within one generation (30 to 50 years). He believed that human energy use and resource flows in their current form are

> **Critical objectives of eco-efficiency**
>
> 1 A reduction in the material intensity of goods and services.
> 2 A reduction in the energy intensity of goods and services.
> 3 Reduced dispersion of toxic materials.
> 4 Improved recyclability.
> 5 Maximum use of renewable resources.
> 6 Greater durability of products.
> 7 Increased service intensity of goods and services.
> (WBCSD 2000)

destructive to the environment and ultimately unsustainable. Specifically, he proposed that energy use must *decrease* by a factor of ten and resource productivity and efficiency must *increase* by a factor of ten to remain within sustainable limits. His ideas were captured in the Factor 10 Manifesto (Schmidt-Bleek 2000).

The Ecological Rucksack is a unit of measure that has emerged from the Factor Ten approach (see box). Similar to the idea of Ecological Footprints coined in 1995 by William Rees and Mathis Wackernagel, this metric quantifies resource intensity for a specific product, service or system within a defined boundary. The Factor Ten approach does not directly address issues of eco-toxicity or similar issues. Instead, it focuses on the efficiency with which materials and energy are used to achieve desired outcomes. The ecological footprint metric, unlike ecological rucksacks, does include calculations to account for the ecological burden placed on ecosystems by waste from the product or system being evaluated.

The eco-effectiveness approach was articulated by William McDonough and Michael Braungart in their 2002 book *Cradle to Cradle: Remaking the Way We Make Things*. This paradigm fundamentally differed from eco-efficiency in removing the focus on ever-increasing efficiency of ultimately unsustainable processes. Instead, eco-effectiveness takes a biomimetic approach that designs industrial processes to emulate natural systems. One of the key differences between the two perspectives is their treatment of waste. In eco-efficiency, waste is a necessary by-product of production that should be minimized to the extent possible. Under the eco-effectiveness paradigm, waste from one system should be thought of as food for another. In this way, design solutions are created and evaluated from cradle to cradle (C2C), with material and energy flows cycling indefinitely within the overall system and being recharged with solar input to ecological systems. McDonough and Braungart subsequently developed the C2C product certification system to evaluate products in terms of these criteria, discussed further in Chapter 4. This certification system evaluates a product's material health, level of material reutilization, energy required for production, water usage and discharge quality, and social responsibility.

The differences between these two perspectives are made evident in the concept of regenerative design. Formalized by landscape architecture professor John Tillman Lyle in his 1994 book *Regenerative Design for Sustainable Development*, the idea of regenerative design is to create human systems that are able to regenerate and repair damage caused to the natural world. Lyle identified two different types of systems corresponding to the eco-efficiency and eco-effectiveness paradigms (Figure 2.12). The so-called Paleotechnic or linear throughput paradigm best matches the eco-efficiency paradigm, with the aim being to gain as much as possible from the minimum quantity of resources through the system and minimize the amount of waste generated at the downstream end. The Regenerative or cyclical system corresponds with the eco-effectiveness paradigm, based on the assumptions that resources are not

Ecological Rucksack:
'The total amount of natural material input for manufacturing a product minus the weight of the product itself.'
(Schmidt-Bleek 2000)

Ecological Footprint:
'The amount of land and water area a human population would hypothetically need in order to provide the resources required to support itself and to absorb its wastes, given prevailing technology. In order for humanity in total to be sustainable, the total net area of all ecological footprints should be less than or equal to the total land/water area of the Earth.'
(Rees and Wackernagel 1995)

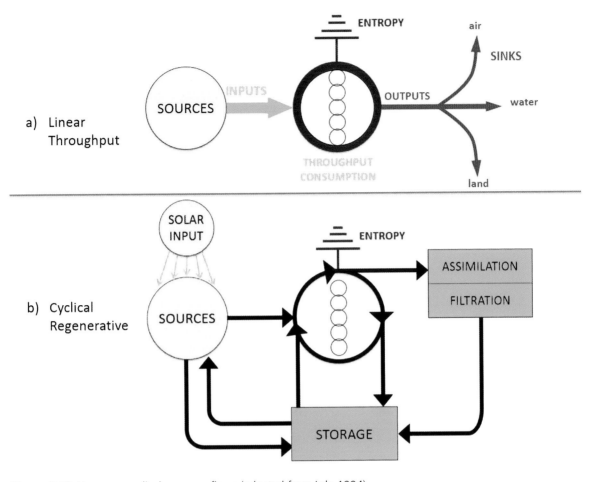

Figure 2.12 Linear vs. cyclical resource flows (adapted from Lyle 1994)

unlimited, should be cycled repeatedly through the system, and are ultimately regenerated through inputs of solar energy to biological systems.

Twelve strategies guide the regenerative design process (see *Strategies for regenerative design*). Over several decades beginning in the 1970s, Lyle worked with students, faculty and experts from multiple disciplines to employ these strategies in the design of an institute at his university California State Polytechnic Institute (Cal Poly) in Pomona, California. The facility was finished around the same time as Lyle's book in 1994. Although Lyle died in 1998, the Lyle Center for Regenerative Studies remains today as a vital part of the Cal Poly College of Environmental Design, offering a graduate degree and undergraduate programs in regenerative studies. The facility itself houses the students who study at the centre as a living laboratory, offering them an opportunity to explore regenerative design principles in the context of living systems.

Strategies for regenerative design

1 Letting Nature do the work.
2 Considering Nature as both model and context.
3 Aggregating, not isolating.
4 Seeking optimum levels for multiple functions, not the maximum or minimum level for any one.
5 Matching technology to need.
6 Using information to replace power.
7 Providing multiple pathways.
8 Seeking common solutions to disparate problems.
9 Managing storage as a key to sustainability.
10 Shaping form to guide flow.
11 Shaping form to manifest process.
12 Prioritizing for sustainability.

(Lyle 1994)

Both eco-efficiency and eco-effectiveness have value to add to the quest for sustainability, and a combination of both philosophies is essential for the pursuit of sustainability in designed systems. Critical to the success of these philosophies is the influence they have on human choices and behaviours, discussed next.

Influencing human decisions and behaviours

The evolution of sustainability science has included many insights about how natural systems and stocks of natural resources require constraints on human behaviour if sustainability requirements for future generations are to be maintained. In parallel to these ideas, thinking about how to influence human decisions and behaviours toward sustainability goals has also advanced. This section describes some of the relevant theory about human behaviour that influences the overall sustainability equation.

Case study: Lodging Complex, City of Knowledge, Panama City, Panama

Architect Rossana Merida and Dr. Rodolfo Valdes-Vasquez

The Lodging Complex is located in the City of Knowledge former military base in Clayton, Panama. The City of Knowledge is a non-profit organization committed to becoming a sustainable city promoting research, innovation and human development as well as creating economically viable and environmentally friendly facilities.

This project was developed from an international competition and consists of four building modules, each with two levels with 24 rooms, and all united by closed corridors with louvers that allow natural ventilation and daylighting. The ground corridor allows for occupants' engagement where they find a cafeteria, sitting area, meeting rooms, and laundry. Each building module is surrounded by beautiful tropical gardens. In addition, the outdoor area has protected large trees that are significant for wildlife habit. Walking and biking paths and access to public transportation encourages

Table 2.1 LEED Scorecard – The Lodging Complex

LEED BD+C: New Construction v3.2009	
Sustainable sites	24/28
Water efficiency	7/10
Energy and atmosphere	22/35
Material and resources	7/14
Indoor environmental quality	10/15
Innovation	6/6
Regional priority credits	4/4

continued

Figure 2.14 Between building modules, the connecting corridors have louvers which allow both natural ventilation and daylighting

Figure 2.13 The exterior of the complex features the ground corridors and the building modules

occupants to use alternative modes of transportation, aiming to decrease CO_2 emissions and connecting to nature.

The Lodging Complex is the first green residential complex in Panama to achieve LEED Platinum, the highest level of certification awarded by the USGBC. The project achieved 80 points and the certification was awarded in May 2013. Table 2.1 shows the respective LEED scorecard. In addition, the complex complies with the environmental strategies of the City of Knowledge. LEED Certification was based on multiple sustainability strategies, including:

- Implementing an erosion control and sedimentation management plan during construction.
- Using approximately 20 per cent of the materials with recycled content.
- Implementing waste management plan during construction, including on-site waste recycling dumpsters 70 per cent of the site with native plants, which require little maintenance and irrigation.
- Using approximately 70 per cent of wood from certified forests.
- Taking full advantage of natural lighting and the inclusion of occupancy sensors, which helps to reduce energy consumption by 23 per cent.
- Applying waterproofing on the roof to decrease the heat island effect and lower the temperature in the internal spaces, saving on energy consumption.
- Targeting 75 per cent of the occupied spaces to have natural lighting and 90 per cent to have an exterior view.
- Including low-flow fixtures and toilets that save 35 per cent of water consumption.

continued

- Using a rainwater collector system for cleaning and irrigation.
- Filtering 90 per cent of water coming out of the building using rainwater channels around the perimeter of each module.
- Using biodegradable materials for cleaning purposes during operation, and creating an educational program for occupants.

Sources: Photos courtesy of Architect Rossana Merida and Dr. Rudolfo Valdes-Vasquez

The Jevons Paradox and the rebound effect

One critique of the eco-efficiency paradigm of sustainability is its failure to consider the unintended consequences of increasing the efficiency of technology. In economics, the Jevons Paradox describes a phenomenon in which resource use actually *increases* following an improvement in the efficiency of a technology or process that uses the resource. Economist William Stanley Jevons was the first to observe this effect in 1865, while studying the use of coal in British industry. Jevons observed that, despite expectations to the contrary, increasing efficiency of technologies using coal actually *increased* the total amount of coal used due to growing demand for products produced with that coal, since the cost of producing these products decreased due to improved efficiency. He also observed that increased efficiency tends to accelerate economic growth, which further increases demand. Ultimately, Jevons concluded that technological progress to improve efficiency is unreliable as a way to conserve resources in a market economy (Jevons 1866).

Since Jevons' original observations, the Jevons Paradox has been explored in the context of a variety of other resources, including fuel efficiency, energy efficiency and water efficiency, and similar effects have been observed. For instance, economists have observed that consumers tend to travel more when their cars are more fuel efficient, causing an increase in overall fuel use that is nevertheless more productive for the consumer due to the lower unit cost of fuel per unit of benefits it delivers (Small and Van Dender 2005). Similarly, increases in road capacity have been noted to increase rather than decrease congestion, an effect known as the Lewis-Mogridge Position (Mogridge 1990). This phenomenon is also known as the rebound effect, where increased efficiency or capacity leads to increased demand that offsets the amount of resource or capacity saved through efficiency. When the amount of resource consumed due to increased demand exceeds the amount saved by efficiencies driving that demand, the Jevons Paradox occurs (Alcott 2008).

Tragedy of the Commons

The Tragedy of the Commons is a second theory describing human behaviour that must be considered in designing approaches to achieve sustainability. Originally coined in an 1833 essay by economist William

Forster Lloyd, the Tragedy of the Commons was brought to the fore by ecologist Garrett Hardin in 1968. Both authors used a hypothetical story to illustrate what would happen when individually owned cows are allowed by their owners to graze unregulated on shared land. In the story, each animal is owned by an individual member of the community, and the land on which they are allowed to graze is owned collectively by the community. Although each owner may understand conceptually the consequences of overgrazing the land, individually each seeks to act in his own best interests to add additional cows to the herd, rationalizing that one small additional demand placed by an additional animal will not be enough to destabilize the system.

In situations where the carrying capacity of the field is much greater than the demands placed by the herd, individual choices made to promote self-interest have notable benefits for the individual but negligible consequences for the community as a whole. However, as the field approaches its capacity to support the size of the current herd, each additional burden placed by a self-interested individual pushes the community resource closer to collapse. The tragedy also applies to other finite but renewable resource bases such as fisheries, water sources and forests, as well as non-renewable resources such as oil and coal. The rate at which the resource is depleted depends on the number of users wishing to consume the resource, the intensity with which each individual consumes, and the overall resilience of the resource base in response to consumption (Daniels 2008).

Hardin argued that individuals must find solutions to such problems through their relationship with society and the fellow individuals that comprise it. He explicitly argued against the use of altruism and conscience as a reliable means of managing the commons, where individuals are trusted to take no more than their fair share. Altruism, he argued, is inherently self-extinguishing in that it allows advantages to selfish individuals who ignore the rules, known as free riders, compared to those who behave altruistically. Ultimately, the Tragedy of the Commons highlights the need for social policies and interventions to guide and control individual behaviours in the market.

Market-based instruments (MBIs)

Both the Jevons Paradox and the Tragedy of the Commons occur in economic systems where individuals make rational choices to optimize individual outcomes at the expense of society as a whole. Along with other challenges of free market economics discussed further in Chapter 9, individual choices, be they rational or irrational, can often cause unanticipated consequences in systems behaviour that result in undesirable consequences for sustainability. Market-based instruments (MBIs) are policy interventions used in free markets to control and influence these effects.

MBIs influence the behaviour of the market by incorporating costs that are often externalized to others or to society as a whole. For instance, the cost of pollution is ultimately borne by society as a whole, which must suffer from degradation of environmental quality and individual health as a result. The benefits of the economic processes causing that pollution, on the other hand, accrue to the individual or organization that causes the pollution. As long as the societal costs of pollution are not borne by the individual polluter, that individual has no incentive to control the pollution to reduce societal costs. Instead, the rational economic choice is to continue the polluting behaviour in order to realize the benefits to the individual.

The notion of 'polluter pays' or emissions trading is one type of MBI in which artificial markets are established where individuals must pay for the right to emit pollutants as part of economic activities. These markets impose externalized costs of pollution onto polluters, incentivizing them to reduce pollution to avoid those costs. Within these markets, individuals who are able to reduce their own levels of pollution at lower cost may sell credits for their reductions to others for whom it is more expensive. Often referred to as 'cap and trade', these markets establish maximum permissible levels of pollution (caps) for each individual that must be achieved. These caps can either be achieved by technological process improvements or through trading of pollution rights with other individuals. Other types of MBIs include environmental taxes on negative or undesirable behaviours, and deposit-refund systems where penalties can be refunded upon remediation of undesirable behaviour.

Some argue that irrational systems-level behaviour such as the Jevons Paradox and its analogues mean that eco-efficiency strategies will ultimately not be able to achieve desired sustainability outcomes. However, conservation policies such as cap and trade have been found to control the rebound effect leading to the Jevons Paradox by artificially maintaining the cost of using resources at previous levels (Wackernagel and Rees 1997; Freire-González and Puig-Ventosa 2015).

Other types of policy mechanisms to influence behaviour include voluntary agreements and regulatory instruments. Generally, MBIs are implemented with participation and enforcement using these other mechanisms. Labelling requirements for producers and products are an example of an indirect market mechanism that can help consumers in the market make informed purchasing and investment choices. Overall, these mechanisms have helped to even the playing field by internalizing the negative costs to society of unsustainable behaviour. However, policy can also be used to negative ends in designing solutions to societal problems, discussed next.

Product semantics and the politics of artefacts

Product semantics refers to the meaning of a product to users and others affected by its use. Policies are one means by which both positive and negative semantics can be imposed upon the design of technological solutions, and the semantics of a product can influence the sustainability impacts of its use.

In a 1980 article 'Do Artifacts Have Politics?' political scientist Langdon Winner discussed the ways in which technologies can influence the behaviour and social outcomes of their users and others who are affected by their use. Illustrating the power of policy to influence design and subsequently social and environmental outcomes of that design, Winner presents an example of tunnels on New York City's Long Island Parkway. Urban planner Robert Moses, designer of the tunnels, purportedly chose the clearance height of the tunnels to be too low to permit public transit buses to use the parkway. This design decision deliberately excluded access by transit and the poor and minority people who used it from being able to enjoy the beaches of Long Island. From the standpoint of one group of stakeholders, the outcome of this design decision could have been considered positive, whereas for others, it was arguably negative (Winner 1980). In both cases, the parameters selected to define the design had undeniable social consequences.

In a similar fashion, design decisions have consequences for both societal and environmental outcomes, leading to sustainability impacts that result from their use. For example, the design of American communities to accommodate motor vehicles has led to neighbourhoods in which alternative modes of transportation such as walking or bicycling are impractical, dangerous or impossible. Approaches to designing walkable neighbourhoods are now being rediscovered and implemented as part of comprehensive planning policy in many neighbourhoods, not just to promote more environmentally friendly alternatives to travel by car, but also to improve the health and increase healthy choices by residents.

With its role in shaping and regulating the design of technological solutions, policy has the power to differentially reward some interests while disadvantaging others. When encapsulated as policy, design parameters, prescriptive requirements and even performance goals can have unanticipated consequences not only for the beneficiaries of those artefacts but also other humans and ecosystems. Such outcomes may be difficult to predict and are often impossible to identify from within the socioeconomic system driving their development. Challenges such as this are one driving factor behind the Precautionary Principle, discussed next.

The Precautionary Principle

The Precautionary Principle arose from the idea that under conditions of uncertainty, it is often best to determine that a course of action will not cause irreversible problems before deciding to implement it. The principle resonates with common aphorisms in multiple cultures, including the ancient medical principle of 'first, do no harm' and the more contemporary 'better safe than sorry'.

The idea gained wider attention within the sustainability movement as a result of the United Nations Rio Conference on Environment and Development (Earth Summit) in 1991, and was mentioned explicitly in

Agenda 21, Principle 15. The 1998 Wingspread Statement on the Precautionary Principle states the principle in this way:

> When an activity raises threats of harm to human health or the environment, precautionary measures should be taken even if some cause and effect relationships are not fully established scientifically.

The Precautionary Principle places the burden of proof on the proponent of an action to provide reasonable evidence that the action is not likely to cause significant harm. This idea is contradictory to the notion of 'innocent until proven guilty', which has widely guided economic behaviour and requires proof of harm to demand that an action be stopped. R.B. Stewart has articulated the Precautionary Principle in terms of four key elements (see *Four operational elements of the Precuationary Princinple*).

The Precautionary Principle brings to light the idea that potential social and environmental outcomes of technological design must be considered as a critical part of design. Also critical is the idea that design can deliberately be developed to influence the behaviour of humans that use designed artefacts, discussed next.

Four operational elements of the Precautionary Principle

1 **Non-preclusion**: Scientific uncertainty should not automatically preclude regulation of activities that pose a risk of significant harm.

2 **Margin of safety**: Regulatory controls should incorporate a margin of safety; activities should be limited below the level at which no adverse effect has been observed or predicted.

3 **Best available technology**: Activities that present an uncertain potential for significant harm should be subject to best technology available requirements to minimize the risk of harm, unless the proponent of the activity shows that they present no appreciable risk of harm.

4 **Prohibition**: Activities that present an uncertain potential for significant harm should be prohibited unless the proponent of the activity shows that it presents no appreciable risk of harm.

(Stewart 2002)

Design, behaviour change and community-based social marketing (CBSM)

Herbert Simon, widely known for his work in human decision making and artificial intelligence, offered a theory of design as early as 1969 that acknowledged its capacity to encourage 'courses of action to change existing situations into preferred ones' (Simon 1969). The idea that a designed object or environment determines the type of actions that can be undertaken with it was initially proposed by James Gibson in the 1970s (Gibson 1977). Donald Norman built further on this idea in the 1980s with his work in environmental psychology and industrial design.

Norman explored ideas from behavioural psychology and human factors as they apply to designed artefacts, including the behaviours or functions an artefact affords its user (Norman 1988). Norman believed that an artefact's design should intuitively facilitate correct use to achieve its desired function, based on inherent attributes of the design itself. For instance, an emergency exit bar placed at hand height on a door affords pushing, making it easy to use the exit door correctly when trying to escape in an emergency, and preventing the possibility of panicked people crowding against an inward-opening door that cannot be opened (Figure 2.15).

The idea of design for behaviour change further built on these ideas, as did the notion that the design of artefacts and technologies can help people achieve a better understanding of what mechanisms allow them to work (see *Learning from energy efficient home design*). With these ideas came increased awareness that design could have both desirable and undesirable consequences that could be brought about not only intentionally but also unintentionally, particularly in complex systems. The notion of socially responsible design sought to deliberately create an intended user experience with the design to encourage desirable behaviour and discourage undesirable behaviour (Tromp et al. 2011).

Figure 2.15 An emergency door push-bar affords correct actions to successfully open the door

Learning from energy efficient home design: saving energy through improved energy literacy

Frederick Paige, Leidy Klotz and Julie Martin

Energy is an invisible and complex form of matter, making it a concept that many struggle to understand. Over the past 60 years, overall U.S. energy consumption has tripled (EIA 2016), despite an abundance of efficiency improvements. Occupants are driving up residential energy consumption by purchasing an increased number of appliances, electronics and lighting devices which have offset heating and cooling efficiency gains (EIA 2013).

Previous studies support the notion that occupant education is a powerful tool to help reduce energy consumption (Janda 2011; Laitner and Ehrardt-Martinez 2009; McMakin et al. 2002; Zografakis et al. 2008). The Department of Energy (DOE) explicitly states that efficient home design requires a whole-house systems approach, in which designers' and occupants' energy literacy plays a role (DOE 2016). Energy literacy is an understanding of the nature and role of energy in the universe and in our lives (DOE 2012). Green materials, decentralized renewable energy sources and sustainable construction practices will all fall short of their full potential when underutilized due to a lack of public understanding. By increasing societal energy literacy, more people will have the knowledge necessary to reduce their own consumption through changing personal behaviours (Ehrhardt-Martinez and Laitner 2009).

To be engaging and relatable, education on energy should be based in the real world (DeWaters et al. 2014), and the home provides concrete and simple lessons that can be utilized to improve

continued

energy literacy. With the assistance of home-related examples, lessons on energy become personally relevant, easy to understand and applicable. In a recent exploratory case study, different ways in which energy efficient homes can influence the energy education of occupants and visitors were discovered. Two cases (a model high efficiency home for affordable sustainable living, and a combination of three weatherization retrofit homes) were selected to investigate a wide range of home features, visitors and occupants. Ten interviews with visitors and home occupants provided rich detail in how the energy literacy of study participants was influenced by their interactions with these energy efficient homes and the technologies they contain.

Energy efficient features like renewable energy sources provided important insights about the effect visibility and interactivity can have on an individual's energy literacy. For example, solar panels are a very publicized renewable energy source, but have varied levels of interactivity. Commercial installations are frequently publicized but are usually located on high inaccessible rooftops. Residential installations have shown to be effective at educating neighbourhoods but they can be easily confused for objects with similar design features such as solar water heaters, or blend into roofs with black shingles. Importantly, a combination of publicity and real-world distribution was found to make solar panels an effective educational tool. When investigating participants' understanding of the various ways electricity is generated, distant yet large and important power plants were talked about as obscure objects.

> When I lived out in California, they got those things that they look like windmills . . . I'm talking about those things make me look like my eyes would cross because they had thousands. It was like a field of them . . . we may not have the wind source to power those type of things.
>
> (Julie sharing how wind farms are a very visible example of how humans can generate electricity in a variety of ways)

In contrast, engaging micro generators were discussed with a confident understanding:

> It was a washing machine that ran on a bike, so instead of using electricity to spin, you spin. You would spin and then the washing machine would spin, but you can also just pedal a bike to create electricity that runs an appliance. I would love to see that at a home.
>
> (Patty supporting the power of micro generators have to improve the energy literacy of their users)

In the future, in-home experiences will not replace traditional formats of energy education, but instead will complement it. When presented with the option of passive vs. active opportunities to share lessons on energy, although participants in the case studies respected the positive aspects of traditional passive learning processes, they favoured more active processes. For example, textbooks, videos and historical documentation were considered to have the ability to cover a topic with great breadth and depth. Yet at the same time, participants felt as if the passive lessons would not be as inviting or memorable as active learning processes resulting from interaction with the built environment:

> I think it is hard to show people, we're getting electricity from that power line. I mean it's kind of just obvious and somewhat boring. I think it would be interesting to show people and explain to people how the energy is getting into that power line and where it's coming from and that makes them realize, the environmental sources that we're using.
>
> (Cathy explaining how energy is transmitted in a visually boring way that hides the interesting phenomenon)

continued

Figure 2.16 Windfarm
Source: Iaer Ireis (CCO)

With the Earth's limited and constrained energy sources, engineering the world of the future will require some indirect human engineering to accompany technological advances. Surrounded by transparent infrastructure, from which people can learn, more individuals will be able to make informed energy decisions. The future of the built environment is in the hands of a growing and energy-hungry population. From an ethical standpoint, we must not only design what people want, but also integrate what they need to survive.

References

DeWaters, J.E., Andersen, C., Calderwood, A., and Powers, S.E. (2014)."Improving climate literacy with project-based modules rich in educational rigor and relevance." *Journal of Geoscience Education* 62(3), 469–84. (1 August, 2014): doi:10.5408/13-056.1

DOE (2012, July). Energy Literacy: Essential Principles and Fundamental Concepts for Energy Education. U.S. Department of Energy. Retrieved from www1.eere.energy.gov/education/energy_literacy.html.

DOE (2016). "Whole-House Systems Approach | Department of Energy." http://energy.gov/energysaver/whole-house-systems-approach (accessed 12 July 2016).

Ehrhardt-Martinez, K., and Laitner, J.A. (2009). "Breaking out of the economic box: Energy efficiency, social rationality and non-economic drivers of behavioral change." Proceedings, ECEEE. http://eceee.org/library/conference_proceedings/eceee_Summer_Studies/2009/Panel_1/1.350 (accessed 2 November 2013).

EIA. (2013). "Energy use in homes – Energy explained, your guide to understanding energy." http://eia.gov/Energy Explained/?page = us_energy_homes#tab2 (accessed 7 July 2016).

EIA. (2016). *U.S. Energy Information Administration – Monthly Energy Review September 2016*. http://eia.gov/total energy/data/monthly/pdf/sec1_3.pdf (accessed 10 October 2016).

Janda, K.B. (2011). "Buildings don't use energy: People do." *Architectural Science Review* 54(1), 15–22.

Laitner, J.A., and Ehrhardt-Martinez, K. (2009). "Examining the scale of the behaviour energy efficiency continuum."

McMakin, A.H., Malone, E.L., and Lundgren, R.E. (2002). "Motivating residents to conserve energy without financial incentives." *Environment and Behavior* 34(6), 848–863.

Zografakis, N., Menegaki, A.N., and Tsagarakis, K.P. (2008). "Effective education for energy efficiency." *Energy Policy* 36(8), 3226–3232.

Figure 2.17 Properly shaped and located recycling bins encourage desired recycling behaviour

In more recent years, the idea of community-based social marketing has emerged as a way to influence behaviours that benefit both individuals and communities for the greater social good. The aim of this approach is to reduce barriers to desired actions and amplify benefits to those who take such actions (McKenzie-Mohr 2011). Tactics used to achieve this aim include prompts for desired behaviour, social norms, incentives, commitments, communication and removing barriers to action. For example, signs reminding people to recycle solid waste, coupled with properly placed receptacles to allow them easily to do so help achieve the social goals of waste diversion. Receptacles with shaped openings that make putting an item in the wrong bin impossible are an example of an affordance that further supports the goal (Figure 2.17).

Even more recently, choice architecture has emerged as a relevant way to influence human behaviour. Choice architecture is the design of ways in which options can be presented to decision makers to influence their decisions. Factors such as the number of choices presented, the order in which they are given, the way in which attributes are described and the presence of a baseline or default choice all have been demonstrated to influence choices of decision makers. In the sustainable infrastructure field, Shealy and Klotz (2015) have explored the ways in which choice architecture can help encourage more sustainable choices with regard to green rating systems and overcome irrational decision biases (see *The choice architecture of sustainability rating systems*). Critics of this approach argue that choice architecture can artificially limit the choices of rational decision makers. However, these ideas are being widely applied through policy and design to encourage healthier and more sustainable behaviours such as choosing healthier food, saving for retirement or installing energy saving technologies.

The choice architecture of sustainability rating systems: an approach to encourage more sustainable infrastructure design decisions

Tripp Shealy and Leidy Klotz

Rating systems like Leadership Energy and Environmental Design (LEED), Energy Star and others are often used to guide design decisions to encourage more sustainable buildings and infrastructure. Embedded within any such rating system is choice architecture, which refers to the way information is presented to a decision maker. Just as there is no 'neutral' building architecture that does not influence in some way how people navigate a building, there is also no neutral choice architecture: some options must be first, attributes are or are not presented, and, just as in other domains, these factors are likely to influence decisions in sustainable infrastructure development. Choice architecture matters because how a choice is presented affects the reasoning process; even when two presentation formats are formally equivalent, each may give rise to different psychological processes.

continued

For example, when presented with a mile-per-gallon (mpg) fuel efficiency metric, consumers wrongly assume that increases in efficiency have a linear effect on fuel use and CO_2 emissions, suggesting that an increase from 10 to 20 mpg has the same benefit as going from 40 to 50 mpg (Larrick and Soll 2008). This, however, is untrue: the shift from 10 to 20 mpg reduces fuel use by 50 per cent, whereas from 40 to 50 mpg, fuel use is reduced by only 20 per cent. Consumers make this mistake because either they do not understand the metric or cannot do the calculations. But, when presented with fuel efficiency information using a linear metric, such as gallons per mile, their ability to pick the most beneficial change improves. In part because of findings like these, recent EPA revisions to car energy use labels support better decisions in two ways: by providing an easy to understand metric of total gasoline costs, and by providing fuel usage as gallons per mile (Ungemach et al. under review).

Rating systems that do not account for systematic biases in the reasoning process when designing and constructing infrastructure may unintentionally contribute to a process that leads away from more sustainable outcomes. For example, LEED may inadvertently set goals that are too low and thus discourage the ambition needed to achieve energy use performance that is technically and economically feasible (Klotz et al. 2010). Conversely, intentionally crafted choice architecture can use differences in the way decisions are posed to help decision makers achieve desired outcomes.

The Envision rating system for sustainable infrastructure

Envision is a leading U.S. rating system for sustainable infrastructure. While LEED has been used mostly for buildings, Envision is meant for a range of infrastructure projects (i.e. roads, bridges, pipelines, railways, airports, dams, levees, landfills and water treatment systems). Envision is similar to LEED because both are appropriate for project planning to inform goal setting and early design considerations. Envision, like LEED, is also used voluntarily by construction and design firms, but it can also be mandated by local governments and municipalities.

The Envision rating system is composed of 60 questions divided into five categories: Quality of Life, Leadership, Resource Allocation, Natural World, and Climate and Risk. Each question, or credit, is associated with a series of points. Engineers use Envision's guidance manual to decide the number of points achievable for their project. Levels of achievement are ranked from lowest to highest: Improved, Enhanced, Superior, Conserving and Restorative. The scale of points varies for each credit, but all points accumulate moving from the Improved through Restorative levels. For example, Quality of Life question 1.3 asks: 'How will the project team develop local skills and capabilities?' The Improved level (1 point) is achieved by hiring a local workforce whereas the Conserving level (12 points) is achieved through a training program for minorities and disadvantaged groups. The training program must also leave a competitive local workforce in place for future projects. To meet Conserving and Restorative levels means that the project provides sustained benefits to the community, economy and local environment after the construction phase is complete (i.e. a trained, diverse workforce is more competitive for future projects in the community). The goal of Envision is to move project teams from the conventional construction standards (zero points) to the highest possible levels of sustainability (defined by Envision as Restorative).

Choice architecture modifications to the Envision rating system

A series of experimental studies was conducted to explore the role of choice architecture in affecting the choices of users of the Envision rating system, with the aim of determining how Envision users might be motivated to consider higher levels of sustainability in their choices. The studies varied

continued

the way in which points were earned within the system and explored the use of a 'role model' project as a reference point. The research team also studied the effects of combining the two interventions.

In the first study, the research team explored the Endowment Effect by shifting the rating scale to an endowed scale where points were taken away for lesser levels of sustainability than Conserving, rather than being added for each gain in level (Table 2.2). This intervention was based on reference dependence, which means the default frames other outcomes as a loss or gain, and this framing subsequently impacts the decision (Dinner et al. 2010). When tested with professional engineers, the group given the modified version with a potential to lose points (n = 32) scored an average of 66 per cent of the total possible points compared with the control group's (n = 33) 51 per cent. In responses to a post-task survey, 95 per cent of participants believed their scores were realistic and achievable (Shealy et al. 2016). In other words, simply restructuring the point system significantly improved engineers' sustainability goals.

The second study tested the Role Model Effect, where providing a similar role model that performs well in a seemingly attainable way acts as an example to encourage similarly high goals (Lockwood and Kunda 1997). In this study, each credit on the Envision rating system had a specific example to explain how a peer firm achieved a high score on a reference project. When tested with engineering professionals, the group given the role model version (n = 27) achieved 74 per cent of available points, which was 20 per cent more points than the control group (n = 26) (p = 0.003) (Harris et al. 2016). This difference is equivalent to two levels of certification in Envision, from a silver certification to platinum. As with the endowed points study, the results from the role model study indicate that simply restructuring infrastructure planning decisions can influence the outcome, at least for initial goal setting, which is one of the purposes of Envision.

The third study combined the Endowment and Role Model effects to determine whether combinations of interventions could result in even better performance. Endowing users with points in order to restructure choices as a loss or gain in value is a passive intervention because decision makers are likely unaware of the changes in framing (Bovens 2009; Loewenstein et al. 2014). By comparison, the role model project is a more explicit endorsement intended to draw decision makers' attention to a preferred option. This study was implemented using engineering students in their final year of study. The combined group was able to achieve 79 per cent of points while the control group achieved 56 per cent. The combined intervention group outperformed the control group on every credit by an average of 4 points. This is more than either the endowed intervention or role model project separately. The difference between the combined intervention compared to the control was significant (p = 0.0001).

Across the studies, each choice intervention (Table 2.3) improved engineers' decision making for higher achievement for sustainability. The endowment and role model project similarly improved Envision users' goals for achievement towards sustainability. When added together the results were

Table 2.2 Modifications to Envision rating scale

Levels of Achievement	Current Scale	Endowed Scale
Industry Convention	0*	(–12)
Improved	1	(–11)
Enhanced	2	(–10)
Superior	5	(–7)
Conserving	12	12*
Restorative	15	(+3)

* Indicates number of starting points.

continued

Table 2.3 Choice architecture interventions lead to higher goals for sustainability

Intervention	Control Points Achieved (%)	Intervention Points Achieved (%)	Difference (%)	p	n
Endowment	51%	66%	15%	<0.01	65
Role Model	54%	74%	20%	<0.01	54
Combined	56%	79%	23%	<0.001	56

more significant than either version alone. In this case, the combined choice architecture approach appears to elicit complementary cognitive process that together improve decision making.

Conclusions

The objective of Envision and other rating systems is to increase the consideration for sustainability during infrastructure planning. This research provides a method to better meet this objective. Structuring the decision processes to align with tested behavioural science theories can improve multi-stakeholder decision making about infrastructure. Both engineering professionals and student engineers achieved more points when endowed points and provided with a role model project, and the combined effect was significant.

The potential impacts of this small change in choice architecture are immense. Suppose role models led to 30 per cent better performance on the Envision credit 'Reduce Greenhouse Gas Emissions'. Applied to all U.S. infrastructure, this represents a reduction of over 4 billion tons of CO_2 (estimate based on a per-capita carbon footprint of infrastructure of 53 tons and a U.S. population of 316 million). Considering the successful cash-for-clunkers program invested roughly $3 billion dollars and saved an upward estimate of no more than 30 million tons of CO_2 (Li et al. 2013), adding positive role models to an infrastructure rating system appears promising, even just for one of the 60 Envision credits. Positive role models or the combined intervention might offer similar gains in the 59 other sustainability outcomes defined by Envision. While Envision is only used currently on a fraction of infrastructure projects, it is rapidly expanding, as are similar rating systems.

Better understanding the influence of tools like Envision (and countless other similarly structured sustainability rating systems) on decision making will help avoid scenarios where unanticipated behavioural or decision barriers limit achievement. Compared to the costs of infrastructure itself, simply restructuring choices is a relatively inexpensive approach to support more informed decisions. These types of interventions are also less intrusive than legal responses. Through more empirical studies and field experiments, researchers can begin to predict decision outcomes based on these and other cognitive biases and better improve decision making for infrastructure project planning and the stakeholders these projects serve.

References

Bovens, L. (2009). "The ethics of nudge." In *Preference change: Approaches from philosophy, economics, and psychology.* T. Grune-Yanoff and S.O. Hansson, eds. Springer Netherlands.

Dinner, I.M., Johnson, E.J., Goldstein, D.G., and Liu, K. (2010). *Partitioning Default Effects: Why People Choose Not to Choose* (SSRN Scholarly Paper No. ID 1352488). Social Science Research Network, Rochester, NY. http://papers.ssrn.com/abstract = 1352488.

continued

Harris, N., Shealy, T., and Klotz, L. (2016). "'HOW exposure to 'role model' projects can lead to decisions for more sustainable infrastructure." *Sustainability* 8(2), 130.

Klotz, L., Mack, D., Klapthor, B., Tunstall, C., and Harrison, J. (2010). "Unintended anchors: Building rating systems and energy performance goals for U.S. buildings." *Energy Policy* 38(7), 3557–3566.

Larrick, R.P., and Soll, J.B. (2008). "The MPG illusion." *Science* 320(5883).

Li, S., Linn, J., and Spiller, E. (2013). "Evaluating 'Cash-for-Clunkers': Program effects on auto sales and the environment." *Journal of Environmental Economics and Management* 65(2), 175–193.

Lockwood, P., and Kunda, Z. (1997). "Superstars and me: Predicting the impact of role models on the self." *Journal of Personality and Social Psychology* 73(1), 91–103.

Loewenstein, G., Bryce, C., Hagmann, D., and Rajpal, S. (2014). *Warning: You Are About to Be Nudged* (SSRN Scholarly Paper No. ID 2417383). Social Science Research Network, Rochester, NY. http://papers.ssrn.com/abstract = 2417383.

Shealy, T., and Klotz, L. (2015). "Well-endowed rating Systems: How modified defaults can lead to more sustainable performance." *Journal of Construction Engineering and Management* 141(10), 04015031.

Shealy, T., Klotz, L., Weber, E.U., Johnson, E.J., and Bell, R.G. (2016). "Using framing effects to inform more sustainable infrastructure design decisions." *Journal of Construction Engineering and Management* 0(0), 04016037.

Ungemach, C., Camilleri, A., Johnson, E., Larrick, R., and Weber, E. (Under review). "Translated attributes: Aligning objectives and choices through signposts." *Management Science*.

The future of sustainability theory

What will society say of sustainability in the future? Sustainability theory has been influenced by many thought leaders and global trends in the past 50 years or more, and has been operationalized as its own science as well as applied to problems in many specific disciplines and domains. Over time, the many attempts to define the term 'sustainability' have led some to conclude that the term itself is meaningless, given that it has been defined differently for seemingly every different situation. In fact, some people have even stopped using the term altogether.

Others have brought forth the idea that sustainability is not enough, that we as a society should be striving for more regenerative practices, not just net zero. Architect William McDonough is well known for his quip that no one would want just a 'sustainable' marriage – everyone ought to aspire for more than that (McDonough and Braungart 2013). Likewise, designer Janis Birkeland has pointed out that current approaches to green best practices and rating systems ultimately reward designs that are still net-negative. These initiatives that are meant to drive more sustainable development ultimately backfire in unexpected ways.

Birkeland has advocated strongly for what she calls 'Positive Development', in which our built environment increases quality of life, health and safety, and functionality without costing more or requiring more resources (Birkeland 2008). She points out that many decision makers have used the idea of sustainability to encourage development while accepting that ecological and social tradeoffs are necessary for economic growth. Birkeland advocates that, rather than trying to live within ecological limits of an already seriously degraded planet using tactics that further degrade its carrying capacity, we design systems that *expand* those limits and repair some of the damage human development has caused. She challenges the notions of conventional sustainability that ecological losses can be offset with social benefits in a sustainable system. Her focus is specifically on retrofit and repair of existing buildings and cities to increase their ability to contribute positively to sustainability goals.

Many leaders in the applied sustainability domain have begun to explore the concept of resilience as a natural extension of sustainability science. With strong roots in biology and ecology, resilience is of growing concern as it becomes apparent that climate change may lead to stresses our current human communities and societies are not well prepared to sustain. Key concepts from resilience will be a large part of the future of sustainability in the construction discipline, including resilient inter-dependent infrastructure, resilient communities, and the Food-Energy-Water Nexus. Each of these concepts is discussed in more detail in Chapter 10.

> If we labeled cigarettes the way we label buildings, people might start smoking more 'light' cigarettes to get healthier.
> (Janis Birkeland, *Positive Development*, 2008)

Discussion questions and exercises

2.1   What was your first exposure to the idea of sustainability in your personal or professional life? What sources of information did you find most impactful? What convinced you that sustainability was worthy of your attention?

2.2   Map the evolution of sustainability in your region or country. What were the major milestones in time? How do they correlate with other significant historical, social or political events occurring in your country at that time? Were any associated with significant natural or human-made disasters? How have these external events shaped your society's reactions to the tenets of sustainability and sustainable development?

2.3   How have advances in sustainability in other countries impacted its evolution in your region or country? What elements of sustainability have derived from actions in other countries? For example, was the prevalent green building rating system used in your country developed elsewhere? How did it spread to your country? How has it evolved since then?

2.4   How do the principles of Agenda 21 apply to your country or region? Which principles are easiest for your country to achieve? Which are most difficult? Why? Which principle seems most critical to you for achieving sustainability? Why?

2.5   Principle 15 of Agenda 21 mentions the precautionary approach. Restate this principle in your own words. What arguments can be raised against the precautionary principle? What arguments can be made in favour of it? Choose an example of something about which there is debate regarding safety and health of humans and/or the environment. Who is in favour, and what is their motivation? Who is opposed? Why?

2.6   How large is your personal carbon footprint? Using an online calculator (search 'carbon footprint calculator'), determine how many pounds of carbon your activities generate on an annual basis. What is the most significant source of carbon among your activities? What could you do to reduce your individual carbon footprint?

2.7   Think about the most recent significant purchase you have made. What information did you consider in making the decision? What additional factors from the Triple Bottom Line could you have considered? What information would you have needed to incorporate these factors into your decision?

2.8   Choose one of the four scenarios for energy descent and climate change (Brown Tech, Green Tech, Earth Steward and Lifeboats) that most closely represents your personal beliefs about the likely future of society. What evidence can you find to support the likelihood of this scenario vs. the others? What actions would need to happen to mitigate its most negative effects?

2.9   Consider the different approaches used to define what sustainability means, including the Brundtland definition, Munasinghe's triangle, three pillars/Triple Bottom Line and Heinberg's axioms. Which of these definitional approaches resonates the most with you? Which is easiest to explain to others? Which is most useful in making decisions? Which is most commonly referenced in your country as a way to understand sustainability?

2.10   Compare the lists of goals from Agenda 21, the Millennium Development Goals and the Sustainable Development Goals. Which set of goals seems easiest to understand? If you were in charge of developing a plan for your community to achieve sustainability, which of the sets of goals would you find most helpful? Why?

2.11   Consider Ben-Eli's five core principles of sustainability. Why is the spiritual dimension included? How could this principle be applied in the kinds of decisions made by design and construction professionals?

2.12   Compare the different applied sustainability frameworks in the chapter, including the Natural Step, the Hannover Principles, places to intervene in a system, eco-efficiency, and eco-effectiveness. Which do you think would be easiest to apply as a practicing designer or builder? If you owned a company and had to choose one of these as a yardstick for your company, which one would you choose, and why?

2.13   What examples of the Jevons Paradox or rebound effect have you seen in your life? Has there been a case where you changed your behaviour in response to some technological change that improved efficiency? What did you consider in making that change? What would happen if everyone thought that way?

2.14   What are examples of common resources that are impacted by construction projects? Consider not only materials consumed, but also ecosystem services. How has society attempted to moderate the use of those resources? What has worked, and what hasn't?

2.15   Consider the building in which you are located. Are there any examples of design in that building that make particular types of behaviours more difficult, or that make use of the building more difficult for specific groups of users? How could your building be changed to increase the likelihood of healthy and sustainable behaviours by its occupants?

References

Alcott, B. (2008). "Historical overview of the Jevons Paradox in the literature." In *The Jevons Paradox and the Myth of Resource Efficiency.* J.M. Pollmeni, K. Mayumi, and M. Giampietro, eds. Earthscan, London, UK.

Bartlett, A.A. (1998). "Reflections on sustainability, population growth, and the environment – revisited." *Renewable Resources Journal* 15(4), 6–23.

Commoner, B. (1971). *The closing circle.* Knopf Publishing, New York NY.

Commoner, B. (1972). "Response to critique on *The Closing Circle.*" *Bulletin of the Atomic Scientists* 17, 42–56. (May).

Daly, H., and Cobb, J. (1989). *For the common good: Redirecting the economy toward community, the environment, and a sustainable future.* Beacon Press, Boston, MA.

Daniels, B. (2008). "Emerging commons and tragic institutions." *Environmental Law* 37(515), 517–519.

DuBose, J.R., and Pearce, A.R. (1997). "The natural step as an assessment tool for the built environment." *Proceedings of the 1997 CIB Conference on Green Building,* Paris, France.

Ehrlich, P.R. (1968). *The population bomb.* Sierra Club/Ballantine Books, New York, NY.

Ehrlich, P.R., and Holdren, J.P. (1972). "Critique on *The Closing Circle.*" *Bulletin of the Atomic Scientists* 16, 18–27. (May).

Freire-González, J., and Puig-Ventosa, I. (2015). "Energy efficiency policies and the jevons paradox." *International Journal of Energy Economics and Policy* 5(1), 69–79.

Gibson, J.J. (1977). "The theory of affordances." In *Perceiving, Acting, and Knowing: Towards an Ecological Psychology.* R. Shaw and J. Bransford, eds. John Wiley & Sons, Hoboken, NJ, 127–143.

Hardin, G. (1968). "The tragedy of the commons." *Science* 162(3859), 1243–1248.

Heinberg, R. (2008). "Five axioms of sustainability." *Museletter #178.* (February). http://richardheinberg.com/178-five-axioms-of-sustainability (accessed 13 January 2017).

Holdren, J.P., Daily, G.C., and Ehrlich, P.R. (1995). "The meaning of sustainability: biogeophysical aspects." In *Defining and Measuring Sustainability: The Biogeophysical Foundations,* M. Munasinghe, and W. Shearer, eds. The United Nations University and The World Bank, Washington, DC.

Holmgren, D. (2009). *Future scenarios: How communities can adapt to peak oil and climate change.* Chelsea Green, White River Junction, VT. http://futurescenarios.org.

Hubbert, M.K. (1956). "Nuclear energy and fossil fuels." Shell Development Company, Exploration and Production Research Division, Publication Number 95.

Jackson, P.M. (2006). Why the "Peak oil" theory falls down – myths, legends, and the future of oil resources. CERA – Cambridge Energy Resources Associates.

Jevons, W.S. (1866). *The coal question. 2nd ed.* Macmillan & Company, London, UK.

Lyle, J.T. (1994). *Regenerative design for sustainable development.* John Wiley & Sons, New York, NY.

McDonough, W., and Braungart, M. (1992). *The Hannover Principles: Design for Sustainability.* William McDonough & Partners, Charlottesville, VA. http://www.mcdonough.com/writings/the-hannover-principles/.

McDonough, W., and Braungart, M. (2002). *Cradle to cradle: Remaking the way we make things.* North Point Press/Farrar, Straus, and Giroux, New York, NY.

McDonough, W., and Braungart, M. (2013). *The upcycle: Beyond sustainability – designing for abundance.* North Point Press/Farrar, Straus, and Giroux, New York, NY.

McKenzie-Mohr, D. (2011). *Fostering sustainable behavior: An introduction to community-based social marketing.* 3rd ed. New Society Publishers, Gabriola Island, BC.

Meadows, D.H. (1999). "Leverage points: Places to intervene in a system." The Sustainability Institute, Hartland, VT. Archived at *http://donellameadows.org.*

Meadows, D.H., Meadows, D.L., Randers, J., and Behrens III, W.W. (1972). *The limits to growth: A report for the club of rome's project on the predicament of mankind.* Universe Books, New York, NY.

Meadows, D.H., Meadows, D.L., and Randers, J. (1992). *Beyond the limits: confronting global collapse, envisioning a sustainable future.* Chelsea Green Publishers, White River Junction, VT.

Meadows, D.H., Randers, J., and Meadows, D.L. (2012). *Limits to growth: The 30-year update. 3rd ed.* Chelsea Green Publishers, White River Junction, VT.

Mogridge, M.J.H. (1990). *Travel in towns: Jam yesterday, jam today, and jam tomorrow?* Macmillan Press, London, UK.

Munasinghe, M. (1993). *Environmental economics and sustainable development.* World Bank, Washington, DC.

Munasinghe, M., and Shearer, W., eds. (1995). *Defining and measuring sustainability: The biogeophysical foundations.* The United Nations University/World Bank, Washington, DC.

Nattrass, B., and Altomare, M. (1999). *The natural step for business.* New Society Publishers, Gabriola Island, BC.

Norman, D.A. (1988). *The psychology of everyday things.* MIT Press, Cambridge, MA.

NRC – National Research Council. (2011). *Sustainability and the U.S. EPA.* Committee on incorporating sustainability in the U.S. environmental protection agency, science and technology for sustainability program, policy and global affairs division, National Research Council. National Academies Press, Washington, DC.

Robèrt, K.H., Holmberg, J., and Eriksson, K.E. (1994). "Socio-ecological principles for a sustainable society – scientific background and Swedish experience." *Ecological Economics*.

Schlosberg, D. (2007). *Defining environmental justice: theories, movements, and nature*. Oxford University Press, London, UK.

Schmidt-Bleek, F. (2000). "Factor 10 manifesto." Working paper, Factor 10 Institute (January). http://factor10-institute.org/files/F10_Manifesto_e.pdf. (accessed 14 August 2016).

Shealy, T., and Klotz, L. (2015). "Well-endowed rating systems: How modified defaults can lead to more sustainable performance." *Journal of Construction Engineering and Management* 141(10), 04015031.

Simon, H.A. (1969). *The science of the artificial*. MIT Press, Cambridge, MA.

Small, K.A., and Van Dender, K. (2005). "The effect of improved fuel economy on vehicle miles traveled: Estimating the rebound effect using U.S. state Data, 1966–2001." *Policy and Economics*, University of California Energy Institute, U.C. Berkeley, Berkeley, CA.

Stewart, R.B. (2002). "Environmental regulatory decision making under uncertainty." *Research in Law and Economics* 20, 76.

Tainter, J. (1988). *The collapse of complex societies*. Cambridge University Press, Cambridge, UK.

Tromp, N., Hekkert, P., and Verbeek, P. (2011). "Design for socially responsible behavior: A classification of influence based on intended user experience." *Design Issues* 27(3), 3–19.

Turner, G. (2008). "A comparison of 'the limits to growth' with thirty years of reality." Commonwealth Scientific and Industrial Research Organisation (CSIRO). https://ideas.repec.org/p/cse/wpaper/2008–09.html (accessed 13 January 2017).

Turner, G. (2014). "Is global collapse imminent?" MSSI Research Paper No. 4. Melbourne Sustainable Society Institute, The University of Melbourne, Australia. http://sustainable.unimelb.edu.au/sites/default/files/docs/MSSI-ResearchPaper-4_Turner_2014.pdf (accessed 13 January 2017).

Von Bertalanffy, L. (1969). *General system theory*. George Braziller, New York, NY.

Wackernagel, M., and Rees, W. (1997). "Perceptual and structural barriers to investing in natural capital: Economics from an ecological footprint perspective." *Ecological Economics* 20(3), 3–24.

WBCSD – World Business Council for Sustainable Development. (1992). *Changing course*.

WBCSD – World Business Council for Sustainable Development. (2000). *Eco-efficiency: Creating more value with less impact*.

Waite, M. (2013). "SURF framework for a sustainable economy." *Journal of Management and Sustainability* 3(4), 25–40.

Wingspread Conference on the Precautionary Principle. (1998). "Wingspread statement on the precautionary principle." (January 12). http://sehn.org/wing.html (Accessed on 14 August, 2016).

Winner, L. (1980). "Do artifacts have politics?" *Daedalus* 109(1), Winter.

Chapter 3
Sustainability policies and programmes

Sustainability improvement involves two complementary perspectives: developing more sustainable technological solutions that can work *in concert* with or *despite* the spectrum of likely human behaviours, and changing human behaviour to increase the uptake of those more sustainable strategies and technologies. A hybrid approach including both perspectives is likely to be more effective than either extreme, namely, a combination of technological innovations that improve sustainability coupled with active approaches toward diffusing those innovations throughout the population targeted to adopt it (Gardner and Stern 1996; Rogers 2003). A broad variety of technological options are available to those who would increase the sustainability of their built facilities, and these are the focus of subsequent chapters of this book. This chapter, however, focuses on the behavioural interventions that can facilitate the adoption of sustainability technologies, strategies and practices in the built environment in the form of policies and programmes at the organizational level.

With growing interest in adopting sustainability innovations in both the public and private sectors, the options available to encourage sustainability best practice are many and varied. As a result, the types of policy and programme options also vary across different implementation contexts. In the context of this chapter, a sustainability policy is a plan or specified course of action that guides the decisions and behaviours of individuals with regard to the goal of improving sustainability within an organization. Sustainability policies are typically supported by specific programmes, or packages of actions and associated resources directed to achieve specific objectives within the larger policy. This chapter explores the types of policy and programme mechanisms that can be used to promote the adoption of sustainability practices and technologies in practice. It defines, compares and contrasts policy and programme options to provide key information for designing effective sustainability policies that have beneficial effect in different contexts.[1] The chapter also explores the evolution of sustainability policies and programmes in the corporate sector, including six examples of corporate sustainability policies and programmes from different sectors of the construction industry.

Sustainability as an innovation in the AEC industry

Unlike other less permanent technologies, built facilities represent significant one-time investments on the part of their owners, and many are designed uniquely to meet the specific needs, goals and aspirations of the owner (Brand 1994). In this case, the actions and values of a single individual, namely the building owner, can strongly influence which technologies are included in the design and what are the overall design goals of the facility (NAHB 1989).

However, in many projects, the entity that fills the role of owner and for whom a built facility is designed and built may not be the primary occupant, user, operator or maintainer of the building. While design of built facilities and the technologies that comprise them often includes input at some level from building occupants, users and operators, these important stakeholders generally entrust the actual design and construction decisions to professional agents such as architects, engineers and builders. These construction professionals have the necessary expertise and resources to develop and implement a solution on behalf of a project's owner, occupants, users and operators. Effective integrative design and project delivery requires both active involvement of critical stakeholders in formulating the scope and goals of the project as well as performance feedback after the project is complete, so that construction professionals can adapt the design intent (and subsequent design realization) to better accommodate the reality of use environments in future projects.

From a human behaviour standpoint, multiple tactics can be used to facilitate the adoption and diffusion of sustainability innovations in the design and delivery of buildings and infrastructure systems, including policy (laws, regulations and incentives), education and other programmes, and moral, religious and/or ethical appeals. One way to consider the goal of increasing built environment sustainability is to think of sustainability as an innovation that can be adopted by built environment stakeholders. An innovation is anything 'new or perceived as new by the unit of adoption' (Rogers 2003). Rogers describes the innovation-decision process as the sequence of events between an individual's (or decision-making entity's) first knowledge of an innovation, continuing through evaluation of the merits of the innovation, a decision to adopt or reject, implementation of the innovation if a decision is made to adopt, and confirmation of the adoption decision by either sustained use or subsequent rejection. The process is preceded by prior conditions including perceived needs or problems that may drive the quest for new solutions. The available knowledge of innovations is influenced by factors ranging from socioeconomic and personality variables of the potential adopter to typical communication behaviour that may inform adoption choices. Adoption decisions are influenced by characteristics of the innovations themselves, ranging from perceived relative advantage of the new solution over the status quo, to compatibility with existing norms, complexity, trialability and observability of outcomes (see *Attributes of innovations affecting their adoption*).

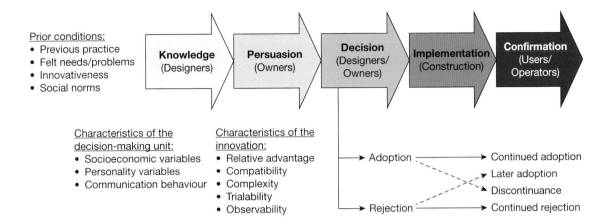

Prior conditions:
- Previous practice
- Felt needs/problems
- Innovativeness
- Social norms

Characteristics of the
decision-making unit:
- Socioeconomic variables
- Personality variables
- Communication behaviour

Characteristics of the
innovation:
- Relative advantage
- Compatibility
- Complexity
- Trialability
- Observability

Figure 3.1
Innovation-decision process
for the construction industry
(adapted from Rogers 2003)

Figure 3.1 shows Rogers' innovation-decision process in the context of the construction industry and delivery of capital projects, including the different stakeholders involved in each stage of the process. Not shown are the designers of facility technologies who supply the industry with technological innovations in the first place. The process is typically shepherded by a design professional (an architect or engineer) who brings knowledge of state of the art in facility technologies and strategies and provides expertise and counsel to facility owners who do not themselves possess such specialized expertise. Owners are responsible for approving design decisions and are therefore the key players in the persuasion stage. Together with designers, owners make adoption or rejection decisions in the decision stage of the process as part of completing the facility design.

After design is completed, the design itself (still an idea on paper) is turned over to yet another stakeholder – the constructor – who may or may not have been involved in prior stages. The job of the constructor is to bring to the process the knowledge necessary to implement any

Attributes of innovations affecting their adoption

- **Relative advantage** – the degree to which the innovation offers some actual or perceived benefit compared to the solution being displaced by it.
- **Complexity** – the degree to which an innovation is perceived to be difficult to understand or use.
- **Compatibility** – the degree to which an innovation is consistent with existing values, past experiences, needs, and existing technological and organizational infrastructure of potential adopters.
- **Observability** – the degree to which the results caused by an innovation (or the innovation itself) are visible to others, and the degree to which cause and effect are apparent to the observer.
- **Trialability** – the degree to which a potential adopter can experiment with an innovation without commitment, changing current behaviors, or significant investment.

(adapted from Rogers 2003)

innovations incorporated into the design, which may also require additional innovation decisions. At this point, however, the nature of the original innovation is largely fixed, and any changes to that innovation are likely to only become more costly as the project progresses. The final stage, confirmation, involves another set of stakeholders: the facility users and operators, who then face decisions on an ongoing basis regarding whether to continue, adapt or reject innovations built into the facility based on other stakeholders' decisions. Owners of multiple buildings must also determine whether or not to employ successful innovations in future projects based on their experiences with those innovations in earlier projects. Repeated use of innovations across multiple cases is called *routinization,* where an innovation becomes part of the organization's routine and is no longer considered innovative.

In general, the ability to influence innovation decisions cost-effectively diminishes markedly as the project moves from left to right along the innovation-decision process, although there are ample opportunities for informal changes or 're-invention' as the process progresses. For instance, while a designer has control of the initial decision to adopt an innovation, the responsibility for the continued operation is often in the hands of the facility manager or even the tenant or occupant of the facility.

Although 'decision' is indicated as a discrete step in the generic process, in actuality each step offers an opportunity to reject an innovation (Rogers 2003). For instance, two types of rejection are possible at the confirmation stage: *active rejection,* which consists of considering adoption of the innovation but then deciding not to adopt it, and *passive rejection* (also called non-adoption), which consists of never really considering the use of the innovation. Re-invention is generally perceived as positive by adopters of innovations in that it affords an opportunity to customize the innovation for a better fit with its context of application. In contrast, re-invention is typically perceived as negative by the designers of innovations, although innovations have been shown to be more persistent when re-invention is implemented to increase compatibility of the innovation with adopters' beliefs, past experiences and context of application (Rogers 2003; DuBose 1994).

How can organizations wishing to reap the benefits of sustainable facility technologies most effectively promote the implementation of these innovations in their capital projects? How can such organizations maximize the benefits of lessons they learn in early adoption and effectively apply those lessons to subsequent implementation of the innovations? Existing theory suggests that organizations perceiving a net benefit from the adoption of an innovation will have an increased likelihood of choosing to implement that innovation on subsequent projects through routinization (e.g. Klein and Sorra 1996; Rogers 2003). However, multiple factors influence whether or not a particular innovation will be assimilated into the standard operating procedure for the organization. The development of both formal and informal standard operating procedures for an organization is the realm of policy, discussed next.

The owner's perspective: how policy can be used to promote sustainability

Designing a successful sustainability policy and associated programmes within the complex context of an organization is difficult by any standard. Not only may the actions and solutions denoted in a policy be innovative to members of the organization, but also the new policy itself will be an innovation that may require significant changes to current behaviours in its adoption. Multiple stakeholder perspectives must be considered and aligned behind a common vision and plan of action. Clear benefits that outweigh potential risks must be shown from social, environmental and economic perspectives to support the social and business case for changing the status quo. Finally, the implementability of the specific programme elements themselves must be carefully considered in designing the programmes and policies to ensure their sustainability and effectiveness in the long term, despite the constantly evolving and shifting political nature of organizations.

Stakeholders of a sustainability policy

One of the most important challenges in designing a sustainability policy is considering the perspectives and interests of each of the categories of stakeholders who will be affected by such a policy. These stakeholders include:

- Owner organizations, including organization leaders and facilities staff who will be responsible for interpreting the policy, implementing its requirements and evaluating the results with regard to capital projects.
- Supporting personnel, such as government energy offices, property offices or departments of general services in the case of public agencies, or finance departments in private organizations, who may be responsible for managing funding to implement sustainability programmes.
- Management, leadership or elected officials, whose endorsement of a sustainability policy exposes them to the potential political risks and rewards the policy might bring as it is implemented.
- Facility occupants, who will benefit from high-performing buildings but may suffer if facility scope must be reduced to achieve high performance, or if building technologies do not perform as anticipated.
- Shareholders or taxpayers who will provide some or all of the funding to support sustainability programmes and who will benefit from increased environmental quality and enhanced productivity of facility users, as well as lower operating expenses.
- The architecture, engineering and construction industry, which provides capital project-related services for facilities and will be required to comply with policy requirements and deal with associated implications for how business is conducted.
- Lobbies who represent key industries or non-profits in the jurisdiction and are sensitive to the potential influence of policies on their constituencies.

All of these stakeholder groups have the potential to either offer support or inhibit the implementation of a sustainability policy or programme as it is put forth. Owners and supporting personnel in particular will play a strong role in the ongoing success of and ability to sustain a sustainability programme for built facilities.

Potential sustainability policy and programme risks and rewards

A variety of potential risks and rewards are associated with sustainability policies and programmes, and must be considered when designing a policy or programme configuration. Since the ultimate objective of a sustainability policy for built facilities is to increase the sustainability of facilities in the organization or jurisdiction, the Triple Bottom Line of sustainability (social, environmental and economic criteria) provides a balanced basis to evaluate and compare potential programme designs (Elkington 1997; Munasinghe 1993 and others).

Social considerations

From a social standpoint, multiple factors contribute to the success of a sustainability policy or programme. Since this chapter is primarily focused on sustainable construction programmes undertaken by public or private-sector organizations, socio-political factors are considered here from the standpoint of organizational leadership or elected officials who will mandate or endorse sustainability policies, the users and operational staff of buildings and infrastructure subject to the policies, and the share-holders, taxpayers and supporting industries whose interests those organizations serve. Construction stakeholders also have obligations to consider social sustainability elements as part of capital projects.

From the government standpoint, including both elected officials who make policy and agency stakeholders who implement or experience the direct impacts of that policy, social considerations focus on how well these agents are able to meet their obligations to the public toward achieving greater good. Similar concerns exist for the shareholders of private-sector organizations, except that they are defined by how well the overall organization is able to meet its individual social goals. Potential rewards of sustainable construction programmes and policies include (e.g. Kats 2003, Portland Energy Office 1999, Romm and Browning 1995, USDOE 2003, US Green Building Council 2004, Wilson 2005):

- increased effectiveness of facility users/occupants, potentially result-ing in better services provided by those organizations to their customers or constituencies.
- improved image/reputation of environmental leadership both within the organization and with respect to other organizations.

From the standpoint of the constituencies whose interests the organ-ization serves, effective sustainability programmes and policies can indi-rectly offer, in terms of potential social rewards (same sources as above):

- support for economic development for local industries and resultant increase in wealth and quality of life.
- increased health and productivity due to improved environmental quality.
- availability of funding to enhance other programmes that is no longer needed for facility operations and maintenance.
- better service from more efficient and effective employees.

These potential rewards are likely to result from effective implementation of sustainability policies and programmes, but there are also social risks if programmes do not perform as expected. These risks are primarily associated with the opportunity costs of investments in sustainable facilities if they do not result in performance improvements (e.g. Athena 2002, Best 2005, Bray and McCurray 2006, Greenspirit Strategies 2004, Johnston 2000, Myers 2005, Winter 2004). If programmes do not perform well, the organizations who have implemented those programmes are accountable to shareholders or taxpayers for how funds have been spent. With programmes that are new or not well understood, the perceived political risk associated with endorsing them may outweigh the promise of benefits (Pearce 2001, Pearce et al. 2005a).

Environmental considerations

The second category of considerations includes the natural environment and the impacts on ecological systems and resource bases that can be mitigated or improved by implementing sustainability programmes. Potential rewards may include (Wilson 2005):

- reduction in resource consumption, including water, energy and materials.
- reduction in waste, destruction and pollution that lead to ecosystem degradation and biodiversity loss, including solid waste, waste-water/water pollution, air pollution and site disturbance.
- increase in sustainable site development practices and improvements in transportation efficiency.

Policy alternatives can be compared environmentally in terms of likely numbers of projects or facilities affected by each proposed scenario over time and the scale of change in those facilities or projects. As with social rewards, potential environmental rewards must also be weighed against potential risks if policies and their programmes do not behave as expected. Environmental risks include considerations such as unproven or unfamiliar construction materials and technologies that fail in operation and must be replaced, requiring additional resources to address the problem. While the potential for innovative technologies and practices to backfire is inherent in all innovation (Rogers 2003), it is especially ironic in the case of environmental technologies, where replacement potentially entails more waste and resource consumption than installing a traditional alternative in the first place (Pearce 2001).

Economic considerations

The third category of evaluation criteria focuses on economics. This includes both direct costs and benefits associated with sustainability policies and programmes, and indirect costs and benefits that occur as a result of better-performing capital construction processes, the facilities that result from them, and the impacts of those facilities on their occupants. The most obvious economic cost impacts of sustainable construction programmes and policies are the direct costs of implementation. Categories of programme implementation costs include (Bennett and James 1998, Pearce 2004):

- programme administration costs.
- cost of project registration/documentation/certification.
- increased first cost of projects due to improved systems, additional design and construction requirements, additional documentation and certification, and building commissioning.
- increased lifecycle costs of projects for maintenance of unfamiliar systems.
- programme marketing costs.
- training costs.
- technical assistance costs.
- evaluation/compliance costs.

These costs are coupled with the risk of remediation costs if unknown sustainable technologies do not perform as expected. Potential quantifiable areas of savings resulting from effective sustainable construction programme implementation include (Bennett and James 1998, Pearce 2004, Wilson 2005):

- Savings in operating and maintenance costs, such as energy costs, system replacement, water/wastewater treatment and waste disposal.
- savings in first cost due to system optimization, design right-sizing, reduced waste generation and recycling revenues.
- reduced liability, for instance for human health risks.
- reduced environmental management and compliance costs.
- improved productivity and employee retention.

Many indirect economic benefits can also stem from sustainable construction programmes, including value of resources saved for future use, value of environmental image, and value of environmental quality due to avoided negative impacts, although these kinds of costs are not typically included directly in decision making since they are difficult to quantify or attribute to specific project decisions (Bennett and James 1998). Nevertheless, all of these factors should be considered at least conceptually in terms of evaluating the costs and benefits of green building programmes and designing policies that optimize benefits for stakeholders (see also Chapter 9).

Together, social, environmental and economic impacts of a policy and its programmes provide a balanced measure of the ultimate performance of that policy in terms of its influence on the sustainability of capital facilities for an organization. These three categories of impacts serve as the fundamental basis for evaluating policy and programme designs, along with programme implementability considerations.

Implementability considerations

In addition to the likely social, environmental and economic impacts that may result from a sustainability policy, the design of the policy itself and its accompanying programmes contributes to its potential to succeed within its organizational and political context. From an organizational standpoint, implementability considerations focus on the compatibility of potential policies and programmes with the standard operating procedures, constraints and conventions of implementing organizations. Specific considerations for implementability include (Rogers 2003, Vanegas and Pearce 2000):

- Compatibility with statutory requirements and funding processes.
- availability of a trigger to establish urgency of need for the programme, such as an energy crisis or environmental disaster.
- degree of change required of individuals within the organization who are affected by the policy or programme.
- level of additional burden imposed by the policy or programme on implementing agents.
- existence of enthusiastic change agents and support networks with appropriate stature and resources within the organization.
- existence and observability of rewards or benefits for policy and programme achievement.
- absence or ability to mitigate potential risks associated with policy and programme implementation.
- likelihood of strong political or leadership endorsement or, conversely, significant political opposition to the programme by leadership, lobbies or other constituencies.

These factors influence the degree and rapidity with which individuals and organizations adopt new or unfamiliar technologies and practices that have the potential to improve their existence. Together with the potential risks and rewards, implementability considerations can help predict which policy and programme designs will take hold and be successful in achieving their full potential while avoiding risk of failure.

The possible elements of a sustainability policy or programme for an owner organization are diverse and must be evaluated in terms of social impacts, environmental impacts, economic impacts and implementability. These variables form a basis for constructing potential policy approaches that can serve as paths for action by an organization toward greater sustainability.

Elements of a sustainability policy or programme

In defining the elements that could be incorporated as part of an overall sustainability policy or programme for an owner organization, particularly with respect to capital facilities, three basic categories of options emerge:

policy options, whereby formal guidance is put in place to require or encourage sustainability-related activities; **programme options**, which may provide funding, information or other needed resources to make sustainability easier to achieve; and **evaluation options**, which serve to measure the outcomes of the programme and evaluate its success. The next three subsections describe potential options within each of these three categories, along with their pros and cons in terms of the evaluation criteria identified in the previous section.

Policy options

Policies can come in the form of an executive order signed by a CEO, governor or public leader, a bill passed by legislature, or even an internal organizational directive issued by the executive staff. The following options (Table 3.1) could be made in any of these forms:

- **Require a specified level of performance** – A policy can mandate compliance with a specific set of sustainability guidelines or achievement of a measurable, specified level of performance. A straightforward method would be to require that a project or development meet some specified level of rating under a standard such as the US Green Building Council's (USGBC's) LEED rating system or a similar applicable standard. This option can be implemented with varying degrees of rigour ranging from simply meeting the standard on an honour basis vs. a formal third-party certification process. The scope of the mandate can also be varied by limiting the policy to specific facility types; limiting it to facilities or projects of a certain scale, such as over $1 million or greater than 25,000 sq ft; applying it to all projects for a specific subset of the organization or agency; or only requiring each subset of the organization or agency to do one pilot project under the policy (DuBose et al. 2007).
- **Endorse and encourage performance** – Another policy-based approach is to issue a policy which does not mandate any specific action but instead officially endorses sustainability as a priority for the organization's facilities and encourages members of the organization to voluntarily adopt sustainable construction practices. Such a policy could make specific mention of a rating system and/or a level within a rating system that it is desirable to achieve.
- **Encourage sustainable construction activity** – In addition to being used to show support for sustainability, policies can also be used to create programmatic elements that provide inspiration and support for implementation of sustainability practices by organizations and agencies (see 'Programme options'). While these programmes can be created outside of official policy documents, in some cases a policy such as an executive order gives leaders an opportunity to officially endorse a programme. This endorsement can be beneficial to the leader and may encourage others in the organization to take advantage of the programmes because they know those programmes have support from the top. Creating supporting programmes can be done in conjunction with a requirement to meet a specified sustainability performance standard.

Table 3.1 Sustainability policy options – pros and cons

	Require specified level of performance	Endorse and encourage performance	Encourage sustainability in general	Create working group to set standards
Social	May increase jobs if done externally. Occupants may be healthier and happier in their workplaces. Increased productivity is possible.	Same as requiring performance, but with likely fewer participants and subsequent reduced costs and benefits.	Benefits and costs are programme-specific.	Long-term impacts may result from eventual greater uptake of subsequent policies. Resulting network can provide inspiration for independent action and opportunity to share lessons and experiences.
Environmental	Reduced environmental impact and demands on infrastructure.	Reduced environmental impacts and demands on infrastructure, although lower impact than mandate due to lower uptake.	Benefits and costs are programme-specific.	Long-term impacts may result from eventual greater uptake of subsequent policies.
Economic	Increased costs because of certification fees and other costs associated with each standard, but expected lifecycle operational savings.	Lower cost of implementation than mandate due to lower uptake; potentially lower lifecycle cost savings as well.	Could be costly to implement; depends on the nature of the programmes specified. Benefits are also programme-dependent.	There may be direct implementation costs for meeting support and documentation.
Implementability	Greater personal load for implementers if done in-house. Some oppose particular standards for various reasons. Often constitutes an unfunded mandate. Provides temptation to do things that may not be cost-effective or appropriate just to get points.	Potentially lower uptake, but those who do adopt are less resistant. Can be used as a springboard for a future requirement while building support. Gives formal political endorsement to people inside agencies who already aspire to do green building. Lower probability of 'point-mongering'.	Gives formal political endorsement to people inside agencies who already aspire to do green building, and gives specific direction on how to proceed. Does not offend people by imposing requirements without support. Provides tools and resources to support the end goals.	Works well to achieve broad consensus and buy-in as long as all key parties are represented. Adds additional workload to volunteer participants. Requires strong leadership and effective facilitation. Recommendations from the council may carry more weight than simple political mandates. Gives alternative factions the chance to present themselves to the council and be carefully evaluated. Provides a mechanism for figuring out the most appropriate way to get things done.

- **Create a working group to develop standards or plans** – In a situation where there is support for sustainability but not a consensus about what standards to use or how far-reaching a policy needs to be, a policy creating a working group to take up the issue is a possible option. This option is a useful first step and gives a clear signal of a leader's support for the issue. Creating a working group in the form of a council or task force gives leadership the opportunity to have input from critical stakeholders in a transparent process. This approach also avoids some of the political risks of mandating a specific set of guidelines to a resistant population of potential adopters.

Programme options

A variety of programme options have been developed by agencies and organizations for increasing uptake of sustainability-related techniques and practices in the built environment. Programmes can be established through formal policies or can be created on an ad-hoc basis. Critical to the success of all programmes is a source of funding and support to build, promote and sustain the efforts undertaken by those programmes. Potential programme options include (Table 3.2):

- **Technical support** – Providing technical support can be useful to create capacity for sustainable construction. It can also help overcome ignorance about new and innovative sustainable construction techniques that are different from conventional practice. Technical assistance can be provided directly by the specific organization designated to promote sustainability, or indirectly by providing funding for technical assistance by external providers.
- **Training** – Unlike technical support which provides assistance for specific projects, training opportunities can be used to inform facility stakeholders on topics including general sustainability, sustainable construction principles, rating system requirements and technical details of specific technologies. Organizations may not need to implement training internally but instead may find ways to increase the number of personnel attending existing sustainable building training events through subsidizing the cost of training, providing release time to attend or merely encouraging attendance at training seminars.
- **Guidance documents** – Many implementing organizations have found that it is useful to create guidance documents to distil the wide variety of available information on sustainability into a more concise format that contains information relevant for their specific context. These documents can be a tailored version of third-party rating systems highlighting strategies that have been most successful for similar projects by that organization (Pearce et al 2005c). They also include tailored checklists to address attributes of that organization's projects and guidebooks containing more detailed information (Bosch and Pearce 2003).

- **Demonstration projects** – Demonstration or pilot projects are implemented to show what sustainable construction entails and demonstrate its benefits in the context of a real project, without necessarily committing to an ongoing policy of constructing in this way for future projects. A successful pilot project can help dispel the fears and objections of opponents, and incurs a much lower political cost and risk than formally putting a policy into place. If successful, it can be a precedent toward establishing a policy later on.
- **Incentives/subsidies** – Instead of mandates, another approach is to reward organizations or parts of the organization that are 'ahead of the curve' in already pursuing these practices and provide motivation for other parts of the organization to follow their lead. Incentives might include reimbursing the cost of rating system certification, leadership awards for most sustainable new projects, improved performance reviews for involved individuals and positive press coverage.
- **Modified organizational practices** – Established procedures can sometimes make achieving sustainability difficult. Programmes to modify internal organizational constraints can remove procedural barriers to sustainability-related actions. Examples include contract vehicles for commissioning or energy savings performance contracting, waiving 1-year contract limit requirements and prequalification of contractors or products (DuBose et al. 2007).

Evaluation options

Evaluation involves determining the compliance of policies and programmes to requirements and their effectiveness, both at the individual project level and overall. Sometimes but not always, policies explicitly specify how compliance should be measured or demonstrated, and they sometimes specify reporting and accountability requirements that programme implementers must follow to document compliance (Pearce et al. 2005a). These measures also provide baseline data to evaluate the effectiveness of the programme as a whole. Voluntary programme evaluation is also possible. Options include (Table 3.3):

- **Third-party certification or evaluation** – This method is effective for ensuring that specific projects follow prescribed sustainability guidelines. This mechanism creates a clear metric with reduced administrative burden on the organization for ensuring compliance, but it does put responsibility on the agents managing the construction process. Certification through third parties also involves expense that some organizations have found to be a barrier when they would rather invest that funding directly into the building.
- **Regular reporting requirements** – This requirement can be combined with third-party certification. Some state governments in the United States, for instance, have set periodic reporting requirements for their agencies to report back to a central agency, committee or council on their sustainable construction accomplishments and

Table 3.2 Sustainability programme options – pros and cons

	Technical support	Training	Guidance documents/ tools	Demonstration projects	Incentives/ subsidies	Modified organizational practices
Social	May increase confidence to try innovations; adds capabilities to teams that don't already have them; can make sustainable construction seem like something that requires expert assistance and disempower individuals.	May increase confidence to try innovations; may provide networking opportunity and peer interaction; may build greater internal capacity and support since it empowers individuals who receive training.	May increase confidence to try innovations. Individually empowering, but essentially an individual effort; no specific opportunities for networking.	Occupant benefits can derive from sustainable design. Long-term impacts may result from eventual greater uptake of sustainability best practices that are effectively demonstrated on these projects.	Impacts may result from greater uptake of sustainability best practices.	Impacts may result from greater uptake of sustainability best practices.
Environmental	Impacts may result from eventual greater uptake of sustainability best practices.	Impacts may result from eventual greater uptake of sustainability best practices.	Impacts may result from eventual greater uptake of sustainability best practices. Depending on the nature of the document, can help to tailor efforts to those most effective for the specific context, e.g. what works best in a given climate.	Environmental benefits from sustainable design of demonstration facility.	Long-term impacts may result from eventual greater uptake of sustainability best practices that are effectively demonstrated on these projects.	Impacts may result from greater uptake of sustainability best practices.

	Technical support	Training	Guidance documents/ tools	Demonstration projects	Incentives/ subsidies	Modified organizational practices
Economic	Have to pay for implementation, but ultimately will speed learning curve and build broadly applicable capacity that can result in long-term savings.	Have to pay for implementation, but ultimately will speed learning curve and build broadly applicable capacity that can result in long-term savings.	Less expensive because it is generated once but used many times. Relatively minimal ongoing costs for updating, and dynamic options that are self-updating are possible.	Could capitalize on existing sustainability projects by designating them as demonstration projects and promoting them. Access to different funding sources and donations is often possible. Funding commitment is project by project, not an ongoing commitment.	Direct first cost of implementation varies by programme type, e.g. paying for rating system third-party certification or commissioning. Should be phased out over time.	No cost outside normal operating costs for organization, unless feasibility studies or similar are required.
Implementability	Technical assistance does not ensure uptake, but at least it applies to an immediate real project situation. Potential exists to capture and transfer lessons learned via centralized tech support provider. Depending on who provides the support, can generate or suppress market capacity.	Providing training does not ensure uptake. The next relevant project may not happen soon.	Requires individual adaptation to specific cases. May have to provide dissemination and training to ensure widespread effective use. Can be tailored to meet the culture, constraints and needs of the organizational context, and therefore be more easily adopted. Does not ensure uptake.	Lower risk of perceived failure on these projects (since they are designated as pilots/ demonstrations) encourages greater innovation. Improves the implementability of future projects due to ability to learn from these special cases.	Greatly reduces the most significant barrier to implementation: perceived increased first cost.	One effort can be used multiple times by multiple agencies and projects. Examples include contract vehicles for commissioning, waiving 1-year contract limit requirements, prequalification of contractors or products, etc.

Table 3.3 Sustainability evaluation options – pros and cons

	Third-party certification	Regular reporting requirements	Performance monitoring and reporting	Post-construction evaluation
Social	Pride in certification outcomes; external validation; visible reward for achievement.	Introduces accountability. Can also introduce a spirit of competitiveness and motivation to excel. Visible and public acknowledgement of achievements.	Introduces accountability. Opportunity for feedback and action may empower facility manager to proactively deal with problems, resulting in greater occupant satisfaction and productivity.	Most likely option to give good information on true social impacts that can be applied as lessons to future projects. Can be empowering to occupants.
Environmental	Ensures that basic standards are met, but doesn't necessarily guarantee environmentally beneficial outcomes during operation.	Can build on third-party certification and encourage positive environmental outcomes.	Encourages meeting environmental performance goals during operation rather than just meeting standards upfront.	Encourages meeting environmental performance goals during operation rather than just meeting standards upfront.
Economic	First cost of implementation can be considerable; risk that certification is not achieved.	Can be minimal cost to implement; work imposed on existing personnel.	Ongoing programme costs can be considerable, but afford the opportunity for operational adjustments that can result in savings.	Ongoing programme costs can be considerable, but afford the opportunity for operational adjustments that can result in savings. Likely to employ a third-party to perform.
Implementability	Generally a limited-time event per project with standardized milestones. Considerable opposition may exist regarding the level of effort and cost required. Many stakeholders have to provide data to meet third-party certification requirements, making documentation complex and difficult. Risk that certification is not achieved.	May require a centralized person to continually pester agencies for data. Can be piggybacked on other data submittal requirements such as annual performance or status reporting.	Requires initial investment in monitoring equipment or possible employment of a third-party. Benefits from ongoing data analysis and interpretation. Agency maintains control and can take immediate action to remedy defects as they are discovered. Proactive.	Fear of identifying problems that are otherwise not obvious. Could be embarrassing or reflect poorly on the project team or building. Perception of less control than with performance monitoring. Likely to be a one-time event, not ongoing.

whether or not they followed any policies that have been established (Pearce et al 2005a). Reporting requirements may create a greater sense of accountability, which ultimately results in greater action.

- **Operational performance monitoring and reporting** – If the goal is to achieve better-performing projects that consume fewer resources, requiring that organizations monitor their capital projects and facilities and regularly report this data to a central authority may be effective. A requirement that organizations develop and implement an action plan for how they will remedy any performance deficiencies can be even more powerful.
- **Post-construction evaluation** – In addition to rating systems that help to guide design and operationalize sustainability during a project, evaluation of post-construction performance is a useful tool to ensure that projects are indeed meeting their design intent and thereby making progress toward the underlying or driving goals of the sustainability policy such as energy savings. Post-construction evaluations may range from a simple walk-through to intense investigative studies using a variety of research methods to correlate physical factors with user-related outcomes. Post-construction evaluation may include (Federal Facilities Council 2002):
  - utility studies, including power and water consumption
  - employee productivity studies
  - absenteeism studies
  - indoor environmental testing for buildings
  - user satisfaction evaluations
  - acoustical studies.

Whatever the mechanism or mechanisms for evaluation, measuring the impacts of sustainability policies and programmes is essential to remain accountable to the shareholders or taxpayers who ultimately support those programmes and benefit from their existence.

Phases of a sustainability programme for owners of capital projects

Sustainability programmes generally evolve through four basic phases (DuBose et al. 2007, Pearce et al. 2005a), as shown in Figure 3.2. These phases are essential to the success of any sustainability programme, although they may be implemented in different ways by different stakeholders in different contexts. The phases mirror the innovation-decision process (Rogers 2003), beginning with an Inspiration phase that includes knowledge, awareness and persuasion that move a person or organization to adopt a sustainability policy or practices. Inspiration may, at some point, be followed by Motivation, the stage in which a formal or informal policy is developed to shape subsequent organizational actions toward meeting sustainability goals. Motivation is followed by Implementation, where programmes are developed to support the activities needed to meet the goals of the policy, followed by Evaluation, where compliance with policy requirements and assessment of programme

Figure 3.2
Phases of a sustainability programme

performance is undertaken. These four phases provide a structure for mapping the different ways in which owner organizations have approached each step in creating a sustainability programme.

Phase 1: Inspiration

Inspiration can be defined as any activity that increases awareness about sustainability goals. It includes formal and informal mechanisms by which stakeholders become informed about the benefits and opportunities of sustainable design and begin to consider its application in their own situations. Exposure of stakeholders to sustainability concepts can come through a variety of channels, including:

- reading about sustainability in trade publications or journals;
- attending meetings with speakers who address the subject;
- interacting with colleagues who have sustainability interest or experience;
- participating in formal training or conferences on sustainability;
- having discussions with sales/marketing people from sustainability-related product manufacturers.

Exposure can be intentional or unintentional on the part of any given stakeholder, but even uninterested stakeholders have an increasing likelihood of hearing about sustainable construction concepts just by virtue of doing their jobs on a daily basis. Coverage of sustainability and sustainable development concepts has increased considerably over time, as evidenced by the growing number of articles in general trade publications and sustainability-related tracks in traditional industry conferences. Publications, courses and conferences are emerging that are dedicated specifically to sustainable construction issues, and even coverage in popular press and television helps to ensure exposure to sustainability concepts by project stakeholders.

Some stakeholders may become inspired to pursue sustainability by virtue of being faced with an external threat or crisis that requires a change in the status quo, such as energy or water shortages, volatility in fuel or building material prices, specific environmental constraints associated with project sites, or human-induced or natural disasters that offer significant opportunities to rebuild. With their focus on resource efficiency and reduced environmental impact, sustainable capital projects offer a chance to invest in solutions that reduce demands on resource bases and ecosystems both initially and over their substantial lifecycle. This benefit means that sustainable solutions can emerge as leading contenders in crisis situations.

Prevailing conditions

As with the adoption of most policies or programmes, a variety of triggers create a climate in which new ideas are accepted and allowed to flourish, even in the midst of opposition. The presence of a crisis cannot be underestimated as a way to establish support for sustainability policies and programmes. For example, in the United States, Chesapeake Bay, Maryland had been suffering from many years of pollution, causing several administrations to work toward improving the quality of the bay. In Arizona, Phoenix experienced brownouts in the summer of 2004 due to the failure of a large transformer. In Maine, extremely high energy costs made sustainable construction more attractive. Other conditions, such as the recognition that high operation and maintenance costs of facilities could be reduced through sustainable design, also help to create a climate that is conducive to sustainability policy and programme development. Organizations with rapid growth recognize the importance of implementing sustainability policies and programmes to ensure that the capital projects they will add to their inventory are well designed, provide a better working environment, and are more efficient to operate.

Leadership

Whether initiated by organizational leadership or workers, sustainability-related activities are more likely to occur in an organization if there are champions to promote them. Whether motivated by personal conviction, political gain, economic benefits or some other impetus, someone (or a group of persons) becomes inspired to launch new programmes or policies. Advancing a more sustainable approach to building design, construction and operation/maintenance from an occasional occurrence to standard operating procedures within an organization requires dedication, determination and tenacity on the part of sustainability champions. They will encounter stakeholders who are opposed to changing existing practice because of fear, a lack of understanding about sustainability strategies or an unwillingness to part with the old way of doing things. Typically, it takes a larger number of champions working from the bottom up within organizations compared with fewer champions at higher levels with greater decision- and policy-making authority to make significant changes in a relatively short period of time.

Government activities to encourage private sector sustainability

Sustainability activities in the private sector may advance the rate of adoption of sustainable design and construction practices in the market at large. The existence of prominent private-sector sustainable projects increases awareness of the importance of sustainability among the general population. As the private sector builds more sustainable facilities, members of the design and construction community gain valuable knowledge and experience in creating sustainable facilities and infrastructure systems, and therefore build capacity among professionals.

Rewarding organizations and individuals that choose to build sustainable facilities is an effective method to motivate the adoption of best practices in design, construction and operations while simultaneously improving health, prosperity and quality of life for all. Thus, many governments have begun to implement different types of incentive programmes, along with structural and financial incentives. Structural incentives such as density bonuses and expedited permitting can be implemented at low or no cost to government authorities. These incentives encourage developers to build more sustainably by making healthy, efficient and high-performance development an even more attractive option.

One example of a procedural incentive is simple modification in zoning permissions and review processes that can yield impressive dividends for developers and building owners alike. The city of Dallas, Texas adopted a green building ordinance requiring energy and water efficiency improvements for new residential and commercial buildings. In Dallas, expedited permitting is available if the organization planning new residential construction submits a residential green building checklist (LEED for Homes, GreenPoint Rated, Green Communities, GreenBuilt North Texas or others). For commercial construction greater than 50,000 sq ft, expedited permitting is available if a project attempts a number of priority LEED credits.

Financial incentives are also a highly successful means of encouraging developers to follow sustainable design and construction practices. While these incentives necessarily require a financial investment in more sustainable projects, state and local governments are finding that these investments pay dividends to a community's Triple Bottom Line: ecology, economy and equity.

Case study: Songdo IBD, South Korea's sustainable city

Songdo International Business District (Songdo IBD) is a sustainable development located on the waterfront of Incheon, South Korea covering 1500 acres. The development team of Songdo IBD shows its commitment to sustainability for this new city with six development goals: open and green space; transportation; water consumption, storage and reuse; carbon emissions and energy use; material flows and recycling; and sustainable city operation. Songdo IBD includes an international business district area, 80,000 apartments, and can handle 300,000 commuters daily. As a smart city, sensors, networks and controls have been built into the streets and buildings to facilitate ubiquitous computing. Occupants can control environmental conditions remotely, be instantly informed about alerts and crimes, or learn when the next bus will arrive. To achieve the goals of this sustainable development, the development team has incorporated the LEED rating systems for new construction and neighbourhood development. Sustainable strategies to achieve six goals are:

Open and green space
- Songdo IBD has been designated with 40 per cent open space – 600 acres – to maximize the connection to nature within the city for residents, workers and visitors.
- A 100-acre Central Park (which is modelled after New York City's Central Park) provides an open space for residents, workers and visitors.
- All blocks connect pedestrians to open space, walking/biking corridors and public park.
- Native and adapted species have been planted to reduce water needs in its open space.

continued

Transportation

- Incheon subway line runs through the centre of Songdo IBD.
- A 25 km network of bicycle lanes within the project facilitates safe, carbon-free transportation.
- 5 per cent of parking capacity within each project block is set aside as parking for fuel-efficient and low-emitting vehicles. Office and commercial blocks reserve an additional 5 per cent of parking capacity for carpool vehicles.
- Parking is primarily located underground or under a canopy to minimize the urban heat island effect and maximize pedestrian-oriented open space above ground.
- Infrastructure for electrical vehicle charging stations is integrated into parking garage designs to facilitate the transition to low emissions transportation.

Water consumption, storage and reuse

- The Central Park canal uses seawater instead of fresh water, saving thousands of litres of potable water per day.
- Irrigation-based potable water use targets a 90 per cent reduction vs. international baseline, reduced through the use of efficient landscape design, water-saving irrigation systems, reclaimed stormwater and reuse of treated greywater from a city-wide central system.
- Potable water consumption in plumbing fixtures targets a 20–40 per cent reduction based on the use type of the project.
- Stormwater runoff is reused to the maximum extent possible given the project's climate zone and annual rainfall pattern.
- Vegetated green roofs can reduce stormwater runoff, mitigate the urban heat island effect and promote biodiversity and species habitat preservation.

Figure 3.3 Songdo IBD has been designated with 40 per cent open space – 600 acres – to maximize the connection to nature within the city for residents, workers and visitors

continued

Carbon emissions and energy use

- All buildings in Songdo IBD integrate energy-saving strategies to reduce energy consumption and carbon emissions.
- A central city-wide co-generation facility fuelled by natural gas provides clean power and hot water to the project.
- Energy-efficient LED traffic lights and energy efficient pumps and motors are used throughout Songdo IBD.
- A centralized pneumatic waste collection system is used to collect wet and dry waste, eliminating the need for waste removal vehicles.

Material flows and recycling

- 75 per cent of construction waste is targeted to be recycled.
- Recycled materials and locally produced/manufactured materials are utilized to the maximum extent possible.
- Low-VOC materials are incorporated into all buildings.

Sustainable city operations

- Sustainable procurement goals and recycling guidelines are integrated into the operational structure of the city through the facilities management digital interface.
- Facilities management and maintenance contracts mandate environmentally friendly (low/zero VOC, EcoLabel, Good Recycled designations or equivalent) products.
- Smoking is prohibited in public areas and office buildings except for specially designated areas.

The Songdo IBD project is an exemplary sustainable development that sets a high standard for sustainable design for large-scale city development across the globe.

Figure 3.4 A 100-acre Central Park located in the middle of Songdo IBD provides an open spaces for residents, workers and visitors

Source: All photos courtesy of POSCO E&C

Figure 3.5 Songdo Convensia has incorporated many sustainable features including public transportation systems, energy-saving lighting systems, water-saving fixtures, a construction waste management programme and low-VOC materials to achieve the goals of sustainability in the building. Songdo Convensia is the first LEED certified convention centre in Asia

Major financial incentives for sustainable construction include tax credits and abatements, fee reductions or waivers, grants and revolving loan programmes. For example, the US state of New Mexico created legislation that provides tax credits based on the square footage of a green facility. For commercial buildings, the tax credits range from US $3.50/sq ft for a building that achieves LEED-NC Silver certification, to US $6.25 for a building that achieves LEED-NC Platinum certification by the USGBC. For residential buildings, the tax credits range from US $5.00/sq ft for buildings that achieve LEED for Homes Silver certification to US $9.00/sq ft for a building that achieves LEED for Homes Platinum certification.

Facing opponents

Some organizations or people are more difficult to inspire to change their practices than others. Organizations typically find opposition from those who believe that sustainability best practices will increase design and construction costs more than is reasonable. Other opponents represent industry trade associations (e.g. forestry, vinyl and refrigerant manufacturers) that do not agree with a particular credit or point within a rating system, such as credit in the LEED Green Building Rating System to reward use of lumber certified as sustainably harvested by the Forest Stewardship Council. Building owners sometimes oppose specific prerequisites such as commissioning in rating systems like LEED because they feel as though they already pay for proper design and in-stallation, making commissioning unnecessary. In order to pass sus-tainable construction policies, it is important to identify, acknowledge and address the questions and concerns of opponents. It is not necessary, and in fact is nearly impossible, to appease everyone, however. Some organizations have effectively addressed some opponents by allowing exemptions and providing allowances that address their concerns.

Inspiration may come from a variety of sources encouraging sus-tainability best practices. When an organization finally decides that sustainable design and construction is important enough to develop a policy, this phase is called Motivation. The next section describes this phase of the sustainability programme development process.

Phase 2: Motivation

The second phase, Motivation, refers to formal and informal drivers that compel people to actually try sustainability techniques on their projects. Internal motivation helps people become convinced through repeated exposure to sustainability information during the Inspiration stage that they would like to try sustainable construction tactics on their own projects. Alternatively, formal, external motivators such as execu-tive orders, legislation or internal policies may encourage or require sustainability action. At the Motivation stage, some stakeholders may have already had exposure to sustainability through the Inspiration process to be receptive, but others may be suddenly subject to new requirements without prior exposure, particularly in the case of mandates such as executive orders or legislation.

Sustainability policies in this stage fall along a spectrum of permanence corresponding to the mechanism used to institute them. For instance, at the state level in the United States, policies can be issued as executive orders by a governor, or as legislation formally passed by the legislature. The former is less permanent since it can be rescinded or superseded by subsequent governors when the issuing governor leaves office, or it can lose urgency even if it remains in effect when a new governor with different priorities takes office. Executive orders typically have no formal enforcement mechanism, although they may include metrics or formal evaluation and reporting procedures to encourage compliance.

Legislation, on the other hand, is considerably more permanent since it is passed by governing bodies of elected officials and is therefore unaffected by changes in administration. It is also typically much more difficult to put in place, since it must receive majority support across multiple elected officials through the formal legislative process, and is subjected to more scrutiny and possible opposition from lobbying groups. Legislation also has penalties to encourage compliance, since failure to comply is actually breaking the law. While the nature of penalties depends on the specific legislation, a legislation-based sustainability policy has a much higher likelihood of being successful in the implementation phase due to the broader level of sustained support it must achieve in order to be issued in the first place.

While some organizational sustainability policies are focused solely on establishing sustainability standards for capital projects, many policies also include a wide array of additional issues such as transportation fleets, operating procedures and investment criteria. For example, in the United States:

- Some policies reach beyond buildings alone. For instance, in California, the state policy requires that the Teacher Retirement System seek investments in green buildings and green technologies.
- In order to equip future building professionals to design and construct sustainable facilities, Nevada's legislation requires that the university system provide students with the essentials of sustainable design and construction to help them prepare to become LEED accredited professionals.
- New York's executive order policy includes some very specific operating requirements to save energy such as turning off lights in unoccupied areas, shutting off unused equipment and adjusting thermostats. New York's policy also includes goals for improving the fuel economy of state-owned vehicle fleets.

Sustainability policies have also established programmes to support organizations as they implement sustainability. Specific programmes to support sustainability implementation are broad and varied, and the next section of this chapter, Implementation, discusses examples of such programmes in greater detail.

Phase 3: Implementation

The third phase of the sustainability programme framework is Implementation. This is the step in which parties responsible for turning policy into practice decide what programmes and actions will be needed to meet policy goals. It also involves executing those programmes and actions to achieve the goals. Examples of Implementation programmes and actions include:

- technical assistance programmes, both internal and privatized;
- education and outreach programmes, both internal and privatized;
- incentive programmes, including subsidies and grants to offset project costs;
- award programmes;
- organization-specific guidelines or application guides;
- modified operating practices.

Which programmes are most effective in a given context depends on the level of market expertise on sustainability, the structure and culture of the organization in which the programmes are being implemented, and available financial resources including third party funding such as grants. Programmes that are specifically mentioned and established as part of formal policy typically have more success in motivating change, since they are formally endorsed by leaders and have resources explicitly provided for them. Some activities to assist organizations in implementing a sustainability policy include technical support, training, the development of guidance document, and demonstration projects. The following sub-sections describe these programmes in greater detail.

Technical support

In order to increase sustainable design and construction capacity and ensure that sustainability principles are followed, some organizations in the public sector offer technical assistance, either directly using their own experienced staff or indirectly by providing funding for technical assistance by other entities. For example, in the United States, the New York State Energy Research and Development Authority (NYSERDA) administers an innovative and successful technical assistance programme called FlexTech. Through this programme, NYSERDA approves a list of qualified providers in various fields of expertise and makes these contractors available to its clients to perform customized technical assistance on a cost-share basis. In Colorado, the Office of Energy Management and Conservation provides free technical assistance to all public agencies in the state of Colorado through its Rebuild Colorado programme to support energy performance contracting. In Pennsylvania, the Governor's Green Government Council has an experienced engineer on staff that provides direct assistance, primarily to state agencies but also to private sector clients, depending on availability.

Training

Training informs facility stakeholders on topics ranging from general sustainability awareness to technical details of sustainability best practices. In the US state of Maine, for example, state government provided funding for several training courses on green building techniques, including official LEED training conducted by USGBC trainers, as well as other courses taught by volunteers with the knowledge and experience to train the building community. In California, the Division of the State Architect, Department of General Services (DGS) has developed and participated in the development of sustainability-related training materials, including a series of videos, to advance green building in California. Pennsylvania has also produced a series of videos that is widely available via a website. By making these resources publicly available on their websites, organizations can also help a countless number of others with their sustainable construction efforts.

Guidance documents

Organizations sometimes develop guidance documents to assist their staff in implementing sustainability. In the US state of Washington, for example, school districts adopted the Washington Sustainable School Design Protocol and were allowed use it instead of LEED as a rating system. NYSERDA worked with a coalition of other agencies to fund the development of University at Buffalo's High Performance Building Guidelines. In Colorado, the State Energy Office produced a guidebook for how to implement LEED based on strategies that have worked given the unique climate, utility costs, typical payback, construction styles and other factors unique to that state. The Commonwealth of Pennsylvania has produced several well-known guidance documents not only on sustainable design, but also on requirements for leased facilities to meet sustainability objectives.

Demonstration projects

There is perhaps nothing like the success of others to inspire one to try something innovative, including sustainable design and construction. For instance, in the US state of Maryland, a long-abandoned warehouse building was converted into offices for the Maryland Department of Environment and several other state agencies. This project, which included a 30,000 sq ft green roof, was considered to be quite successful and has been used as a demonstration facility to encourage others to try similar tactics on their own projects. The state of Arizona Department of Environmental Quality (DEQ) building in Phoenix, completed in July 2002, was designed to be LEED certified. The building is very energy efficient – utility bills have been about US $1.16/sq ft vs. about US $1.50/sq ft for a conventional Phoenix-area building built in the same year. In addition, the local utility installed a 100 kW photovoltaic array on the roof of the parking garage to demonstrate renewable energy. The building has performed very well as an efficient building, a useful showcase and an educational tool.

Organizations with sustainability policies can demonstrate a commitment to assisting their staff through programmes like those described in this chapter. How does an organization know whether or not it is complying with requirements and effectively implementing appropriate actions? The phase of a sustainability programme designed to measure compliance with sustainability policies, Evaluation, is discussed next.

Phase 4: Evaluation

The fourth key phase of a sustainability programme is Evaluation. Evaluation can cover programme compliance and/or the effectiveness of the policy, either at the individual facility level or over an entire portfolio of facilities or functional areas. Sometimes but not always, policies explicitly specify how compliance should be measured or demonstrated and specify reporting and accountability requirements that programme implementers must follow. In the United States, many organizations either reference or explicitly incorporate the USGBC LEED Green Building Rating System as an evaluation mechanism and/or support tool. Some organizations require that all of their projects become certified by the USGBC at a particular level. For example, California governor Arnold Schwarzenegger signed Executive Order S-20–04 in 2004 requiring LEED Silver certification for all state-funded 'significant' (50,000 sq ft, prototype or highly visible buildings with an educational purpose) new and renovation projects. Other organizations do not explicitly require certification but simply say that buildings 'shall meet the requirements for certification'. This is done to avoid the costs associated with formal certification, to make the policy more amenable to opponents of the referenced rating system, or to encourage a diversity of sustainability programmes and approaches within the organization.

While many policies do refer to rating systems such as LEED, they do not all require formal certification. In addition to the prevalent rating system within a particular context, some policies provide for the possibility of using an alternative rating system instead. One alternative to LEED referenced by some organizations in the United States is Green Globes, described further in Chapter 4. Other policies specify that it is acceptable to use a particular rating system 'or equivalent'. Many states in the United States, as well as federal and local government entities, have realized that LEED is not necessarily a perfect match for their building types and contexts, and have developed application guides or variants on the basic LEED framework to better meet their goals.

The most straightforward evaluation method is to require certification by a third party for all projects covered by a policy directive. This creates a clear metric with little administrative burden on the organization for ensuring compliance, but it does put considerable responsibility on the individuals managing the capital project process. Whatever the evaluation mechanism selected, the importance of evaluation as part of the policy is paramount to ensuring that the goals of the policy are being met. Careless specification and use of a rating system to get points for the sake of points can lead to counterproductive project decisions.

For example, increasing evidence suggests the possibility that highly rated projects under the LEED system may not always perform as expected. Some project teams are so eager to obtain points under LEED that they may pursue technological solutions that are not relevant or appropriate for their projects (Bray and McCurray 2006, Myers 2005).

Many organizations require their staff to regularly report sustainability-related initiatives and activities as a way to measure compliance and overall programme effectiveness. In many cases, this also entails periodically reporting specific data consistently, which is helpful in monitoring progress. For example, in the United States, Colorado's policy contains a requirement that all state agencies give an annual report to the Greening Government Council enumerating all their accomplishments related to the state's Executive Order, including data on the savings from those measures. In other states and at the federal government level, the reporting requirement is given to one agency tasked with overseeing the programme. For instance, the Maryland Green Building Council must produce an annual report to the governor and General Assembly describing the efforts of state agencies, including green buildings, clean energy procurement and greenhouse gas reduction. Likewise, Pennsylvania's policy tasks the Department of Environmental Protection with compiling an annual report on the progress of all the commonwealth's agencies in reaching their goals set forth in their green plans.

Evaluation of policies themselves can include provisions for periodic review and updating of the policy over time and requirements for periodic reporting and accountability, as mentioned in the previous subsection. Although funding may not be specifically provided for these activities in the policy, other resources such as grants or partnerships with universities can often be tapped to conduct these studies and monitor progress toward policy goals. Together, these elements combine to achieve change toward the goal of increased sustainability through best practices.

The corporate perspective: how to transform a construction company to deliver more sustainable projects

From the perspective of a government agency trying to incentivize more sustainable practices in the built environment, there are many ways to promote and encourage owners to seek to adopt sustainability as they develop and operate their capital facilities. However, this perspective can differ drastically from the perspective of the architecture, engineering and construction (AEC) industry that must respond to the needs of owners to actually deliver more sustainable projects. From the perspective of construction as a production process, contractors and subcontractors represent the general and specialized units within the process that accomplish the actual production of the desired product (Gil et al. 2000). With design–build procurement and pre-construction services on the rise, these players also have an increasingly important role in the planning and design phases of capital projects (Horman et al. 2006), and their input has the potential to shape how projects are realized in multiple ways (Errasti et al 2007), including influencing planning and design decisions that have cost implications for the project. Together with the supply chain that provides materials and products for the production process, these actors constitute the constructor subsystem within the larger socio-enviro-technical (S-E-T) system that comprises a capital project (Rohracher 2001).

The stakeholders that collaborate in the delivery of green capital projects must deal with a number of unique qualities that characterize more sustainable projects, including (Pearce 2010):

- Tightly coupled designs and multifunction materials and systems (Riley et al. 2003; Rohracher 2001).
- Procurement of unusual products with limited sources (Klotz et al. 2007; Pulaski et al. 2003; Syphers et al. 2003).
- Existence of incentives and resources not available to other projects (Grosskopf and Kibert 2006; Pearce 2008; Rohracher 2001).
- Requirements for additional information and documentation (Lapinski et al. 2005, 2006; Pulaski et al. 2003).
- Greater involvement of later stakeholders in earlier project phases along with greater integration of their input (Cole 2000; Gil et al. 2000; Matthews et al. 1996; Pulaski and Horman 2005c; Pulaski et al. 2006; Reed and Gordon 2000; Rohracher 2001).

Considered from the standpoint of contractors and subcontractors, sustainability as an innovation may not always seem to offer immediate benefits over the status quo, especially for later adopters who miss the opportunity to use sustainability expertise as a market differentiator. In terms of relative advantage, the benefits of sustainable construction tend to accrue to other stakeholders or even non-stakeholders, particularly the owners and occupants of the facility and future generations who may benefit from reduced environmental impacts and resource consumption (Taylor and Wilkie 2008). These benefits may also be difficult for the constructor to see since they are typically spatially and temporally distant from decisions made during construction (Gardner and Stern 1996, Khalfan 2006), thereby reflecting poor observability. From the standpoint of trialability, integrated design and delivery tactics require contractors to jump in with both feet rather than being able to try sustainability concepts and practices at their own pace (Horman et al. 2006, Lapinski et al. 2005). The use of new technologies and products may require deviation from established subcontractor and supplier networks, thereby reflecting poor compatibility with contractor assets that traditionally afford a source of competitive advantage (Kale and Arditi 2001). Finally, all of these factors combined with the demands of extensive new documentation, product qualification requirements and additional general requirements such as waste management, indoor air quality best practices and commissioning lead to a disadvantage in terms of complexity (Klotz et al. 2007). From this perspective, it should come as no surprise that not all members of the constructor subsystem have yet universally embraced sustainability (Panzano et al. 2004). How, then, to align a construction company with the goals of sustainability and address those goals to strategic advantage?

The evolution of corporate sustainability for the construction sector

Despite the challenges associated with diffusing sustainability throughout the construction industry, the concept has a history in this sector that evolved from some of the fundamental trends that have shaped construction over the last 50 years or more. This section describes some of the major trends that were precursors to sustainability in the construction industry, leading to today's approaches for strategic sustainability in corporate operations in the construction industry.

Environment, safety and occupational health (ESOH)

Although ideas pertaining to safety in the construction sector have been around since the time of Hammurabi, corporate attention to protecting the environment, safety and occupational health in the workplace has evolved comparatively recently. Driven by both regulation and corporate liability, these notions emerged most strongly in the 1970s with the establishment of the Occupational Safety and Health Act in the United States and numerous environmental regulations discussed in Chapter 2. Since that time, construction along with many other industries has invested significant attention and resources to improving the safety and occupational health of its workers, and reducing impact on the natural environment. Many construction companies have special departments or personnel within the organization devoted specifically to managing and improving safety and occupational health within the organization and establishing a safety culture within the company.

Increasingly, companies have personnel or resources specifically dedicated to management of environmental compliance as well. For example, managing stormwater on construction sites in the United States has gained significant importance since the development of the National Pollutant Discharge Elimination System (NPDES) permit programme under the Clean Water Act. This programme requires permitting and enforcement of stormwater controls on construction sites, and imposes site shutdowns and/or fines for violations. As requirements to comply with the Clean Water Act have become increasingly more stringent, construction companies have begun to realize the real impacts noncompliance will have on project budgets and schedules, and they have taken actions accordingly. Some companies manage these issues internally through special departments or functionalities within the project team, while others outsource specific responsibilities for stormwater management to outside firms.

In either case, the importance of systematic managing ESOH issues for company success and competitive advantage is becoming more universally recognized. Although these issues have historically been driven by regulation, legal liability, or both, in today's market they form the underpinning for a company's reputation and success in the marketplace. Some industry experts have suggested that eventual adoption and internalization of sustainability principles within the construction industry will follow a similar path.

Total quality management (TQM)

TQM is another trend that has been applied across multiple industry sectors before being embraced by the construction sector. The concept originated in the early twentieth century as an application of statistical theory to product quality control, and saw widespread evolution and adoption in domains such as the automotive industry in the 1940s and 1950s, especially in Japan and the United States. Eventually, TQM evolved from a narrow focus on improving the quality of products and began to be applied more broadly to the task of improving all aspects of a company's operations, with a focus on 100 per cent customer satisfaction. By the 1990s, TQM broadly included concepts addressing multiple aspects of corporate management, including customer focus, employee involvement, continuous improvement and integration of quality management into the total organization.

Often called 'business excellence' today instead of TQM, the concept of systematic, holistic management of an organization's operations based on consistently and competitively meeting customer needs eventually gained attention from other organizations such as service industries, government agencies and charitable organizations or nonprofits. Today, it has been more widely adopted by companies in the construction sector, especially outside the United States (Pheng and Teo 2004). Benefits of adoption cited by construction firms who have adopted TQM include reduction in rework costs, better employee job satisfaction because employees do not need to attend to defects and client complaints, recognition by clients, work carried out correctly from the start, subcontractors with proper quality management systems, and closer relationships with subcontractors and suppliers.

TQM is a significant precursor to sustainability in the construction industry because of its similarities to the concept during corporate adoption (Fust and Walker 2007). Both concepts have initially been ambiguous with regard to implementation, and both have been often perceived by firms as a cost or risk rather than an opportunity for competitive advantage. As with TQM, the firms most likely to reap competitive advantage from early adoption of sustainability principles are those that see the opportunities it presents for improvement rather than 'an added cost to absorb, another risk to manage, or one more regulation with which to comply' (Fust and Walker 2007). There are a number of parallels between the two concepts based on a review of early leaders in multiple sectors (Fust and Walker 2007), and four important attributes characterize successful programmes in TQM that also apply to sustainability initiatives (see *Characteristics of successful corporate TQM and sustainability programmes*).

Characteristics of successful corporate TQM and sustainability programmes

- A CEO champion.
- Carefully chosen initiative leaders.
- Multidisciplinary teams.
- Dual focus on risk and opportunity.

(Fust and Walker 2007)

Environmental management systems (EMS)

Key elements of an EMS

- Purpose and scope.
- Policy statement.
- Identification of significant environmental impacts.
- Development of objectives and targets.
- Implementation plan/programme.
- Training plan/programme.
- Review/update plan.

Along with TQM, the trend of EMS represents an influence in the construction sector that has promoted a more systematic, deliberate approach to addressing both need and opportunity in how business is conducted. While TQM focuses its systematic approach on ensuring 100 per cent customer satisfaction with a company's products or services, EMS focuses instead on reducing unwanted impacts on the environment resulting from a company's operations, and increasing environmental benefits as an outcome. While each implementing firm is likely to have its own unique approach to managing environmental impacts, some key elements are common across most plans (see *Key elements of an EMS*).

Most EMS initiatives follow the ISO 14001 standard, which incorporates a four-step, cyclical approach to continual improvement of environmental performance in operations (see Figure 3.6). This problem-solving process, also known as the Deming cycle after its originator, originally evolved from the ideas of TQM, and is also a part of the ISO 9000 standard for TQM. The cycle begins with a planning step, where objectives are set and a course of action is developed to achieve those objectives. This is followed with a 'Do' step in which the course of action is implemented, often on a small scale at first. Following this step, a 'Check' process is used to evaluate the results of the action and compare them with the original objectives. Any differences between planned and actual outcomes are noted and analysed in an 'Act' process, where changes to the course of action are made to address any identified problems. The process then begins again, and continues until no need to further improve is noted. At this point, the process is standardized, shown as a wedge in Figure 3.6, to stabilize the improved results and prevent backsliding. Then, a new, refined scope is selected to continue the improvement process.

An EMS is generally developed as a written plan and then implemented in practice. Several common issues have been noted in practice that must be addressed for the successful implementation of an EMS (Burden 2010):

- Full commitment of management to the EMS.
- User friendliness.
- Clear identification of environmental compliance requirements.
- Degree of complexity.
- Adequate financial, physical and human resources.
- Measurable objectives and targets that facilitate continuous improvement.
- An organizational culture of continuous improvement.

In Europe in particular, there has been widespread adoption of EMS by organizations in the construction industry under the ISO 14001

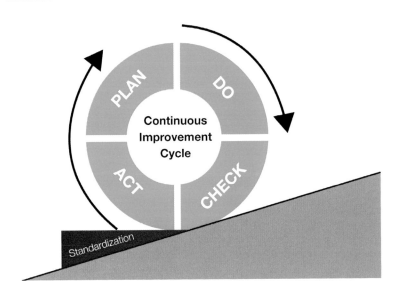

Figure 3.6
Cycle of continuous
improvement

standard, whereas in the United States and many other countries, adoption has been slower. However, EMS and TQM have set the stage for the next step in the evolution of corporate sustainability for construction: sustainability management systems.

Sustainability management systems (SMS)

Similar in form to EMS, SMS plans take a broader focus to include social and economic goals relevant for the business being managed. With this broader scope, companies can move beyond simple conformance with standards or compliance with regulations into managing the effectiveness and strategic sustainability of their enterprises as a whole.

PricewaterhouseCoopers (PWC) has identified four key questions companies should ask as part of developing sustainability management systems (PWC 2004):

- What are the key sustainability risks and challenges relevant to our business goals, and are we effectively managing these issues?
- Are our sustainability objectives complementary with, or competing with, our other business objectives?
- Are our division-level SMS concordant with our corporate-level systems, and are these systems aligned with our overall business strategies?
- Are our SMS designed to deal with emerging sustainability risks that could affect the company's long-term success?

To address these questions and create a framework for evaluating effectiveness of SMS, PWC has developed an approach for assessment called ORCA (see *The ORCA framework*). The ORCA approach is

> ### The ORCA framework
>
> - **Objectives**: what are the company's key business and sustainability objectives?
> - **Risks**: what are the main sustainability risks and challenges the company faces?
> - **Controls**: what management practices have been implemented to manage these risks and challenges?
> - **Alignment**: to what extent is the SMS aligned with overall business goals?
>   (Price Waterhouse Coopers 2004)

organized around relevant corporate objectives for growth, production levels, cost control, reputation, environment, health and safety, and external stakeholders.

As mentioned in earlier sections, alignment of business and sustainability objectives is an essential feature of companies that can successfully pursue sustainability goals to strategic advantage. The next section addresses this and other key issues that are part of an effective corporate sustainability strategy.

Organizational strategies for corporate sustainability in the construction sector

Given the evolution over time of increasingly holistic and outcome-oriented approaches for managing corporations, how should a firm proceed in developing a plan or management system for corporate sustainability? Developing sustainability objectives, forming an effective sustainability team, finding opportunities within the firm to increase sustainability and choosing the right strategies for implementation are key aspects of a strategy to pursue corporate sustainability. The following subsections describe these issues in greater detail.

Setting effective sustainability goals and objectives

There are many ways to set sustainability goals and objectives, and many levels at which these goals and objectives can apply. At the overall corporate level, for example, sustainability goals for companies in the construction sector may deal with overall impacts of the business across all of its projects and operations. Other goals may be developed for individual projects and pertain to the specific context and requirements for that project. Some projects pursuing formal certification under a recognized rating system may choose to use the framework of the rating system as a starting point. Other projects may have goals outside any particular framework. The details of specific rating systems and standards at both these levels are covered in more detail in Chapter 4. However, no matter what external points of reference are employed, the principles of effective goals and objectives remain the same.

Goals are broad statements of what should be achieved for or by the company or project as a whole. Objectives are more specific, and have the characteristics of being both measurable and time-bound. Both objectives and goals for a project should be attainable and relevant to the company or project and issues at hand. The SMART acronym provides a way to remember what are the characteristics of effective objectives: they should be Specific, Measurable, Attainable, Relevant and Time-bound (see *SMART characteristics of effective objectives*).

Measurable means that it is possible to definitively say whether or not an objective has been achieved at a certain point in time. Some objectives can be

---

SMART characteristics of effective objectives

The five characteristics of effective objectives are:

S   Specific
M   Measurable
A   Attainable
R   Relevant
T   Time-bound

(Duran 1981)

expressed in terms of quantifiable performance – for instance, 'Reduce water use by 50 per cent'. These objectives require modelling or measurement to determine whether the specified level has been met. Others may be measurable in terms of 'yes' or 'no' – for instance, 'Use bio-based form release oil for all concrete forms', or 'Develop an education plan to ensure that 50 per cent of all field employees have at least one external sustainability accreditation'.

Time-bound means that the point at which the objective will be evaluated is defined at the beginning. For instance, the objective 'Reduce water use by 50 per cent' does not explicitly specify the measurement period. If developed at the project level, must this objective be met during construction, or is it meant to be evaluated for a typical year of operation? The objective only makes sense if the time period for evaluation is specified.

The objective 'Reduce water use by 50 per cent' must also be made more *specific* to be useful. The scope of the objective should be clear. What water uses will be counted? What is the baseline against which the reduction should be measured?

To be *relevant*, the objective should have a scope that is meaningful for the time period and project being considered. For instance, reducing water use may be meaningful for a building during operations, but is not likely to be relevant for a highway project during operations. However, water use might be relevant for certain aspects of the construction of a highway project, depending on the means and methods involved.

Lastly, effective objectives should be *attainable*, given the resources and parameters of the organization. Objectives are most useful when they require additional effort beyond common practice. However, they should be reasonably achievable when considering the additional costs and efforts of achieving them. Determining what makes an objective attainable is a balance between the cost of meeting the objective and the benefits that can result. Setting unattainable objectives may result in unwise investment of resources. It may also be bad for morale when the unattainable objective is inevitably not met.

In general, sustainability goals for a company or project can be organized into three overlapping categories corresponding to the Triple Bottom Line (Figure 3.7). Goals and objectives should focus on the 'what' rather than the 'how'. For instance, specifying a goal of diverting 50 per cent by weight of construction waste from landfill disposal retains flexibility for the project team to determine the most effective way to achieve the objective in practice, rather than requiring them to use a particular means to achieve the objective.

Forming effective sustainability teams

Concurrently with the development of sustainability objectives at the corporate level, a key element of corporate sustainability is forming an effective corporate sustainability team. Various approaches exist within a corporation to develop and roll out such a team, including horizontal approaches, vertical approaches and Special Weapons and Tactics (SWAT) team approaches.

Figure 3.7
Triple Bottom Line factors

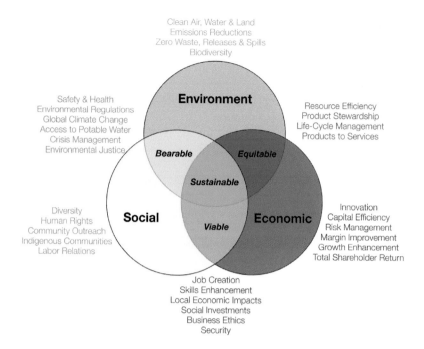

The horizontal approach to sustainability rollout involves training everyone at a particular level or key function within a firm to have basic skills and capabilities pertaining to sustainability. For example, a general contractor firm may elect to train all of its site superintendents on sustainability principles and encourage them to obtain credentials under a relevant accreditation programme such as the Sustainable Construction Supervisor programme managed by the National Center for Construction Education and Research in the United States. In this way, the company can be certain that on every project, there will be at least one person to serve as a sustainability champion and identify opportunities that can be pursued.

The vertical approach to sustainability teams involves developing deep expertise about every relevant aspect of sustainability for project delivery by educating all of the players involved with 'greening' a particular project. In this approach, each player from project manager to labourer receives basic sustainability training followed by additional training relevant for their particular project role. For firms just getting started with sustainable construction, this type of training may happen when the firm becomes involved with an owner or design organization that demands sustainability best practices for a capital project. In this case, members of the project team will learn on the job rather than in any organized fashion. The vertical approach can also be achieved through corporate-wide sustainability awareness or training programmes for everyone in the firm, not just a particular project, although it may be difficult to provide information at a level that can be used by everyone in their day-to-day activities using this approach. If done in a focused way for a particular project, the result may be a sustainability SWAT team that takes on a specialized role within the firm to provide sustainability services across multiple projects.

The SWAT team approach is so named for the special weapons and tactics teams commonly found in law enforcement organizations. These elite teams receive special training and resources to perform high-risk operations, and are called upon for difficult tasks that are beyond the capabilities of regular police officers. In similar fashion, a corporate sustainability team formed using the SWAT team approach has the specialized capabilities to address sustainability challenges internally from any aspect of the company's operations, and may be deployed anywhere within the company to solve specific sustainability problems. Such teams are often part of the resources available within a corporate sustainability office, and may be led by the corporate sustainability director if such a role exists within the firm.

In any of these approaches, the organization should take advantage of expertise from all aspects of the company to have the most holistic perspective (Table 3.4). If internal expertise does not exist within the firm to address relevant issues, it may be brought in through the use of external consultants or new hires, or internally cultivated via education or training of existing personnel.

As mentioned earlier, the support of upper management correlates highly with the overall success of a corporate sustainability policy, especially in firms with centralized organizational structures (Accenture 2010a). A different approach may be preferable in decentralized

Table 3.4 Expertise on the corporate sustainability team (adapted from Fust and Walker 2007; Porter 1985)

Function	Expertise
Board of directors	Shareholder value impact; outside knowledge of emerging best practices and opportunity initiatives
Research and development	Perspectives on how current and future products and practices relate to sustainability
Marketing	Understanding how sustainability supports marketing strategy and provides differentiation with target markets
Supply chain	Understanding of sustainability issues related to material and product inputs and natural resource utilization
Operations	Insights and ideas on process and product improvements with sustainability benefits
Environmental/safety/occupational health	Historical compliance perspective to expand initiatives *beyond* compliance; incubation of new sustainability-related revenue opportunities such as carbon trading
Finance	Evaluation of the costs and benefits of sustainability programs; risk management expertise
Human resources	Performance management links and recruitment support
Legal/compliance	Knowledge of current and pending laws and regulations; understanding of how sustainability risks affect regulatory risks and vice versa
Public relations	Communications to external stakeholders; corporate social responsibility

organizations. In any event, the design and governance of sustainability initiatives in an organization should be structured in a way that makes the most sense for a given business environment. A strong change management orientation, coupled with a clear understanding of how sustainability can support business objectives, is also essential.

Finding opportunities for increasing sustainability in the firm

As the organization begins to consider how it can improve the sustainability of its operations and business practices, management frameworks such as the ISO standards or the British standard BS 8900–1:2013 – Managing Sustainable Development of Organizations provide a useful starting point. In general, each organization will identify things it can do from the standpoint of the primary product or service it provides to the market, and the internal support practices it uses to deliver those products or services. Figure 3.8 shows a model of corporate functions that can serve as a starting point for inventorying sustainability improvement opportunities. In the construction industry, the ultimate product is capital buildings and infrastructure that comprise our built environment, and all members of the industry contribute to that outcome according to their own functional capabilities. Chapters 5 to 9 of this book explore the range

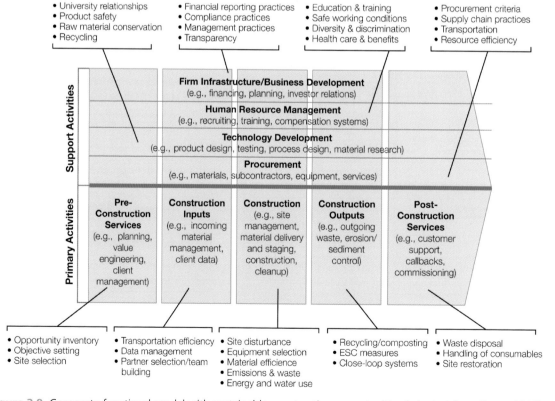

Figure 3.8 Corporate functional model with sustainable construction opportunities (adapted from Porter 1985)

of best practices that are used in the market today to improve the sustainability of capital projects, and can be a resource for firms seeking ideas to improve their own projects.

To improve internally, a variety of best practices can help to increase the sustainability of internal operating practices. Table 3.5 provides a checklist of some of the most common practices adopted by companies in general in their quest to increase operational sustainability. Additional examples are provided in the last part of this chapter as case studies of corporate sustainability.

Of these practices, some of the most foundational deal with increasing the level of knowledge of stakeholders within and outside the firm about sustainability opportunities and practices. The next section describes various approaches to corporate education, training and awareness which are an essential part of creating a corporate sustainability culture.

Building sustainability knowledge and competencies

Ultimately, improving corporate sustainability requires both increasing the levels of knowledge about sustainability within the firm, and creating a sustainability culture that values and rewards sustainable choices at all levels of organizational decision making. The following subsections describe the critical skills and competencies necessary for stakeholders of built environment sustainability, the many different approaches to training programme delivery and knowledge acquisition within the firm, opportunities for obtaining credentials and knowledge assessment, and recommendations for creating a sustainability culture.

Critical sustainability skills and competencies

Before a plan for training, education and awareness can be designed, the desired competencies, skills and educational outcomes must be identified that will be achieved by the plan. Multiple competencies are necessary to achieve sustainability in the built environment, and the competencies required of different stakeholders in the capital projects delivery process differ in both functional role in the project and level of management in each organization. At the highest levels of management, a basic understanding and awareness of sustainability principles is necessary, as is the ability to see the firm within its larger environmental and societal context and forecast how the firm will evolve in the future. At the line level where technical decisions are being made, more discipline-specific knowledge is required, as well as an understanding of how each decision contributes to the bigger picture. At the worker level, a sound knowledge of basic sustainability principles will help ensure that sustainability is achieved even when novel circumstances arrive. Even supporting functions such as finance, marketing and other areas are essential to the success of the organization and critical to its sustainability objectives. A framework such as the one shown in Figure 3.8 can be a useful basis for systematically identifying what critical skills are required and what opportunities exist for each key function within the organization.

Table 3.5 Checklist of organizational best practices for sustainability

	Practice
	A formal policy on sustainability
	Green team(s) or sustainability working groups
	An orientation programme that includes sustainability or social/environmental issues
	A training programme that includes sustainability or social/environmental issues
	A formal process for capturing lessons learned
	Access to external sustainability specialists/partners/consultants
	Employees with sustainability expertise
	An employee handbook/guide on sustainability
	A website or newsletter on sustainability programmes
	Marketing information on the company's green practices
	Technical information on the company's green practices
	Listings in major green guides/directories for products/services
	A programme for customer/client education on sustainability
	A programme for supply chain education on sustainability
	Formal sustainability metrics/evaluation programme
	Regular sustainability audits of products/processes
	A sustainability accounting system (Triple Bottom Line)
	A strategic plan for sustainability implementation
	A recycling programme for office waste
	A green housekeeping programme
	A programme for greening the company's capital facilities
	A policy of sustainable/green purchasing/procurement
	A programme for reusing resources (e.g. coffee mugs)
	A telecommuting/flextime policy
	Alternative fuel fleet vehicles
	Bicycle facilities/carpool incentives
	A teleconference-instead-of-travel policy
	A policy of paperless operations
	A community/social service/philanthropy programme
	Public transport passes for employees
	A green investment programme for retirement/pension plans
	Lunch'n'learn programmes on sustainability
	A policy of letting employees participate in external committees
	An annual sustainability report to shareholders/stakeholders
	A green travel policy (e.g. environmentally friendly hotels)
	An environmental code of conduct for employees
	Energy-efficient office equipment

Pearce and Ahn (2010) have identified a range of sustainability competencies developed by professional societies and documented in the literature that are thought to be important for achieving sustainability in engineering and construction, especially among technical disciplines:

- Ability to communicate and solve problems effectively and professionally with people from other disciplines and cultures.
- Ability to decide and competence to act in ways that favour sustainable development; having an attitude of care or stewardship; self-efficacy.
- Understanding the influence of culture and context on attitude toward sustainability, being able to contextualize knowledge, and valuing diversity.
- Ability to understand the complexity of real-world problems, differentiate between problems and symptoms, tolerate uncertainty and ambiguity, and resolve conflicts.
- Knowledge and tolerance of disciplinary perspectives that are not one's own.
- Ability to think holistically, comprehend interrelatedness and search for integrated solutions.
- Ability to challenge dominant ideologies.
- Awareness of the role of humans within a larger systems context, and humility regarding current state of knowledge.
- Ability to expand the scale of thinking in spatial, temporal, biological and intellectual terms; breakthrough or lateral thinking in the context of complexity.
- Ability to evaluate impacts and manage tradeoffs between technological, ecological, human and economic elements.

In addition to these core sustainability competencies, an array of more fundamental skills are also essential for achieving sustainability in the construction industry (Pearce and Ahn 2010):

- Knowledge of people and how to motivate action, especially in collaboration with other disciplines.
- Ability to work in teams, undertake inclusive visioning, implement stakeholder management and communicate effectively.
- Ability to cope with novel situations, analyse requirements, identify resources, develop solutions, monitor progress and learn from the process.
- Ability to filter, interpret and integrate information and evidence, evaluate the testimony of experts, and situate and explain one's own perspective in this context.
- Performance under constraints.
- Ethical judgement.
- Leadership, change management, project/process management and life-long learning.

From the standpoint of domain-specific knowledge, the Construction Industry Council (CIC) in the United Kingdom has used an industry-based forum to develop its sustainability skills matrix, which identifies critical sustainability requirements and associated functional skills required to achieve them. These skills are mapped against an array of functions representing all phases and steps in the capital project delivery function, and the role of each function is identified in terms of required skills it should have to achieve sustainability. Skills are identified in the three core dimensions of sustainability, as shown in Table 3.6.

At the most basic level, everyone in the organization should understand how the company has operationalized sustainability for its own operations, and how their job functions and decisions affect sustainability overall. They should also be able to follow a process for sustainable decision making that enables them to identify opportunities to improve sustainability and make sound decisions about actions to take when addressing those opportunities (see Table 3.7).

Approaches for knowledge acquisition

Given the broad array of desired characteristics required of construction stakeholders pursuing sustainability, how should a firm design a sustainability awareness, education and training plan that can move its employees to master these skills and abilities, and apply them in the workplace? Designing a plan for acquiring sustainability knowledge requires identifying potential sources of knowledge and means of delivering it, then integrating those sources of knowledge as part of a training plan to support corporate operations.

Sustainability knowledge can come from a variety of sources. It can be brought to the firm through new hires with outside training or experience, by hiring consultants, or by partnering with other organizations that have complementary sustainability expertise or experience. It can be cultivated among the firm's existing employees through training, self-study or participation in outside events such as conferences, local green building councils or standards development. It can also be brought into the firm through investment in physical resources such as a sustainability library or resource centre, or access to online sustainability databases and tools. Finally, sustainability knowledge may also be provided at no cost to the firm through the outreach and educational efforts of other stakeholders in the capital projects industry such as product manufacturers wishing to educate the firm about their green products, or clients sponsoring training for project team members involved in their products.

If a firm decides to pursue formal training or education options for its employees, the array of possible choices is vast. Firms that expect to have ongoing training needs due to factors such as acquisitions or employee turnover, or firms that are large in size with significant support departments, may elect to develop their own in-house training programmes. External consultants or universities can be employed to assist with designing and rolling out these programmes. Firms with less intensive training needs may elect to bring in outside trainers from

Table 3.6 Essential sustainability skills for the built environment (adapted from UK Sector Skills Council)

Social sustainability

Optimize opportunities and social benefits:
- Create usable public and private space to deliver successful communities.
- Improve health, well-being, accessibility and security of community.
- Enhance employment and skills development opportunities for the local community.

Promote sustainable communities through planning and design:
- Meet requirements of local, regional and national development and regeneration strategies.
- Ensure appropriateness of development to needs of the community, including multiple use and adaptability.

Engage stakeholders:
- Consult with public authorities, the general public and other stakeholders including end users, and respond accordingly.
- Involve and manage expectations of stakeholders in the development process, from concept to commissioning.
- Consult and manage expectations of stakeholders on changes to ongoing use and operation.

Minimize negative impacts:
- Plan for effective public and private transport use.
- Control nuisance (noise, dust, light, etc.).
- Ensure a secure site during construction.
- Ensure health and safety of site workers and local community.
- Protect, enhance and maintain appropriate social access to environmentally sensitive areas.
- Assess and mitigate flood risk.

Environmental sustainability

Take account of natural capacity:
- Assess and mitigate wider environmental impacts, such as water supply, sewerage, transport, waste.
- Respond to projected impacts of climate change.

Optimize environmental benefits:
- Minimize energy demand and meet it efficiently, aiming to achieve carbon neutrality.
- Minimize water demand and aim to maintain water sufficiency from public supply.
- Optimize efficiency of materials use.
- Maximize range of environmental benefits in design.
- Maintain and enhance biodiversity.

Minimize negative impacts:
- Reduce, reuse, recycle, recover waste.
- Reduce emissions to air, land and water.
- Reduce transport impacts.
- Protect ecological resources.
- Minimize taking of environmentally valuable land.
- Minimize pollution of air, land and water.
- Manage and control in situ contamination of land.
- Protect archaeological and historically valuable resources.

Economic sustainability

Ensure economic viability and improve processes:
- Use technologies and materials consistent with sustainability principles.
- Keep up to date with advances in construction/technology.
- Establish cost and benefit on the basis of whole life value.
- Manage the supply chain effectively.
- Keep up to date on regulatory and planning requirements.
- Operate effective project management and contingency planning procedures.
- Maximize range of economic benefits including flexibility of use.
- Achieve cost-effective outperformance of statutory requirements.

Enhance business opportunities:
- Meet requirements of national, regional and local economic strategy.
- Capitalize on funding/grants available for more sustainable development.

Source: CIC (2008).

Table 3.7 Steps in a sustainable decision process

To improve the sustainability of a system, stakeholders must:
• recognize opportunities to improve the sustainability performance of a system in terms of key dimensions including the environmental, social and economic dimensions.
• identify a range of feasible and contextually appropriate best practices that could be used to address those opportunities.
• evaluate and compare these practices in terms of their likely impacts according to traditional qualitative and quantitative criteria such as first- and lifecycle cost, performance, time and quality and in terms of their relative impacts on system sustainability.
• design a recommended course of action to increase the sustainability of the system that takes into account the context of implementation.
• support recommendations with convincing evidence and well-organized analysis delivered in a professional fashion.
• plan and execute the implementation of recommended actions within the system.
• evaluate the impacts of implementing those recommendations on specific system attributes in terms of sustainability.

various sources for a one-time kick-off training event, then integrate key sustainability concepts throughout other existing training programmes to provide a mechanism for ongoing updates. Outside trainers and their training programmes can be offered to cover generic course content, or customized to address the specific needs of the firm. It is even possible to integrate training with real projects to provide authentic experiences with project decision making. Some sustainability consultants routinely provide basic training on sustainability as part of their services when they have been employed to facilitate pursuit of project certification. This helps to ensure that everyone understands certification requirements and is on the same page with regard to project goals and objectives.

One decision that will have to be made in designing a training plan is who will receive what training when. In other words, will different functional groups and management levels in the firm be trained separately, or will multiple functions be integrated as part of a single training event? Although more technical sustainability training is difficult to design when multiple functions (such as procurement, design, construction safety, financing and marketing) are included, this type of training also offers the benefit of allowing trainees to become more familiar with what other functions in the company do, and what challenges they face with regard to implementing sustainability. Designing training to incorporate active learning and discussions can also serve the strategic purpose of identifying potential implementation barriers and designing remedies to overcome them.

Another decision that must be made is to decide what topics and skill sets will be covered in different elements of the training. Often, training may begin with an awareness-level event for everyone together that introduces basic sustainability principles and the company's goals and objectives for corporate sustainability. This general training is then followed by more detailed training for each functional unit that covers issues specific to that unit. Integration events can be included in the training design to allow different functions to interact as part of active learning exercises. Periodic refresher or update training is also a good idea as new practices, standards and technologies emerge.

It is often easy to engage external sources of knowledge in training programmes because organizations with such knowledge have a desire to promote their products or services. For example, it is quite common for product manufacturers to offer seminars or 'lunch and learns' at no cost to the firm to introduce new products or systems. These activities are a valuable way to expand a company's knowledge base, but care should be taken to ensure that a variety of competing firms are represented so that employees receive a balanced perspective on the state of the art in industry.

Another key part of knowledge acquisition is a formal mechanism for sharing lessons learned across project teams or throughout the firm. Some companies already have formal data systems in place to capture this knowledge. Other possibilities are to assign mentors from experienced project teams to guide novice teams during initial pilot testing of sustainability. Firms often develop case studies of their projects as part of marketing or annual reporting, and these case studies can be formally presented or made available across the firm as a means of sharing lessons learned. Field trips to active projects, or even other projects of current clients or partners, can also be a useful way to learn in practice.

Not all sustainability training has to be exclusively sustainability-oriented. Many firms seek to incorporate sustainability ideas throughout the training programmes already in place in the firm. For example, including sustainability concepts as part of safety training makes a great deal of sense when discussing topics like management of hazardous materials and spill prevention. Such training could emphasize the benefits of using products that are less hazardous in the beginning, and the reduced intervention and control requirements for non-toxic or low-emitting products.

Finally, many opportunities exist for a firm to support formal and informal learning among employees outside formal training programmes. Some firms may choose to support employees in obtaining advanced degrees or specialized training outside the company in areas that support the sustainability mission. Many technical programmes are available for graduate study in sustainable construction topic areas, as are sustainability-related business and MBA programmes. Firms can also create opportunities for learning by sending employees to sustainability-related conferences, seminars and trade shows. Employees can also learn a great deal by becoming involved in external service activities such as serving on standards development committees or the boards of local green building non-profit organizations. Obtaining subscriptions to relevant publications, databases and online tools also supports the transfer of knowledge. Many trade publications with relevant information are even available at no cost from industry associations. Firms can further encourage the sharing of knowledge by facilitating opportunities to discuss sustainability knowledge in formal or informal seminars, symposia or 'lunch and learn' sessions.

Obtaining credentials and knowledge assessment

For many firms, having employees earn formal credentials as evidence of their sustainability knowledge is useful for professional credibility and market recognition. Firms often set goals to have some percentage of their technical staff obtain credentials under relevant rating systems, for example. Credentials come in many forms and may be associated with rating systems, technical specialties or completion of formal education programmes offered through colleges or universities.

Many major rating systems such as LEED and BREEAM (see Chapter 4) have formal accreditation or licensing requirements for individuals who participate in the project certification process. For example, the LEED rating system has associated credentials at three levels of expertise: the Green Associate, LEED Accredited Professional and LEED Fellow levels. Green Associate is the entry certification, and requires either LEED project experience or other evidence of green expertise such as a university class in order to sit for the exam. LEED Accredited Professionals must provide a history of project experience as well as additional expertise or training to take the LEED AP exam, and are accredited with regard to a particular specialty within the LEED rating system, including Building Design + Construction, Operations + Maintenance, Interior Design + Construction, Neighbourhood Development and Homes. Different exams exist for each of these specialties. LEED Fellows are selected by the US Green Building Council (USGBC) in recognition for their extraordinary expertise or contributions to the field of green building. All of these credentials levels must be maintained through documented maintenance activities such as taking or teaching continuing education courses, attendance at green building conferences or formal events or participating in events with a local USGBC chapter.

Other sustainable construction-related organizations have developed credentials for various stakeholders in the project delivery process. For example, in the United States, two different certifications have been independently developed for builders, craft workers and trade workers. The Green Advantage certification is targeted at builders, and offers two different certification exams: the GA Certified Associate exam, which is targeted at workers, and the GA Certified Practitioner exam, targeted at supervisors. Green Advantage Certified Associates (GACAs) and Practitioners (GACPs) have passed their respective exams and thereby demonstrated their knowledge of green building principles, materials and techniques. There is precedent under the LEED rating system for obtaining innovation credits if all construction personnel on a job have received green building training and successfully passed the GACP exam, for example.

The Sustainable Construction Supervisor credential is targeted at site superintendents responsible for supervising craft workers on site. The Sustainable Construction Supervisor exam is administered by the US National Center for Construction Education and Research (NCCER), and identifies individuals who, through training or on-the-job experience, are capable of managing the on-site aspects of administering a sustainable construction project. NCCER has also developed training to support this certification, and it is expected to be formally endorsed by the Green Building Certification Institute, which manages credentials associated with the LEED rating system.

In addition to licensing or credential requirements associated with rating systems, some universities or colleges offer degrees or certificates associated with sustainable construction. Multiple Masters of Science degree programmes are available in sustainable construction from a variety of reputable universities, both in person and as distance learning degrees. There are also more technically specialized certifications available from various licensing boards that deal with various aspects of the built environment (see Table 3.8).

Table 3.8 Examples of professional credentials for sustainable construction

Credential	Source
Certified Green Building Engineer (GBE)	Association of Energy Engineers
Certified Sustainable Development Professional (CSDP)	Association of Energy Engineers
Certified Energy Manager (CEM)	Association of Energy Engineers
Certified Water Efficiency Professional (CWEP)	Association of Energy Engineers
Certified Carbon and GHG Reduction Manager (CRM)	Association of Energy Engineers
Building Analyst	Building Performance Institute
Healthy Home Evaluator (HHE)	Building Performance Institute
BPI Rater	Building Performance Institute
Certified EcoBroker/Ecosociate	Association of Energy and Environmental Real Estate Professionals
Envision Sustainability Professional (ENV SP)	Institute for Sustainable Infrastructure
Certified Energy Rater	Residential Energy Services Network
Building Energy Modeling Professional	American Society of Heating, Refrigeration and Air Conditioning Engineers
Building Biology Practitioner	Institute for Bau-Biologie and Ecology
Building Biology Environmental Consultant	Institute for Bau-Biologie and Ecology
Green Globes Professional (GGP)	Green Building Initiative
Guiding Principles Compliance Professional (GPCP)	Green Building Initiative
Certified Commissioning Professional (CCP)	Building Commissioning Association

New credentials are being established at a very rapid pace, and not all credentials are equally rigorous. Before pursuing a credential in an area of interest, individuals should carefully review the requirements and testing standards, and choose credentials that are offered and maintained by reputable organizations in the field. Firms should work with their employees to make informed choices about what credentials may have the greatest long-term impacts for the organization when considering which to sponsor. It is also important to be aware that many reputable credentials require ongoing maintenance costs, either directly to maintain registration as an individual with credentials (such as with a professional engineering license), indirectly in the form of ongoing training requirements, or both.

Creating a sustainability culture

Creating a culture of sustainability within a firm is an important step toward achieving corporate sustainability goals. How can sustainability be incorporated as part of the day-to-day activities of everyone in the firm? What can be done to raise awareness? What is the most effective way to communicate about sustainability to influence behaviour?

These questions and others are important for ensuring that sustainability truly becomes part of the corporate culture and not just another management fad. The Golden Thread principle (Slaveykova 2011; UK Dept. for Communities and Local Government 2012) is one way to infuse sustainability as part of the basic business infrastructure of the firm. This principle highlights the importance of making explicit links between the external drivers of the business enterprise, corporate and divisional strategic and operational plans, and individual work plans and performance appraisal. Aligning sustainability goals with performance incentives helps to ensure that employees are not faced with the challenge of resolving trade-offs between sustainability and some other competing objective. Instead, they can direct their energy toward achieving desirable outcomes.

Another tactic is to make sustainability real by relating actions to easily understood outcomes. The Carbon Trust, a non-profit organization dedicated to reducing carbon emissions, recommends using posters with easily understood facts, such as 'Lighting an office overnight wastes enough energy to heat water for 1,000 cups of tea'. These facts should be based on the specific company in question to further increase the relevance to employees.

The Prince of Wales Accounting for Sustainability Project recommends focusing communications in a way that ensures each person receives the messages most relevant to their own opportunities, rather than taking a blanket approach and saturating people's attention. Overwhelming individuals with sustainability messages can lead to what the authors call 'communication fatigue', which results in important messages being obscured or ignored. The authors also recommend using sustainability champions in each business or functional unit in the firm to disseminate sustainability information in a way that is relevant to each particular business area. Combined with a corporate sustainability team, this tactic helps sustainability team members internalize and endorse corporate sustainability goals as they communicate and teach those goals to others. Having a green team with rotating representation from different parts of the firm will allow a continuous flow of new ideas and ongoing institutional learning.

One key aspect of a sustainability culture is a tolerance for failure in the quest for innovation. Many firms have found success in using pilot projects as a way to provide a 'safe' environment for trial and error. By designating a project as a pilot, there is an implicit understanding that new, unfamiliar tactics will be tried that are outside the realm of the firm's current practice. This understanding protects employees who innovate and often results in a greater willingness to try new ideas that can lead to competitive advantage.

Ultimately, a firm's sustainability culture and knowledge should be formalized via corporate guidelines for sustainability and performance goals at both the corporate and project levels. Individuals are more likely to make the most sustainable choices in a given situation if it is clear to them what that choice is, and if they believe that their efforts will be recognized and/or rewarded.

After an overall education and training strategy and plan has been developed for the firm, the next step is to plan and implement the necessary operational actions that will allow the company to achieve its sustainability goals. Two sides of the coin exist for companies in the construction sector to obtain competitive advantage from implementing sustainability principles. First, a company embarking on this quest must first carefully evaluate its own operational practices and actions to ensure that how it does business is as aligned with sustainability principles as possible, from an internal perspective. Second, the company should evaluate opportunities for increasing the sustainability of the products and services it offers, thus aligning the company from an external perspective with sustainability principles. The next section explores these two areas of opportunity using case studies of leading firms in the design, construction and product manufacturing components of the construction sector.

Corporate sustainability policies in AEC industry firms

When deciding to adopt sustainability principles as a guiding force for a company, organizational leaders have a wide variety of potential actions to take. The concept of sustainability is no longer unfamiliar to the construction industry, and most companies have at least an awareness of the term, if not some basic competencies in delivering more sustainable capital projects. However, comparatively few firms have thoroughly internalized the concepts of sustainability both internally within the firm and externally with respect to the projects or products that firm delivers. The next sections offer case studies of leading companies with notable sustainability philosophies, policies or programmes that can serve as models for others in the construction industry in the areas of engineering and design, construction and project management, and product manufacturing.

Design firms

Increasingly, firms engaged in designing and engineering solutions to challenges in the built environment are including sustainability as a driving criterion for their projects. Whether due to client demand from outside the company or strategic choice from within, design firms increasingly look for ways to ensure that their projects meet core sustainability objectives. Beyond what they do on their projects, however, some firms are also taking active steps to increase the sustainability of their own operations, provide resources that are useful and groundbreaking to the industry at large and weave principles of sustainability into their core values and practices for everything they do. The following case studies provide examples of companies who are leading the way in sustainability among engineering and design firms worldwide.

Example 1: Foster + Partners

Foster + Partners is a design firm headquartered in London, UK. With 21 offices worldwide, the firm employs more than 1100 staff from more than 70 different countries. Their practice is organized into six design studios, each with a senior executive partner as leader. Design groups are not limited by geographic area or building type; instead, each group has a rich cross-section of projects of various sizes around the world to promote diversity in thinking, creativity, innovation and motivation. Team members maintain a close relationship with clients, and move with each project to the building site, maintaining a local office until the project is complete. Designs are regularly reviewed by a company-wide design board bringing expertise and perspectives from multiple disciplines to guide the direction of each design. Supporting elements of the organization include departments for research and development, information technology, contract management and construction, and business development.

The firm has actively been investigating sustainable materials and technologies to incorporate into design since the 1970s, long before sustainability became a significant industry trend. The firm aims to establish a sustainability profile at the beginning of every project which includes specific sustainability objectives and methods for achieving them. The profile serves as a system to monitor the sustainability agenda of individual projects and to promote a strong sustainable design ethic. Each team is also encouraged to record methods used on a project for sustainable design that are then collected into a central database that can be accessed to inform future projects.

While not every project meets every sustainability goal set in its profile, the firm believes it has an obligation to try to persuade clients to adopt sustainable strategies on each project. The firm has created a Sustainability Forum with the purpose of promoting the use of

Foster + Partners

Sustainability practices

- Design review
- Use of sustainable materials and technologies
- Project sustainability profiles
- Capture of lessons learned
- Sustainability Forum
- Formal and informal training
- Partnering with industry to develop new technologies

Ten Themes of Sustainability

- Energy and carbon
- Mobility and connectivity
- Materials and waste
- Water
- Land and ecology

- Community impact
- Well-being
- Prosperity
- Planning for change
- Feedback

Operational policies

- Company procedures
- Anti-corruption policy
- Sustainability policy
- Health and safety policy
- Diversity and equality policy
- Data protection policy
- Maternity policy
- Paternity policy
- Grievance policy
- Capability policy

sustainable technologies and methods throughout the practice. The Forum, which is part of the research and development group, is an interdisciplinary working group that has representatives from each of the six design studios, the information group, communications, training and research departments. With representation from all key parts of the company, the Forum provides a means to integrate sustainability knowledge across the company and its projects.

In its 2014 Corporate, Social, and Environmental Responsibility (CSER) report, the firm set formal targets for 2020 in 10 key theme areas (see box). These themes were developed by the firm based on influences from existing frameworks such as the UN Global Compact, One Planet, LEED and BREEAM. These themes are the basis for evaluating the firm's sustainability performance using metrics such as per capita CO_2 emissions, total waste recycled, per capita water consumption, level of charitable donations, number of employees trained in sustainability and others. Their work is guided by corporate commitments to ethics, health and safety. Policies in 10 key areas help to ensure that these commitments are met (see *Foster + Partners*).

Foster + Partners ensures ongoing sustainability expertise by providing formal and informal training to staff on a range of issues including renewable energy, sustainability criteria and assessment, environmental analysis and visualization techniques. It has a goal to train at least 10 per cent of its technical staff as LEED Accredited Professionals. It has a strong commitment to research, actively scanning the industry for new knowledge and techniques and systematically evaluating their relevance and appropriateness for individual projects. The firm counts among its projects some of the most well-known examples of sustainable design worldwide, including the Commerzbank Headquarters in Frankfurt, Germany, the German Parliament at the Reichstag in Berlin and the master plan for the Masdar Initiative in Abu Dhabi (see *Case study: Masdar City*). The firm has also partnered with industry in the development of new integrated systems for renewable energy, including super-efficient wind turbines and new cladding systems that can harvest solar energy, as well as a design for the fuel station of the future in association with Nissan Motors for the Geneva Motor Show in 2016.

Case study: Masdar City, Abu Dhabi, United Arab Emirates

Masdar City, located in the United Arab Emirates' capital city of Abu Dhabi, was envisioned as a clean technology research hub to rival Silicon Valley while being entirely carbon neutral and waste-free. Home to the Masdar Institute of Science and Technology and commissioned by the Abu Dhabi Future Energy Company in 2006, the city's 5.5 sq km was envisioned to be home to 50,000 people, 1500 businesses and some 40,000 commuters, striving to be entirely self-sufficient for all its energy needs. Approximately 80 per cent of water would be recycled and reused. In addition to solar and biomass conversion, other renewable sources of power would include geothermal, hydrogen and wind. Biological waste would be used to create fertilizer, and industrial waste would be recycled or reused.

Initial phases of the project involved thousands of individuals and included a range of innovative building and infrastructure systems. 6D Geographic Information Systems (GIS) models were used to track costs, project schedules and carbon emissions as part of the collaborative project delivery process, and were incorporated into an automated, paperless asset management system to optimize the performance of the city over its life.

Figure 3.9 Construction in Masdar City was carefully managed to work toward carbon neutrality

continued

During the first phase of construction, which began in 2008, approximately 5000 workers were housed on site in a workers' village in harsh desert conditions, which posed a significant challenge for maintaining carbon-neutral objectives for construction due to the needs for water, power and transportation of workers. Worker and material logistics were monitored and carefully managed to maintain the goal of carbon neutrality (Figure 3.9). Project goals for worker health, safety and quality of life were also established under the Estidama Pearl green building rating system. A minimum level of three pearls is a goal for all buildings in the development, which is comparable to LEED Gold. Some of the innovative features of the city's development included:

- Planning construction lay-down areas to be as close to where the work occurs as possible.
- Managing the more than 100 different contractors on site.
- Undertaking clash detection and alignment at the whole facility scale.
- Managing change orders and tracking environmental impacts, especially compliance with carbon neutrality requirements.
- Combining building information models (BIMs) for individual assets into accurate representation of spatial networks.
- Visualizing energy and water use for the city as a whole during optimization to maintain carbon neutrality.

Masdar City has a variety of innovative design features that were developed to be more sustainable than their conventional counterparts. Key sustainable design strategies include:

- Optimal orientation of the city and street grid on a southeast-northwest axis to provide shading at street level throughout the day, minimize thermal gain on building walls and facilitate the flow of cooling breezes through the city.
- Optimizing utility substation layouts, network routes and facility placement for water and sewage treatment plants, recycling centres, a solar farm, geothermal cells and plantations of different tree species for production of biofuels.
- Integrated, mixed uses to keep residential areas close to business, industry, cultural and university/educational facilities.
- Low-rise, high-density urban design to reduce energy use for transportation and reduced heating and cooling loads.
- Incorporation of public spaces as part of the urban fabric to facilitate interaction and engagement among residents, commuters and visitors.
- Pedestrian-friendly design, including overshading to provide a cooler street environment (Figure 3.10).
- Convenient public transportation through a network of electric buses, a personal rapid transit (PRT) system, and integration with the light rail/metro system of Abu Dhabi.
- Traditional Arabian city design that takes into account indigenous strategies to deal with the harsh desert climate and reduce energy consumption while promoting socially diverse environments and lively public spaces.

Building design, urban form and infrastructure design were all undertaken using careful prioritization to achieve the biggest environmental gains with the least financial investment, especially the city's orientation and overall form (Figure 3.11). At the building scale, responsive shading, natural lighting and passive ventilation were incorporated to achieve comfort and performance goals with minimal use of energy. At the very top level of investment, active controls and renewable energy systems were proposed to minimize energy consumption toward the goal

continued

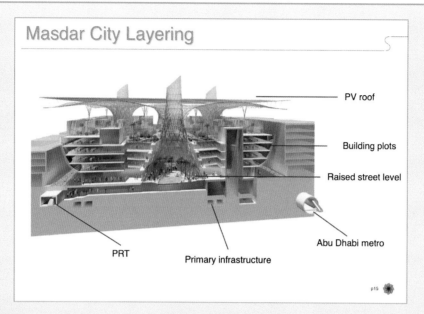

Figure 3.10 A layering approach was used in designing the city to provide critical infrastructure and occupied spaces in shaded areas between buildings and beneath photovoltaic canopies

Figure 3.11 Masdar City incorporates the Masdar Institute of Science and Technology as part of a unique, carbon neutral urban environment inspired by indigenous Arabic design strategies

continued

of carbon neutrality for the city (Figure 3.12). Key infrastructure features of Masdar City's design included:

- An international photovoltaic (PV) competition testbed begun in 2008 that tests technology from more than 35 suppliers for energy yield, efficiency, ambient temperature effects, sand effects and other factors.
- The Beam Down Project, a new type of concentrated solar power (CSP) plant using mirrors to focus the sun's rays onto a receiver where it heats a heat-transfer fluid (such as molten salt, oil or water). The concentrated heat then can be used to generate steam to power a turbine and generate electricity. This CSP design (Figure 3.13) is more efficient than conventional designs because the receiver is located at ground level, thus reducing the energy ordinarily used to pump the heat-transfer fluid to the top of an elevated receiver tower.

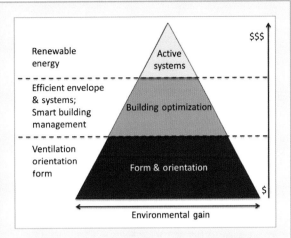

Figure 3.12 Sustainable design strategies were carefully prioritized to achieve the maximal efficiency for the least investment

- Exploratory geothermal testing to locate optimal geothermal resources for thermal heating and cooling, domestic hot water production, power production and desalination. A significant portion of the city's cooling loads were expected to be served using absorption chillers continuously supplied with geothermal heat.

Figure 3.13 A ground-level concentrated solar power (CSP) plant uses mirrors to focus the sun's rays to generate electrical power

continued

- A 10 MW solar photovoltaic farm comprising over 87,000 polycrystalline and thin-film PV modules on the outer boundary of the city, which is the largest grid-connected solar plant in the Middle East. Cleaning dust from the panels in the desert environment represents a significant cost, especially in terms of water use, so ongoing testing is critical to find improved solutions for this purpose.
- A material recycling centre, which diverted on the order of 96 per cent of construction waste from the construction of the city for recycling or reuse. At the centre, wood is segregated and stockpiled for reuse or processed in a wood chipper. Other materials are separated and sent offsite for recycling. Excavated sand and crushed concrete/masonry are retained on site for fill. Waste that cannot be recycled is captured for fuel in a waste-to-energy plant.

Figure 3.14 The city's personal rapid transit system transports people, freight and waste in driverless vehicles through parts of the city

- A joint parking garage and rapid transit station at the gateway to the city (Figure 3.14), where commuters and visitors can transfer to the city's personal rapid transit (PRT) system. Sustainable design features of the station include an efficient radiant cooling system, backlit recycled glass walls, low-carbon concrete benches and recharging berths for PRT vehicles waiting in the station.
- A personal and freight (FRT) rapid transportation system, which consists of driverless vehicles controlled by an advanced navigation system using magnets embedded in each corridor to determine position, onboard sensors for obstacle detection and a wireless connection to a central control system to coordinate operation among all vehicles. Three of the vehicles are flatbeds that are dedicated to transport deliveries throughout the Masdar Institute and transport waste for sorting, reuse or recycling.
- An undercroft area for the PRT system to separate pedestrian traffic from vehicular traffic, and a utility trench below that area to allow for easy access to the city's complex utility infrastructure of power and sewerage lines, piping for potable, grey and black water, wiring for IT and communication systems, waste and recycling networks, and other plant and equipment.
- The Masdar Institute PRT/FRT station, featuring light tubes for natural daylighting, and prominent staircases and hidden elevators to encourage people to make more sustainable choices about vertical conveyance.
- A student reception area featuring a seasonally adaptive system to allow open-air ventilation during cooler months, and closed areas during hot summer months to allow for air conditioning. The master plan for the development dictated that at least as much attention be paid to the spaces between buildings as to the building themselves. Unique fractal patterns constructed from reconstituted stone are used for walkways, and shading, planting and water features lower the perceived temperature from that of open desert or conventional city areas.
- An iconic wind tower to capture prevailing breezes and add moisture to create a low-energy evaporative cooling system for personal comfort.
- A large urban square at the base of the wind tower to take advantage of conditioned air. The square will contain cafés and retail outlets surrounding a public space as well as other services

continued

such as a gym, prayer room, organic grocery and bank. A raised platform beneath the wind tower can serve as a performance stage.

- Building envelopes using best practices such as high insulation levels, strict air-tightness standards and air-filled cushions backed with reflective foil cladding to minimize solar gain and reflect light to the street below. Residential buildings feature red, sand-coloured glass-reinforced concrete screens based on traditional Arab mashrabiya screens that provide shade, reduce solar gain and provide visual contact with the street while protecting residents' privacy.
- The Masdar Institute Laboratory Building which features active and passive energy management technologies, an open plan, column-free floor plan to facilitate interdisciplinary collaboration, overhead service carriers to provide plug-and-play access for utilities in the labs, and easily reconfigurable furniture to facilitate interaction of researchers.
- The Masdar Institute Residential Building, which contains both men's and women's dormitories with studio apartments, and family dormitories with two-bedroom units. The units were constructed using modular, prefabricated bathrooms to minimize fabrication waste and ensure consistently high quality.
- A Family Square to serve as a meeting place and point of interaction for families, containing cafés, services and a water feature that uses a thin layer of flowing water to provide efficient cooling during the summer.
- A Knowledge Center that features PV energy harvesting, self-shading overhangs, FSC- and PEFC-certified glulam (glued laminated) timber roof and a zinc-cladded roof selected for its overall lowest environmental footprint.
- 'Green Finger' linear parks oriented to take advantage of prevailing winds and channel cooling breezes into the heart of the built environment. The parks serve as shaded oases for residents, workers and visitors, and incorporate walking, jogging and bicycle trails for recreation.
- A ready-mix concrete batch plant used to produce concrete using supplementary cementitious materials such as flyash and blast furnace slag, with a lower concrete footprint than conventional concrete.
- A membrane bioreactor that uses suspended, growth-activated sludge and microporous membranes to separate liquids and solids instead of secondary clarifiers. This technology performs well in warm climates, has a compact footprint, is easily scalable and produces an effluent suitable for reuse applications such as toilet flushing, district cooling and landscape watering.

Other projects were also planned to result in a unique 'innovation ecosystem' that facilitates the discovery and development of new technologies for solar energy, green building and urban sustainability. Signature tenants for the city include the General Electric (GE) Ecomagination division, Lockheed Martin, Mitsubishi Heavy Industries and Siemens. In 2015, the International Renewable Energy Agency (IRENA) established its first permanent headquarters in Masdar City, in the first building in the United Arab Emirates (UAE) to receive the highest level of certification – four pearls – under the Estidama Pearl rating system.

Despite its ambitious goals, the city's development overall has not met its original timeline of completion by 2016, and only about 5 per cent of the total footprint has been completed to date. Due in large part to the global economic recession beginning in 2008 and subsequent drop in oil prices, the timeline for completion has been extended to 2030, and the original goals of carbon neutrality have yet to be achieved. The primary investor for the city is Abu Dhabi's state-owned investment company Mubadala, for whom a major source of revenue has been oil production. Overall, the UAE recognizes that oil has a finite future, and is working to expand its portfolio to include renewable energy and sustainable technologies.

continued

Full time residents of the city are presently limited to students at the Masdar Institute of Science and Technology, and about 2000 people commute to the city to work on a daily basis. Despite the challenges in realizing the overall vision for the city, Masdar remains one of the most holistic and ambitious examples of urban sustainability and sustainable infrastructure developed to date.

Sources

ESRI (UK) Ltd. (2009). *Masdar City: The world's first carbon-neutral city.* https://gisandscience.com/2009/05/15/masdar-city-the-worlds-first-carbon-neutral-city/ (accessed 18 September 2016).

Goldenberg, S. (2016). "Masdar's zero-carbon dream could become world's first green ghost town." *The Guardian*, 16 February. https://theguardian.com/environment/2016/feb/16/masdars-zero-carbon-dream-could-become-worlds-first-green-ghost-town (accessed 18 September 2016).

IRENA – International Renewable Energy Agency. (2015). "IRENA opens doors on new permanent headquarters in Masdar City." http://irena.org/News/Description.aspx?NType = A&mnu = cat&PriMenuID = 16&CatID = 84&News_ID = 411 (accessed 18 September 2016).

Masdar City (2011). *Exploring Masdar City.* http://thefuturebuild.com/assets/images/uploads/static/1745/masdar_city_exploring1.pdf (accessed 18 September 2016).

Masdar City (2016). *Masdar City Fact Sheet.* http://masdar.ae/assets/downloads/content/270/masdar_city_fact_sheet_310716.pdf (accessed 18 September 2016).

Source: All images courtesy of MASDAR

Example 2: HOK

Founded in 1955, Hellmuth, Obata + Kassabaum (HOK) is a firm that provides planning, design and delivery solutions for the built environment. With 24 offices in Europe, Asia and North America, HOK employs more than 1800 professionals to complete its award-winning projects in multiple sectors including education, government, hospitality, residential, commercial, health care, institutional, retail and transportation. HOK brings what it calls a 'whole world design ability' to built environment challenges, with disciplines serving the entire facility lifecycle including architecture, engineering, construction services, interiors, planning, urban design and landscape architecture.

HOK as a company began to focus on sustainable design beginning around 1990. In 1993, the company formally established sustainable design as one of its core values. As early players in the USGBC, leaders from HOK helped with the development of the LEED Green Building rating system. More recently, HOK has become a leading design practice through efforts such as an alliance with the Biomimicry Guild 3.8 that enables a new focus on living buildings and the use of living systems as a model for the built environment. Also known for its exemplary work in community service, the company recently issued a 24/SIXTY report describing its effort for each office to devote a minimum of 60 hours of design services to support non-profits and community projects in honour of its 60th anniversary in 2015. The firm is also an industry leader in integrating building information modelling (BIM) as part of the process of sustainable design and project delivery.

Notable elements of HOK's policy on sustainability (see *HOK sustainability policies and programmes*) include a firmwide Sustainable

Roadmap and Operations Plan adopted in 2009 that comprehensively addresses opportunities for the firm to green its operations, including LEED certification of office spaces, purchasing, model shop practices, recycling and waste reduction. The firm has taken numerous steps to address its carbon footprint, in both its projects and its own operations. Annual carbon footprint analyses are conducted for every HOK office worldwide, which include energy use, air travel, employee community and other factors. Each employee receives an electronic copy of the report, and the firm has committed as a company to meet the American Institute of Architects' 2030 Challenge to achieve carbon neutrality by 2030. It also has a goal for all design professionals within the firm to become LEED accredited.

> **HOK sustainability policies and programmes**
>
> - Sustainable Operations Plan:
>   - LEED certified office spaces
>   - Purchasing
>   - Model shop practices
>   - Recycling and waste reduction
> - Carbon neutral as a firm and for client projects by 2030
> - LEED accreditation of all design professionals within the firm
> - Sustainable design services and industry-recognized resources

HOK has been engaged by multiple clients including the US Department of Defense, the UK Building Research Establishment, and the US General Services Administration to develop educational programmes, web tools and design guidelines to support sustainability in public sector facilities. In 2009, its sustainable development guide for the US federal government, *The New Sustainable Frontier: Principles of sustainable development,* won several awards for its contributions to sustainability in the public sector. The firm has also played an industry-wide leadership role in sustainability by developing some of the first widely used information resources for the construction industry. Authored by two professionals in the firm, *the HOK Guidebook to Sustainable Design* (Mendler, Odell and Lazarus 2005) is now in its second edition and is a widely used resource based on extensive databases and resources developed within the company.

One notable feature of this book especially reflects HOK's unique design philosophy with regard to sustainability: its coding of best practices based on whether or not they require owner approval to implement. In the book, the authors acknowledge that while some sustainability best practices require involvement of the owner in the decision to move forward because of additional cost or other constraints, some practices should be included in every project without asking. This perspective highlights the important role of the architect in moving sustainability into the industry, and emphasizes the fact that many practices for sustainability in construction represent the best choice from a technical and economic standpoint as well as an environmental or social standpoint.

In 2016, HOK signed a partnership with the International Finance Corporation (IFC), a member of the World Bank Group, to use the new Excellence in Design for Greater Efficiencies (EDGE) green building certification system and software (see edgebuildings.com) in at least five of its projects in developing countries. HOK already regularly uses green building certification systems on many of its projects, but the EDGE rating system was specially developed to apply to projects in developing nations where rating systems such as LEED or BREEAM may be difficult to apply. HOK will serve as a pilot tester for the new rating system and is committed to help create awareness of its benefits throughout the design and construction industry.

Construction and project management firms

Sustainable innovation has also resulted in significant evolution of firms playing the roles of general contractor, project manager or programme manager. The following cases present two exemplary companies whose sustainability programmes and initiatives have been recognized internationally for the leadership they represent to the industry.

Example 1: Hyundai Engineering and Construction

Hyundai Engineering and Construction Co., Ltd. (Hyundai E&C) is the fifteenth largest construction company in the world, and has delivered a range of construction and engineering projects worldwide. Since its formation in 1947, the company has focused on its goals of 'promoting the happiness of humankind, creating new value for human life, and building a bright and promising future by connecting people to people, culture to culture and land to land through construction'. With offices in Europe, Africa, North America, and throughout Asia and the Middle East, Hyundai E&C specializes in industrial construction, civil works, building works and power plant construction, including nuclear power plants, substations and renewable energy plants. In the industrial construction sector, the firm specializes in desalination plants, power plants, hydrocarbon processing plants and industrial plants of many types.

Hyundai E&C has made sustainability a core value of company operations, and issued its first sustainability report in 2009. Since then, it has been consecutively selected as a global leader in sustainability management by the Dow Jones Sustainability Index (DJSI). One of Hyundai E&C's main efforts for sustainability was to establish the Green Smart Innovation Center (GSIC), an experimental green building research facility, in its Research & Development (R&D) centre located in Yongin, Gyeonggido, South Korea in 2014 (see *Case study: Hyundai E&C's Green Smart Innovation Center*).

Case study: Hyundai E&C's Green Smart Innovation Center

Finished in 2014, Hyundai E&C's Green Smart Innovation Center (GSIC) (Figure 3.15) has achieved LEED platinum certification by the US Green Building Council and G-SEED Certification (Green 1 level) in South Korea.

The GSIS is a four-storey research and experimental testing facility of 26,586 sq ft (2,470 m^2) that includes (Figure 3.16):

- a *residential research facility* designed to verify comfort and effectiveness of energy reduction while taking residents' lifestyles into account;

continued

Figure 3.15 Hyundai E&C Green Smart Innovation Center (GSIC), Yongin, Gyeonggido, South Korea

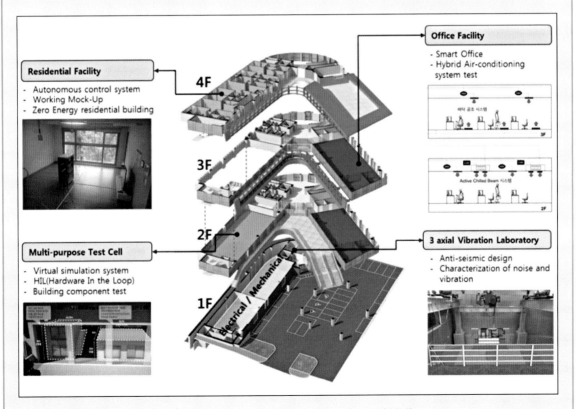

Residential Facility
- Autonomous control system
- Working Mock-Up
- Zero Energy residential building

Office Facility
- Smart Office
- Hybrid Air-conditioning system test

Multi-purpose Test Cell
- Virtual simulation system
- HIL(Hardware In the Loop)
- Building component test

3 axial Vibration Laboratory
- Anti-seismic design
- Characterization of noise and vibration

Figure 3.16 Layout of the Hyundai E&C Green Smart Innovation Center (GSIC)

continued

- an *office research facility* that delves into cutting-edge workplace environments to which the Smart Façade system and a complex combined system are applied;
- a *three-axis vibration laboratory* aimed to assess vibration characteristics of architecture and plant equipment; and
- a *multi-purpose test cell* intended to experiment with building assemblies under various climatic environments from tropics to polar regions.

An integrated design approach was used for the GSIC that incorporated multiple methods and tools to encourage and enable design and construction specialists in different areas to work together to produce an energy efficient integrated design. For this project, the integrated project team including all stakeholders undertook serious charrette meetings to develop a design concept, including passive and active energy saving strategies and technologies and renewable energy. In addition, the project team also utilized multiple analysis and simulation tools for passive design, including energy, ventilation, daylighting and insulation tools (Figure 3.17).

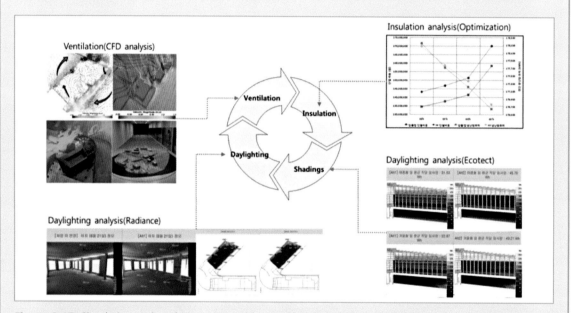

Figure 3.17 Simulation and analysis tools used for integrated design in the GSIC

The integrated project team also included many innovative building systems to improve performance and increase the use of renewable energy, including a radiant floor heating system, building integrated photovoltaics (BIPV), a chilled beam system, underfloor HVAC, geothermal heating and cooling, a heat recovery system, a wind turbine, an Energy Storage System (ESS), a panel-based PV system, and others.

Through implementing passive and active energy saving strategies, the GSIC is projected to reduce energy consumption of nearly 60% in the office and common use spaces in the building. The passive and active design approaches were expected to reduce total energy use by over a third, with renewable energy generating nearly 26% of total energy. Figure 3.19 illustrates how energy strategies were used to reduce energy consumption and generate renewable energy in the building.

continued

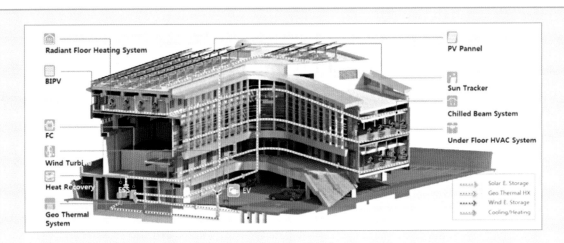

Figure 3.18 Active and renewable strategies and technologies at the GSIC

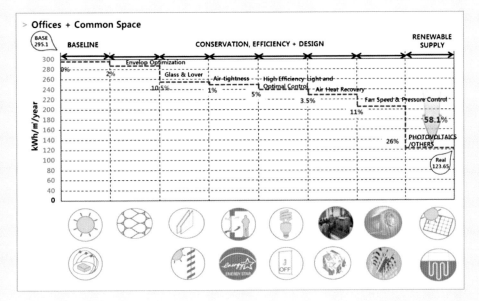

Figure 3.19 Energy saving opportunities using passive and active strategies and renewable energy for office and common space sections

The GSIC was designed to demonstrate how the residential building section could achieve zero energy housing. Based on the team's study and assessment, the residential section of the GSIC could reduce 46.6% of energy compared to a baseline design by implementing passive and active strategies. The remainder of energy needs would be generated from renewable sources including the geothermal heat pump system, solar PV system and the energy storage system. Figure 3.20 illustrates how energy saving strategies were used to achieve the zero energy residential part of the building.

The GSIC incorporates a Smart Building Energy Management System (Smart BEMS) for smart building, efficient operation, and high indoor environmental quality within the building (Figure 3.21).

continued

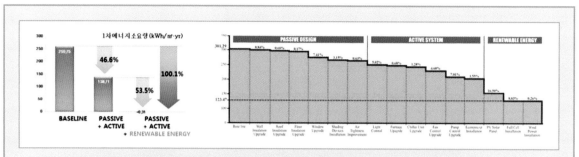

Figure 3.20 Energy saving opportunities for a zero energy residential building

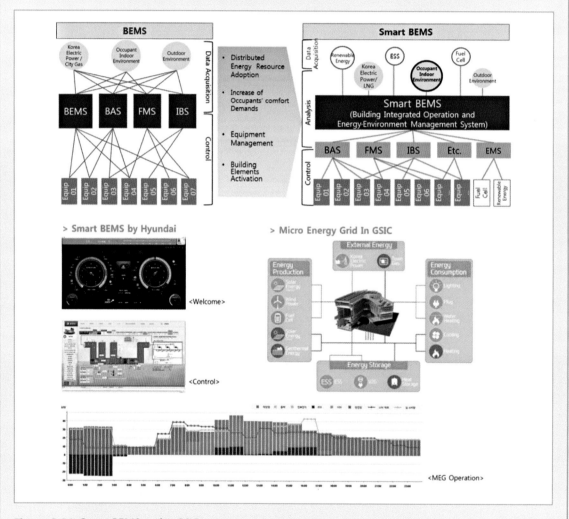

Figure 3.21 Smart BEMS at the GSIC

continued

The Smart BEMS at the GSIC is a computer-based system that helps to manage, control and monitor building technical services (HVAC, lighting, PV, etc.) and the energy consumption of devices used by the building. It also provides the information and the tools that building managers need both to understand the energy usage of their building and to control and improve its energy performance. The Smart BEMS system also takes into account indoor air quality so that it is possible to provide a better indoor environment to occupants.

After completion of the GSIC in November 2014, the researchers at the R&D centre of Hyundai E&C monitored energy consumption of the building to validate the predicted performance of their energy saving strategies and technologies. The actual energy consumption of the building has been measured to be 120.89 Kwh/m² per year, less than the design target of 123.47 Kwh/m² per year. With these data, designers were able to prove that the GSIC actually achieves its goals of energy saving.

Overall, the GSIC demonstrates the dedication of the Hyundai E&C sustainability initiative as a leading sustainable contractor in the world. Using this facility, Hyundai E&C has been able to improve and validate the green strategies and technologies used in their project buildings through innovative design, construction and performance measurement. The GSIC allows Hyundai E&C to demonstrate their capacity for green and zero energy buildings, which further improves their competitiveness in the global market and recognition as a leader in sustainability.

For more information

Hyundai E&C. (2016). *Corporate Sustainability Report.* http://en.hdec.kr/EN/Sustainability/Global.aspx
(accessed 4 September, 2016).

Source: All images courtesy of Hyundai E&C

In April 2010, Hyundai E&C also publicly committed to join the United Nations Global Compact, an initiative between private industry, institutions and other groups, to master the challenges of globalization and achieve sustainable development. As part of this commitment, it has committed to meeting the Global Compact's ten principles, centred around respecting human rights, promoting social and environmental standards and fighting corruption (see *United Nations Global Compact – ten principles for business*). These principles were derived from the Universal Declaration of Human Rights, the International Labor Organization's Declaration on Fundamental Principles and Rights at Work, and the Rio Declaration on Environment and Development. The commitment has been honoured through annual reporting on activities and performance relevant to each of the ten principles through the company's annual sustainability report.

In its 2016 corporate sustainability report, Hyundai E&C identifies four global megatrends it believes will shape the construction industry between now and 2030: urbanization, water scarcity and pollution, energy and fuel, and material resource scarcity. Hyundai E&C has focused its efforts on these trends in particular to identify future opportunities to transform the industry. It has also closely linked activities in each of these four areas to the UN's Sustainable Development Goals (SDGs), including goals 6 – Clean Water and Sanitation, 7 – Affordable

United Nations Global Compact – ten principles for business

1   Businesses should support and respect the protection of internationally proclaimed human rights.
2   Businesses should make sure that they are not complicit in human rights abuses.
3   Businesses should uphold the freedom of association and the effective recognition of the right to collective bargaining.
4   Businesses should uphold the elimination of all forms of forced and compulsory labour.
5   Businesses should uphold the effective abolition of child labour.
6   Businesses should uphold the elimination of discrimination on respect of employment and occupation.
7   Businesses should support a precautionary approach to environmental challenges.
8   Businesses should undertake initiatives to promote greater environmental responsibility.
9   Businesses should encourage the development and diffusion of environmentally friendly technologies.
10 Businesses should work against all forms of corruption, including extortion and bribery.

and Clean Energy, 9 – Industry, Innovation and Infrastructure, and 11 – Sustainable Cities and Communities. The company's sustainability report meets the Global Reporting Initiative's requirements for corporate social responsibility reporting in the areas of economic, environmental and social activities. In addition, it also presents data relevant for the ISO 26000 standard on Social Responsibility (see *ISO 26000: Social Responsibility core subjects*).

In response to these challenges, Hyundai E&C has declared its sustainability mission to be 'creating sustainable habitat for humans' (Hyundai E&C 2016). The company identifies five kinds of corporate capital (Financial, Manufactured and Natural, Intellectual, Human and Social and Relationship) that organize its aim of creating shared value (Figure 3.22). These capitals are then monitored in terms of a framework for an environment-friendly construction process used by the company to deliver value to its customers, focusing on six main functions over the project lifecycle:

- **Engineering**, including expansion of green certification efforts using LEED, Green Mark and other rating systems.
- **Procurement**, including assessment of materials' environmental friendliness using the Hyundai Environmental Goods Standard (HEGS).
- **Transportation**, including introduction of environment-friendly vehicles and efficient operation.
- **Construction**, including on-site resource and environmental management, on-site inspection and protection of ecosystems.
- **Deconstruction**, including integrated environmental waste management systems, environment-friendly deconstruction guidelines, and improved waste management through recycling.
- **Operation**, including increased green certification for the operational phase of the lifecycle.

ISO 26000: Social Responsibility core subjects

- Organizational governance
- Human rights
- Labour practices
- The environment
- Fair operating practices
- Consumer issues
- Community involvement and development

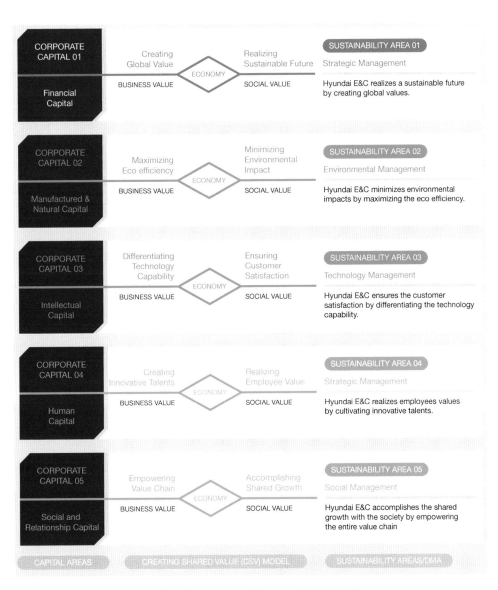

Figure 3.22 Hyundai E&C's 'creating shared value' model (Hyundai E&C 2016)

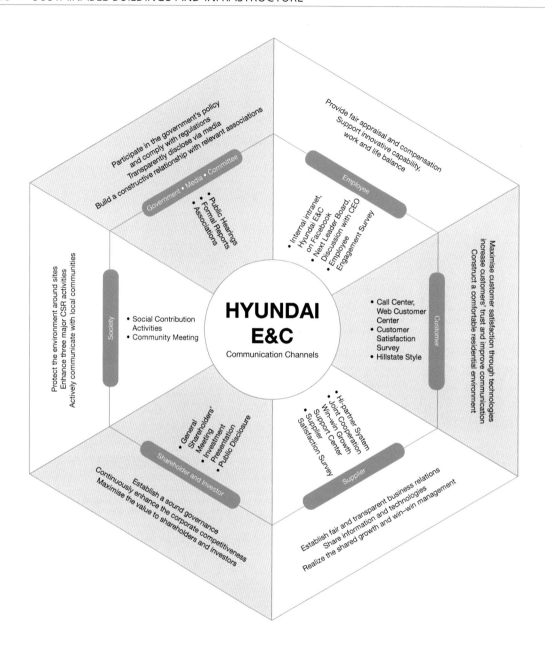

Participate in the government's policy and comply with regulations
Transparently disclose via media
Build a constructive relationship with relevant associations

Government • Media • Committee

- Public Hearings
- Formal Reports
- Associations

Provide fair appraisal and compensation
Support innovative capability,
work and life balance

Employee

- Internal Intranet, Hyundai E&C on Facebook
- Next Leader Board, Discussion with CEO
- Employee Engagement Survey

Protect the environment around sites
Enhance three major CSR activities
Actively communicate with local communities

Society

- Social Contribution Activities
- Community Meeting

HYUNDAI E&C

Communication Channels

Maximise customer satisfaction through technologies
increase customers' trust and improve communication
Construct a comfortable residential environment

Customer

- Call Center, Web Customer Center
- Customer Satisfaction Survey
- Hillstate Style

General Shareholders' Meeting
Investment Presentation
Public Disclosure

Shareholder and Investor

- Hi-partner System
- Joint Cooperation Win-win Growth
- Support Center
- Supplier Satisfaction Survey

Supplier

Establish a sound governance
Continuously enhance the corporate competitiveness
Maximise the value to shareholders and investors

Establish fair and transparent business relations
Share information and technologies
Realize the shared growth and win-win management

Figure 3.23
Hyundai E&C's Communication Channels (adapted from Hyundai E&C 2016)

Hyundai has also established customer satisfaction as a priority, with an emphasis on enhancing modes and opportunities for communication with all involved stakeholders (Figure 3.23). It has established specific mechanisms and goals for communicating with six different stakeholder groups (Employees, Customers, Suppliers, Shareholders/investors, Society and Government/media) and it monitors and evaluates those channels to improve their effectiveness over time.

Hyundai E&C has been striving to meet the expectations of the DJSI's sustainability criteria and incorporate them into its practices.

In 2014, Hyundai E&C reduced waste emissions by 26.3 per cent, posting 452,021 tons, down from 613,302. In the case of sewage emissions for wastewater, the company reduced the amount by 2.1 per cent, recording 1,093,787 tons, down from 1,117,667 tons the previous year. The company is also actively engaging in outreach programs. Since 2009, Hyundai E&C has been locally involved in six outreach programs, including one that supplies free lunches to undernourished children. The company is also extending the same kind of activities abroad by engaging in 18 outreach programs in 13 countries including the Philippines, Colombia and Kenya.

Through these and multiple other activities, Hyundai E&C has emerged as not only a leader within the construction sector, but also an acknowledged corporate leader across the whole set of sectors that are part of the DJSI. Its sustainability report (Hyundai 2016) is also an excellent example of comprehensive and transparent reporting of sustainability-related issues and accomplishments.

Example 2: Skanska

Among the top ten largest construction firms worldwide, Skanska is a leading international project development and construction company. Founded in 1887 and headquartered in Stockholm, it employed 43,000 employees in 2015 and nearly four times as many subcontractors in markets throughout Europe and North America on an average of 10,000 projects annually. Within these markets, Skanska is well known for its building and civil engineering projects as well as projects in the oil, gas and energy sector. In the United States in particular, Skanska targets projects undertaken through public–private partnerships (PPP). Its offerings also include single-family and multi-family housing projects; property investment, planning, development and management; and large infrastructure projects such as bridges, tunnels and roads.

Skanska began its quest for sustainability in 1995 by joining the World Business Council for Sustainable Development. In the next several years, it obtained ISO 14001 certification for various business units, with all operations worldwide being certified by 2000. The firm produced its first environmental report in 1997, which evolved into its first corporate sustainability report in 2002, organized around the Global Reporting Initiative (GRI) elements. It was first listed in the DJSI in 1999, and in the FTSE4Good social responsibility index in 2003. In 2005, it was the only construction firm listed in the initial launch of the Global 100 Most Sustainable Corporations list. In 2007, it ranked #1 in Engineering News Record's first ever Top Green Contractors listing, and was ranked in the top six in 2015. Some of its notable projects include the first LEED Certified McDonald's restaurant in the United States, a LEED-CI Platinum office in the Empire State Building in New York City, the 100 GWh El Totoral Wind Farm in Chile, Boston Logan International Airport Terminal A, the State of California Central Utility Plant and the Swiss Reinsurance Headquarters building in London. Its Swedish headquarters is certified LEED Platinum.

Skanska's Six Zeros of Deep Green

- net zero primary energy for buildings and net positive primary energy for civil/infrastructure projects
- near zero carbon in construction
- zero waste
- zero hazardous materials
- zero unsustainable materials
- net zero water for buildings and zero potable water for construction in civil/infrastructure

As a firm, Skanska has incorporated sustainability as part of its core values through a code of conduct and sustainability goals in five key areas: Safety, Ethics, Green, Corporate Community Investment, and Diversity and Inclusion. These goals reflect the economic, environmental and social goals of sustainability and include key stakeholders both within and outside the firm. Skanska encourages both internal and external stakeholders to hold it to these goals, providing a means on its website to report breaches in its code of conduct. It further integrates core values in its business practices by linking incentives to performance at various levels within the company. It also integrates sustainability management at the line level in each of its core business units. The decentralized nature of the organization has led to a matrix organizational structure for sustainability at the level of each business unit.

As part of its trademarked 'Journey to Deep Green', Skanska seeks to go beyond compliance to set aggressive sustainability goals in the areas of energy, carbon, water, material selection and waste in accordance with its Skanska Color Pallette (Skanska 2016). The colour pallette is a strategic tool to measure and communicate the company's progress toward achieving Deep Green, which it hopes will 'Future-proof' its projects. It has three basic levels: Vanilla, which reflects basic compliance with all applicable laws, regulations and standards; Green, which are projects that go beyond compliance at least by 25 per cent in terms of energy use as well as other goals for carbon, materials selection, waste or water use; and Deep Green, defined as having a near-zero impact on the environment (see *Skanska's Six Zeros of Deep Green*).

One of Skanska's notable accomplishments resulting from its green initiatives is its publicly available book *Green Thinking: There's more to building a green society than just building* (Skanska 2010). The book includes an overview of ways in which the firm is taking steps to contribute to a green society in terms of energy conservation, carbon reduction, water conservation, material use, lifecycle planning, community involvement and local impact. It also ends with a section on things you can do, targeted specifically to key stakeholders such as politicians and public officials, along with developers, planners, financiers, investors, tenants and clients (see *Skanska's top ten green tips*).

Skanska also began issuing annual sustainability reports in 2002, and has published a series of reports on Green Urban Development highlighting development trends for a more sustainable society in the areas of energy, materials, water and carbon (Skanska 2015). The company has also contributed to several notable international reports on green building practices, including a comprehensive report on *Building the Business Case for Green Building* in collaboration with the World Green Building Council (WGBC 2013). This report identifies the business benefits of green building in terms of design and construction costs, asset values, operating costs, workplace productivity and health, and risk mitigation. It also contains a section on scaling up these benefits from green buildings to green cities, including case studies from New York City, Abu Dhabi and other cities.

Skanska's top ten green tips

1 Ask yourself if there's any reason not to build green.
2 Involve affected parties early in the process, so you can incorporate their opinion and concerns into the solution, rather than try to find remedies afterwards. Make 'suppliers' into 'partners' so you can take advantage of their knowledge, experience and enthusiasm.
3 Take the entire lifecycle of the building or structure into account when planning it. What's interesting isn't the cost of construction, but the total cost over the structure's lifetime. Also, you avoid future environmental issues.
4 Think long-term flexibility to avoid the high costs and environmental impact generated by complicated rebuilding when having to respond to evolving needs.
5 Look at new financial planning models that support long-term gains rather than short-term low costs.
6 Be open to innovative solutions. Consider the local conditions and utilize the best technology. Use local materials to avoid unnecessary transport, and use local renewable energy sources as much as possible.
7 In the planning and design process, keep the four Rs in mind – reduce, reuse, recycle and recover.
8 Define your own plan for the Deep Green journey and make it visible. Start now. Communicate the strategy and operations so everyone involved understands their necessity and what they need to do. Inspire others to think greener.
9 Consider the positive effects of green building on the brand equity of your company.
10 Again, ask yourself if there's any reason not to build green.

Construction product manufacturers

In the manufacturing sector, sustainability initiatives among con-struction materials and systems producers are as diverse as the materials and systems themselves. The following two case studies illustrate different approaches to sustainability from one of the very early leaders in the sustainability arena (Interface) and one of today's leading inter-national sustainability advocates (LafargeHolcim).

Example 1: Interface global

Founded in 1973, Interface is known as a world leader in soft-surface modular floor coverings, also known as carpet tiles. The company produces primarily commercial carpets, although it entered the residen-tial carpet market in 2003 with the production of FLOR carpet tiles. Since its founding, Interface has grown into a global enterprise with annual sales over US$1 billion in 100 different countries, from manufacturing facilities in Australia, China, the Netherlands, Thailand, the UK and the United States.

In 1994, founder, chairman and CEO Ray Anderson transformed the company's business strategy and practices with the aim of achieving sustainability in its manufacturing enterprises without sacrificing business goals. The story of this transformation, based on Anderson's personal awakening to environmental concerns, is well-known in the sustainability world and documented along with his business model for sustainability in his 1999 book, *Mid-Course Correction* (Anderson 1999).

In 1997, Interface set a goal to eliminate any negative impact the company has on the environment by the year 2020, dubbed 'Mission Zero'. The company worked to achieve this goal through three primary paths: innovative solutions for reducing its footprint, new ways to design and make products, and an inspired and engaged culture. Its efforts included developing new processes for recycling old carpets, creating a leased carpet programme, utilizing the work of indigenous peoples, switching to solar and other alternative energy sources, and reducing water use and contamination. In 2007, the company achieved third-party certified net negative greenhouse gas emissions. During this time, Interface concurrently grew its profits, not just in the United States but internationally as well. Interface characterized its quest as 'climbing Mount Sustainability'.

Following principles of the Natural Step (see Chapter 2), the company bases its industrial processes on an understanding of natural systems. One of its most well known contributions to a paradigm shift in the industry was Entropy, a carpet tile inspired by forests and the random patterns of sticks and leaves that are found on the forest floor. Entropy carpet tiles were designed to be laid randomly without any need for colour matching or orientation. This saved significant installation time and product waste, and allowed for easier replacement of small sections over the lifecycle as needed. The Entropy product line currently represents nearly 40 per cent of Interface's total sales.

Among Interface's most fundamental challenges is the reality that carpet has been a product of materials derived from non-renewable fossil fuels, such as nylon. To address this challenge, Interface developed its ReEntry programme as a means to recover reusable materials from among the 5 billion pounds of carpet that ends up in landfills each year and displace the use of virgin non-renewable resources in its products. It has also led the way in achieving carbon neutrality for its products, beginning in 2003 for its first carbon neutral product, Cool Carpet. This product is a result of a systematic process to inventory and reduce greenhouse gas emissions resulting from the product's lifecycle, then offset the remaining emissions through the purchase of carbon offsets. Greenhouse gases that cannot be eliminated from the product's lifecycle are balanced by buying and retiring carbon offsets on the carbon market, or sponsoring other projects that reduce, avoid or sequester carbon dioxide to prevent it from entering the atmosphere.

In 2016, Interface announced a new mission called 'Climate Take Back', which represents a major step forward from its original Mission Zero goals (Makower 2016). The new programme includes four basic commitments (see *Interface's Climate Take Back commitments*). The basic aim of the programme is to demonstrate that a company can help to reverse the impact of climate change while remaining profitable, by actually using carbon dioxide molecules as a beneficial ingredient to build useful things such as plastics and other materials. Interface is also incorporating principles of biomimicry by designing its factories to behave like forests, providing ecosystem services such as nitrogen cycling, water storage and purification, temperature cooling and wildlife habitat. The company is pilot testing these ideas at its facilities in Minto, New South Wales, Australia and LaGrange, Georgia in the United States.

Perhaps the most notable influence of Interface within the sustainability movement has been from an organizational culture standpoint. Before his death in 2011, Ray Anderson widely shared his vision and models with peer organizations worldwide, serving as a model and inspiration for those who believed that sustainability was incompatible with business success. Within the firm, employees are encouraged and rewarded to present new ideas that can contribute to the firm's goals, and many of these ideas have been adopted with positive results. Employees are infused with the company's philosophies and then take these ideas into their local communities through volunteering and philanthropy. Interface is also known for pioneering the idea of an Eco Dream Team, where influential thought leaders in the sustainability domain are regularly brought into the firm to provide fresh perspectives, learning and ideas.

Interface intends to be a completely sustainable company from an environmental footprint perspective, serving a restorative function within its economic, social and environmental contexts worldwide. The company acknowledges that being truly restorative involves not only reducing its own impacts, but also returning more than it takes to the global ecosystem. The overall goal is to expand beyond internal improvements to affect the whole supply chain that interacts with the company. By 2020, the company predicts its internal efforts will result in sourcing 95 per cent of all materials from recycled or bio-based sources and reduce total energy use by half, water intake by 90 per cent, greenhouse gas emissions by 95 per cent and waste to landfills by 100 per cent. These efforts are significant, but the company hopes that they can ultimately eclipse their internal efforts through facilitating and encouraging change throughout the industry.

Example 2: LafargeHolcim

In 2015, construction materials giants Lafarge and Holcim merged to become the leading provider of cement, concrete and aggregate products worldwide. The resulting company, LafargeHolcim, has plants in 90 countries, over 100,000 employees, and over 2,500 plants including ready mix concrete, aggregate, cement and grinding plants. Lafarge originally opened its doors in France in 1833 as an operator of limestone quarries, winning the 'contract of the century' to provide the hydraulic lime used to construct piers for the Suez Canal in 1864 and quickly becoming an international leader in the cement industry. Founded in Switzerland in 1912, the Holcim company was also a leading supplier of cement and aggregates worldwide, supplying ready-mix concrete and asphalt along with related services to markets on every continent. Together, the employees of LafargeHolcim provide products, solutions and services for the building, infrastructure, distribution and retail, oil and gas and housing sectors.

Interface's Climate Take Back commitments

- We will bring carbon home and reverse climate change.
- We will create supply chains that benefit all life.
- We will make factories that are like forests.
- We will transform dispersed materials into products and goodness.

To achieve its vision of 'creating shared value with society', LafargeHolcim has integrated sustainable development at the fundamental core of its business strategy. Two of its key mindsets, sustainable environmental performance and corporate social responsibility (CSR), underlie its basic business strategies, priorities and ultimate goal of value creation. The company's focus areas for its sustainability efforts lie in five major areas (see *LafargeHolcim's sustainability priorities*). Its 2030 plan sets out a vision of an 'innovative, climate-neutral construction sector that embraces the circular economy' (LafargeHolcim 2015a). Among its environmental goals for 2030 are a 40 per cent reduction in CO_2 emissions per tonne of cement vs. 1990 figures as well as significant conversion to waste-derived resources, recycled aggregates and reclaimed asphalt pavement. Social goals include zero fatalities both on and off-site, increased gender diversity, and extensive stakeholder engagement efforts.

Like Interface, LafargeHolcim aspires to reach beyond its own immediate operations to influence the industry at large, increasing the innovativeness and sustainability of the entire construction value chain (ibid.). Examples of efforts 'beyond the fence' include offering end-of-life solutions for concrete and asphalt products, contributing positively to water impacts in water-scarce areas, and developing initiatives to benefit 75 million people and combat bribery and corruption, particularly in high-risk countries.

From a social sustainability standpoint, integrity is a key issue for LafargeHolcim. The company operates under a Code of Business Conduct (CoBC) that ensures all business is conducted with integrity. A global Integrity Line is available to enable employees to confidentially disclose any integrity concerns they may have, in 36 different languages,

LafargeHolcim's sustainability priorities

Our people:
- Keeping people safe
- Championing diversity
- Respecting human rights

Our communities:
- Affordable housing solutions
- Inclusive business models
- Social investments
- Stakeholder and community engagement
- Promoting clean and fair business practices
- Acting for responsible sourcing in our supply chain

Climate:
- Leadership in net CO_2 emissions per tonne of cement

- Reducing CO_2 emissions from buildings and infrastructure
- Managing energy
- Our advocacy

Circular economy:
- Transforming waste into resources
- Providing end-of-life solutions for our products

Water and nature:
- Reducing freshwater withdrawal
- Showing a positive impact on water resources in water-scarce areas
- Safe water, sanitation and hygiene at the workplace
- Positive change for biodiversity
- Managing other environmental impacts

through telephone or online. Compliance training and transparent community engagement also support the ethical conduct of the company.

LafargeHolcim uses an Integrated Profit and Loss (IP&L) statement to express its corporate performance using the Triple Bottom Line, including quantifying externalities such as social and environmental impacts not ordinarily included in corporate accounting (LafargeHolcim 2015b). Some of the key performance areas tracked include socio-economic metrics (stakeholder value, strategic social investments, inclusive business, occupational injuries and health, human rights, and employee education) and environmental metrics (CO_2 emissions, air quality, water, biodiversity, waste, secondary resources and environmental incidents).

An area of particular importance to the firm is sustainable construction, since its products represent the very foundation of the construction industry and contribute significantly to its impact on energy use and climate change. Toward that end, the firm continues to support the LafargeHolcim Foundation for Sustainable Construction, which uses its resources to support efforts addressing five target issues important for sustaining the human habitat (see *LafargeHolcim Foundation for Sustainable Construction*). Its activities include sponsoring an international competition for excellent projects in sustainable construction, which awards US$2 million in cash prizes in each round to its winners. The competition celebrates innovative, future-oriented projects from around the world, and is open to architects, planners, engineers, project owners, builders and construction firms. Students in the final year of undergraduate study or in graduate programmes are also eligible to apply for the 'Next Generation' category of the award. The competition is presently in its fifth round. Over 200 awards have been made to date.

The foundation also provides seed funding for key efforts outside the competition, such as setting up a new Center of Excellence in Sustainable Housing and Rural Infrastructure in India, supporting a post-tsunami master planning effort in Constitución, Chile, and supporting microfinance and capacity building in developing nations. It currently spends on the order of SF36 million (US$38.5 million) annually on community initiatives, donations and in-kind support for its corporate social responsibility initiatives, representing more than 1 per cent of net income before tax. Community initiatives are evaluated using a social engagement scorecard which facilitates the selection of projects that are aligned with the company's CSR priorities.

As part of its sustainability programmes, LafargeHolcim has undertaken a materiality review to identify the most important areas of concern for the company with regard to sustainability, including carbon emissions, water use and raw material supply. From an operational standpoint, LafargeHolcim acknowledges that the ongoing sustainability

---

LafargeHolcim Foundation for Sustainable Construction

Target Issues to Sustain the Human Habitat for Future Generations

- **Progress** – Innovation and transferability
- **People** – Ethical standards and social inclusion
- **Planet** – Resource and environmental performance
- **Prosperity** – Economic viability and compatibility
- **Place** – Contextual and aesthetic impact

of its enterprise relies on long-term access to raw materials acquired through quarrying. Accordingly, the firm pays special attention to the ecological and biodiversity impacts of quarrying. Nearly all of its quarries have rehabilitation guidelines and plans that have led to conservation and restoration projects resulting in new habitats such as wetlands, forests and natural grasslands on previous quarry sites. The use of water in quarrying, aggregate production and cement production is also a top priority emerging from the stakeholder materiality assessment, and a new water management scheme for all business units was implemented in 2013.

Common themes

What can we learn from these and other examples of leaders in sustainable construction? There are many opportunities at the corporate and project levels to reduce an organization's impacts on the planet and increase its positive contribution to society. No matter what portfolio of strategies a company chooses to adopt, several common themes emerge in common across the leaders examined here.

First, the organizations that are most successful in becoming sustainability leaders have a clear picture of the firm within its larger global and societal contexts. They understand how their actions have the potential to influence that context both positively and negatively, and they pay careful attention to integrate their sustainability actions within that context in a way that capitalizes on their individual strengths.

Second, the leading firms examined here make special effort to internalize sustainability within the firm and integrate it with core business principles and practices. These companies are not content to add sustainability as just one more requirement they must meet to achieve compliance, or one more fad in which the market is interested today. Instead, they look for sustainability impacts in what they already do – their core strengths and liabilities – and weave the concept throughout their core values, standard operating procedures and policies.

Third, leading firms are recognizing their power to influence the behaviour of others outside the boundaries of their own enterprises. From design to construction to product manufacturing, these leading companies understand that they are part of a larger value creation effort with supply chains upstream and beneficiaries downstream that can be influenced by each firm's decisions and choices.

Fourth, many of the firms that are leaders in sustainability have become recognized as such through their stature in the sustainability community. One key way they achieve this is through active participation in the development of industry resources, organizations and systems such as green building councils, rating systems, scorecards, books and guidelines. By giving their time and expertise to support the development of resources that can benefit everyone, they increase both visibility and credibility among their peers.

Finally, the firms explored in this chapter clearly understand the relationship between credibility in the green market and their own corporate behaviour. They have taken steps to improve not only their products

> ## An effective corporate sustainability policy:
>
> - clearly states the company's vision statement and presents core values and principles
> - contains content relevant to the company's existing culture and operation
> - is signed and dated by either the owner or a company official representing the executive level, demonstrating personal commitment at the top of the company
> - is made public, both on the company's website and in the facility where all employees and visitors can see it
> - is reviewed and renewed annually to ensure the policy is current and applicable
> - has supporting documents and resources
> - is audited by a third party for conformance.
>
> (Gary Jones, director of environment, health, and safety affairs, PIA International)

and contributions to value in the market, but also the operations and internal practices used to generate their company's value added. These companies contribute to the sustainability of the world in which we all live by truly walking the talk of sustainability.

Ultimately, the tactics that are most effective for achieving change in a given context depend on many factors, including the level of expertise and experience with sustainability already existing in that context, the structure and culture of the organization in which the programmes are being implemented, and the financial and other resources and that can be applied to the programme. While observations made in this chapter are based on a synthesis of data collected across multiple organizations, the most effective approach for a given organization will depend where that organization is in its process of sustainability adoption. Stakeholders interested in developing sustainability policies and programmes in their organizations can draw from these lessons and ideas in constructing programmes that fit the context and requirements of their situations (see *An effective corporate sustainability policy*). The matrices and criteria established in this chapter can be adapted contextually and developed in more detail as needed to support the most effective selection of programme elements to maximize the chances of success of those programmes within their individual context.

Sustainability policy in the future

What will be the relationships between and among stakeholders in the construction industry in the future? How will business-driven corporations be held accountable for impacts of their actions that are borne by society at large? Much controversy exists about the best way to achieve sustainability goals that are seemingly at odds with the interests of business, and even at the national level, some nations still place the economy ahead of the environment. In reality, the economy cannot and will not exist without the environment, so ultimately this way of thinking must change.

Reports on industry sustainability leaders by Accenture (2010a, 2010b) found that companies considered to be leaders in their fields were finding new ways to drive value delivered by their firms. First, firms that are successful sustainability leaders grow their revenue through new products and services that leapfrog current thinking about today's solutions. Second, they are reaping benefits to their own bottom line by reducing costs through efficiency gains in energy and water use, raw material use, waste generation and human productivity. Third, these firms' proactive approaches to sustainability enable them to manage operational and regulatory risk more effectively. Finally, by considering sustainability as a core value for their business, they are able to build intangible assets such as their brand, reputation and collaborative networks that will ultimately serve them well in the market.

In the coming years, we will see an evolution of current thinking about value in the corporate world. Companies that do not align themselves with sustainability principles will be less competitive in a climate of increased resource scarcity, more stringent public policy and greater transparency. At the same time, greater success will come to firms that are early movers in their domains with regard to operationalizing sustainability for their industry and implementing it through effective policy. Carbon in particular will be a major driver of industry opportunity, with poor performers subject to penalties while sustainability leaders reap benefits from carbon trading. Figure 3.24 shows a continuum of corporate maturity with regard to sustainability developed by the Accenture Corporation (2008). In the future, more firms will gravitate

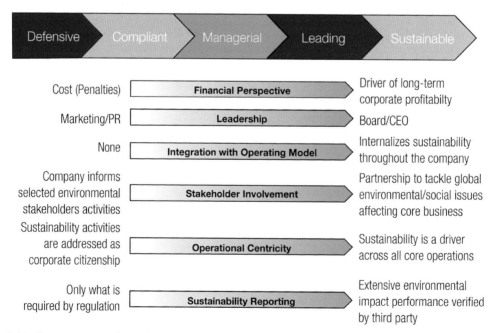

Figure 3.24 Corporate maturity with respect to sustainability
Source: Accenture 2008

toward the right side of the continuum to their continued advantage. From the standpoint of public policy, it is difficult to predict how government will structure initiatives to promote greater sustainability. To date, sustainability leaders of industry have increasingly employed self-regulation as a way to preempt government mandates. Although today's trends suggest that the future will hold more self-regulation, voluntary compliance and market-based policy, increases in economic instability, political unrest, undeniable evidence of climate change, and other factors may lead to a more regulatory approach by many government entities. Overall, there is likely to be increasing focus on impacts of the developing world by developed nations as they enter world markets and compete for finite resources. Hopefully, such attention will result in a just and equitable distribution of resources, and effective policy initiatives to ensure ongoing global stability and prosperity for all.

Discussion questions and exercises

3.1   Consider the last major decision you made to adopt an innovation. What attributes were most important for you in making the choice you did? Now consider an action you might personally take to improve your own sustainability. How does it compare to what you do now, in terms of the five attributes of innovations affecting their adoption?

3.2   Inventory the existing sustainability policies of your organization and/or for the built facilities with which you interact. What are the major components of these policies? What components are missing?

3.3   What social risks would be associated with a sustainability policy for your organization's facilities? If you already have a sustainability policy, talk with the people responsible for putting that policy in place. What challenges did they face in implementing the policy? How did they overcome those challenges? If you do not already have a policy, what challenges do you imagine you would face in implementing one?

3.4   If your organization has a sustainable facilities policy, how does that policy reduce negative impacts on natural ecosystems and minimize resource consumption? If you do not already have a policy, what would you recommend including in a new policy to address these issues?

3.5   If your organization has a sustainable facilities policy, what costs are associated with implementing that policy? Consider both direct costs such as enforcement and training, and indirect costs such as the potential for the failure of unfamiliar technology. What economic benefits might result from implementing the policy?

3.6   Consider your organization's structure and standard operating procedures. Who would be involved in implementing a sustainability policy? What arguments might be most effective in getting the support of those stakeholders?

3.7   Which of the four policy options – require a specified level of performance, endorse and encourage performance, encourage sustainability in general, or create a working group to set standards – would be most effective in the context of your organization's current views on sustainability? Why?

3.8   If your organization already has a sustainable facilities policy, which programme options have been developed as part of its implementation? Which additional programme options would be beneficial for your organization beyond what already exists?

3.9   If your organization already has a sustainable facilities policy, how is the effectiveness of the policy evaluated? If no policy exists, which approach to evaluation would be most successful in the context of your organization?

3.10 What building or infrastructure-related policies have been put in place by your local, regional or national government? Which of them apply to the facilities with which you interact? Identify specific features in the buildings you occupy that are a result of those policies.

3.11 Does your organization have sustainability goals and objectives? If yes, evaluate those goals and objectives using the SMART characteristics described in the chapter. If no sustainability goals and objectives exist, draft a set that meets SMART criteria.

3.12 Does your organization have a sustainability team? If yes, identify the members of that team and the roles they play. Where are they located on the organizational chart? If no team presently exists, where would a team best fit on the organizational chart if one were created? Why?

3.13 Adapt the corporate functional model shown in Figure 3.8 to your organization. What opportunities exist in your organization to increase sustainability for each functional area of the model? Consider each of the major areas of the Triple Bottom Line. How specifically can you affect organizational sustainability within the responsibilities of your present position or job?

3.14 Review the checklist of organizational best practices for sustainability shown in Table 3.5. Which practices is your organization already following? Which should be considered for future implementation? Are there any sustainability practices your company implements that are not already on the checklist?

3.15 Which of the core sustainability competencies discussed in the chapter do you possess? Which of these critical skills and abilities should be further developed? Think of examples where you have demonstrated these competencies, skills and abilities in your personal and professional life.

3.16 To what extent does your organization have a sustainability culture? How can your organization's sustainability culture be enhanced?

3.17 Where on the spectrum of corporate maturity with regard to sustainability (Figure 3.8) does your organization fall? Discuss the spectrum with people in your organization responsible for sustainability functions and document the history of your firm in progressing through the spectrum.

Notes

1   Portions of this chapter have been adapted from work previously published by the author and colleagues (DuBose et al. 2007; Pearce et al 2005a; Pearce et al. 2007) and are included with permission from the *Journal of Green Building*.

References

Accenture. (2008). "Compatible aims: Sustainability and high performance." https://microsite.accenture.com/NonSecureSite CollectionDocuments/CompatibleAims.pdf (accessed 4 January 2011).

Accenture. (2010a). "Driving value from integrated sustainability: High performance lessons from the leaders." https://microsite.accenture.com/sustainability/Documents/Driving_Value_from_Integrated_Sustainability_High_Performance_Lessons_from_the_Leaders.pdf (accessed 4 January 2011).

Accenture. (2010b). "Sustainability and its impact on the corporate agenda." https://microsite.accenture.com/NonSecure SiteCollectionDocuments/By_Subject/Strategy/PDF/Accenture_POV_Management_Consulting_Sustainability_and_its_Impact_on_the_Sustainabiliy_Agenda.pdf (accessed 4 January 2011).

Anderson, R. (1999). *Mid-Course Correction*. Peregrinzilla Press, White River Junction, VT.

Athena Sustainable Materials Institute. (2002). "U.S. Federal and Military Applications of LEED: Lessons Learned." Report prepared for the Department of National Defence and Public Works and Government Services, Canada, March.

Australia Building Code. (2008).

Bennett, M. and James, P. (eds). (1998). *The green bottom line: Environmental accounting for management.* Greenleaf Publishing, Sheffield, UK.

Best, A.M. (2005). "Building on common ground: Green buildings and the University of Washington." MS degree project final report, Public Administration Programme, University of Washington, Seattle, WA.

Bosch, S.L., and Pearce, A.R. (2003). "Sustainability in public facilities: Analysis of guidance documents." *Journal of Performance of Constructed Facilities, Special Issue on Facilities Design and Management* 17(1), 9–18.

Brand, S. (1994). *How buildings learn.* Viking Press, New York, NY.

Bray, J., and McCurray, N. (2006). "Unintended consequences: How the use of LEED can inadvertently damage the environment." *Journal of Green Building* 1(4).

Burden, L. (2010). "How to up the EMS ante." www.environmentalmanagementsystem.com.au/ems.html (accessed 1 January 2011).

CIC – Construction Industry Council. (2008). "Sustainability skills matrix for the built environment functions." www.cic.org.uk/activities/sustainabilityskillsmatrix.pdf (accessed 31 December 2010).

City of Portland Office of Sustainable Development (2003). "ReThinking development: Portland's strategic investment in green building." City of Portland Office of Sustainable Development, Portland, OR.

Cole, R.J. (2000). "Cost and value in building green." *Building Research and Information* 28(5/6), 304–9.

Doran, G. (1981). "There's a S.M.A.R.T. way to write management's goals and objectives." *Management Review* 70(11).

DuBose, J.R. (1994). *Sustainability as an inherently contextual concept: Some lessons from agricultural development.* M.S. Thesis, School of Public Policy, Georgia Institute of Technology, Atlanta, GA.

DuBose, J.R., Bosch, S.J. and Pearce, A.R. (2007). "Analysis of state-wide green building policies." *Journal of Green Building* 2(2), 161–77.

Elkington, J. (1997). *Cannibals with forks: The Triple Bottom Line of twenty first century business.* Capstone, Oxford, UK.

Errasti, A., Beach, R. Oyarbide, A. and Santos, J. (2007). "A process for developing partnerships with subcontractors in the construction industry: An empirical study." *International Journal of Project Management* 25(3), 250–6.

Federal Facilities Council. (2002). *Learning from our buildings: A state-of-practice summary of post-occupancy evaluation.* National Academies Press, Washington, DC.

Foster + Partners. (2014). "Corporate, social, and environmental responsibility report." http://fosterandpartners.com/media/967039/foster_and_partners_2014_cser_report_online.pdf (accessed 4 September 2016).

Fust, S.F., and Walker, L.L. (2007). "Corporate sustainability initiatives: the next TQM?" Executive Insights, Korn/Ferry International. www.kornferryinstitute.com/files/pdf1/KFsustainability.pdf (accessed 1 January 2011).

Gardner, G.T., and Stern, P.C. (1996). *Environmental problems and human behavior.* Allyn & Bacon, Boston, MA.

Gil, N., Tommelein, I.D, Kirkendall, B. and Ballard, G. (2000). "Lean product-process development process to support contractor involvement during design." Proceedings, ASCE 8th International Conference on Computing in Civil and Building Engineering, August 14–17, Stanford, CA, 1086–93.

Greenspirit Strategies. (2004). "The cost of green: A closer look at State of California sustainable building claims." Engineered Wood Association, www.apawood.org/level_b.cfm?content = pub_ewj_arch_f04_green (accessed 2 August 2005).

Grosskopf, K.R., and Kibert, C.J. (2006). "Developing market-based incentives for green building alternatives." *Journal of Green Building* 1(1), 141–7.

Horman, M.J., Riley, D.R., Lapinski, A.R., Korkmaz, S., Pulaski, M.H., Magent, C.S., Luo, Y., Harding, N., and Dahl, P.K. (2006). "Delivering green buildings: Process improvements for sustainable construction." *Journal of Green Building* 1(1), 123–40.

Hyundai E&C. (2016). Corporate Sustainability Report. http://en.hdec.kr/EN/Sustainability/Global.aspx (accessed 4 September 2016).

Johnston, D.R. (2000). "Actual costs – is building green too expensive?" *Building Green in a Black and White World*, New Society, Saint Paul, MN, 59–62.

Kale, S., and Arditi, D. (2001). "'General contractors' relationships with subcontractors:Aa strategic asset." *Construction Management and Economics* 19(5), 541–9.

Kats, G. (2003). *The costs and benefits of green buildings: A report to California's Sustainable Building Task Force.* California's Sustainable Building Task Force, Sacramento, CA.

Khalfan, M.M.A. (2006). "Managing sustainability within construction projects." *Journal of Environmental Assessment Policy and Management* 8(1), 41–60.

Klein, K.J., and Sorra, J.S. (1996). "The Challenge of Innovation Implementation." *Academy of Management Review* 21(4).

Klotz, L., Horman, M. and Bodenschatz, M. (2007). "A lean modeling protocol for evaluating green project delivery." *Lean Construction Journal* 3(1), 1–18, April.

LafargeHolcim. (2015a). "Building for Tomorrow: Sustainability Report 2015." http://lafargeholcim.com/sites/lafargeholcim.com/files/atoms/files/06132016-press-lafargeholcim_sustainability_report_online_2015.pdf (accessed 4 September, 2016).

LafargeHolcim. (2015b). "LafargeHolcim Integrated Profit and Loss Statement 2015." http://lafargeholcim.com/sites/lafargeholcim.com/files/atoms/files/06132016-press-lafargeholcim_integrated_profit_loss_statement_2015.pdf (accessed 4 September 2016).

Lapinski, A., Horman, M., and Riley, D. (2005). "Delivering sustainability: Lean principles for green projects." Proceedings, 2005 ASCE Construction Research Congress.

Lapinski, A.R., Horman, M.J., and Riley, D.R. (2006). "Lean processes for sustainable project delivery." *Journal of Construction Engineering and Management* 132(10), 1083–91.

Makower, J. (2016). "Inside Interface's bold new mission to achieve 'Climate Take Back'." GreenBiz. https://greenbiz.com/article/inside-interfaces-bold-new-mission-achieve-climate-take-back (accessed 4 September 2016).

Matthews, J., Tyler, A., and Thorpe, A. (1996). "Pre-construction project partnering: Developing the process." *Engineering, Construction and Architectural Management* 3(1/2), 117–31.

Matthiessen, L.F., and Morris, P. (2004). *Costing Green: A comprehensive cost database and budgeting methodology.* Davis Langdon SEAH International, Boston, MA. www.davislangdon.com/upload/images/publications/USA/2004%20Costing%20Green%20Comprehensive%20Cost%20Database.pdf (accessed 10 October 2011).

Mendler, S.F., Odell, W., and Lazarus, M.A. (2005). *HOK guidebook to sustainable design.* 2nd ed. Wiley, New York.

Munasinghe, M. (1993). *Environmental economics and sustainable development.* World Bank, Washington, DC.

Myers, T. (2005). *Should the state follow LEED or get out of the way?* Washington Policy Center, Washington, DC. www.washingtonpolicy.org/publications/opinion/should-state-follow-leed-or-get-out-way (accessed 10 October 2011).

NAHB – National Association of Homebuilders Research Center. (1989). *Diffusion of innovation in the housing industry.* U.S. Department of Energy, Washington, DC.

Northbridge Environmental Management Consultants (NEMC). (2003). *Analyzing the cost of obtaining LEED certification.* American Chemistry Council, Arlington, VA. www.cleanair-coolplanet.org/for_communities/LEED_links/AnalyzingtheCostofLEED.pdf (accessed 10 October 2011).

Packard Foundation. (2002). *Building for sustainability report: Six scenarios for the David and Lucile Packard Foundation Los Altos project.* David and Lucile Packard Foundation, Los Altos, CA.

Panzano, P.C., Roth, D., Crane-Ross, D., Massatti, R., and Carstens, C. (2004). "The innovation diffusion and adoption research project (IDARP): Moving from the diffusion of research results to promoting the adoption of evidence-based innovations in the Ohio Mental Health System." *New Research in Mental Health*, Vol. 15, ed. D. Roth, Ohio Department of Mental Health, Columbus, OH.

Pearce, A.R. (2001). "Sustainable vs. traditional facility projects: A holistic cost management approach to decision making." Paper for US Army Forces Command, Fort McPherson, GA.

Pearce, A.R. (2003). "An online knowledge base for sustainable military facilities and infrastructure." Final Project Report to Region IV Department of Defense Pollution Prevention Partnership, University of South Carolina.

Pearce, A.R. (2004). "Sustainability design cost differential analysis for constructed projects using the LEED Green Building rating system." Project report to US Air Force Reserve Command Headquarters, Robins AFB, GA.

Pearce, A.R. (2008). "Sustainable capital projects: leapfrogging the first cost barrier." *Civil Engineering and Environmental Systems* 25(4), 291–301.

Pearce, A.R. (2010). "Costing sustainable capital projects: The human factor." Proceedings, Transitions to Sustainability Conference, New Zealand Society for Sustainability Engineering and Science, 1–3 December, Auckland, NZ.

Pearce, A.R., and Ahn, Y.H. (2010). "Strategic entry points for sustainability in university construction and engineering curricula." Proceedings, Transitions to Sustainability Conference, New Zealand Society for Sustainability Engineering and Science, 1–3 December, Auckland, NZ.

Pearce, A.R., DuBose, J.R., Bosch, S.J., and Carpenter, A.M. (2005a). "Greening Georgia facilities: An analysis of LEED requirement impacts." Final project report to the Georgia Environmental Facilities Authority, Atlanta, GA, 30 September.

Pearce, A.R., Bosch, S.J., DuBose, J.R., Carpenter, A.M., Black, G.L., and Harbert, J.A. (2005b). "The Kresge Foundation and GTRI: the far-reaching impacts of green facility planning." Final project report to the Kresge Foundation, Troy, MI, 30 June.

Pearce, A.R., DuBose, J.R., and Bosch, S.J. (2007). "Green building policy options for the public sector." *Journal of Green Building* 2(1), 156–74.

Pearce, A.R., DuBose, J.R., Bosch, S.J., and Carpenter, A.M. (2005c). "Sustainability and the state construction manual: Georgia-specific voluntary guidelines.' Final project report to the Georgia State Finance and Investment Commission, Atlanta, GA, December.

Pearce, A.R., and Fischer, C.L.J. (2001a). "Systems-based sustainability analysis of Building 170, Ft. McPherson, Atlanta, GA." Summary analysis report to the Army Environmental Policy Institute, Atlanta, GA, May.

Pearce, A.R., and Fischer, C.L.J. (2001b). "Resource guide for systems-based sustainability analysis of Building 170, Ft. McPherson, Atlanta, GA." Report to the Army Environmental Policy Institute, Atlanta, GA, May.

Pheng, L.S., and Teo, J.A. (2004). "Implementing total quality management in construction firms." *Journal of Management in Engineering* 20(1), 8–15.

Porter, M.E. (1985). *Competitive advantage: Creating and sustaining superior performance*. Free Press, New York.

Portland Energy Office. (1999). *Green building options study: The city's role in promoting resource efficient and healthy building practices*. City of Portland Office of Sustainable Development, Portland, OR.

PWC – PricewaterhouseCoopers. (2004). "Sustainability Management System Assessments." www.pwc.com/en_CA/ca/sustainability/publications/sustainability-management-system-assessments-2004-en.pdf (accessed 1 January 2011).

Pulaski, M.H., and Horman, M.J. (2005c). "Organizing constructability knowledge for design." *Journal of Construction Engineering and Management* 131(8), 911–19.

Pulaski, M.H., Horman, M.J., and Riley, D.R. (2006). "Constructability practices to manage sustainable building knowledge." *Journal of Architectural Engineering* 12(2), 83–92.

Pulaski, M., Pohlman, T., Horman, M., and Riley, D. (2003). "Synergies between sustainable design and construction at the Pentagon." Proceedings, 2003 ASCE Construction Research Congress.

Reed, W.G., and Gordon, E.B. (2000). "Integrated design and building process: What research and methodologies are needed?" *Building Research and Information* 28(5/6), 325–37.

Riley, D., Pexton, K., and Drilling, J. (2003). "Procurement of sustainable construction services in the United States: The contractor's role in green building." *UNEP Industry and Environment* 26(2/3), 66–9.

Rogers, E. (2003). *Diffusion of innovations*. 5th ed. Free Press, New York, NY.

Rohracher, H. (2001). "Managing the technical transition to sustainable construction of buildings: A socio-technical perspective." *Technology Analysis and Strategic Management* 13(1), 137–50.

Romm, J.J., and Browning, W.D. (1995). *Greening the building and the bottom line: Increasing productivity through energy-efficient design*. Rocky Mountain Institute, Snowmass, CO.

SBW Consulting. (2003). *Achieving silver LEED: Preliminary benefit-cost analysis for two city of seattle facilities*. City of Seattle, WA, www.seattle.gov/dpd/cms/groups/pan/@pan/@sustainablebliding/documents/webinformational/dpds-007573.pdf (accessed 10 October 2011).

Skanska. (2008). "Our sustainability agenda." www.skanska.com/upload/About per cent20Skanska/Sustainability/Skanska_Sustainability_Agenda_Overview_Master_Copy_Release_1_2008_05_15.pdf (accessed 3 January 2011).

Skanska. (2010). "Green thinking: there's more to building a green society than just building." http://group.skanska.com/sustainability/reports-publications/green-thinking-the-book/ (accessed 4 September 2016).

Skanska. (2015). "Green urban development reports." http://group.skanska.com/sustainability/reports-publications/green-urban-development-reports/. (accessed 4 September 2016).

Skanska. (2016). "How we define green." http://group.skanska.com/sustainability/green/how-we-define-green/ (accessed 4 September 2016).

Slaveykova, K.G. (2011). *Sustainable development's golden thread: The principle of integration*. Master Thesis, Utrecht University, Ultrecht, Netherlands.

SCTG – Sustainable Construction Task Group. (2002). "Reputation, risk, and reward: the business case for sustainability in the uk property markets." www.dti.gov.uk/construction/sustain/rrnr.pdf/ (accessed 31 December 2010).

Syphers, G., Baum, M., Bouton, D., and Sullens, W. (2003). *Managing the cost of green buildings*. State of California's Sustainable Building Task Force, Sacramento, CA.

Tan, Y., Shen, L., and Yao, H. (2010). "Sustainable construction practice and contractors' competitiveness: A preliminary study." *Habitat International* 1–6.

Taylor, B., and Wilkie, P. (2008). "Briefing: Sustainable construction through improved information flows." *Proceedings of the Institution of Civil Engineers – Engineering Sustainability* 161(ES4), 197–201.

UK Department for Communities & Local Government. (2012). "National Planning Policy Framework." http://planning guidance.communities.gov.uk (accessed 17 September 2016).

UNEP – United Nations Environment Programme. (2010). "Carrots and Sticks – Promoting Transparency and Sustainability: An update on trends in Voluntary and Mandatory Approaches to Sustainability Reporting." http://unep.fr/shared/publications/pdf/WEBx0161xPA-Carrots per cent20& per cent20Sticks per cent20II.pdf (accessed 6 November 2010).

UNWCED – United Nations World Commission on Environment and Development. (1987). *Our common future*. Oxford University Press, Oxford, UK.

USDOE – US Department of Energy. (2003). "The Business Case for Sustainable Design in Federal Facilities." Energy Efficiency and Renewable Energy Programme, Federal Energy Management Programme, USDOE, Washington, DC.

USDOE. (2010). "Database of State Incentives for Renewables and Efficiency (DSIRE)." www.dsireusa.org/. (accessed 31 December 2010).

USGBC – US Green Building Council. (2003). *Building momentum: National trends and prospects for high-performance*. USGBC, Washington, DC.

USGBC. (2004). "Making the Business Case for High Performance Green Buildings." USGBC, Washington, DC. USGBC. (2010). Public Policy Search www.usgbc.org/PublicPolicy/SearchPublicPolicies.aspx?PageID = 1776 (accessed 31 December 2010).

Vanegas, J.A., and Pearce, A.R. (2000). "Drivers for change: An organizational perspective on sustainable construction." *Proceedings, Construction Congress VI*, 20–22 February, Orlando, FL.

Wilson, A. (2005). "Making the case for green building." *Environmental Building News* 14(4), 1 ff.

Winter, S. (2004). *GSA LEED cost study*. United States General Services Administration (GSA), Washington, DC http://wbdg.org/ccb/GSAMAN/gsaleed.pdf (accessed 10 October 2011).

WGBC – World Green Business Council. (2013). "The Business Case for Green Building: A Review of the Costs and Benefits for Developers, Investors, and Occupants." http://group.skanska.com/globalassets/sustainability/reporting-publications/reports-on-green-building/business_case_for_green_building_report_web_2013–03–13.pdf (accessed 4 September 2016).

Chapter 4
Green rating systems

Objective measurement is critical to evaluate progress toward sustainability. A variety of rating systems, standards and information sources exist that can help decision makers evaluate sustainability-related attributes at multiple scales. Rating or measuring the sustainability of capital projects and their components serves a variety of purposes, including:

- baselining – establishing an initial measurement against which to calibrate future performance.
- benchmarking – providing a basis for comparison with competitors and identifying what is the state-of-the-art in a given practice.
- prioritization, decision support or selection – establishing a basis to choose and implement solutions with the objective of maximizing benefits.
- documentation – capturing evidence to support conformance with standards, compliance with regulations, or progress being made toward improvement.

Various tools have been developed to accomplish these purposes, at scales ranging from raw materials and individual building products through assemblies, buildings, developments, cities and business enterprises. Table 4.1 shows examples of such systems.

Threshold measurement tools establish a single result, value or outcome to represent the sustainability of a material, product or facility. For instance, forest products certified by the Forest Stewardship Council (FSC) either do or do not meet FSC certification requirements. Likewise, buildings receive a single certification level under the Leadership in Energy and Environmental Design (LEED) rating system (Certified, Silver, Gold or Platinum) based on the number of points achieved in the rating system. It is possible to review the specific credits and points awarded to a project under LEED. However, certified buildings are generally known by their certification level rather than the specific profile of points they have achieved.

Profile tools, in contrast, provide values for multiple indicators of sustainability. For instance, in the Building for Environmental and Economic Sustainability (BEES) tool, building materials are evaluated in terms of both economic and environmental variables, and both sets of information are presented as a profile. Likewise, the Athena lifecycle

Table 4.1 Examples of assessment tools at various scales

Threshold tools	Profile tools	Scale
Raw material	Forest Stewardship Council Certification (www.fsc.org)	BEES (www.bfrl.nist.gov/oae/software/bees/)
Product or assembly	GreenSeal (www.greenseal.org), GreenLabel Plus (www.carpet-rug.org)	Athena (www.athenasmi.ca)
Building	LEED (www.usgbc.org), GreenGlobes (www.greenglobes.com)	SB Tool (www.iisbe.org)
Development, city or region	LEED-ND (www.usgbc.org), Ecological Footprint (www.gdrc.org/uem/footprints/index.html), Carbon Footprint (various)	ICLEI Profile (www.iclei.org)
Enterprise	Ranking in Dow Jones Sustainability Index (www.sustainability-indexes.com), Carbon Footprint (various)	Global Reporting Initiative's Triple Bottom Line (www.globalreporting.org), SAM Corporate Sustainability Assessment (www.sam-group.com)

analysis tool provides an inventory of information about the lifecycle impacts of building assemblies and whole buildings, including embodied energy and energy consumption; global warming potential; air, water and land emissions; and weighted and absolute resource use. This chapter explores a sample of both types of green rating systems at the material or product, whole facility, site and organizational scales, including prevalent systems used in countries around the world.

Evaluating material and product sustainability

Green product rating and labelling systems are one way to make sense of the many material attributes affecting product sustainability. Many sources have developed information that helps building stakeholders understand how materials will perform and what impacts they have. The following subsections describe major information sources for green building materials and products, common labelling systems and logos, and overarching principles of lifecycle assessment for building.

Sources of green product information

Green product information is available in a variety of forms. For some attributes such as recycled content or sustainable harvest certification systems are available that allow third parties to review and objectively confirm a manufacturer's claim. Certification systems usually have multiple criteria that must be met in order for a product to be certified.

Multiple directories of green products are also available. Some directories may allow product manufacturers to suggest products to be listed and do not necessarily have any third party review or qualification of what products are listed. Other directories such as the GreenSpec database of green building products (www.buildinggreen.com) provide screening of products before they are listed. A product's qualifications are explicitly listed and verified as part of its listing in the GreenSpec database.

Handbook of Sustainable Building (Anink et al. 1996), *Green Building Handbook* (Howard et al. 1998) and *Environmental Resource Guide* (AIA 1990) all contain more general information about the environmental performance and lifecycle impacts of building materials in general. They do not provide specific manufacturer or brand information, but they are useful in understanding and comparing the environmental and social impacts of different types of construction materials such as steel vs. aluminium, for example.

Product labelling systems also provide various kinds of information about a product's green attributes. Labels may provide values for specific qualities about a product, such as grams per litre of volatile organic compounds or percentage of recycled content. They may display a logo that indicates the product has met some requirement. Lifecycle assessment is a still more detailed set of information about a product. It is used as the basis for some green product certifications. Resources such as the *Environmental Resource Guide* contain lifecycle assessment information about building products from their manufacture to end-of-lifecycle. This can also be called 'cradle to grave' analysis.

Lifecycle assessment (LCA)

The lifecycle of a construction material can be divided into three phases: upstream of use, the use of the material itself and downstream of use. Upstream impacts are all of the side-effects a product has before it is actually used. Depending on the product, these may include the effects of:

- harvesting the raw materials for the product from the natural environment.
- transporting those raw materials to factories for processing.
- processing the raw materials into finished components.
- transporting components to other factories for assembly.
- assembling the components into the finished product.
- packaging the finished product for transportation.
- transporting the finished product through the supply and distribution network to the project.

Complex products such as window and door assemblies have many steps along the way. Products such as stone or aggregate may have far fewer steps. Each of these steps consumes materials and energy, and each step also produces waste. Keeping track of all of the materials, energy and waste associated with the product is known as lifecycle assessment, and tools such as the Athena Impact Estimator for Buildings (Athena Institute 2016) or the BEES database (NIST 2010) are available to facilitate this analysis at the material, assembly or building level. Ecological Footprint Analysis (described further in UEM 2010) is a variant on lifecycle assessment applied at the community or development scale that expresses sustainability ratings in terms of the amount of land area it would take to provide the necessary resources and absorb the waste generated by a community. This area is then compared with the actual area taken up by the development as a basis to evaluate the sustainability of its continuing existence.

In addition to the materials and energy consumed to make a product, waste is also produced during harvesting of raw materials, manufacturing and transport. A product's emissions, including solid waste, liquid waste and air pollutants, are also tracked in a lifecycle analysis. Depending on how and where the product is manufactured, it may have significant emissions from the energy used in the manufacturing process. If it is complex with many components, the transportation required to bring those components together may be a significant factor. Emissions such as sedimentation from timber operations may be tracked as soil loss. All of these factors are considered when evaluating the ecological history of a building material. Building materials and products also have impacts while they are in use in a building, and at the end of their lifecycle. A lifecycle assessment takes all of these phases into account.

Lifecycle assessments can be used as a basis for comparing products and materials during the design process. Some building-scale rating systems incorporate this process as part of credits pertaining to the products and processes used to construct built facilities. For instance, the LEED rating system and the Living Building Challenge rating system, discussed in the next section, both refer to Red List materials (see *The Red List*) as products that should be avoided as part of the design and construction of a building. The International Living Future Institute has developed a product declaration format and database for information about products that can be used to determine whether or not a product meets Red List limitations (see www.declareproducts.com).

The Red List

Many materials used in buildings contain chemicals that are known to be harmful to humans and the natural environment. The International Living Future Institute's Red List (ILFI 2016) contains a list of harmful chemicals that must be avoided in projects to achieve Living Building Challenge certification. Other 'Red Lists' of building materials have also been developed, including the Banned Chemicals List developed for Cradle to Cradle certification and the Perkins and Will Transparency List. The GreenScreen List translator (www.greenscreenchemicals.org/) has been developed as a way to cross-reference red lists or other hazard screening lists across different countries and contexts.

The U.S. Green Building Council incorporates this concept as part of its *Building Product Disclosure and Optimization – Material Ingredients* credit in LEED v.4, referencing the GreenScreen Benchmark, Cradle to Cradle Certified, and the European Union's REACH Optimization third party standards (USGBC 2014). Of particular concern under the LEED Rating system are persistent bioaccumulative and toxic chemicals (PBTs) and persistent organic pollutants (POPs), both of which cannot be metabolized by organisms or easily broken down. Even small doses of these chemicals can accumulate over time and cause harm.

continued

Table 4.2 Potentially harmful chemicals and their uses in buildings

Chemical	Use(s) in buildings
Alkylphenols	Cleaning products, thermoplastics
Asbestos	Wall insulation, vinyl floor coverings, paint compounds, roofing, heat-resistant fabrics
Bisphenol A (BPA)	Polycarbonate plastics (high-impact glazing and coatings) and epoxy resins (used for pipe lining, coatings, adhesives and hard-surface substrates)
Cadmium	Batteries, pigments for paints and primers, zinc coatings for corrosion resistance, plastic stabilizer, cadmium telluride solar panels
Chlorinated Polyethylene and Chlorosulfonated Polyethylene (CSPE)	Geomembranes, roofing materials
Chlorobenzenes	Solvents, herbicides, rubber
Chlorofluorocarbons (CFCs) and Hydrochlorofluorocarbons (HCFCs)	Insulation, refrigeration systems, fire suppression systems
Chloroprene (Neoprene)	Gaskets, weatherstripping and other sealant products
Chromium VI	Metal plating/finishes, stainless steel, anti-corrosion agent, some pigments and dyes, wood preservatives, paints and primers
Chlorinated Polyvinyl Chloride (CPVC)	Hot and cold water piping, sprinkler piping, water storage
Formaldehyde	Glues and binders, particularly in engineered and composite wood and agrifibre products; insulation, furniture, adhesives and binders, paints and coatings
Halogenated Flame Retardants (HFRs)	Flexible and rigid foam, elastomers, wire coatings, some natural fibre fabrics and furniture, polyisocyanurate insulation
Lead	Piping, solder, plumbing fittings and connectors, roofing, flashing, pigments, dyes, leaded glass and PVC
Mercury	Thermostats, switches, fluorescent lamps, certain batteries
Polychlorinated Biphenyls (PCBs)	Old electrical transformers and devices; coolants, lubricants and insulators for electrical equipment
Perfluorinated Compounds (PFCs)	Water- and stain-repellent coatings, semiconductors
Phthalates	Plasticizers used in PVC flooring, upholstery, wall coverings, wire jacketing
Polyvinyl Chloride (PVC)	Plumbing pipe, carpet backing, textile coatings, wire and cable jacketing, flooring and wall coverings
Polyvinylidene Chloride (PVDC)	Filters, screens, outdoor furniture, artificial turf, underground materials
Short Chain Chlorinated Paraffins (SCCP)	Lubricants and coolants for metal cutting and forming
Wood treatments containing Creosote, Arsenic or Pentachlorophenol	Lumber/timber rated for exterior use and ground contact
Volatile Organic Compounds (VOCs) in wet applied products	Paints, adhesives, sealants

continued

Of these lists, the International Living Future Institute's list is one of the most well-known. Containing over 800 toxins as of its most recent update in July 2014, this list includes 22 categories of potentially harmful chemicals. Table 4.2 lists these major categories of chemicals along with typical building materials that may contain them (synthesized from ILFI 2016, Atlee 2010, Melton 2012, and Wikipedia).

Some building materials appear on one or more red lists but not all. Examples of building materials not included on the International Living Future Institute's Red List (shown in Table 4.2) include:

- Copper, which can create runoff toxic to aquatic species when used as exterior roofing or flashing, particularly in area with acid rain.
- Boric acid, which has been linked to reproductive toxicity in humans and is used as a natural flame retardant and biocide in cellulose and cotton insulation.
- Polyurethane foam, used for building insulation and air sealing as well as furniture and appliance applications, which contributes significant greenhouse gas emissions due to its blowing agents.

Concerns have been raised about the widespread use of antimicrobials in building materials to prevent mold and bacterial growth, as well as the increasing number of applications using nanotechnology. Biocides used for agriculture and landscaping, including pesticides, herbicides and fungicides, may also make their way into the built environment through their use on crops destined for biobased materials. Advances in green chemistry lead some to hope for a day when there will be 'green lists' of environmentally positive options instead of 'red lists' that should be avoided (USGBC 2014). Experience with Red Lists has shown that sometimes materials selected to replace known hazards can create unforeseen health hazards themselves (Melton 2015). Meanwhile, being aware of and avoiding red list materials where possible can help reduce risk of harm to both current and future generations.

For more information . . .

Atlee, J. (2010). "Chemistry for designers: Understanding hazards in building products." *Environmental Building News/Buildinggreen.com* (1 March).

ILFI – International Living Futures Institute. (2016). "The Red List." International Living Futures Institute web reference. http://living-future.org/redlist (accessed 10 August 2016).

Melton, P. (2012). "Why's that on the Red List?" *Environmental Building News/Buildinggreen.com* (30 April).

Melton, P. (2015). "The Problem with Red Lists." *Environmental Building News/Buildinggreen.com* (1 February).

USGBC – U.S. Green Building Council. (2014). *LEED Reference Guide for Building Design and Construction, v. 4.0.* USGBC, Washington, DC (1 July).

Another important consideration in evaluating a product is its durability and fitness for purpose. Choosing a product because its upstream environmental impacts are reduced is a suboptimal choice if the product does not perform well in use and has to be replaced frequently. Lifecycle costing is a method used to take into account a product's expected service life. Service life is the length of time a product is expected to remain in use before it fails or is otherwise unusable. Lifecycle costing of products and systems, discussed further in Chapter 9, takes into account the product's initial purchase price and cost of construction, plus the cost of operations, repairs and replacement over the lifecycle of the building. Products with longer service lives may cost more to purchase initially, but they can have a lower lifecycle cost when taking into account the need to purchase fewer products over time. The cost of operation is also a significant factor. While more durable or efficient systems may cost more initially, the savings in energy or other operational costs over their lifecycle may make them a better choice.

During operations and maintenance, building materials and products have differing requirements. Some products may require considerable maintenance using special cleaning products. Other products may require energy to operate and maintain. As mentioned earlier, choosing products with fewer upstream impacts is not the best choice if significant negative impacts are required during operations and maintenance. Although some products may cost more to purchase in the first place, their lifecycle maintenance requirements may be much less. The whole lifecycle of a product should be considered in making material selection decisions.

The end-of-lifecycle requirements of a product are also important in lifecycle assessment. Some products are difficult to dispose due to the materials they contain. For example, products containing multiple types of plastic may be difficult to recycle and dangerous to incinerate. Many types of plastics generate dangerous chemical compounds when burned, such as dioxins. These compounds are released into the air and can cause diseases such as asthma and certain types of cancer when breathed. They can also fall out and be absorbed into the soil or groundwater, creating more long-term pollution problems.

Other products, such as structural elements of steel or aluminium, may have a high recycled value at the end of their life. The length of the product's service life, the difficulty of disassembly and recycling, and the hazardous effects of disposal all should be considered as part of life-cycle assessment. lifecycle assessment can thus be useful in choosing products that are most sustainable for a building.

Product evaluation, rating and labelling systems

A variety of green product labels can be found on materials and products used in green building. Figure 4.1 shows the country of origin of some product ecolabels from various parts of the world, and a listing of these ecolabels and their attributes and requirements follow. Most are based to some degree on individual criteria that would be part of a larger lifecycle assessment, and some labels are based on a full lifecycle assessment of the product.

Blue Angel Certification (Germany)

The Blue Angel ecolabel was developed in 1978 by the German government to recognize products and services that are more environmentally friendly than their conventional counterpoints in terms of health, climate, water and/or resource use. Conformance with label requirements is established by the Environmental Label Jury, an independent

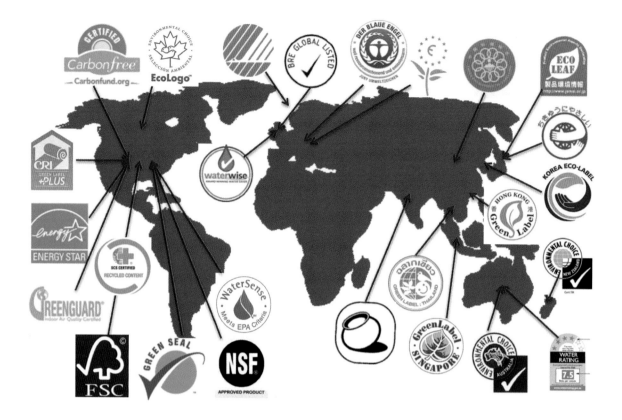

Figure 4.1
Selected product ecolabels and their countries of origin

decision-making body consisting of a variety of product stakeholders. It applies to building products, transportation, waste management and a variety of other products and services.

BRE Certified Environmental Profile (UK)

The Building Research Establishment Environmental Profiles Certification Scheme was developed to compare the environmental performance of building materials, products and systems. Environmental profiles developed under this system allow designers to compare products across a variety of parameters. Profiles are reviewed annually by an independent third party and recalculated every 3 years to ensure ongoing validity. They focus primarily on the production and end-of-lifecycle phases of the product lifecycle, and cover social attributes such as human rights, labour relations, training and education, and safety, along with environmental attributes including biodiversity, carbon/greenhouse gas emissions, energy, chemical and material use, toxics, solid waste and recycling, water quality and use, and wastewater.

CarbonFree Certification (US/UK)

The CarbonFree product certification label is awarded to products that have eliminated or offset all carbon emissions associated with the production of the product. It covers carbon and greenhouse gas emissions preceding the use phase of the product. The protocol for this certification was originally developed jointly by organizations in the United States and United Kingdom. It is presently administered by Carbonfund.org in the United States.

China Environmental Label (China)

Initiated by the Ministry of Environmental Protection of the People's Republic of China in 1993, this label provides environmental standards for construction materials, packaging and other products. Fifty six individual standards exist under this labelling system. It tracks environmental performance over the whole lifecycle of a product in terms of chemicals and raw materials used, energy type and quantity, solid waste and recycling, pesticides/herbicides/fungicides, water use and wastewater generation. It carries reciprocity with the Environmental Choice New Zealand label. Conformity is established by an independent third-party organization.

Eco-Leaf (Japan)

Developed in 2002, the eco-leaf label is awarded to building products, services and whole buildings based on lifecycle analysis consisting of (1) a product environmental aspects declaration, (2) a product environmental information data sheet and (3) a product data sheet. Organizations applying for this certification self-certify conformance with the ecolabel criteria, including carbon and greenhouse gas emissions, energy production and use, chemical and material use, toxics, water use and quality, solid waste and wastewater generation. The primary recipients of this label to date are electronic devices, but standards also exist for water metre boxes, grid electricity and insulation materials.

EcoLogo (Canada)

Originally developed by the government of Canada in 1988 but now recognized worldwide, EcoLogo now appears throughout North America, Mexico and the United Kingdom. It is applied to energy, buildings, waste management, building products and a variety of other products and services. The EcoLogo programme examines the entire lifecycle of products in developing product standards, although not all phases are included for all standards. Issues considered include biodiversity, carbon and greenhouse gas emissions, energy, genetically modified organisms, pesticide/herbicide/fungicide use, recycling and solid waste, water quality and use, and wastewater. Ninety-one product standards presently exist, with additional standards under development. Conformity is determined by an independent third party, and ongoing audits are conducted randomly/by surprise to ensure ongoing authenticity.

EcoMark (Japan)

Developed in 1989 by the Japan Environment Association (JEA), this standard applies to building products, cleaning products and other types of goods and considers the whole product lifecycle in terms of energy, material use, impacts on natural resources, pesticides/herbicides/fungicides use, recycling and solid waste, toxics and water quality/use. It is considered equivalent to the Environmental Choice New Zealand label and the Korean Ecolabel. Conformity is assessed by the JEA and audited annually.

Energy Star (USA)

Energy Star is a voluntary certification program developed by the US Environmental Protection Agency (USEPA). Certification is awarded for meeting standards with respect to energy use and associated impacts during the manufacturing and use phases of a product's lifecycle. It is applicable to appliances, building products, home electronics, office equipment and whole buildings. USEPA provides technical reviews of product attributes before awarding permission for a product to display the Energy Star logo. Products must perform in the top 25 per cent of all products in their class in order to display the Energy Star logo.

Environmental Choice (New Zealand)

This labelling programme provides voluntary environmental specifications for many different types of products, including construction materials, cleaning products, office products and services, furniture and fittings and others. It covers the whole lifecycle of a product in terms of carbon and greenhouse gas emissions and offsets, chemicals, energy production and use, material use, solid waste and recycling, toxics, water use and wastewater generation. It carries reciprocity with multiple other ecolabels including the Korean Ecolabel, certain Green Seal standards and the EcoMark label from Japan. Conformity with the standard is established by independent third parties.

EU Ecolabel (European Union)

This voluntary standard was developed in 1992 by the European Commission and applies to building products, cleaning products, retail goods, tourism, and other products and services. It is found throughout the European Union. It considers a variety of environmental issues throughout the whole product lifecycle, including chemical and material use, energy use, recycling and solid waste, toxics, water quality and use, and wastewater generation.

Forest Stewardship Council (FSC) Certification (USA)

The Forest Stewardship Council standards for certified sustainably harvested wood and wood products are used to indicate that labelled products are made from wood that is harvested in a sustainable way. Scientific Certification Systems is the organization that provides third-party certification of claims under FSC's standards. Products meeting FSC requirements must have chain-of-custody certification from initial harvest of wood products through final manufacture.

Good Environmental Choice (Australia)

Australia's Good Environmental Choice rating was developed in 1991 to meet ISO 14024 standards and is applicable to a wide variety of products including building materials and retail goods. Standards exist in 21 different categories ranging from adhesives and carpets to cleaning services and products, construction materials, agricultural products and sanitary products under this programme, which considers both social factors such as labour relations, work safety and environmental factors such as chemical and material use, energy, recycling, toxics and waste-water generation. This certification is equivalent to a number of standards in Asia and Oceania, and conformity is assured by third-party verification. The standard is recognized by the Green Building Council of Australia and the Infrastructure Sustainability Council of Australia.

GREENGUARD (USA)

The GREENGUARD and GREENGUARD Gold ecolabels indicate that a product meets strict chemical emissions standards and therefore contributes to healthier indoor air quality. GREENGUARD Certification is awarded by the GREENGUARD Environmental Institute, an independent, third-party certification organization. Products certified by the GREENGUARD Environmental Institute include construction materials such as paint, adhesives, insulation, flooring and furnishings. GREENGUARD Certified products can contribute to points in the LEED Green Building Rating System and satisfy the requirements of hundreds of green building codes, standards and procurement policies. GREENGUARD Gold focuses specifically on the safety of sensitive individuals such as children and the elderly, and provides stricter certification criteria to ensure that products are acceptable to use in environments such as schools and healthcare facilities.

Green Label Plus (USA)

The CRI Green Label and Green Label Plus are standards that apply to carpets, adhesives and cushions. Developed by the Carpet and Rug Institute, a trade association for the carpet industry, this logo indicates that a product has been tested by an independent third party to meet emissions standards for multiple chemicals including benzene and formaldehyde. The program also includes the Seal of Approval program for certification of vacuums as a measure of carpet cleaning effectiveness.

GreenSeal (USA)

The GreenSeal logo is awarded to products that meet requirements developed by the GreenSeal organization for environmental or human health performance. The GreenSeal logo is awarded to paint and other products for meeting indoor air quality standards. GreenSeal also certifies the environmental and energy performance of windows, occupancy sensors, cleaning chemicals and services and other building-related products. Each standard measures a single attribute of products of a particular type.

Hong Kong Green Label (Hong Kong)

The Hong Kong Green Label (HKGL) system is an environmentally preferable product certification scheme applied to a variety of products in Hong Kong including construction materials, computers and electrical appliances, cleaning products and packaging. It can be found throughout China, Hong Kong, India and Taiwan. Factors considered over the whole product lifecycle include carbon/greenhouse gas emissions, energy production and use, material and natural resource use, toxics, waste generation, water use and wastewater generation. Conformity with the standard is verified by independent third parties under the ISO 17025 testing standard, with ongoing random audits to ensure compliance.

Korean Ecolabel (Republic of Korea)

This labelling programme developed by the Korea Eco-Products Institute establishes and manages eco-product standards for building materials, forest products, packaging and a variety of other products. It provides environmental trend information to the public and otherwise supports the diffusion of environmentally friendly products in the market. Over 150 products and services are covered by standards under this labelling scheme, ranging from building materials and consumer products to automobile insurance and car sharing services. The scheme considers the production, use and end-of-lifecycle phases of product lifecycles in terms of chemicals and materials used, toxics, solid waste and recycling, energy use and wastewater generation. Conformance is assessed by independent third-party organizations under ISO 17025 standard testing procedures.

NSF International (USA)

The National Sanitation Foundation, now known as NSF International, develops US national standards and provides third-party certification for building products and other types of food, water and consumer products. Products certified by NSF International display the NSF certification mark on their label or packaging. Founded in 1944, NSF is well known for testing and certifying plumbing system components such as piping, valves and fittings and treatment system components.

The organization also certifies products using methods such as eco-efficiency analysis.

Nordic Swan Ecolabel (Denmark)

The Nordic Swan Ecolabel was established in 1989 and can be found throughout the Nordic countries of Denmark, Sweden, Finland, Norway and Iceland. It addresses the major lifecycle phases of building products, forest products and other retail goods in terms of safety considerations and environmental considerations, including carbon/greenhouse gas emissions, energy, material and chemical use, solid waste and recycling, toxics, water quality/use and wastewater generation. It is recognized as an equivalent to the EU Ecolabel, and is verified by independent third-party certification.

Scientific Certification Systems (USA)

Scientific Certification Systems (SCS) is a nonprofit organization that certifies a variety of environmental claims by product manufacturers, including recycled content, fair trade, legal harvest, carbon neutrality, responsible sourcing, biodegradability and others. Many products make claims of recycled content or other environmentally or socially preferable features. However, these claims should be third-party certified by a reputable organization such as SCS in order to be credible. For recycled content products, they should also indicate the source for the recycled content (pre- or post-consumer/recovered). Most often, re-cycled content is expressed in terms of weight percentage of the product. SCS certification is found throughout the world, including North and South America, Europe and Asia.

Singapore Green Label Scheme (Singapore)

This green labelling scheme applies to buildings and building products as well as services and the companies that provide such products and services. There are multiple standards for products in categories includ-ing building materials, solar powered products, interior finish materials, office supplies and general consumer products in the labelling scheme, which includes whole lifecycle consideration of social and environ-mental attributes including worker health and safety, community, housing and living conditions, animal welfare, biodiversity, greenhouse gas emissions and offsets, energy and material use, toxics, recycling and solid waste, and impacts to soil and water. Conformity is assessed by the Singapore Environment Council on an annual basis.

Thai Green Label (Thailand)

The Thai Green Label is an environmental certification that applies to building products, water, energy, transportation and other products, with 107 standards presently existing and several others under development. Developed in 1994, this standard evaluates products over the whole lifecycle in terms of animal welfare, carbon/greenhouse gas emissions and offsets, chemical and material use, genetically modified organisms, recycling and solid waste, impacts on soil and water, and a variety of other factors. It is considered equivalent to the Environmental Choice New Zealand certification. Initial certification and ongoing audits are provided by an independent third party.

Water Efficiency Labelling and Standards (WELS) Scheme (Australia)

This rating label provides water efficiency information for water-using household products in Australia. It applies to building products including showers, taps, flow controllers, toilets, urinals, hot water heaters and recirculation devices as well as appliances such as dishwashers and clothes washers, and considers water use during the use phase of the lifecycle. Verification is self-performed by the company seeking the WELS label for their products, with random audits by the Australian Federal Government to ensure compliance.

WaterSense (USA)

WaterSense is a programme developed by the USEPA to indicate water-conserving devices. WaterSense applies to residential and commercial plumbing products. All products displaying the WaterSense logo have been third-party certified to meet EPA standards. WaterSense specifications are available for toilets, urinals, faucets for bathroom sinks, showerheads, landscape irrigation controllers and new homes. New specifications are under development for other water-consuming devices such as water softeners, sprinklers and soil-based moisture control technologies. Rebates are also available for WaterSense-labelled products.

Waterwise Recommended Checkmark (UK)

The Waterwise Marque is awarded annually to products that reduce water use/waste and raise awareness of water efficiency in the United Kingdom. Established in 2006, this label is applied to buildings, retail products and building products in the United Kingdom. It focuses on water use during the use phase of the product lifecycle. Waterwise UK is the non-profit organization that manages the label and verifies conformance with standard requirements.

Given the many ecolabels that exist, determining which ones are credible is a key task for professionals making product selection decisions. Third party organizations such as the International Standards Organization (ISO) or the American National Standards Institute (ANSI) provide standards that should be met by ecolabelling organizations to provide credibility and auditability of results. Other things to consider in evaluating product information are discussed next.

Evaluating product sustainability information

As evidenced by the variety of evaluation, rating and labelling systems described in this section, there is no lack of information about sustainability in the marketplace. However, the quality, accuracy and reliability of that information may not always be suitable to support decision making. It is essential to critically evaluate sustainability-related claims before using information as part of decision making for the built environment (see *Reviewing sustainability claims*). The following subsections describe important considerations for evaluating sustainability information for products and with respect to rating systems overall.

Figure 4.2 shows an environment-related claim on a product label. This label is one example of the types of information that is provided to consumers about a product's environmental performance. In this case, the label does not make a product-specific claim but instead quotes a government agency about how general energy-saving practices can influence heating and cooling costs. By reference, the implication is that using this product can provide similar performance, although no specific evidence is provided.

Messages such as this one can be confusing to the uninformed consumer. Since the label does not make any product-specific claims about energy savings, it is technically correct. However, it does not provide the necessary details to evaluate what the product itself can do. When reviewing information about a product, it is important to consider the following critical factors.

First, what is the source of the information, and what is its interest in the product? Information from independent sources (such as universities or government organizations) may be more reliable than

> ### Reviewing sustainability claims
>
> - What is the source of the information, and what is their interest in the product?
> - What is the basis of the information? Is it supported by third-party verification?
> - Are standardized tests and protocols used to produce product data? Are they regularly reviewed and updated?
> - Are ecolabels from recognized, credible sources?
> - Is extraneous information provided that is vague, irrelevant or distracting?

Figure 4.2 Extraneous information on a product label can distract buyers from a product's actual features; this label implies a specific level of energy savings but does not provide enough information to substantiate the claim

Figure 4.3
Ecolabels do not necessarily indicate compliance with a recognized environmental standard. The earth smart label is not based on a recognized standard

information from sources with a vested interest in the product (such as trade associations). Second, what is the basis of the information? Is it supported by third-party verification, or is it provided by the product manufacturer without external review? Is contact or reference information provided for the third-party verifier? Third, are standardized tests from recognized agencies such as ISO or ANSI used to produce data about the product? Using standardized test methods to evaluate products allows them to be compared with other similar products. Fourth, if the product has an ecolabel, is the label from a recognized, credible third-party source? Labels such as the one shown in Figure 4.2 may appear to be supportive of environmental benefits, but there are no limitations on their use and they are not based on an objective standard of review. Finally, is extraneous information provided that is vague, irrelevant or distracting? The label shown in Figure 4.2 is an example of this type of information. It may lead to conclusions about the product that are not substantiated in fact.

Some labels and logos appear on product packaging without any reference to an established standard. For example, the earth smart label (Figure 4.3) has been trademarked as a graphic and is used on the labels of various building products. However, it is not linked with a recognized standard and does not require any testing or verification prior to use. As such, it is not a reliable basis for decision making during product selection.

Obtaining information about the environmental benefits or sustainability of a product, building or service can be challenging, although the sources of such information are growing. In some cases, product manufacturers are being driven to provide this information in the form of Environmental Product Declarations (EPDs) by building-scale rating systems such as LEED. In cases where an EPD is not available, a valuable source of product information is the manufacturers, vendors and

> ## Checklist of questions for product manufacturers
>
> - Where is the product made?
> - Where are the raw materials from?
> - What environmental claims does the manufacturer make regarding:
>   - recycled content?
>   - recyclability?
>   - rapid renewability?
>   - certified sustainable harvest?
>   - toxicity?
>   - biodegradability?
>   - low emissions?
> - Are these claims backed up by third-party certification?
> - What organization certifies the product or verifies manufacturer claims?
> - What documentation is available? Where can it be obtained?
> - Are any special certifications or training required to install the product and maintain the warranty?
> - Are there any special requirements for purchase or shipping?
> - Are there any comparable alternative products? How do they compare?

suppliers from whom the product is obtained (see *Checklist of questions for product manufacturers*).

When collecting product information, the project team should coordinate efforts within their companies. For commonly used products, this information can be collected and archived for use in future projects and green building rating certifications. This can save considerable time in subsequent projects.

When considering rating or measurement systems in general at the product, project, development, corporate, national or other level, it is similarly important to consider issues of accuracy, reliability and quality of information.

Misleading claims about the environmental benefits of a product, company or other entity are known as greenwash. Greenwash is the act of misleading customers regarding the environmental practices of a company or the environmental benefits of a product or service. Common types of greenwash identified by TerraChoice, an organization that specializes in validating environmental marketing claims, include the following (TerraChoice 2009):

- **Hidden trade-offs** – a claim based on a small number of environmental attributes without attention to other important environmental issues. An example is a product advertised as having recycled content packaging while the main content of the product contains highly toxic compounds.

- **Irrelevance** – a claim that may be truthful but is unimportant for decision making. 'CFC-free' is a good example – Chlorofluorocarbons are banned by law, so all products should be CFC-free, and most already are.
- **Lesser of two evils** – a claim that may be true within the product category but distracts the consumer from other negative product attributes. A commonly cited example is organic cigarettes.
- **Fibbing** – a claim that is completely false. Such claims are increasingly infrequent, but may still occur from time to time, especially in sectors where sustainability is a relatively new concept.
- **False labels** – a claim made using a counterfeit or fake ecolabel, or a claim of third-party endorsement where none exists.
- **No proof** – a claim that cannot be supported with easily accessible data or reliable third-party certification.
- **Vagueness** – a claim that is so broad that its meaning is likely to be misunderstood by the consumer. An example is the term 'all natural' – while 'natural' implies an inherently good and safe product, there are many substances such as mercury and formaldehyde that are natural but not safe in products.

As information sources about the sustainability of products, buildings and services become more widely available, so too will the challenges of ensuring that information is valid and reliable. This is also a challenge with regard to rating the sustainability of built facilities and infrastructure systems, discussed next.

Evaluating sustainability at the building scale

At the scale of whole buildings, various rating systems have been developed around the world to evaluate capital project sustainability, starting in the 1990s and continuing through today (Figure 4.4). Green building rating systems offer valuable information to owners and purchasers of buildings such as the building's expected or actual overall environmental performance. Green building rating systems also provide a way to help the construction market improve in terms of meeting green project goals. Most green building rating systems include explicit performance thresholds that buildings must meet in order to be certified. They also typically provide guidelines that help project teams meet or exceed those performance thresholds.

The following subsections describe some of the more well-known systems in more detail, beginning with rating systems originating in specific countries. International rating systems are discussed next, followed by a brief overview of local and regional rating systems for other types of buildings. It is important to note that some of the rating systems such as BREEAM and LEED have been widely applied in countries outside their country of origin. Rating systems relevant for a single country are noted as such in the description.

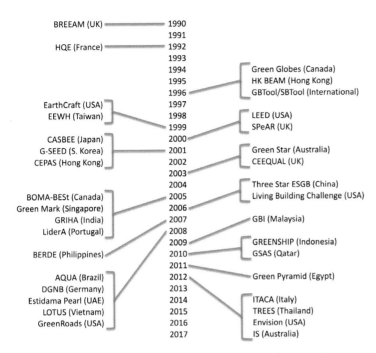

Figure 4.4 Timeline of Building Rating System Development (Pearce and Kleiner 2014)

National and international green building rating systems

Most of the green building rating systems on the market today were developed in a particular country to serve the specific needs of that country's buildings. However, many of these tools have been applied across multiple countries to meet the demand for green building ratings in countries that do not yet have their own rating system. Figure 4.5 shows the green building rating systems discussed further in this section, mapped by country of origin and initial application. The colours in Figure 4.5 are used to differentiate between continents.

Building Research Establishment Environmental Assessment Method (BREEAM)

One of the first assessment methods to be developed for evaluating project sustainability was BREEAM. As of 2017, over 561,700 buildings worldwide have been certified under the BREEAM family of rating systems, and well over two million have been registered for certification in 77 different countries (www.breeam.org). Developed in the United Kingdom, specific versions of this rating system exist for communities, infrastructure, new construction, in-use buildings and refurbishment/fit-out, and it has also been adapted for use in other contexts to take into account environmental weightings; local codes, standards and building

Figure 4.5
Green building rating systems by country of origin

methods; and important local environmental issues. BREEAM has been applied to a wide range of building types, including offices, retail, education, prisons, courts, healthcare and industrial facilities, and housing, and it was a guiding standard for development of venues for the 2012 Olympic Games held in London (see *Case Study: London 2012 Olympic and Paralympic Games Venues*). It awards points or credits for performance above regulation in ten major areas:

- **Energy**, covering both operational energy and carbon dioxide generated as a result of the facility.
- **Management**, which deals with management policy, commissioning, site management and procurement.
- **Health and well-being**, covering both indoor and external issues such as noise, light and air quality.
- **Transport**, including transport-related carbon dioxide and factors related to the location of the project.
- **Water**, including both consumption and efficiency inside and outside the facility.
- **Materials**, including embodied impacts of building materials over their lifecycle such as embodied carbon dioxide.

- **Waste**, including construction resource efficiency as well as operational waste management and minimization.
- **Land use**, dealing with the choice of site, the building footprint, enhancement of ecology and long-term biodiversity management.
- **Pollution**, including external air and water pollution resulting from the project.
- **Innovation**, including sustainability-related actions not covered elsewhere.

Case study: London 2012 Olympic and Paralympic Games Venues, UK

As part of its selection to host the 2012 Olympic Games, the city of London committed to the International Organizing Committee of the Olympics that it would conduct the first sustainable Olympic and Paralympic Games to be held. London promised to be the first summer host city to embed sustainability in its planning from the very beginning and to set the standard for all subsequent games. The guiding principles for the games were to:

- use venues already existing in the United Kingdom wherever possible.
- only make permanent structures that would have a long-term use after the Games.
- build temporary structures for everything else.

The London team planned not only to push the boundaries of what had previously been done, but also to capture lessons learned that could be publicly shared to provide a 'learning legacy' of the experience. The intent was to use the Games to help to regenerate East London and improve quality of life there while encouraging more sustainable living across the whole of the United Kingdom. The vision that guided the development of all venues for the game was to use the power of the Games to inspire change in everyone from athletes to spectators to supporters and organizers of the games, not only with respect to ecological sustainability, but with respect to other critical sustainability factors as well, such as accessibility and attitudes toward disability.

The venues were developed using ten principles of sustainability created by the World Wildlife Federation (WWF) and BioRegional Development Group (see *Ten principles of sustainability*). The British Standard for Sustainable Events (BS 8901) was also used as a tool to ensure that the games were hosted sustainably, and London 2012 developed its own guidelines for corporate and public events (2010) and sustainable procurement and sourcing (2011). With these principles in mind, the Sustainability Plan for the games focused on five key themes (2009):

- **Climate change** – minimizing greenhouse gas emissions and ensuring legacy facilities are able to cope with the impacts of climate change.
- **Waste** – minimizing waste at every stage of the project, ensuring no waste is sent to landfill during Games-time, and encouraging the development of new waste processing infrastructure in East London.
- **Biodiversity** – minimizing the impact of the Games on wildlife and their habitats in and around Games venues, leaving a legacy of enhanced habitats where possible.
- **Inclusion** – promoting access for all and celebrating the diversity of London and the United Kingdom, creating new employment, training and business opportunities.
- **Healthy living** – inspiring people across the country to take up sport and develop active, healthy and sustainable lifestyles.

continued

Ten principles of sustainability

- **Zero carbon** – making buildings more energy-efficient and delivering all energy with renewable technologies.
- **Zero waste** – reducing waste, reusing where possible and ultimately sending zero waste to landfill.
- **Sustainable transport** – encouraging low carbon modes of transport to reduce emissions; reducing the need to travel.
- **Sustainable materials** – using sustainable and healthy products, such as those with low embodied energy, sourced locally, made from renewable or waste resources.
- **Local and sustainable food** – choosing low-impact, local, seasonal and organic diets and reducing food waste.
- **Sustainable water** – using water more efficiently in buildings and in the products we buy; tackling local flooding and water course pollution.
- **Land and wildlife** – protecting and restoring existing biodiversity and natural habitats through appropriate land use and integration into the built environment.
- **Culture and heritage** – reviving local identity and wisdom; supporting and participating in the arts.
- **Equity and local economy** – creating bioregional economies that support fair employment, inclusive communities and international fair trade.
- **Health and happiness** – encouraging active, sociable, meaningful lives to promote good health and well-being.

(WWF/BioRegional Development)

The sustainability plan focused on three core phases of delivery – building the stage for the Games, staging the Games, and building a lasting and sustainable legacy after the Games. As a result of these goals and objectives, the design and construction of venues for the Games incorporated a variety of strategies and practices to move toward greater sustainability.

Regular environmental monitoring and surveys were undertaken throughout to manage the effects of construction on the local environment and the people who lived and worked around the Olympic Park. Considerations included air quality, noise, ecology, water quality, flooding and transport impacts. Detailed archaeological documentation of the history of venue sites was also conducted to preserve the site heritage. Contaminated soil was remediated in some venues such as the Aquatics Centre site, and included remediation of asbestos and other contaminants. Old industrial buildings were demolished on site, with recovery of 98.5 per cent of materials for reuse or recycling.

External standards were used for multiple purposes to guide development. Third-party certification was undertaken to ISO 14001 standards for the Olympic Development Authority's Environment and Sustainability Management System. Third-party rating and certification of development was also conducted with the aim for all Olympic Park venues to achieve a BREEAM rating of 'Excellent', the Athlete's Village to achieve the Code for Sustainable Homes Level 4, and all registered development sites targeting CEEQUAL 'Very Good' rating.

Aggressive measures were incorporated for carbon reduction, including a 31 per cent reduction in carbon emissions for the Velodrome over 2006 building regulations and use of combined

continued

heat/power and biomass systems for energy needs. The Velodrome used compact design, natural lighting, natural ventilation for passive cooling and other strategies to achieve these goals. It also incorporates a rainwater harvesting and supply system and water-efficient fixtures. The design team was able to optimize both the foundation system and cable net system to save 1,000 tonnes of steel in the construction of the facility.

Energy retrofits were undertaken in surrounding communities both to strengthen social sustainability and to meet carbon-reduction goals. A target of 100 per cent use of public transport during the Games was also set to reduce negative impacts and overall carbon footprint. Significant improvements were made to local Underground facilities to support this goal, including a new lighting system dubbed 'spectator-powered lighting' that used electricity generated by human footfalls on surfaces in the station.

Figure 4.6 The London 2012 Olympic Site

Figure 4.7 The Velodrome cycling facility incorporated significant material savings in its construction and is the most energy-efficient building on site: rainwater is harvested from its iconic roof surface

continued

Figure 4.8 Aerial view of the Olympic Village showing the newly developed Parklands alongside the River Lea

A new water recycling treatment works was built to treat and repurpose reclaimed wastewater and black water to a level that exceeds bathing water standards. Water from this plant was used for irrigation and toilet flushing. Careful measures were also taken for stormwater management and optimized design of surface water drainage outfalls to minimize negative impacts on local waterways. A waterways restoration strategy was devised for the whole park that addressed water quality, flood-risk management, navigation, biodiversity and recreation. During construction, discharges of wastewater were made to a special manifold system that incorporated metering and monitoring to ensure that unauthorized discharges could not be made. The manifolds also incorporated break tanks that allowed silt and gravel to settle out rather than being discharged into the sewer network.

A goal of at least 50 per cent by weight was set for delivery of construction materials to the Olympic Park using sustainable transport such as barges. By project completion, 63 per cent of materials had been delivered using this low impact method. Recycled aggregate was incorporated in precast concrete units used in the Olympic Stadium and Aquatics Centre seating terraces, temporary bridge decks and Handball Arena. Recycled aggregate was also incorporated into the cladding of bridge abutments and retaining wall facings in the Olympic park to create a common visual identity. All temporary venues were designed to use bolted steel truss systems that could be reused when the Games were over.

Over 45 hectares of new wildlife habitat was set aside as part of the 250 hectare site, with the potential to become Sites of Importance for Nature Conservation (SINC). Seed stock was collected from the site prior to redevelopment for subsequent use in creating habitat, and local plantings and native species were incorporated throughout. Invasive species such as Japanese knotweed and giant hogweed were eradicated on site, and 675 new bird and bat boxes were added to improve habitat for wildlife.

From a workforce standpoint, promotion of diversity and equity in workforce hiring and contracting was a key goal, including programmes such as the Women into Construction project, training programmes through the National Skills Academy for Construction, and hiring unemployed people from the local borough as part of the project. Over 40,000 people were employed to work

continued

on the project by the time it was completed. Health and safety of workers was an important priority, and a target of zero fatalities was maintained during all Games-related construction. By completion, the accident rate remained well below the industry average. To manage potential safety hazards during excavation, alternative techniques such as suction excavation were employed for small excavations to reduce the risk of cable strikes, and a common standard was developed for the safe use of quick hitches when detaching buckets on excavators. Other hazard mitigation measures were also undertaken throughout the project. A set of Visual Standards was developed through consultation with contractors to establish a strong health and safety culture on the project. These standards presented photographic images of good and bad practices based on actual hazards identified for the project, and a manual was produced and distributed to all contractors on site. The standards were used as part of regular compliance reviews, and workers were encouraged to notify supervisors if working in areas where the standards were not observed.

PVC Policy Implementation

In recognition of the harmful health impacts of phthalates in PVC-based products, a policy was developed for optimizing use of PVC as part of buildings and facilities in games venues. Each PVC-based product, including membrane wraps, flooring, cabling and piping, had to be carefully evaluated to determine whether it was the most appropriate choice. All PVC-based materials incorporated in the project had to be justified in a written report showing that appropriate mitigation measures were used to ensure safe and healthy use.

As a result of the policy, manufacturers developed a new non-phthalate PVC that was used in membrane wraps and roofing for construction of several venues including the Water Polo Arena, Aquatics Centre, Eton Manor and the Royal Artillery Barracks. Wraps also had to have 30 per cent recycled content to meet project goals, which posed a challenge for achieving required engineering properties. The phthalate-free fabric had a 15 per cent cost premium over conventional fabric, and take back and reuse was difficult given the way in which the material was used. While overall the policy was welcomed by the plastics industry as an opportunity to innovate, it was recognized that in some cases, PVC-based materials remain the best choice. Some products were used that ultimately had a higher lifecycle costs and environmental impacts due to shipping, fabrication and installation. However, on a project of this magnitude, these impacts were judged to be worthwhile in light of the positive impacts on industry innovation.

(london2012.com/learninglegacy, ODA 2010/374)

Sources

London 2012. (2009). *London 2012 Sustainability Plan: Towards a one planet 2012*. http://learninglegacy.independent.gov.uk/documents/pdfs/sustainability/1-cp-london-2012-sustainability-plan-2nd-edition.pdf (accessed 28 December 2016).

London Organising Committee of the Olympic Games and Paralympic Games. (2011). *LOCOG Sustainable Sourcing Guide*, 3rd ed. http://learninglegacy.independent.gov.uk/documents/pdfs/sustainability/cp-locog-sustainable-sourcing-code.pdf (accessed 28 December 2016).

London 2012. (2011). *London 2012 Sustainability Report – A blueprint for change*. http://learninglegacy.independent.gov.uk/documents/pdfs/sustainability/2-london-2012-sustainability-report-a-blueprint-for-change.pdf (accessed 28 December 2016).

continued

London 2012. (2012). *London 2012 Pre-Games Sustainability Report – Delivering change.* http://learninglegacy. independent.gov.uk/documents/pdfs/sustainability/3-pre-games-sustainability-report-neutral.pdf (accessed 28 December 2016).

London 2012. (2012). *London 2012 Post-Games Sustainability Report – A Legacy of change.* http://learning legacy.independent.gov.uk/documents/pdfs/sustainability/5-london-2012-post-games-sustainability-report-interactive-12-12-12.pdf (accessed 28 December 2016).

London 2012. (2012). *Sustainability Guidelines – Corporate and public events,* 3rd ed. http://learninglegacy.independent. gov.uk/documents/pdfs/sustainability/cp-london-2012-sustainability-guidelines-for-corporate-and-public-events.pdf (accessed 28 December 2016).

London 2012. (2016). *Learning Legacy.* http://learninglegacy.independent.gov.uk/index.php (accessed 28 December 2016).

Source: All images courtesy of London 2012; used with permission

The total number of points or credits in each category is multiplied by a weighting factor reflecting the relative importance of each represented issue. The total score for the building is then translated into a rating on a scale of one to five stars corresponding to pass, good, very good, excellent and outstanding. Assessment of BREEAM ratings is provided by trained and licensed assessors who produce a report outlining the project's performance and its overall rating.

Comprehensive Assessment System for Building Environment Efficiency (CASBEE)

CASBEE was developed by the Japan Sustainable Building Consortium and Japan Green Building Council in conjunction with several other Japanese government agencies including the Ministry of Land, Infrastructure and Transportation (MLIT) in 2001. The CASBEE system measures both the improvement in living amenities for building users within a property and the negative environmental impacts caused by the building and its construction within and outside the property.

CASBEE projects are rated in five different categories: excellent (S), very good (A), good (B+), fairly poor (B-) and poor (C). Versions of the rating system exist for three different scales: construction (housing and buildings), urban (town management) and city management. Multiple tools exist specifically for housing (new detached, existing detached, housing units, housing renovation and housing health), and project-scale tools include new construction, existing buildings and renovation. There are also tools for market promotion, commercial interiors, temporary construction and community health.

Overall, CASBEE was designed for assessment of buildings over their whole lifecycle, from pre-design through new construction, operations and eventual renovation. The system is based on two main considerations: the environmental load (L) the building places on its environment and quality (Q) of the built environment that is delivered through amenities for building users. CASBEE develops its rating of projects by evaluating the building environmental efficiency (BEE), which is measured as the ratio of quality delivered to environmental load imposed (Q/L). Specific considerations for quality of the built environment (Q) include:

- indoor environment
- quality of service
- outdoor environment on site.

Considerations for environmental load (L) include:

- energy
- resources and materials
- negative impacts to off-site environment.

Since the introduction of CASBEE in Japan, multiple local governments across Japan have mandated its use or have created incentive programmes to help promote its adoption. For example, the cities of Osaka and Nagoya subsidize highly rated projects in their city limits. The city of Kawasaki provides lower interest rate home loans to promote the CASBEE rating system, and other cities also provide some flexibility in obtaining building permits and expediting the review process. CASBEE has been primarily used in the country of Japan, although its principles can also be applied in other contexts. As of July 2016, 330 commercial buildings, 119 houses and 92 other projects have been certified.

Estidama Pearl

The Estidama Pearl rating system was developed in 2008 by the emirate of Abu Dhabi as a way to promote sustainability by addressing the unique environmental, cultural, social and economic needs of the Middle East. The word 'estidama' means sustainability in Arabic, and the Estidama program is the larger initiative of which the Pearl rating system is a part. Five pearls are available under the rating system, with one pearl being the lowest level of certification. As of 2010, all new development applications in Abu Dhabi were required to achieve at least one pearl, and all new private buildings must achieve one pearl. New government-led projects, including schools, mosques and government buildings, must achieve at least two pearls.

The Pearl rating system has multiple versions for different types of projects, including communities, buildings and villas (residential). The Buildings version of the rating system features seven categories of criteria, including:

- **Integrated Development Process (IDP)**, pertaining to post-occupancy considerations, lifecycle costing, worker accommodations, commissioning and others.

- **Natural Systems (NS)**, including assessment, protection, remediation and enhancement of natural ecology.
- **Livable Buildings (LBi and LBo)**, including criteria for enhancing livability both outdoors through outdoor amenities and indoors through indoor environmental quality.
- **Precious Water (PW)**, including indoor and exterior water use reduction, monitoring and management.
- **Resourceful Energy (RE)**, including improved energy performance, renewable energy and management of ozone, and global warming impacts.
- **Stewarding Materials (SM)**, including use of environmentally preferable materials, waste management, design for adaptability, durability, and disassembly and others.
- **Innovating Practice (IP)**, including innovative cultural and regional practices.

The Pearl rating system has been applied to projects as part of the overall Abu Dhabi Vision 2030, including the Masdar case study in Chapter 3 (p.130). The Pearl rating system is being designed to coordinate with municipal building codes and addresses four primary pillars of sustainability: environment, economy, society and culture. The cultural component has been an important distinction of the Estidama system compared to other rating systems, with its aim of preserving local values and regional culture from the very start.

Green Globes

Initially developed in Canada, Green Globes was modelled after the BREEAM rating system as an offshoot of the BREEAM Canada for Existing Buildings Rating System. Presently deployed both in Canada and the United States, Green Globes has been established as an official standard recognized by the American National Standards Institute (ANSI). The primary market for Green Globes is large developers, institutional owners and property management companies.

In contrast with BREEAM, Green Globes uses an online questionnaire-based approach to rating, with information provided by the project team. The rating system also provides feedback on the environmental impact of project decisions and suggests advice and resources to improve performance. One of the primary advantages claimed by the developers of Green Globes is that it is written in plain language and can be used by a wide range of people with different experience levels. This applicability to laypeople contrasts with BREEAM and other rating systems such as LEED, where expertise and formal training, while not required, are necessary to apply the rating system effectively to a project.

The system can be applied to both new and existing buildings and is suitable for projects including offices, multi-family facilities and institutional buildings such as schools, universities and libraries. New versions are also available for sustainable interiors and green and productive workplaces. The existing building standard in Canada is licensed to and administered by the Building Owners and Managers Association (BOMA), where it is administered as BOMA BESt (Building Environmental Standards). It can also be applied across portfolios to compare the performance of multiple buildings of a single owner. Green Globes can be used throughout the design process to evaluate the impacts of project decisions on point scores. Third-party verification and certification is provided by trained regional verifiers who audit the project information provided by the team.

Ratings under Green Globes are expressed as one, two, three or four globes, indicating increasing levels of environmental performance. Projects are evaluated based on a 1000-point scale including variables in seven categories (management, site, energy, water, resources, emissions and indoor environment). Data submitted online by a project team is verified both through document review and an on-site walk through assessment. The on-site verification by a third party is another major difference between rating systems such as LEED and Green Globes. Green Globes has been applied primarily in the United States and Canada.

Green Rating for Integrated Habitat Assessment (GRIHA)

The GRIHA standard for buildings was developed in 2005 by The Energy and Resources Institute (TERI) in India. It is administered by the GRIHA Council headquartered in New Delhi. The GRIHA rating system family addresses multiple project types, including:

- **Pre-certification** – available for all projects seeking to demonstrate intent and applicability for subsequent rating under the full GRIHA rating system.
- **Small Versatile Affordable (SVA)** – applies to projects less than 2500 sq m in built-up areas and is a streamlined rating tool to meet the needs and capabilities of smaller projects.
- **Rating** – the full GRIHA rating system applicable to all projects.
- **Large development** – applies to development-scale projects larger than 50 hectares such as residential townships, educational and institutional campuses, medical colleges and hospital complexes, special economic zones and hotels/resorts.
- **Prakriti** – applies to existing schools in India.

The full GRIHA rating system consists of 34 criteria, which are either fully mandatory, partly mandatory or optional. One to five stars may be achieved based on the number of points earned in each criteria. A minimum of 50 points must be earned to receive certification. Major categories of criteria include:

- **Site planning**, including site selection, low-impact design, minimizing urban heat islands and site imperviousness.

- **Construction Management**, including air and water pollution control, landscape preservation during construction, and construction management practices.
- **Energy**, including energy efficiency, use of renewable energy, and low ozone-depleting potential materials.
- **Occupant comfort and well-being**, including indoor comfort (visual, thermal and acoustic), maintaining indoor air quality and use of low-VOC materials.
- **Water**, including use of low-flow fixtures, reducing landscape water demand, water quality, on-site reuse and rainwater recharge.
- **Sustainable building materials**, including recycled materials, reduction in embodied energy, and use of low-environmental impact materials.
- **Solid waste management**, including avoiding post-construction landfill use and treating organic waste on site.
- **Socio-economic strategies**, including labour safety and sanitation, design for universal accessibility, dedicated facilities for service staff, and increased environmental awareness.
- **Performance monitoring and validation**, including smart metering and monitoring, O & M protocols, post-occupancy performance assessment.
- **Innovations**.

GRIHA also maintains an online Product Catalogue to provide information on green products in 32 different categories that can be used to develop GRIHA-compliant projects. Product categories include a variety of construction materials, interior finishes and building systems and controls.

Green Star Australia

The Green Star environmental rating system for buildings was developed by the Green Building Council of Australia (GBCA) based on existing systems and tools in overseas markets including the British BREEAM system and the North American LEED system. Green Star was Australia's first comprehensive rating system for evaluating the environmental design and performance of Australian buildings based on a number of criteria, including management, indoor environmental quality, energy, water, materials, land use and ecology, emissions and innovation. These categories are divided into credits, each of which addresses an initiative that improves or has the potential to improve environmental performance. Points are awarded under each credit for actions that demonstrate that the project has met the overall objectives of Green Star. Once all claimed credits in each category are assessed, a percentage score is calculated and Green Star environmental weighting factors are the applied. The Green Star rating tool uses six stars to predictively assess environmental performance of buildings. Projects that obtain a predicted rating of one, two or three stars are not eligible for formal certification. Projects that obtain a predicted four-star rating or above are eligible to apply for formal certification, as follows:

- four-star Green Star certified rating recognizes and rewards 'Best practice'.
- five-star Green Star certified rating recognizes and rewards 'Australian excellence'.
- six-star Green Star certified rating recognizes and rewards 'World leadership'.

To date, Green Star has been applied to multiple different types of building including education, healthcare, multi-unit residential, industrial, office, office interiors, retail centre, office design and office as-built. The system presently has four different tools corresponding to different scales and lifecycle phases, including precinct planning and development (Communities), building design and construction (Design and As-built), fitout design and construction (Interiors), and operations and maintenance (Performance).

Since its inception in 2003 by the Green Building Council Australia, 1322 projects have been certified as of September 2016. Green Star Australia complements the National Australian Built Environmental Rating System (NABERS) managed by the New South Wales Department of Environment and Climate Change (DECC), which rates existing buildings based on their energy performance and climate change impact during operations. Green Star has also been adopted by other nations such as South Africa and New Zealand, where it has been adapted to fit local contexts.

Green Star New Zealand

Based heavily on the Australian version of Green Star, the first Green Star New Zealand rating system was launched in New Zealand in 2007, focusing on office design, and has since evolved to cover multiple building types and multiple phases of the project lifecycle. The Green Star NZ rating system is designed as a three-stage system including design, build and use/performance. This structure reflects the process used by industry to deliver a built facility and also captures unique opportunities to realize environmental benefits at each stage.

In the design phase, the tool rates the design of the building as a reflection of what will ultimately be built. This phase is important because decisions made during design greatly influence subsequent practices for sustainability during later lifecycle phases. Achievement of the design phase certification generally happens prior to construction. Documentation includes submittal of design documents, reports from the design team, and statements from owners demonstrating that Green Star requirements have been met. Green Star Accredited Professionals (GSAPs) are typically involved in compiling documentation for the projects.

In the built phase, the focus is on confirming environmental initiatives proposed in the design phase and measuring and assessing what has actually been built. The credit aims and criteria are the same as for design, but in the built phase, documentation includes actual evidence from the construction phase of the project, including supplier and subcontractor submittals, commissioning reports, as-built drawings, product data sheets and other information. Projects can be certified under Green Star NZ during the built phase without having previously completed design phase certification, and they can also apply for different credits than pursued in the design phase if changes occur during construction.

Requirements are grouped into nine major categories, as follows:

- **Management**, covering issues related to commissioning, use of a Green Star Accredited Professional, development of a building user's guide, and environmental and waste management planning.
- **Indoor environmental quality**, covering indoor air quality, ventilation, daylighting, thermal comfort, views, noise levels and other issues.
- **Energy**, addressing overall energy performance, peak demand reduction, lighting power density and zoning, sub-metering and carbon dioxide emissions.
- **Transport**, including provision of car parking, cyclist and pedestrian facilities, public transport and small parking spaces.
- **Water**, including potable water efficiency, water meters, landscape water efficiency and cooling tower water consumption.
- **Materials**, including recycling, reuse of building components, recycled content, minimization of PVC, sustainable timber, carpets, paints, thermal insulation and floor coverings.
- **Land use and ecology**, including ecological value of site, reuse of land, reclamation of contaminated land, change in ecological value and topsoil issues.
- **Emissions**, including ozone depletion and global warming potential, recovery of refrigerants, pollution of watercourses, light pollution and cooling towers.
- **Innovation**, covering strategies not already addressed in other categories, greatly exceeding Green Star benchmarks, and other environmental design initiatives.

As with Green Star Australia, points are awarded under each credit for actions that demonstrate that the project has met the overall objectives of Green Star NZ. After all claimed credits in each category are assessed, a percentage score is calculated and Green Star environmental weighting factors are applied. The Green Star NZ rating tool uses six stars to measure environmental performance of buildings. Projects that obtain a predicted rating of one, two or three stars are not eligible for formal certification. Projects that obtain a predicted four-star rating or above are eligible to apply for formal certification on the same basis as Green Star Australia.

In 2015, Green Star New Zealand released a new version of its rating tool (v3), with changes to credits within materials, water and energy categories and a new approach to innovation credits. In this new version of the rating system, a new series of innovation challenges were developed that can be pursued by a project team, with additional challenges under development (see *Green Star New Zealand innovation challenges*). Teams can also obtain credit for pursuing innovation credits in the conventional way, for leading innovation in the marketplace, exceeding benchmarks in existing Green Star credits, or undertaking activities outside the scope of the existing tool.

> **Green Star New Zealand innovation challenges**
>
> - Adaptation and resilience
> - Contractor education
> - Culture, heritage and identity
> - Financial transparency
> - Market intelligence and research
> - Marketing excellence
> - Material lifecycle impacts
> - Social return on investment
> - Environmental product declarations
> - Earthquake resilience
> - Circular economy model office

Leadership in Energy and Environmental Design (LEED)

Of all the tools in the market today, the LEED rating system is one of the most well known worldwide for rating and evaluating green buildings, and has been applied to projects in both its home country, the United States, and abroad. Initially modelled after the BREEAM system, the LEED rating system applies to a wide variety of project types, and over 99,000 projects have received certification as of September 2016. LEED is designed to be applicable in different climates and contexts throughout the United States and has been awarded to buildings in other countries as well. LEED has four levels of certification (Certified, Silver, Gold and Platinum). It is administered by the Green Building Certification Institute (GBCI). Presently in version 4.0, the LEED rating system consists of a series of performance goals and requirements in nine categories, as follows:

- **Integrative process (IP)**, focusing on the involvement of multiple stakeholders and disciplines during project planning, design and delivery.
- **Location and transportation (LT)**, covering issues related to the selection of the project site, and amenities it features that affect transportation and mobility.
- **Sustainable sites (SS)**, covering issues related to the characteristics of the project site, impacts to the site during construction and impacts resulting from building operations.
- **Water efficiency (WE)**, which deals with water consumption and wastewater generation by the building in operation.
- **Energy and atmosphere (EA)**, which addresses all aspects of the building's energy performance, energy source(s) and atmospheric impacts.
- **Materials and resources (MR)**, which pertains to the sources and types of materials used on the project, the amount of waste generated and the degree to which the project makes use of existing buildings.

- **Indoor environmental quality (EQ)**, which covers aspects of the building's indoor environment ranging from ventilation to air quality to daylight and views.
- **Innovation in design (ID)**, which rewards the project for going beyond the minimum credit requirements and for using a LEED accredited professional.
- **Regional priority (RP)**, where credits from other categories of special relevance to the project's location are given 'extra credit'.

Each category consists of a series of credits and points that can be earned by a project. Most categories also have prerequisites that the project must meet to be considered for certification. A project must meet all prerequisites in all categories in order to pursue certification. It also must obtain a minimum number of energy performance credits to exceed energy code requirements. Individual versions of LEED for five major project types are now available:

- **Building Design + Construction (BD+C)** – focuses on buildings being built new or undergoing a major renovation, including new construction, core and shell, schools, retail, hospitality, data centres, warehouses and distribution centres and healthcare.
- **Interior Design + Construction (ID+C)** – applies to interior fit-out projects, including commercial interiors, retail and hospitality.
- **Operations + Maintenance (O+M)** – applies to existing buildings undergoing improvement work or under normal operations, including existing buildings, schools, retail, hospitality, data centres and warehouses/distribution centres.
- **Neighborhood Development (ND)** – applies to new land development or redevelopment projects containing residential, non-residential or a mix of uses.
- **Homes** – applies to single family homes and low- or mid-rise multi-family housing. Larger projects are covered under BD+C.

Some of these systems are complementary and reinforce each other. For instance, tenant spaces seeking certification under ID+C can obtain points for locating in buildings that have received BD+C certification. The same applies to LEED-Homes and LEED-ND. Homes seeking certification can receive extra points if the neighbourhood in which they are developed has received ND certification. Likewise, additional points can be earned under ND if the buildings in the development are also certified under an appropriate rating system such as LEED-Homes or BD+C.

During the certification process, the project team can submit documentation to the GBCI for review at two points in time: at the end of the design process, and after construction is complete. The project team may also opt to submit everything at once when the project is complete. Design phase review is useful for the project team to get an idea of how many points they are likely to obtain based on the design of the project. It provides a basis for deciding how hard to work for additional points during construction to achieve overall certification goals or requirements.

MOHURD Three Star Rating System (China)

In China, the Ministry of Housing, Urban and Rural Development (MOHURD) has developed a voluntary, context-specific rating system to encourage the development of green buildings beyond what has already occurred due to the use of international rating systems such as LEED. The Three Star Rating System, developed in 2006, is a green building labelling system that evaluates a building's performance and publishes relevant information about the building to verify the performance. The system targets both residential and public buildings, and awards certifications at one-star, two-star, and three-star levels. The rating system considers building performance in six major areas: land saving, energy saving, water saving, material saving, indoor environment and operations. Buildings can be evaluated at the completion of the construction document phase, at which point a Green Building Design Label is awarded that has a 2-year validity. This evaluation provides a path for subsequent award of the Green Building Label, which is awarded after operation for at least 1 year and is valid for 3 years following certification.

The rating system contains a checklist to guide performance and a composite index for evaluating performance. It was developed to reflect the unique practices in the Chinese construction supervisory system and the unique attributes of Chinese standards and regulations. Management and enforcement of rating requirements and the labelling process are undertaken by the Center of Science and Technology of Construction (CSTC) within MOHURD, with technical support from several Chinese universities specializing in building science, construction and performance.

South Korea Green Building Certification System (G-SEED)

South Korea has experienced dramatic development over the last 40 years, often with little regard for the environment. Since it became a member of the Organization for Economic Co-Operation and Development (OECD), along with endorsing other international environmental movements including the Kyoto Protocol, the Korean government has been interested in comprehensive environmental action plans to achieve the goals of sustainability in the construction industry. One of South Korea's initiatives as part of this effort has been to develop and implement a green building rating system called Green Building Certification System (G-SEED), which began in 2001. The purpose of the G-SEED in Korea is to:

- Reduce the environmental impacts of buildings by incorporating green building principles and practices into every aspect of a building's lifecycle, from design and construction to use, maintenance and even demolition.

- Provide comfortable living and working environments for occupants and visitors.
- Raise the awareness of the importance of the natural environment.
- Encourage the development and use of environmental friendly technologies and materials.
- Promote environmental research and development.

Since its introduction in 2001, G-SEED has expanded to different types of building including multi-family housing units, office buildings, mixed-use residential buildings, schools, retail markets and lodging facilities. G-SEED in Korea includes four major categories to achieve the goals of sustainability in the building sector including:

- land use and commuter transportation
- energy resources consumption and environmental loads
- ecological environment
- indoor environmental quality.

Within the rating system, G-SEED has four different certification grades including:

- first grade of green buildings (above 80 out of 100)
- second grade of green building (above 70 – below 80)
- third grade of green building (above 60 – below 70)
- fourth grade of green building (above 50 – below 60).

The Korean government also offers real estate acquisition and registration tax exemption (5 per cent to 15 per cent) for first and second grades of green buildings to motivate the adoption of G-SEED in Korea. Since G-SEED in Korea is mainly implemented and enforced by the government, it has been very effectively spread throughout the construction industry within a short period of time.

Living Building Challenge (International Rating Standard)

Developed by the International Living Building Institute, the Living Building Challenge is a standard that can be used to rate built facilities in terms of the degree to which they restore the natural and social environment and function effectively as contributors to, not parasites of, the context in which they are built. As described on the Living Building Challenge website, the aim behind the tool is to ask:

> What if every intervention resulted in greater biodiversity; increased soil health; additional outlets for beauty and personal expression; a deeper understanding of climate, culture and place; a realignment of our food and transportation systems; and a more profound sense of what it means to be a citizen of a planet where resources and opportunities are provided fairly and equitably?

Indeed, 'Living Building Challenge' is not a merely a noun that defines the character of a particular solution for development, but more relevant if classified as a series of verbs – calls for action that describe not only the 'building' of all of humanity's longest lasting artefacts, but also of the relationships and broader sense of community and connectivity they engender. It is a challenge to immerse ourselves in such a pursuit – and many refer to the ability to do so as a 'paradigm shift'.

The Living Building Challenge is comprised of seven performance areas, or 'petals', which are subdivided into 20 'imperatives' focusing on a specific sphere of influence, as follows:

- **Site**, including limits to growth, urban agriculture, habitat exchange and car-free living.
- **Water**, including net zero water and ecological water flow.
- **Energy**, including net zero energy.
- **Health**, including civilized environments, healthy air and biophilia.
- **Materials**, including avoiding Red List materials, embodied carbon footprint, responsible industry, appropriate sourcing and conservation, and reuse.
- **Equity**, including human scale and humane places, democracy and social justice and rights to nature.
- **Beauty**, including beauty, spirit, inspiration and education.

In each petal, the intent is clearly articulated, along with a vision of ideal conditions under that category and current limitations faced by market and technological conditions. Imperatives are required conditions that must all be met in order to achieve certification requirements. The system can be adapted to many different project types, grouped into four major typologies:

- renovation
- landscape/infrastructure
- building
- neighbourhood.

Projects are encouraged to look for synergies with regard to their natural and built environment context. This is known as 'scale jumping', and encourages multiple buildings or projects to operate in a cooperative fashion to achieve sustainability goals. Not all of the typologies are required to achieve all 20 imperatives due to differences in the nature of different projects, and many imperatives have temporary exceptions that acknowledge current market limitations. The expectation is that these exceptions will be eliminated over time as the market evolves.

Certification under the Living Building Challenge is based on actual, rather than modelled, project performance. Accordingly, projects cannot be certified until at least 12 consecutive months of operation have been completed. Documentation must also be compiled during planning, design and construction to demonstrate compliance with certification requirements.

Case study: Center for Sustainable Landscapes, Pittsburgh, PA

Phipps Conservatory and Botanical Gardens

The Center for Sustainable Landscapes (CSL) is located at Phipps Conservatory and Botanical Gardens in Pittsburgh, Pennsylvania. Developed originally as a place for workers in Pittsburgh's heavy steel industry to escape the drudgery and pollution of the urban environment in their everyday lives, Phipps Conservatory began construction in 1892 as part of Pittsburgh's Schenley Park. Over the years, its mission evolved to a more contemporary recognition of the relationship between humans and the natural environment, serving as a foundation for the major renovations begun to the conservatory grounds and facilities in 2000.

Phipps Conservatory Mission

'To inspire and educate all with the beauty and importance of plants; to advance sustainability and promote human and environmental well-being through action and research; and to celebrate its historic glasshouse.'

Three major updates were planned: a new welcome centre and entrance to the conservatory (Phase 1); a new tropical forest conservatory and production greenhouses to supplement the original glasshouse (Phase 2), and the Center for Sustainable Landscapes, which would house the administrative, educational and research functions of Phipps' mission (Phase 3). The 12,465 sq ft Welcome Center was completed in 2005 and certified to a level of LEED Silver. It is constructed partly underground to mitigate extreme temperatures, and a glass dome provides extensive daylight throughout the main atrium. A vegetated roof surrounds the dome and contributes both to the building's energy efficiency and provides a location for additional exterior plantings that feature native species to the Pittsburgh area (Figure 4.9).

The second phase 12,000 sq ft Tropical Forest Conservatory was completed in 2006, along with a new 36,000 sq ft production greenhouse next to the original glasshouse. This project did not originally have any certification goals, due to the project team's belief that the all-glass buildings would not be able to meet energy performance requirements. However, the production greenhouses subsequently were awarded LEED Platinum certification under the Existing Buildings Operations + Maintenance (EBOM) rating system due to the use of innovations including:

- Earth tubes for passive cooling (Figure 4.10).
- insulated glass and a shading system that serves as an 'energy blanket' (Figure 4.11).
- open-roof technologies to allow passive ventilation.
- thermal massing in the walls to absorb excess heat during the day and reradiate the heat at night to maintain required temperatures.
- a fuel cell located in the greenhouse to produce electricity.

Phase 3 of the project, the Center for Sustainable Landscapes, was registered for version 1.3 of the Living Building Challenge (LBC) as the ambitious culmination of the Phipps update plan. Unlike rating systems used for other parts of the project, LBC required a different way of thinking about certification. First, all elements of the rating system are mandatory, although exceptions may be granted on a case-by-case basis to reflect what is possible to achieve using existing technologies. The aim of this requirement is to reflect the essential balance required of living systems to remain alive. In mimicking these organisms, living buildings must observe all the requirements of living systems operating within their ecosystem constraints; they are not allowed to selectively ignore requirements based on other project constraints such as cost. Secondly, living buildings must prove that they are meeting performance constraints by providing at least 12

continued

Figure 4.9 Phipps Welcome Center dome is surrounded by native species planted in its living roof

Figure 4.10 The Earth tubes used in Phase 2 incorporate long runs of tubing buried below the foundation which allow outside air to either shed or gain heat before providing fresh air to the conservatory space through vents hidden in the garden's plantings

months of actual performance data before certification is awarded. These requirements are significantly different from those of other rating systems such as LEED, where rating can happen based on predicted or modelled performance during design, and project teams can choose which requirements their design will target as goals.

Due to the challenges of achieving its ambitious requirements, the CSL was only the sixth project registered under LBC to pursue Living Building certification. The project team also had a goal to achieve a Platinum rating under the LEED rating system and a certification under the Sustainable Sites Initiative pilot rating system for its sustainable landscape features. To achieve their performance goals under these rating systems, the project incorporated a variety of features including:

- Natural ventilation systems, including operable windows, retractable doors, and an expansive central atrium to promote the stack effect (Figure 4.12).
- Orientation of the building to minimize energy requirements.

Figure 4.11 A custom-designed shading system along with insulated glass helps to mitigate unwanted heat gain through the building's façade

continued

Figure 4.12 The building's operable windows and retractable doors connect with the outdoors and provide accessibility as well as natural ventilation

Figure 4.13 The roof of the CSL is the main entrance to the building and incorporates a vegetated roof garden

Figure 4.14 Light shelves and energy efficient glazing prevent unwanted heat gain and support natural daylighting

continued

Figure 4.15a, b The building's orientation along the site's steep topography provides an opportunity for sustainable landscaping

- An extensive green roof system used for demonstration gardens (Figure 4.13).
- High performance glazing and light shelves/shades to allow natural daylighting (Figure 4.14).
- Extensive integration of the building as part of the site topography, including placement and orientation of the building along a hillside (Figure 4.15).
- A restorative natural landscape, including a lagoon to collect and purify runoff (Figure 4.16), permeable paving to promote groundwater recharge (Figure 4.17), and constructed wetland to treat runoff from parking areas.
- A rainwater harvesting system to capture water from the production greenhouse roofs for use as potable water in the CSL.
- Use of filtered graywater for non-potable uses in the building.
- Solar distillation technology to process effluent from wastewater for use in the production greenhouse to water orchids (Figure 4.18).
- Photovoltaics and a wind turbine to provide necessary power.
- Geothermal heating, thermal mass and desiccant heat exchange to maximize efficiency of space conditioning.
- A robust building envelope using phase change materials and high efficiency glazing to reduce conditioning needs.

The site for the CSL was formerly a City of Pittsburgh Department of Public Works facility classified as a brownfield due to its legacy of leaking underground storage tanks, salt storage for deicing, vehicle maintenance and other environmental degradation. Both known and unanticipated challenges along with the difficult access to and small size of the project site made restoration and

continued

Figure 4.16 A lagoon captures excess stormwater produced on site and treats it using natural processes

Figure 4.17 Multiple types of pervious paving are used throughout the site

Figure 4.18 Solar distillation technology is used to treat wastewater effluent to levels usable for watering orchids in the production greenhouse

remediation a significant challenge (Figure 4.19). Although LBC requires habitat exchange for all projects (i.e. setting aside an equivalent area to the project site that will be protected from development for at least 100 years), the CSL was exempt from this requirement since it was remediating a damaged site and significantly restoring the landscape on it. CSL was also able to clean, seal, and repurpose the leaking underground fuel storage tanks for water storage and treatment, thereby offsetting significant embodied energy and resources that would have been required to install new tanks.

The CSL is also extensively sensored and metered, allowing it to be carefully monitored for managing energy performance as well as understanding the human behaviour aspects of the innovative technologies employed. In partnership with local universities, the building serves as a

continued

Figure 4.19 The CSL is located on a steep hillside overlooking one of Pittsburgh's many bridges and Schenley Park

living laboratory that can help validate building performance models and improve the design of future buildings. Sensors measure occupancy patterns, temperature, humidity, air quality and other factors. Dashboards are incorporated throughout the property to provide feedback to occupants as well as data used by operations personnel to manage the building's performance.

The team's experience in avoiding Red List materials was one of the more difficult parts of the project. A 60-day specification period was implemented to allow contractor Turner construction to complete the entire material submittal process before construction began, rather than continuously during construction. The purpose of this unusual step was to avoid the temptation to proceed without vetting materials to keep construction activities going. At the time, few sources of information existed to provide the necessary information to vet materials for the project, and significant time was invested by project team members to obtain all necessary information. Materials had to be vetted in terms of multiple requirements, including toxicity, site of origin and other factors. Some products such as the phase change materials used in the wall could not be purchased without some levels of halogenated flame retardants, which have significant human health hazards. The team ultimately received an exemption allowing them to use these materials since they would be embedded within walls and would not likely come into contact with humans. As a result of interacting with the team on this project, the manufacturer of the phase change material pledged to remove the dangerous

continued

chemical from future versions of the product, further extending the impact this project has had on other buildings using this technology in the future.

Overall, the CSL project has raised the bar on how infrastructure systems can be integrated into a project site in a beautiful way. Experiences with on-site water systems have greatly expanded the perspective of local code enforcement officials about what is possible, and have begun to change the mindset both locally and beyond about how human needs can be met on a building site. The project is integrated as part of a facility whose primary mission is educating over

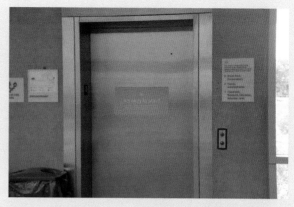

Figure 4.20 Signage encourages people to use stairs instead of the elevator to reduce energy use

300,000 members of the visiting public each year, and as such, it has very high exposure to people for whom the visit is their first experience with many of the innovative technologies it employs. The CSL is an exemplary case of how innovations can be incorporated as part of a project, both to achieve the project's performance goals, as well as to educate those who occupy and visit the building.

Opened in December 2012, the CSL finally received its LBC certification in March 2015 after demonstrating its net zero energy and water performance in operation. Original energy projections proved to be more optimistic than shown in practice, and the key culprit was more extensive use of elevators than expected by visitors to the building. Signage and educational campaigns were used to inform visitors of this unanticipated impact, leading to significant reduction in elevator use (Figure 4.20). The resulting changes in energy performance enabled the building to meet its net zero requirements. The building has also been certified LEED Platinum under LEED version 2.2, and received the highest level of certification possible under the Sustainable Sites pilot (four stars). It also is the first building certified platinum under the new WELL building pilot standard.

Acknowledgements

Special thanks to Mr. Ron Rademacher and Mr. Jason Wirick of the Phipps Conservatory and Botanical Gardens for their stories and tours of the facilities that were instrumental in the preparation of this case study.

Sources

International Living Future Institute. (2016). "Center for Sustainable Landscapes." http://living-future.org/phipps-conservatory-center-sustainable-landscapes (accessed 19 September 2016).

International Well Building Institute. (2016). "The Center for Sustainable Landscapes." www.wellcertified.com/projects/center-sustainable-landscapes (accessed 19 September 2016).

Phipps Conservatory. (2016). "Center for Sustainable Landscapes." https://phipps.conservatory.org/green-innovation/at-phipps/center-for-sustainable-landscapes-greenest-building-in-the-world (accessed 19 September 2016).

Thomas, M.A. (2013). *Building in Bloom: The making of the Center for Sustainable Landscapes at Phipps Conservatory and Botanical Gardens.* Ecotone Publishing, Portland, OR.

EDGE Rating System (International Rating Standard)

The Excellence in Design for Greater Efficiencies (EDGE) green building certification system was developed by the International Finance Corporation and the World Bank as a new approach to rating residential and commercial projects in emerging markets. The need for such a product became apparent as other rating systems available in the market became increasingly complex and expensive to administer. The aim of EDGE is to provide a cost-effective approach to improving building energy and water performance to achieve mass market transformation. The EDGE software application is available for free, allowing designers to quickly and efficiently evaluate possible strategies to optimize their return on investment. Formal certification is available from GBCI for a modest cost depending on the country in which the project is located and attributes of the building. As of 2016, EDGE certification was available in 125 countries worldwide (see edge.gbci.org). To be certified, EDGE projects must achieve a projected minimum reduction of 20 per cent in energy and water use as well as embodied energy in materials compared to a standard building. The rating system applies to a variety of building types ranging from affordable housing to hotels, offices, hospitals and retail establishments.

WELL Building Standard (International Rating Standard)

The WELL Building Standard (WELL) was launched in 2014 by the International WELL Building Institute, and is administered by the Green Building Certification Institute. WELL is based on scientific and medical peer-reviewed literature and research on the key factors in building design and management that affect human health (wellcertified.com). The aim of the WELL Building Standard is to create a common foundation for measuring wellness in the built environment. As such, the WELL standard is a complementary tool that can be applied to apply evidence-based design to built facilities. As of July 2017, over 480 projects have been registered or certified under the WELL Building Standard.

The standard is designed to measure attributes of buildings that influence occupant health in seven basic concept groups:

- **Air** – optimizing indoor air quality.
- **Water** – optimizing water quality and promoting accessibility.
- **Nourishment** – encouraging healthy eating habits.
- **Light** – minimizing disruption to the body's circadian rhythms.
- **Fitness** – using building design technologies and strategies to encourage physical activity.
- **Comfort** – creating an indoor environment that is distraction-free, productive and soothing.
- **Mind** – supporting mental and emotional health, and providing regular feedback and knowledge about the environment.

The standard consists of a series of over one hundred features that can be applied to each project, and may be either mandatory (Preconditions) or optional (Optimizations). A project must meet all Preconditions to be certified at a baseline level, and can receive higher levels of certification based on how many Optimizations are achieved. Some Features are prescriptive, while others are performance-based and allow flexibility in how a project achieves desired outcomes. The International WELL Building Institute also supports a professional credential (WELL AP).

Local and regional building rating systems

In addition to the national and international rating systems described in the previous subsections, many local and regional building rating systems have also evolved over time, especially at the level of residential construction. In the United States, many of these local systems were initially developed by local homebuilder associations to highlight high-performance housing features to the residential market, and some of them even preceded the development of national rating systems such as LEED. In many cases, the rating systems evolved over time and were taken on by local non-profit organizations or utility companies who managed and administered the rating system. Examples of such locally developed systems in the United States include the Austin Energy Green Builder Program at the local level (Austin Energy 2016) and the Earthcraft House program at the regional level (Earthcraft 2016).

Infrastructure, regional and country rating systems

Beyond the individual building scale, other rating and assessment schemes have also been developed that address sustainability for infrastructure projects, for project sites, cities, or regions, and business enterprises. The next sections describe these rating systems, including infrastructure and site rating systems as well as measurement and evaluation systems at the regional and corporate scales.

Infrastructure rating systems

This section presents an overview of the most well-known and widely adapted green infrastructure rating: the CEEQUAL system for infrastructure rating, the Infrastructure Sustainability Rating Scheme, the Greenroads Rating System for roadways, the National Green Infrastructure Certification Program for green infrastructure, the Sustainable Sites Initiative for land development and the Envision rating system. Figure 4.21 shows the country of origin of each of these systems. The colours in Figure 4.21 are used to differentiate between continents.

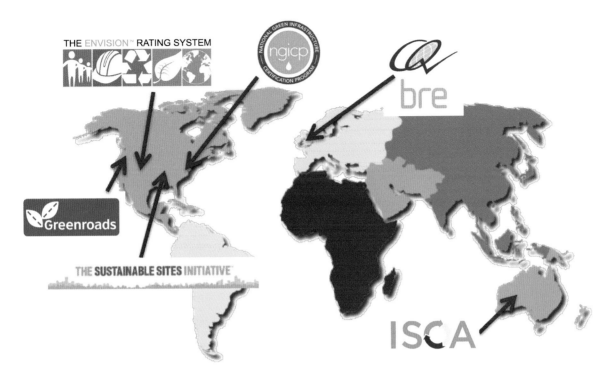

Figure 4.21 Measurement schemes for infrastructure systems

Civil Engineering Environmental Quality Assessment and Award Scheme (CEEQUAL)

CEEQUAL is a rating system targeted toward civil engineering and public works projects. Developed in the United Kingdom by the Institution of Civil Engineers and related organizations, it was formally launched in 2003 and in 2015 transitioned to management by the Building Research Establishment (BRE), the same organization that developed and manages BREEAM. Presently in its fifth version, CEEQUAL has evolved from its initial conception as an environmental rating system into a more comprehensive sustainability assessment and awards scheme. CEEQUAL Version 5 uses a point-based assessment system made up of questions in nine topic areas:

1 Project/contract strategy
2 Project/contract management
3 People and communities
4 Land use (above and below water) and landscape
5 The historic environment
6 Ecology and biodiversity
7 Water environment (fresh and marine)
8 Physical resources use and management
9 Transport

Two different forms of the CEEQUAL Scheme exist: Projects and Term Contracts. CEEQUAL for Projects is applicable to civil engineering, infrastructure, landscaping and public works projects, and exists with guidance for both international and UK and Ireland-specific contexts. Projects can seek awards of six different types based on who applies for the rating: a whole team award, an interim client and design award, a client and design award, a design award, a construction only award or a design and construction award.

The Term Contracts version of the scheme applies to civil engineering and public works projects that are undertaken over a wide geographical or operational area using multi-year contracts. Under this version, two different award types are available: a whole team award and assessment, and a delivery team award and assessment.

Projects are evaluated by an assessor within the project team, whose work is then verified by a certified and trained verifier. The verifier assigns the final rating for the project. Assessment begins by reviewing the list of questions and determining which ones are not relevant for the project. Some questions can be removed from consideration if they are not relevant for the specific type of project, while others are mandatory for all projects.

Four levels of rating are available, including 'pass' (>25 per cent of points), 'good' (>40 per cent), 'very good' (>60 per cent) and 'excellent' (>75 per cent). A variety of projects have received awards under CEEQUAL, including bridges, public buildings, rail projects, river/coastal defence structures, roads and water supply/sewer projects.

Infrastructure Sustainability (IS) Rating Scheme

The IS Rating Scheme (icsa.org.au) is an Australian rating system applicable to a variety of infrastructure projects. Formerly known as the Australian Green Infrastructure Council (AGIC) Rating Scheme, the system was developed by the Infrastructure Sustainability Council of Australia (ISCA), a member-based organization of design and construction professionals from both public and private organizations. The rating scheme applies to projects ranging from roads, rail, bridges, and tunnels, to airports, distribution grids, waterway and port projects, and others. Key themes and associated categories of credits under the IS system include:

- **Management and governance**, including management systems; procurement and purchasing; and climate change adaptation.
- **Using resources**, including energy and carbon; water; and materials.
- **Emissions, pollution and waste**, including discharges to air, land, and water; land; and waste.
- **Ecology**, including ecological functioning.
- **People and place**, including community health, well-being and safety; heritage; stakeholder participation; and urban and landscape design.
- **Innovation**.

The IS scheme has been applied to projects in both Australia and New Zealand. A professional accreditation is also available.

Greenroads Rating System

The Greenroads Rating System (greenroads.org) is a voluntary rating system managed by the Greenroads Foundation in the United States. It targets both new and reconstructed/rehabilitated roads based on sustainable best management practices targeted toward reducing environmental impact, minimizing lifecycle costs and enabling more positive social outcomes. Rating under this system is available at four levels: certified, silver, gold and evergreen. The system has 12 mandatory project requirements (see *Greenroads mandatory project requirements*).

Greenroads also has 45 additional voluntary credits in five categories that can be pursued to achieve a rating (see *Greenroads credit categories*). Four additional credits can be earned in the Creativity and Effort category based on innovations undertaken by the project team including education, innovative ideas, enhanced performance, and local values. Certification levels include Bronze (40 points minimum), Silver (50 points minimum), Gold (60 points minimum) and Evergreen (80 points minimum). Projects are presently being evaluated throughout the United States as well as in Abu Dhabi, New Zealand, Canada, South Africa, Israel and Taiwan.

Greenroads mandatory project requiremens

- Ecological impact analysis
- Energy and carbon footprint
- Low impact development
- Social impact analysis
- Community engagement
- Lifecycle cost analysis
- Quality control
- Pollution prevention
- Waste management
- Noise and glare control
- Utility conflict analysis
- Asset management

Greenroads credit categories

- **Environment and Water**, including habitat and ecological connectivity, vegetation quality, soil management, water conservation, and others.
- **Construction Activities**, including workzone health and safety, equipment fuel efficiency, air emissions, water use, procurement integrity, fair labor, local economic development, and others.
- **Materials and Design**, including preservation and reuse, recycled and recovered content, local materials, long-life design, and environmental/health product declarations.
- **Utilities and Controls**, including utility upgrades, maintenance and emergency access, energy efficiency and alternative energy, electric vehicle infrastructure, travel time reduction, and others.
- **Access and Livability**, including safety audits and enhancements, multimodal connectivity, equity and accessibility, health impact analysis, archaeology & history, and others.

(greenroads.org)

National Green Infrastructure Certification Program (NGICP)

The National Green Infrastructure Certification Program (ngicp.org) was collaboratively developed by the Water Environment Federation, DC Water and other partner utilities and organizations in the United States to set national certification standards for workers involved in the construction, inspection and maintenance of green infrastructure (see *What is green infrastructure?*). While it is not a project certification scheme, the NGICP certifies workers that can then be employed by

What is green infrastructure?

While the term 'green' is often used as a synonym for 'environmentally friendly', in the case of Green Infrastructure (GI) systems, it literally refers to infrastructure that mimics, restores or protects the water cycle as found in natural systems. Often incorporating living systems including trees, wetlands and plants, GI systems use soils, pervious pavement and other materials to capture, direct and treat stormwater before returning it to natural ecosystems or diverting it for uses beneficial to humans.

Types of GI facilities include bioretention structures and areas, permeable pavement, rainwater harvesting, rooftop stormwater management, dry wells and wetlands. In response to increasingly stringent standards for the management of stormwater runoff in the United States, GI systems are being employed more frequently as part of both new construction and post-construction stormwater management.

municipalities and water utilities to provide the knowledge, skills and abilities to properly install and manage green infrastructure projects. Launched in 2016, this program will include a knowledge base, national curriculum and standardized certification examination targeted to construction and maintenance workers.

Sustainable Sites Initiative (SITES)

SITES (sustainablesites.org) was collaboratively developed by the American Society of Landscape Architects (ASLA), the Lady Bird Johnson Wildflower Center and the United States Botanical Garden. It is a voluntary national guideline and performance benchmark system that is based on a fundamental framework of ecosystem services, where sustainable sites provide key ecosystem services such as climate regulation, air and water purification, water retention, pollination and habitat functions, waste decomposition and treatment, human health and well-being, cultural benefits and others. SITES is organized around the idea that there should be priorities in deciding how to develop or influence the landscape. Most importantly, key ecosystem features should be conserved if possible, rather than disturbed and replaced later. Over the project lifecycle, the net health and performance of site ecosystems should increase.

The SITES rating system is based on a set of principles (see *SITES guiding principles*) and performance benchmarks. Version 2 of the rating system requires achieving 18 prerequisites and allocates 200 potential points among 48 credits for each project site. It can be applied to sites both with and without buildings which will be protected, developed or redeveloped for public or private purposes. Credits are weighted based on potential effectiveness in meeting four primary goals, as follows:

- Create regenerative systems and foster resiliency.
- Ensure future resource supply and mitigate climate change.

SITES guiding principles

- Do no harm.
- Apply the precautionary principle.
- Design with nature and culture.
- Use a decision-making hierarchy of preservation, conservation and regeneration.
- Provide regenerative systems as intergenerational equity.
- Support a living process.
- Use a systems thinking approach.
- Use a collaborative and ethical approach.
- Maintain integrity in leadership and research.
- Foster environmental stewardship.
                        (sustainablesites.org)

SITES credit categories

1 Site context
2 Pre-design assessment + planning
3 Site design – water
4 Site design – soil + vegetation
5 Site design – materials selection
6 Site design – human health + well-being
7 Construction
8 Operations + maintenance
9 Education + performance monitoring
10 Innovation or exemplary performance
                        (sustainablesites.org)

- Transform the market through design, development and maintenance practices.
- Enhance human well-being and strengthen community.

Credits and prerequisites are organized into ten different categories or sections based first on project lifecycle phase and second on area of impact (see *SITES credit categories*). The rating system has received considerable attention from the USGBC, which has heavily incorporated elements of the rating system into its sustainable sites credits in version 4.0 of its rating system. SITES certification can be pursued individually for projects without also pursuing LEED certification. Rating under the SITES system is administered by the Green Building Certification Institute.

Envision Rating System for Sustainable Infrastructure

Launched in February 2011, the Envision Rating System (sustainable infra-structure.org/envision) is managed by the Institute for Sustainable Infrastructure in the United States as a tool to guide engineers, owners, constructors, regulators and policy-makers toward the development of more sustainable infrastructure systems. It is based on a set of objective-based goals associated with reliability, resilience, efficiency organizational adaptability and overall project performance. The rating system was collaboratively developed by the American Society of Civil Engineers (ASCE), the American Council of Engineering Companies (ACEC) and the American Public Works Association (APWA).

The rating system consists of 60 credits in five categories (see *Envision credit categories*) in a structured, four-phase process to guide users

Envision credit categories

- **Quality of life**, including purpose, community and well-being.
- **Leadership**, including collaboration, management and planning.
- **Resource allocation**, including materials, energy and water.
- **Natural world**, including siting, land and water and biodiversity.
- **Climate and risk**, including emissions and resilience.

(sustainableinfrastructure. org/envision)

Envision levels of achievement

- **Improved** – Performance that is above conventional; slightly exceeds regulatory requirements.
- **Enhanced** – Sustainable performance that is 'on the right track'; superior performance is within reach.
- **Superior** – Sustainable performance that is 'noteworthy'.
- **Conserving** – Performance that has achieved essentially zero negative impact.
- **Restorative** – Performance that restores natural or social systems (not applicable to all credits).

(sustainableinfrastructure. org/envision)

through the planning and project delivery process in increasing levels of detail. A unique feature of the rating system is the use of rubrics to define a range of achievement levels for each credit, from Improved to Restorative (see *Envision levels of achievement*). This helps project teams to visualize different solutions in terms of their specific project features and constraints. Envision includes a strong focus on social sustainability factors and community interaction to facilitate broad acceptance and support of its recommendations. The system can be applied to infrastructure projects including roads and bridges, transit systems, airports, seaports, water and wastewater systems, energy generation and transmission systems and other physical facilities at the local or regional scale.

The Envision rating system is freely available for use as a tool to support improved project performance by project teams, and an online interactive tool is available to guide users in assessing different levels of achievement for each credit. Projects can also receive third-party evaluation and certification at four different levels: Envision Bronze, Silver, Gold or Platinum. A professional accreditation (Envision Sustainability Professional or ENV SP) is available for professionals interested in being recognized for their expertise with the rating system.

Regional and national sustainability metrics

Other rating systems and schemes exist, and new tools and approaches are continuously being developed to support measurement and decision making for sustainability. Organizations such as the International Council for Local Environmental Initiatives (ICLEI – see iclei.org) have evolved to work hand in hand with local governments to establish local and regional action plans to achieve national and international goals set in Agenda 21, the Rio Conventions, the Millennium and Sustainable Development Goals, and others. At the heart of these efforts lies the task of choosing appropriate metrics to evaluate the issues most critical to each context of application. This context specificity is a goal to which more general rating systems for facilities and products aspire.

At the national scale, multiple approaches including Ecological Footprint (EF), the Environmental Sustainability Index (ESI) and the Environmental Performance Index (EPI) have been developed as a means of benchmarking and comparing the environmental performance of whole countries. These metrics are relevant to the built environment not only because of the role the built environment plays in environmental sustainability, but also because they may provide clues as to areas of future opportunity for new types of development.

Ecological Footprint (EF)

EF, discussed in Chapter 2, is one approach that has been applied at the scale of whole countries as well as at smaller scales. This metric is expressed in terms of the total area of land necessary for the production and maintenance of goods and services required by a defined community. This metric provides a rough estimate of the amount of 'overshoot' or ecological deficit of a given population under the conditions of its current standard of living by comparing the amount of land available to the community vs. the amount of land required.

The Global Footprint Network (www.footprintnetwork.org) maintains estimates of the per capita footprint of 150 countries that is used as a relative ranking of ecological impacts of standards of living in those countries. Based on 2012 data, the countries with the largest ecological footprint per capita were Luxembourg at 15.8 hectares per capita, Australia at 9.3 hectares per capita, and the United States and Canada, tied at 8.2 hectares per capita. The smallest ecological footprints per capita were Haiti at 0.6 hectares per capita, Timor-Leste at 0.5 hectares per capita and Eritrea at 0.4 hectares per capita.

Environmental Sustainability Index (ESI)

The ESI, originally developed by partners Yale and Columbia Universities and recognized by the World Economic Forum, rated 146 countries on 21 elements of environmental sustainability covering natural resource endowments, past and present pollution levels, environmental management efforts, contributions to protection of the global commons and a society's capacity to improve its environmental performance over time (see Key building blocks: environmental sustainability index). ESI rankings were released most recently in 2005.

High rankings under the ESI suggest better environmental stewardship and, coupled with other metrics of economic competitiveness, suggest that economic development need not always come at a high environmental price. Top-ranked countries in 2005 were Finland, Norway, Uruguay, Sweden and Iceland, all of which have substantial

Key building blocks: environmental sustainability index

Environmental systems

A country is more likely to be environmentally sustainable to the extent that its vital environmental systems are maintained at healthy levels, and to the extent to which levels are improving rather than deteriorating.

Reducing environmental stresses

A country is more likely to be environmentally sustainable if the levels of anthropogenic stress are low enough to engender no demonstrable harm to its environmental systems.

Reducing human vulnerability

A country is more likely to be environmentally sustainable to the extent that people and social systems are not vulnerable to environmental disturbances that affect basic human well-being; becoming less vulnerable is a sign that a society is on a track to greater sustainability.

Social and institutional capacity

A country is more likely to be environmentally sustainable to the extent that it has in place institutions and underlying social patterns of skills, attitudes and networks that foster effective responses to environmental challenges.

Global stewardship

A country is more likely to be environmentally sustainable if it cooperates with other countries to manage common environmental problems, and if it reduces negative trans-boundary environmental impacts on other countries to levels that cause no serious harm.

(www.yale.edu/esi)

natural resource endowments, low population density and a history of success in managing development challenges. Lowest ranked countries included North Korea, Iraq, Taiwan, Turkmenistan and Uzbekistan. These countries face numerous issues, both natural and human-induced, and have struggled to effectively manage their policy choices.

Environmental Performance Index (EPI)

The EPI ranks countries based on two key areas of environmental performance: protection of human health and protection of ecosystems. Most recently released in 2016, this metric considers six categories of ecosystem vitality measures (climate and energy, biodiversity and habitat, fisheries, forests, agriculture and water resources) and three categories of environmental health measures (health impacts, air quality, water and sanitation).

EPI data are calculated and maintained by Yale University (www.epi. yale.edu). Individual country scores are calculated based on a 'proximity-to-target' method, where targets are established by international or national policy goals, or established scientific thresholds and country scores are based on how close or far each country is to its target. Countries that have reached their targets receive a score of 100.

An EPI Country Comparison Tool is available that allows comparison of countries' performance for various environmental indicators and across different time periods (visuals.datadriven.yale.edu/country compare/). Countries can be compared according to overall EPI as well as according to performance for each of the nine categories of measures.

Country Sustainability Ranking

The Country Sustainability Ranking is a framework developed by RobecoSAM and Robeco, the organizations that are known for managing the corporate sustainability assessment process used to generate the Dow Jones Sustainability Indices. This ranking is based on countries' strengths and weaknesses in terms of governance, social and environmental factors. These rankings are used by investors to evaluate risks associated with investing in government bonds for different countries. Countries are ranked on a ten point scale based on 17 weighted indicators. Risks are evaluated in five different categories (Economic, Environmental, Geopolitical, Societal and Technological). The most likely risks and the risks with the highest potential impact are estimated (see *Global risks identified by the country sustainability ranking*).

Based on rankings released in October 2016, the highest ranking three countries were Norway, Sweden and Finland, and the lowest ranked countries were Egypt, Venezuela and Nigeria. A total of 62 countries are ranked twice a year under this system (www.robecosam. com/en/sustainability-insights/about-sustainability/country-sustain ability-ranking/index.jsp).

Global risks identified by country sustainability ranking

Top 10 risks in terms of likelihood

1. Interstate conflict
2. Extreme weather events
3. Failure of national governance
4. State collapse or crisis
5. Unemployment or underemployment
6. Natural catastrophes
7. Failure of climate-change adaptation
8. Water crises
9. Data fraud or theft
10. Cyber attacks

Top 10 risks in terms of impact

1. Water crises
2. Spread of infectious diseases
3. Weapons of mass destruction
4. Interstate conflict
5. Failure of climate-change adaptation
6. Energy price shock
7. Critical information infrastructure breakdown
8. Fiscal crises
9. Unemployment or underemployment
10. Biodiversity loss and ecosystem collapse

Corporate rating systems, reporting and strategic actions

In addition to rating projects and building components, a growing area of interest lies in rating enterprises or corporations as a means of increasing market visibility and customer recognition or approval. Such evaluation can be done from two complementary perspectives: (1) evaluating companies in terms of a broad set of metrics that reflect not just conventional economic performance but also social and environmental performance metrics; or (2) rating companies predetermined to be sustainability leaders in terms of conventional performance metrics (such as financial performance). The following subsections describe these two approaches in greater detail.

Measuring financial performance of sustainability-driven companies

The market is a powerful driver for change in the corporate world. Environmentally and socially responsible investing is an approach taken by some market investors to 'steer' the market in general toward more sustainable practices. It is also undertaken by investors who believe that companies pursuing sustainability goals will ultimately be more successful in the marketplace than their conventional counterparts. To effectively guide these types of responsible investments, the market needs a way to monitor the individual and collective financial performance of environmentally and socially responsible companies being traded on the open market.

The Dow Jones Sustainability Indices (DJSI) is a set of metrics that track the financial performance of leading sustainability-driven companies worldwide (sustainability-indices.com). Launched in 1999, this set of indices is a means for asset managers to obtain reliable and objective benchmarks to manage sustainability portfolios. The Robeco SAM Corporate Sustainability Assessment questionnaire (RobecoSAM 2016) is used as a basis to determine which companies are included in the DJSI. The Assessment includes 80–120 industry-specific questions about economic, environmental and social factors relevant for company success. Versions of the questionnaire exist for 60 different industries within the overall DJSI. The family of indices under the DJSI includes multiple groupings of firms based on top-performing firms by geographic region, as well as several international groupings.

Each year, invitations are sent to over 3400 publicly traded companies to participate in corporate sustainability assessment. Of these, the 2500 largest global companies in terms of free-float market capitalization are eligible for possible listing under the DJSI. The corporate sustainability assessment focuses on the long-term development of corporate value. General topics cover such issues as corporate governance, risk and crisis management, human capital development, and the quality of environmental and social reporting. In addition, sector-specific risks and opportunities arising from various sustainability trends are also analysed.

Following the analysis of the corporate data, SAM selects the leading companies from each sector. For example, the top 10 per cent in each industry will be included in the Dow Jones Sustainability World Indexes (DJSI World). The top 15 per cent of companies per industry sector are also featured in the annual SAM Sustainability Yearbook, the reference work on the world's most sustainable companies (see yearbook. robecosam.com). In 2016, the Sustainability Yearbook provided an overview of 59 sectors and 2126 individual companies. The leading companies from each sector are recognized as 'SAM sector leaders'. In addition, companies included in the Yearbook can qualify to the categories of 'SAM gold class', 'SAM silver class' and 'SAM bronze class'. Companies that have made particular progress in their sector are recognized as 'SAM sector movers'.

Measuring the sustainability performance of organizations

Several distinct perspectives exist for measuring sustainability performance at the organizational or corporate level. At one end of the spectrum, corporate sustainability reporting (CSR) is a means for an organization to self-report its goals, accomplishments, and future plans and strategies pertaining to sustainability. Corporate sustainability reports are becoming more commonplace in the market as consumers increasingly base their purchasing decisions on factors beyond cost. These reports, often updated annually, contain key information about an organization's sustainability initiatives, and typically include similar types of information from firm to firm (see *Elements of a corporate sustainability report*).

The Global Reporting Initiative's (GRI's) Triple Bottom Line offers a standardized reporting framework for disclosing corporate sustainability performance in terms of social, economic and environmental performance. The GRI standards (see globalreporting.org/standards) are used for public reporting of performance as well as voluntary performance improvement within corporations. Links to the corporate sustainability reports of hundreds of companies can be found on GRI's website. The GRI's Amsterdam Declaration underscores the importance of having integrated sustainability reporting based on standardized frameworks for performance accounting to allow comparison across industry. According to the Declaration (GRI 2010), an integrated report presents information about an organization's financial performance with information about its environmental, social and governance (ESG) performance in an integrated way. In order for integrated reporting to be a viable and useful activity for companies, it must be underpinned by standardized financial and ESG reporting frameworks. Financial reporting standards, such as International Financial Reporting Standards (IFRS), US Generally Accepted Accounting Principles (US GAAP) and ESG reporting frameworks, principally the GRI Guidelines, act as structural supports for integrated reporting frameworks for corporate sustainability.

Elements of a corporate sustainability report

Executive summary

- Sustainability vision
- Benefits and business case
- Notable achievements
- Barriers and responses

Initiatives

- Current year initiatives
- Rationale for selection
- Progress and achievements
- Impact of initiatives on corporate goals

Accounting

- Inventory of current social and environmental impacts

- Tools and metrics used to measure and reduce impacts
- Goals for impact reduction
- Progress during reporting period

Future plans

- Future plans and initiatives
- Measuring success

Conclusion

- Significant sustainability achievements and benefits
- Reiterate sustainability vision
- Company profile
- External partners

(www.TEDgreenroom.com)

For some metrics of critical significance in the market, special third-party standards are being developed to measure performance at the corporate scale. For instance, carbon footprint analysis and reporting is an approach being adopted by companies to benchmark their performance. One example of such a standard is the Greenhouse Gas Protocol Accounting Standards developed by the World Resources Institute in partnership with the World Business Council for Sustainable Development (WBCSD 2012). The Greenhouse Gas Protocol Standard for products studies all potential contributions to the emissions of a product, including suppliers, transportation, production and disposal. In parallel, the Corporate Value Chain Standard allows corporations to measure and manage their greenhouse gas emissions across their entire supply chain (WBCSD 2010). A variety of third-party organizations such as Climate Neutral Business Network (climateneutral.com) have emerged to provide verification of carbon-related claims at the corporate, not just product, level. As market-based measures for carbon management (cap and trade) become more prominent in the marketplace, these types of evaluation metrics and standards will become more commonplace in the construction industry.

Other organizations such as the Leonardo Academy, a standards developer accredited by the American National Standards Institute (ANSI), are working to develop new standards to evaluate the sustainability of organizations (leonardoacademy.org). Investors, companies and procurement organizations are driving the need for such standards as they seek to evaluate and compare sustainability attributes of organizations beyond what can be evaluated using financially driven indices such as the SAM assessment. The forthcoming ANSI Standard for

Sustainable Organizations will define what a sustainable company or organization is and how its level of sustainability achievements can be measured and documented. The standard is expected to make it easier to evaluate and compare the sustainability of organizations in terms of environmental stewardship, social equity and economic prosperity.

Sustainability evaluation and assessment in the future

Specific challenges abound in measuring the sustainability of the built environment and the products, services and organizations associated with it. At the foundation of the problem is the lack of a widely accepted operational definition of the construct of sustainability. As a context-dependent attribute, the sustainability of a system or artefact will be affected by different factors in different situations. It will also necessarily involve different factors and considerations for different types of products or systems. Thus, there is no 'one size fits all' approach to evaluating sustainability. This is one of the major challenges that will be addressed in the next 10 years with regard to sustainability evaluation and assessment. Other challenges include continuing accountability for ratings, and information accessibility and transparency overall, as discussed in the following subsections.

Project specificity of rating systems

One of the first challenges of rating systems that will be addressed in the next decades is the development of rating systems that can capture specific nuances of very different types of projects while still providing a basis for comparison across those projects. Trends to date in rating system development, especially at the building scale, have shown a tendency to start with a rating system that is applicable to a common situation or case (such as Green Star Offices or LEED for New Construction), then expand and customize that original rating system to be able to address details found in other project types (such as LEED for Schools and LEED for Healthcare). At one end of the spectrum is the notion of having completely customized rating systems for each different type of product or facility. At the other end of the spectrum is the approach of having a single universal rating system that can apply equally to all situations. While frameworks exist that could serve the purpose of a universal measurement tool, they often require information about performance that requires technical expertise to model accurately.

One significant trend that may lead to greater adoption of more universal rating system schemes is the development of new types of user interfaces to those systems. These new interfaces will make complex and powerful performance models accessible to a wider range of users by using case-based reasoning, visual matching, inference, data mining and heuristic methods to make accurate assumptions about a user's specific situation, rather than requiring the user to obtain and manipulate detailed information. In the coming years, the ability to access and manipulate data at multiple levels and quickly develop sophisticated performance models will no doubt far exceed today's standards and lead to rating systems beyond our current ability to envision. Overall, there will be an increased reliance on scientific evidence and systematic methods such as lifecycle analysis to support decisions for built environment sustainability.

Another key trend is the requirement to consider actual vs. estimated performance when rating buildings and infrastructure systems. Many early versions of rating systems assigned credit based on design-phase models of predicted building performance, but data from the population of certified buildings called into question whether this performance was actually achieved in practice. Multiple studies of the energy performance of certified green projects called into question the efficacy of rating systems in achieving desired ends due to the huge variability in actual vs. predicted energy performance (Cotera 2011; Fowler and Rauch 2008; Hughes 2012; Newsham et al. 2009; Scofield 2009; Torcellini 2004; Turner 2006; Turner and Frankel 2008). In more recent iterations of rating systems, efforts have been made to at least require buildings to supply ongoing performance information (e.g. the LEED rating system), and discussions have been ongoing about whether a building might be able to lose its rating if it fails to perform as predicted. The Living Building Challenge rating system requires at least one full year of post-occupancy data showing actual performance before it will certify a building.

Context specificity of rating systems

As with project type, the sustainability of a given solution will vary dramatically based on the context of a given project or solution. For example, incinerating toilets may be tremendously unsustainable where energy is scarce or difficult to produce, but an excellent solution where energy is abundant and water is scarce. Context is a function not only of physical location and constraints, but also socio-economic, political and other factors that can be difficult to define. How can a measurement system incorporate context of use in weighting the relative importance of attributes or variables, without falling prey to the biases and self-interests of the raters?

Newer versions of the US LEED rating system accomplish this goal via a system of region-specific credit requirements applied to buildings based on location. Table 4.3 shows the evolution of approach to this issue over the multiple versions of the LEED rating system. The LEED 4.0 rating system assigns 'bonus points' to existing credits that have been determined by local USGBC chapters to be of particular importance in each location. For example, projects in many parts of the south-western United States receive regional priority credits for achieving higher levels of water efficiency, since water is an increasingly scarce resource in this region. Which credits should receive extra emphasis in a particular area is determined by the collective input of volunteers serving on regional committees.

Table 4.3 Context specificity and the LEED rating system

Approach to context specificity	Version	Launch
Somewhat prescriptive; limited building types – application guides proposed as mechanism for customization	LEED v1	1998
More performance-based; most credits converted; additional application guides developed; additional core rating systems developed for homes, neighbourhood development, existing buildings, core and shell, and commercial interiors	LEED v2.0, 2.1 and 2.2	2001ff
Regional priority credits introduced; credit point values changed to reflect new weightings for relative importance of environmental issues	LEED v3.0	2009

In the future, better integration of existing data and real-time monitoring of local conditions will allow regional credits to be determined using objective algorithms rather than subjective weightings developed by local interests. Weightings may be able to be integrated dynamically as part of rating systems over time, as conditions change. Coupled with trends toward dynamic, environmentally responsive buildings that monitor and self-adjust to changing conditions, measurement systems will be able to monitor the ongoing relationship between built facilities and their environmental contexts, and reward those facilities that best align with their contextual constraints.

Ongoing accountability based on information accessibility and transparency

The ability to have ratings downgraded or revoked is also a growing trend across rating systems at all scales. For instance, British Petroleum was formally removed in 2010 from the Dow Jones Sustainability Index following the massive oil spill in the Gulf of Mexico for which it was responsible early in that year.

A strong movement in the United States advocated in 2010 to revoke LEED ratings for facilities that fail to perform to specified design goals. The USGBC and its founders faced a class action lawsuit claiming that they fraudulently misled consumers and fraudulently misrepresented energy performance of buildings certified under the LEED rating systems, and that LEED was harming the environment by leading consumers away from using proven energy-saving strategies (EBN 2010). Specifically, the suit alleged that USGBC's claim of verifying efficient design and construction is 'false and intended to mislead the consumer and monopolize the market for energy-efficient building design' (EBN 2010). While the lawsuit was ultimately dismissed (Cheatham 2011), it brought to light the challenges associated with rating systems as mechanisms for policy and influencers of market behaviour.

Who is responsible if an owner mandates LEED certification of a building, but the resulting building fails to achieve third-party certification? Are uncertainties in certifying a building so minor that

> [T]ransparency management has become the cornerstone to which all organizations must build in order to guarantee their continued existence.
> (Global Reporting Initiative, Future Trends in Sustainability Reporting)

The new social technologies, media and networks promise – or threaten, depending on your viewpoint – to transform the reporting landscape. They will simultaneously accelerate and deepen market conversations between business and its current stakeholders, and, potentially, bring totally new people and interests into the conversation – with dramatically more powerful information and intelligence resources at their disposal.

(Global Reporting Initiative, *The Transparent Economy*)

designers and builders should be comfortable in guaranteeing a particular rating as part of their contracts? Many policies qualify their mandates through the incorporation of phrases such as 'when cost effective based on lifecycle cost analysis' (Kaplow 2013). Others use incentives such as tax breaks or processing advantages to receive faster approvals to encourage pursuit of certification, rather than mandating certification by law. Regulatory approaches to green building, discussed in Chapter 3, often reference certification and assessment schemes as a way to provide a consistent yardstick for evaluating performance. However, as rating schemes continue to evolve and become increasingly dependent on actual performance of the building in use, these policies may become more contentious.

While the case law involving green buildings pursuing third party certification continues to grow, attorney Stuart Kaplow, who specializes in green building law, reports that most of these cases involve the same kinds of claims brought in conventional building projects, such as construction quality issues. He concludes that there is 'no more or additional liability associated with constructing a green building vs. a similar non green structure' (Kaplow 2016). In the future, better approaches to data tracking, along with improved systems for transparency and accountability, will ensure that public pressure can be brought to bear in a variety of ways against poor sustainability performers.

Advances in sensing and control technologies will bring performance data to the forefront of every decision and make it easy to analyse, track and interpret. Building information modelling is no longer the domain of well-educated domain experts but has instead become ubiquitous, integrating more seamlessly over time with portable devices such as smart phones and tablets to facilitate the construction and facilities management processes. In the future, such integration will enable building management systems to provide moment-to-moment performance data to users of built facilities, offering feedback that can help shape individual choices and behaviours. Individual components and products will be

GRI's key sustainability reporting trends for 2025

- Companies will be held accountable more than ever before
- Business decision makers will take sustainability issues into account more profoundly
- Technology will enable companies and stakeholders to access, collate, check, analyse and correlate data
- Technology will enable companies to operate and report in a highly integrated way
- Ethical values, reputation and risk management will guide decision makers
- New indicators will emerge
- Reports will result both from regulated and voluntary processes
- Sustainability data will be digital.

(www.globalreporting.org/resourcelibrary/Sustainability-and-Reporting-Trends-in-2025-1.pdf)

traceable through their entire lifecycles, with information being immediately available via embedded chips or devices such as is already being done during manufacturing and construction. In addition to more timely, rich data from a variety of new sources, the use of Web 3.0 technologies for social networks will also change who accesses that information and what they do with it (GRI 2010b; GRI 2015a, b). Business leaders will also be accountable to better understand how their decisions affect sustainability challenges, including factors now thought of as externalities (GRI 2015b). Ultimately, real time information will become a basis for adapting human behaviours to improve facility sustainability, even as facilities themselves gain the ability to self-adapt to those behaviours.

Discussion questions and exercises

4.1 Which green rating systems are presently in use by your organization? For what purpose(s) are they used – baselining, benchmarking, prioritization or documentation? Classify each rating system in terms of its scale and whether it is a threshold or profile tool.

4.2 What green product labelling systems are applicable in your country or region? Visit the online ecolabel Index (www.ecolabelindex.com/ecolabels) or a similar guide such as Green Building Alliance's guide (2011) and review the labels that apply to your country. How many of these labels appear on products with which you are familiar?

4.3 Visit a local building material supply store. Inventory product labels to identify any ecolabels shown on product packaging. Which of these labels is based on a recognized third-party standard? Do any product labels contain obvious greenwash?

4.4 Perform an Internet search to identify examples of each of the major types of greenwash discussed in the chapter. Which types are most common? How could you modify the claims made to ensure that they are correct? What additional information would be needed to support each claim?

4.5 Conduct an ecological footprint analysis of yourself (search for one of the calculators freely available online). How many Earths would it take if everyone on the planet lived like you? Compare your results with your peers. What single action could you take to make the biggest change in your ecological footprint?

4.6 Find an example of a type of building product where the options have very different service lives. For example, lamps or light bulbs have vastly different service lives depending on the type of technology employed. How many times must the shortest-lived option be replaced to equal the service life of the longest-lived option? How do the costs compare?

4.7 Choose a building material from a specific manufacturer and obtain the information contained in the Checklist of Questions for Product Manufacturers. How much of this information is available on the manufacturer's website? Contact the manufacturer and obtain the remaining information. How difficult is the information to obtain?

4.8 Which building- or infrastructure-scale rating systems are in use in your country or region? Which buildings have been certified or rated in your area? Locate the online database of projects for the relevant rating system or contact the rating system administrator and identify the closest projects to your location. Visit the project if possible and identify or document the major green features of the project.

4.9   What are the steps necessary to rate or certify a building using your locally applicable rating system(s)? What costs are involved? Do local, regional or national policies require facilities to be rated using this system? If more than one rating system is used in your country or region, which one is most prevalent? Why?

4.10   What rating has your country achieved according to the Environmental Sustainability Index? Ecological Footprint? Environmental Performance Index? Locate the most recent rankings online. Which countries do better than yours? Which do worse? Are you surprised by any of the rankings?

4.11   Does your organization produce a corporate or organizational sustainability report? If yes, locate the most recent report and review the initiatives documented there. Are there any initiatives of which you are unaware? If no report is presently produced, what initiatives within the organization would you recommend including in a future report?

References

American Institute of Architects (AIA). (1990). *Environmental resource guide.* Wiley, New York.

Anink, D., Boonstra, C., and Mak, J. (1996). *Handbook of sustainable building: An environmental preference method for selection of materials for use in construction and refurbishment.* James & James, London.

Athena Institute. (2016). "Athena impact estimator for buildings." www.athenasmi.org (accessed 28 December 2016).

Austin Energy. (2016). "Austin Energy green building." http://greenbuilding.austinenergy.com (accessed 28 December 2016).

Cheatham, C. (2011). "Breaking: Lawsuit against USGBC dismissed." www.greenbuildinglawupdate.com/2011/08/articles/legal-developments/breaking-lawsuit-against-usgbc-dismissed/ (accessed 28 December 2016).

Cotera, P. (2011). "A post-occupancy evaluation: To what degree do LEED certified buildings maintain their sustainable integrities over time?" http://etd.fcla.edu/UF/UFE0042961/cotera_p.pdf (accessed 28 December 2016).

Earthcraft. (2016). "Earthcraft house rating system." www.earthcraft.org (accessed 28 December 2016).

EBN – Environmental Building News. (2010). "USGBC, LEED targeted by class-action suit." www.buildinggreen.com/news-analysis/usgbc-leed-targeted-class-action-suit (accessed 28 December 2016).

Fowler, K. and Rauch, E. (2008). "Assessing green building performance. A post occupancy evaluation of 12 GSA buildings." www.gsa.gov/graphics/pbs/GSA_Assessing_Green_Full_Report.pdf (accessed 28 December 2016).

GRI – Global Reporting Initiative. (2010a). "The Amsterdam Declaration on Transparency and Reporting." www.globalreporting.org/CurrentPriorities/AmsterdamDeclaration/ (accessed 12 November 2010); www.csrwire.com/press_releases/15371-The-Amsterdam-Declaration-on-Transparency-and-Reporting (accessed 28 December 2016).

GRI. (2010b). "The transparent economy." www.globalreporting.org/resourcelibrary/Explorations_TheTransparentEconomy.pdf (accessed 28 December 2016).

GRI. (2015a). "Sustainability and Reporting Trends in 2025 – Preparing for the Future." *Reporting 2025 Project: First Analysis Paper, May 2015.* www.globalreporting.org/resourcelibrary/Sustainability-and-Reporting-Trends-in-2025–1.pdf (accessed 28 December 2016).

GRI. (2015b). "Sustainability and Reporting Trends in 2025 – Preparing for the Future." *Reporting 2025 Project: Second Analysis Paper, October 2015.* www.globalreporting.org/resourcelibrary/Sustainability-and-Reporting-Trends-in-2025–2.pdf (accessed 28 December 2016).

Green Building Alliance. (2011). "Green building product certification and labeling systems." www.go-gba.org/wp-content/uploads/2013/07/Green_Building_Product_Certification__Labeling_Systems.pdf (accessed 28 December 2016).

Howard, N., Shiers, D., and Sinclair, M. (1998). *The green guide to specification: An environmental profiling system for building materials and components* (BRE Report 351). BRE, Garston, Watford.

Hughes, J. (2011). *Comparison of large scale renewable energy projects for the United States Air Force.* M.S. Thesis, Department of Civil & Environmental Engineering, Virginia Tech, Blacksburg, VA.

Kaplow, S. (2013). "King County mandates its green building be the greenest, with a twist." www.greenbuildinglawupdate.com/2013/12/articles/leed/king-county-mandates-its-green-building-be-the-greenest-with-a-twist/ (accessed 28 December 2016).

Kaplow, S. (2016). "Suing but not because it was not a green building." www.greenbuildinglawupdate.com/2016/11/articles/green-building/suing-but-not-because-it-was-not-a-green-building/ (accessed 28 December 2016).

NIST – National Institute of Standards and Technology. (2010). "Building for economic and environmental sustainability." http://ws680.nist.gov/bees/ (accessed 28 December 2016).

Newsham, G., Mancini S., and Birt B. (2008). 'Do LEED-certified buildings save energy? Yes, but . . .' *Energy and Buildings* 41(8), 897–905.

RobecoSAM. (2016). "Corporate sustainability assessment at a glance." www.robecosam.com/en/sustainability-insights/about-sustainability/corporate-sustainability-assessment/index.jsp (accessed 28 December 2016).

Scofield, J. (2009). "Do LEED-certified buildings save energy? Not really . . ." *Energy and Buildings*. doi: 10.1016/j.enbuild.2009.08.006 (accessed 28 December 2016).

Spiegel, R., and Meadows, D. (2006). *Green building materials: A guide to product selection and specification*, 2nd edn. John Wiley, New York.

TerraChoice. (2009). *The Seven sins of greenwashing: Environmental claims in consumer markets*. http://sinsofgreenwashing.com/ (accessed 28 December 2016).

Torcellini, P.A., Deru, M., Griffith, B., Long, N., Pless, S., Judkoff, R., and Crawley, D.B. (2004). *Lessons learned from field evaluation of six high performance buildings*. NREL/CP-550–36290, National Renewable Energy Laboratory, Golden, CO.

Turner, C. (2006). "LEED building performance in the Cascadia Region: A post occupancy evaluation report." www.usgbc.org/resources/leed-building-performance-cascadia-region-post-occupancy-evaluation-report (accessed 28 December 2016).

Turner, C., and Frankel, M. (2008). *Energy Performance of LEED for new construction buildings*. New Buildings Institute. Vancouver, WA.

Urban Environmental Management (UEM). (2010). "Urban and ecological footprints." www.gdrc.org/uem/footprints/index.html (accessed 28 December 2016).

Woolley, T., Kimmins, S., Harrison, P., and Harrison, R. (1997). *Green building handbook: A guide to building products and their impact on the environment*. E & FN Spon, London.

world business council for sustainable development (wbcsd). (2012). "greenhouse gas protocol: a corporate accounting and reporting standard, Revised Edition" www.ghgprotocol.org/standards/corporate-standard (accessed 28 December 2016).

Chapter 5
Project delivery and pre-design sustainability opportunities

In most environmental contexts found on Earth, the built environment is an essential part of the infrastructure necessary for human survival. Buildings provide shelter from adverse climate conditions such as rain and snow, ambient temperature ranges outside human comfort levels and threatening weather conditions. They also afford privacy and security from a variety of dangers, including predatory and pest animals and malevolent humans (Allen 1980). In addition to these roles that contribute to basic human survival, built facilities serve other purposes which help to expand the quality of human life beyond basic biotic survival, including their role as infrastructure for activities such as collection, treatment and/or storage of solid, liquid and gaseous waste, provision and distribution of pure water, processing and distribution of agricultural products into food, and manufacturing and distribution of other products used by humans. This chapter explores the processes used to characterize stakeholder needs when planning a built facility, and the tools and techniques used to make decisions about how to procure the project delivery. The chapter introduces the case of the Trees Atlanta Kendeda Center, which will continue as a running case study through subsequent chapters.

The built environment lifecycle: stakeholders, processes and opportunities

The lifecycle of built facilities typically ranges from 30 to over 100 years (Yeang 1995), and typically consists of the five phases shown in Figure 5.1. External stakeholders are those entities that are based external to the boundary of the facility system, such as contractors, designers, government agencies and others. For these stakeholders, the built facility under consideration represents one of many systems in which they may be involved at any given time. For internal stakeholders, on the other hand, the system under consideration represents a major interest in which they are vested, and may be the only system affecting them at any point in time. These stakeholders, such as owners, tenants, users and clients, have a direct stake and involvement in the facility and the

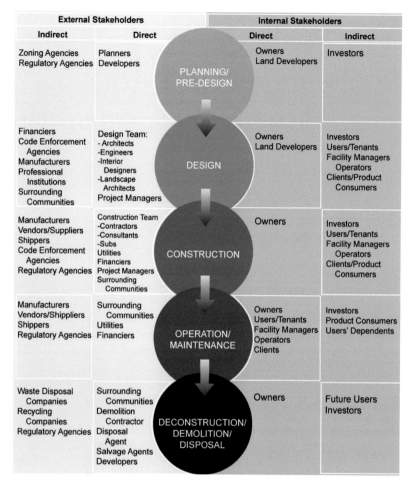

Figure 5.1 Facility lifecycle phases and their stakeholders
Source: Pearce (1999)

functions it serves: it is their needs the facility is designed and constructed to meet.

Direct stakeholders, either internal or external to the system, are those entities whose actions directly bear upon the facility system, who are directly impacted by the behaviour of the facility system, or whose needs are met directly by virtue of their interaction with the facility. Direct stakeholders include users, constructors, designers, owners and surrounding communities. Indirect stakeholders, on the other hand, have no direct impacts on the facility and may have no direct interaction with the facility at all, but nonetheless are indirectly impacted by the existence of the facility system. Indirect stakeholders include the entities that manufacture materials and supplies used to construct the facility, handle waste materials emitted by the facility, invest money in the potential of the facility, and create codes and regulations which must be observed by the facility system. In some phases of the facility lifecycle, stakeholders that were indirectly represented by other stakeholders in earlier phases of the lifecycle become direct stakeholders as their participation in the interactions of the system becomes integral. For example, future users and tenants of the facility system often do not participate directly in the development process, but are represented by the owner and/or developer of the facility during early project phases. After the facility reaches completion and begins operation, these parties become direct stakeholders due to their direct participation in the system operation. The following subsections describe the activities during each of these phases of the lifecycle.

Pre-design phase

The facility lifecycle starts with an idea or concept during the planning or pre-design phase (Halliday 1994, Hendrickson and Au 1989). This phase focuses on defining aesthetic, functional, and physical requirements and constraints for the final building. It also involves identifying any budget, schedule, legal or regulatory constraints that should be taken into account during design (Vanegas 1987). The outcome of the pre-design phase is typically a set of requirements describing the functional expectations the owner has for the facility, also known as a 'programme of requirements' (see *Programme of requirements*). In the LEED rating system, this document is also called the Owner's Project Requirements (OPR) document.

Programme of requirements

Major elements
- Vision and large-scale development goals and objectives for the project.
- Performance criteria by which the resulting project will be evaluated.
- Description of major activities to take place within and around the facility.
- Guiding principles for design and construction.
- Project definition and scope description.
- Design question, issues and restrictions to be resolved prior to or during the design process.
- Summary of referenced standards (if any) and preliminary analysis of applicable codes.
- Design goals and architectural philosophy for the project.
- Functional spaces and square footage summary.
- Description of functional space areas in terms of capacities, activities to take place in each area, general description, relationships to other functional spaces, required equipment and services, and special considerations and other needs.
- Conceptual layout of each functional space area, if known.
- Photos, descriptions, sketches or other information from comparable or similar facilities to show desired outcomes.
- References to site or development master plan elements that will affect design.
- Budget review and analysis.
- Proposed project schedule and constraints.
- Conceptual site and building plans, if available.

The purpose of the programme of requirements document is to consolidate and provide all information needed by the design team to develop a design solution that meets all stakeholder requirements. It is provided to the design firm that is contracted during the design phase to complete the design for the project. If an integrative design and delivery process is intended for the project (see *integrative processes for project design and delivery*), this process may involve stakeholders from other lifecycle phases such as construction, operations and maintenance. It may also include members of the community in which the building is located, as well as direct users of the facility. The People's Place Library in the town of Antigonish, Nova Scotia in Canada is one example of a project that included this participative planning process (see **Case study: The People's Place Library, Antigonish**).

Integrative processes for project design and delivery

The idea of integration is a fundamental change in how projects are designed and delivered. Historically, there have been gaps not only between the professional and trade responsibilities associated with different building systems, but also between the stages of the building delivery process (Figure 5.2). Whether the involvement of non-traditional stakeholders as part of the planning, design and construction process, removing barriers to cooperation among project participants from different lifecycle phases, blending technical and living systems in design solutions, or simultaneously considering the effects of change to one part of a design on other parts, integration is a fundamental idea in delivering successful green projects.

With the release of LEED v4, the green building community in the United States formally endorsed a change in terminology from 'integrated' to 'integrative' to reflect the notion that the process of

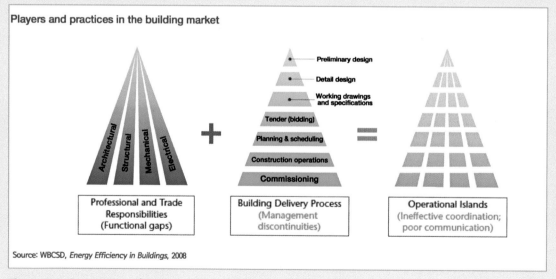

Figure 5.2 Opportunities for integration in the project delivery process
Source: WBCSD (2010)

continued

integrating is ongoing and comprehensive, and never completely finished (Malin et al. 2012). LEED's Integrative Process credit focuses both on process changes that require greater involvement among and between disciplines earlier in the planning and design process, and changes that exploit synergies between different aspects of a building that would not otherwise be caught without this broader perspective. The credit guidance explicitly identifies interdisciplinary opportunities that can be undertaken to improve overall performance at the earliest stages of design, before decisions have been made to constrain the design (USGBC 2013).

According to the USGBC, this practise 'guides the project from visioning to occupancy using a comprehensive, integrative process [that] accounts for the interactions of all building and site systems, relying on an iterative cycle of analysis, workshops, implementation, and performance feedback' (USGBC 2013). The process begins before the development of the Owner's Project Requirements (OPR) or programme of requirements document, based on an integrative design charrette that articulates the project goals, performance targets and mission statement. One of the primary intents of the process is finding solutions that optimize the consumption of water and energy over the building lifecycle through iterative improvements in complementary building systems. Key tools that can be used during the discovery or pre-design phase before the completion of schematic design include:

- **'Simple box' energy modelling analysis** that considers site conditions, massing, orientation, basic envelope attributes, lighting levels and thermal comfort ranges, plug and process load needs, and other programmatic and operational parameters.
- **Preliminary water budget analysis**, including indoor and outdoor water demand, process water demand and supply sources.

Integrated Project Delivery (IPD) is an analogous concept that provides a contractual mechanism for maintaining the focus on project outcomes through the completion of the construction phase and beyond. The American Institute of Architects defines IPD as a delivery approach that 'integrates people, systems, business structures, and practices into a process that collaboratively harnesses the talents and insights of all participants to optimize project results, increase value to the owner, reduce waste, and maximize efficiency through all phases of design, fabrication, and construction' (AIA 2007). An IPD approach to a project removes some of the barriers found in conventional project delivery systems with its focus on open communication, mutual respect and trust, and mutual benefit and reward. Such arrangements include explicit provisions for sharing sensitive or proprietary information, communication and decision making among team members, and managing risk such as legal liability. Various types of contractual agreements can be used, including project alliances or single purpose entities as well as relational contracts that persist over multiple projects. Relational contracts take advantage of the social capital that is developed from teams working together toward a common goal, allowing the associated benefits to be carried forward into other projects rather than relying on competition between firms to provide the best configuration of team members to achieve desired project outcomes.

The American National Standards Institute (ANSI) has developed an Integrative Process Standard for the Design and Construction of Sustainable Buildings and Communities to provide further guidance to the integrative process (ANSI 2012). The standard provides both a step-by-step process for using an integrative process during design and construction as well as a road map for project teams to structure their interactions in a way that results in more creative, holistic solutions (Daley-Peng 2011).

continued

References

AIA – American Institute of Architects. (2007). *Integrated Project Delivery: A Guide.* Version 1. www.aia.org/aiaucmp/groups/aia/documents/pdf/aiab083423.pdf (accessed 29 December 2016).

ANSI – American National Standards Institute. (2012). *Consensus National Standard Guide 2.0 for Design and Construction of Sustainable Buildings and Communities.* ANSI, February 2. ansi.org.

Daley-Peng, N. (2011). 'Clarifying the Integrative Design Process: ANSI Standard gets an overhaul with IP Version 2.0'. http://buildingcapacity.typepad.com/blog/2011/02/clarifying-the-integrative-design-process-ansi-standard-gets-an-overhaul-with-ip-version-20.html (accessed 29 December 2016).

Malin, N., Melton, P., and Roberts, T. (2012). *New Concepts in LEED v4.* BuildingGreen, Inc., Brattleboro, VT.

U.S. Green Building Council. (2013). *LEED Reference Guide for Building Design and Construction.* USGBC, Washington, DC, 1 July 2014.

World Business Council on Sustainable Development (WBCSD). (2010). *Vision 2050.* www.wbcsd.org/Overview/About-us/Vision2050/Resources/Vision-2050-The-new-agenda-for-business (accessed 29 December 2016).

Design phase

The second major phase of the facility lifecycle is design, where the facility is transformed from an idea to a set of buildable construction documents. Design is often divided into four major phases corresponding to the evolution of level of detail in the resulting documents:

- conceptual design
- schematic design
- design development
- construction documents.

A design review is typically conducted at the end of each phase to identify problems or opportunities in the design that could be addressed in the design solution. In an integrative design process, these design reviews will include multiple perspectives and disciplines from the whole team that can help to identify new ideas and solutions to any problems that may emerge. Conceptual design includes developing basic layouts for how the project will be situated on site, plus concepts for building massing and functional areas within each planned facility. This is the phase where the most significant decisions can be made to improve a building's lifecycle performance without negative impacts on cost. It is followed by schematic design, where scaled drawings are produced along with general specifications and relevant perspective and section drawings. Selection of major systems is also part of schematic design. In design development, the architectural design of the facility is undertaken, and engineered systems in the facility are further modelled, sized and specified. Coordination between the design of these systems is also undertaken. The final phase of design, construction documents, involves the production of documents that can be used during construction to actually build the facility, including coordinated architectural and technical drawings, details and construction specifications. These are then used as the basis for project procurement that leads to the construction phase.

Case study: The People's Place Library, Antigonish, Nova Scotia, Canada

Opened in June 2011, The People's Place is a green and socially sustainable regional library serving the Pictou-Antigonish Region of Nova Scotia, Canada. This facility, designed using a community-initiated placemaking process, functions as a civic centre that is an integral part of the community, going beyond the traditional functions of a conventional town library. The place-making process – developed by Project for Public Spaces (PPS), a non-profit planning, design and educational organization headquartered in New York City – was used to involve members of the community in the development and design of the facility and create a vision for what the facility should be.

The resulting facility hosts not only a public library but also other key community functions such as a Community Access Program site, an adult learning association, and Health Connections, which is a local health resource centre that provides a variety of health education and wellness programmes to the local community. Multi-purpose spaces, meeting rooms and even a community kitchen are available for use by non-profit organizations at no cost. The library also provides a venue for public art, including sculptures, woodworking, visual art and textiles. An outdoor patio provides seating and shade for local residents to relax and interact, and a neighbourhood café is also planned to enhance the patio space. Green features of the building include:

- adaptive reuse of an existing facility – a former grocery store – in the downtown area instead of constructing a new building on a greenfield site.
- remediation of environmental contamination found on site, including an underground storage tank and fire debris from an adjacent building.
- a closed-loop ground source heat pump system using twelve 500ft vertical wells located under the parking area.
- heat recovery ventilation, thermal mass and in-floor radiant heating to reduce heating energy required.
- upgraded roof insulation along with a high albedo roof coating to reduce cooling loads.
- envelope with high-performance barrier membrane to reduce infiltration.
- evacuated tube solar hot water heating and photovoltaic power generation.
- high-efficiency LED lighting and low-energy lighting, much of which is controlled by occupancy sensors.
- natural daylighting via skylights on the building's roof.
- high-performance glazing, including a special translucent glazing which is manufactured locally and incorporates a fibrous material that gives it similar insulating value to solid walls while blocking unwanted heat gain and glare during the day.
- energy star rated appliances and equipment.
- rainwater harvesting system used for toilet flushing and watering lawns and plantings.
- 'Stormceptor' catchment drain system to remove oils and sediment from stormwater before it enters local waterways.
- special 'hydration station' located in the facility to encourage the use of refillable bottles.
- main skylight to provide natural ventilation.
- landscaping using local, drought-tolerant species to reduce irrigation requirements.
- use of recycled content or local materials to the extent practicable.

In addition to being a green building that reflects the local community's needs and desires, the library also emphasizes its role as a provider of green knowledge in a green way. Offering free library cards to community members allows the library to offer its books to many different users, thus

continued

Figure 5.3 Green technologies employed in the Antigonish library include photovoltaic and solar hot water heating arrays on the roof, Solera energy-efficient glazing which controls heat gain/loss and glare, indigenous planting, a trellis for shading and Nanawall folding walls that allow inside activities to flow outside to patios when weather is suitable

Figure 5.4 The main entrance to the library features a bicycle rack and dog waterer for thirsty pets, along with exterior artwork to welcome building users and energy efficient windows and photovoltaic panels are also visible

continued

Figure 5.5 Exterior plantings, a sunbreak trellis, a fully opening Nanawall and an artist's bench provide a smooth integration from indoors to outdoors; photovoltaic panels are visible on the roof

distributing these resources over many uses. It also offers a special collection called the 'Green Room' that features a variety of materials on sustainable living and green construction. The Library has also developed a Green Guide to illustrate green features of the building and an Artisan Walking Guide to showcase the artwork incorporated in the building.

Sources

MacIver, M. (2011). 'The People's Place: How placemaking can build today's best libraries.' www.pps.org/blog/the-peoples-place-how-placemaking-can-build-todays-best-libraries/ (accessed 10 October 2011).

People's Place. (2011). 'The People's Place project.' www.peoplesplace.ca/ (accessed 10 October 2011).

Pictou-Antigonish Regional Library. (2011). 'People's Place Green Guide.' www.parl.ns.ca (accessed 10 October 2011).

Source: All photos courtesy of Nikki Dennis

Construction phase

The design phase of the lifecycle is followed by construction, in which workers follow the set of construction documents to build a real building in physical space that meets all of the owner's requirements (Vanegas et al. 1998). If not already involved through integrative delivery and depending on the procurement strategy employed for the project, a construction team may be selected by low bid, best value or some other selection process. The team then works together to deliver the project and hand it off to the facility owner. The outcome of the construction phase is a completely functional building ready to be occupied and used.

Post-occupancy phase

After construction, the post-occupancy phase of the lifecycle begins, during which the building is used to meet the needs for which it was designed. This phase is typically the longest phase of the lifecycle, and is also known as the operations and maintenance (O&M) phase. Operations is the process during which the facility performs its intended functions of use, while maintenance consists of all actions performed on the facility necessary to keep it in proper condition to perform its intended function. Maintenance includes activities such as changing replaceable building components as they fail, cleaning the facility, and minor repairs or replacement of building components that break down (Vanegas et al. 1998). The key stakeholder during this phase is the facilities manager, who coordinates all efforts to keep the building operating properly with respect to the needs of its occupants, and upgrade or repair its systems as necessary.

End-of-lifecycle phase

When a facility no longer meets the requirements of its owners or occupants, one possible choice is to rehabilitate or reconstruct the facility to improve its performance. This process can be even more challenging than the original construction of the building, since buildings can change significantly over their long lifecycles, and those changes may not always be well documented. In some cases, building materials and systems employed in the construction and operations of the facility may be hazardous and require special handling, making the rehabilitation of the building more costly to undertake. In other cases, key materials and systems may themselves be at the limit of their service lives, and it may make more sense to end the lifecycle of the facility. Deconstruction, demolition and disposal are three options for terminating the lifecycle of a facility, ranging from planned, careful disassembly of the facility to destructive, less careful processes and subsequent removal of materials from the site. Owners and facilities managers are key stakeholders who make the decision of how to proceed at the end of the facility lifecycle.

Opportunities for improving sustainability during the facility lifecycle

Built facilities are not independent of other systems; they could not exist without complementary technological and ecological systems to provide sources of matter and energy as inputs and sinks, consumers or storage for system outputs. As such, built facility systems are open systems: that is, systems that exchange matter or energy with their environment (von Bertalanffy 1968, Churchman 1979). The primary links between built facility systems and other technological and ecological systems are via the flows of matter, information and energy across the boundaries of the system. Figure 5.6 shows examples of flows into and out of a built facility, and how they relate to its technological and ecological context.

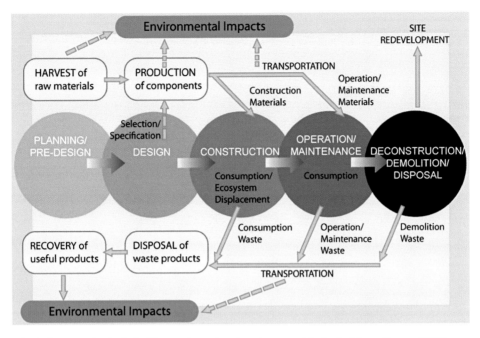

Figure 5.6 The influence of a built facility on the natural environment (adapted from Pearce 1999)

The imports and exports of materials, energy and waste from a building during its lifecycle contribute to the effects that building has on the natural environment, and the greatest opportunity to influence those ultimate impacts is during the pre-design phase of the project, before the facility has begun to take shape as an idea or ultimate product. Systems-based sustainability assessment (see *Systems-based sustainability assessment*) can be used as a technique starting in pre-design as a way to identify and track potential sustainability impacts and improvement opportunities throughout the lifecycle of the facility.

Systems-based sustainability assessment

Systems-based sustainability assessment (Pearce and Fischer 2001a, b; Pearce 1999; Pearce 2016; Pearce and Vanegas 2002) is an approach for identifying sustainability improvement opportunities in the built environment that can be undertaken to reduce a project's unwanted negative environmental and social impacts while improving its outcomes for stakeholders and other humans. Based on techniques such as Material Flow Analysis, this approach requires defining a system boundary not just for project scoping as with LEED and other rating tools, but as a mechanism for tracking and accounting for impacts the project has on other systems with which it interacts. In the context of this approach, a 'facility' is defined as a building, infrastructure system or other type of system designed and built to serve a specific function or afford a convenience or service to its stakeholders. The sustainability of a facility is the attribute which systems-based sustainability assessment seeks to assess.

continued

As discussed in Chapter 2, the sustainability of a facility depends on how well it meets the needs and aspirations of its stakeholders without compromising the ability of non-stakeholders to meet their own needs and aspirations. Non-stakeholders include both future generations as well as other humans on the planet today beyond those who directly influence or are influenced by the facility. The primary ways in which a built facility influences non-stakeholders is through competing with them for available resources and damaging or destroying natural systems that are the primary means of sustainably regenerating those resources (Pearce and Vanegas 2002). Therefore, a facility that can meet the needs and aspirations of its own stakeholders while remaining net neutral or positive with regard to resource use and damage to ecosystems will operate within the major ecological and social constraints of sustainability.

In systems-based sustainability assessment, we compare how a built facility performs in different configurations with respect to the three criteria that define sustainability:

- **Stakeholder Satisfaction** – how well does the facility meet the needs and aspirations of its stakeholders? Does it meet their expectations well enough that stakeholders do not take compensating actions that would reduce sustainability in some other way?
- **Resource Base Impacts** – over its lifecycle, does the facility contribute to the degradation or depletion of resource bases? If so, is that negative impact ameliorated either by natural regeneration of the resource base itself, or through some compensatory positive impact of the facility system?
- **Ecosystem Impacts** – over its lifecycle, does the facility contribute to the destruction or degradation of natural ecosystems? If so, is that negative impact ameliorated either by natural recovery of the ecosystem itself, or through some compensatory positive impact of the facility system?

Measuring the first criterion, stakeholder satisfaction, differs widely based on the functions and affordances the facility was designed to provide and the expectations of the stakeholders that facility was designed to serve. For example, the most basic functions provided by a building are shelter from the external environment and security from threats in that environment. However, stakeholder expectations of contemporary buildings go far beyond these basic functions. For instance, many buildings are expected to provide a comfortable climate for their occupants or users. They often provide connections or interfaces to infrastructure systems which deliver clean water and power to points of use, and convey wastewater and solid waste away from users for treatment or disposal. They may provide spaces in which particular activities can occur, such as working, sleeping, collaboration or recreation. They may even contribute to the social standing, reputation or self-actualization of their owners and communities through aesthetics or other factors. Identifying potential facility configurations that will meet stakeholder expectations is the province of design, and integrative design techniques such as simulation, design review and others can be used to evaluate the degree to which different solutions effectively meet stakeholder expectations.

The second and third criteria, impact of the facility on resource bases and ecosystems, vary based on the attributes of the facility and the ways in which it is used to achieve stakeholder satisfaction. A systematic way to account for a facility's impacts on resource bases and ecosystems is to separate these impacts into those occurring on-site vs. off-site. On-site impacts can be evaluated directly in terms of their impact on ecosystem functionality and stock of resources on site. For example, limiting the extent of building activities to a small portion of a site can reduce damage to existing ecosystems on the site. Likewise, restoring ecosystems can help to create new stocks of resources and new capacity for ecosystem services that improve overall sustainability of the facility.

continued

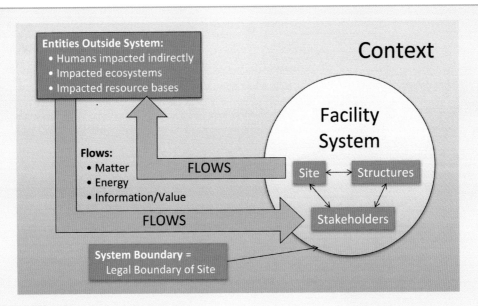

Figure 5.7 Interactions of facilities with context
Source: Pearce (1999); Pearce and Fischer (2001)

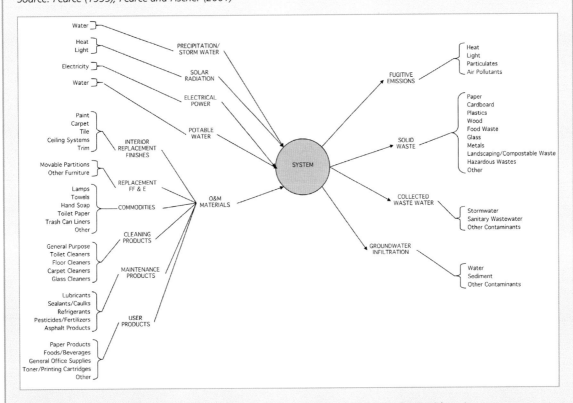

Figure 5.8 Example of cross-boundary flows for the operational phase of a facility lifecycle
Source: Pearce and Fischer (2001)

continued

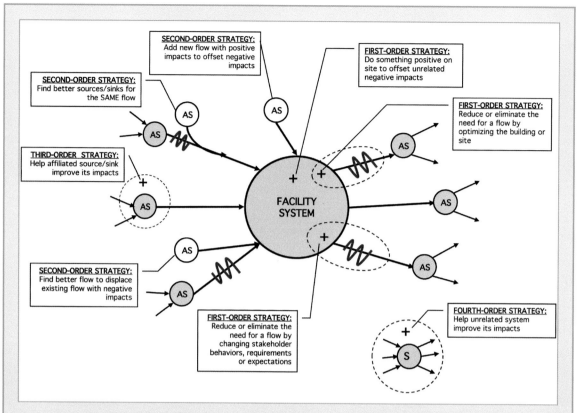

Figure 5.9 The spectrum of strategies for improving facility sustainability
Source: Pearce and Fischer (2001)

Off-site impacts can be identified by inventorying the flows of matter and energy that cross site boundaries over the facility's lifecycle (Figure 5.7), and identifying sources and sinks for those flows where impacts to non-stakeholders, ecosystems and resource bases are likely to occur. The site boundary is a useful demarcation between site and context for buildings, since it represents the legal limit of the facility owner's direct control. Cross-boundary flows are often tracked as part of an organization's accounting system, since there are generally costs associated with procuring materials and energy and disposing of wastes. Inadvertent cross-boundary flows such as stormwater runoff are often explicitly addressed by regulations, and thus must be deliberately managed or mitigated to prevent nuisance and damage to neighbouring sites (Figure 5.8). For infrastructure systems, different system boundaries can be used depending on the purpose of the analysis and what is the locus of control for potential actions to be taken.

In the systems-based approach, a facility is a 'system' of related components whose behaviour in meeting stakeholder needs and aspirations is greater than the sum of its individual parts. Understanding the facility as part of a larger network of supporting facilities (context) provides a basis for identifying specific opportunities to improve sustainability of the facility, both inside and outside the site boundaries (Figure 5.9). The facility including its site, users and structures requires a variety of connections with affiliate systems (AS) in its environment to provide materials and energy for construction and operation, and to absorb waste produced as a result of those activities.

continued

Four different types of sustainability improvement strategies can be identified based on this 'systems' view:

- **First-order strategies** are actions that can be taken directly on-site by the organization and are within the locus of control of the facility owner. They include on-site offsets for negative impacts caused by the facility elsewhere; reducing or eliminating the need for a flow with negative impacts by optimizing the building or site; and reducing or eliminating the need for a flow with negative impact by changing stakeholder behaviours, requirements or expectations.
- **Second-order strategies** can also be influenced directly by the facility owner since they involve the procurement of goods and services related to the facility. These include changing the sources or sinks of a flow to one with better impacts, replacing a flow with a different type that has better impacts, and adding new flows with positive impacts to offset negative impacts.
- **Third-order strategies** involve working with affiliate systems to help them improve their own sustainability, thereby reducing the negative impacts of using the flows those affiliate systems provide. Owners have little or no direct control over these relationships, but they may occur as part of larger strategic initiatives of the organization. Large organizations may have significant influence on the behaviour of their supply chain if the links in that chain wish to maintain good business relationships.
- **Fourth-order strategies** involve finding ways to offset negative impacts that economically or practically cannot be mitigated in other ways. These strategies include the use of offsets to achieve the desired outcome through changes to other unrelated systems, and offer the least degree of control by facility owners.

Developing detailed, quantified inventories of impacts at the whole facility scale (as is done in formal lifecycle assessment) can require considerable investment that is not always necessary to distinguish between options. In some cases, the differences between options can be captured on a relative scale that is sufficient to make informed decisions between 'more' and 'less' sustainable. The merit of the systems-based sustainability assessment approach is that it encourages identification of a range of options at multiple scales both inside and outside the system, some of which would not normally be considered with a focus strictly on the design of the facility itself. By producing a broad range of sustainability improvement opportunities, systems-based sustainability assessment generates a set of possibilities that can be explored in greater detail based on which are most likely to achieve reduced ecological and resource impacts without disrupting the satisfaction of stakeholder needs and aspirations.

References

Pearce, A.R. (1999). *Sustainability and the Built Environment: A Metric and Process for Prioritizing Improvement Opportunities.* Ph.D. Dissertation, Civil & Environmental Engineering, Georgia Institute of Technology, Atlanta, GA.

Pearce, A.R. (2016). Sustainable Urban Facilities Management. *Encyclopedia of Sustainable Technologies,* Elsevier, Oxford, UK.

Pearce, A.R., and Fischer, C.L.J. (2001). *Systems-Based Sustainability Analysis of Building 170, Ft. McPherson, Atlanta, GA.* Summary analysis report to the Army Environmental Policy Institute, Atlanta, GA, May.

Pearce, A.R., and Fischer, C.L.J. (2001). *Resource Guide for Systems-Based Sustainability Analysis of Building 170, Ft. McPherson, Atlanta, GA.* Report to the Army Environmental Policy Institute, Atlanta, GA, May.

Pearce, A.R. and Vanegas, J.A. (2002). "Defining Sustainability for Built Environment Systems," *International Journal of Environmental Technology and Management,* Special Issue on Sustainability and the Built Environment, 2(1), 94–113.

Pre-design opportunities

The pre-design process offers significant opportunity to influence the whole later life of the facility, since decisions made during this phase ultimately constrain subsequent decisions later in the lifecycle. The key steps involved in this activity include definition of the requirements for the facility being constructed, development of the formal programme of requirements and concurrent coordination of these activities with other owner requirements such as master plans to ensure that the facility being constructed is compatible with the development goals for the owner overall.

In this phase, the project team will work closely with owner representatives who will ultimately occupy and use the facility to define objectives, constraints and functional requirements for the resulting facility. These may be based upon the owner's business case for the facility, and may be enhanced and further developed by interviews, charrettes (integrative, collaborative sessions in which stakeholders intensively work together to develop a solution to a design problem), workshops, document review and other sources of information involving facility stakeholders. Interim steps of the process may include visioning and vision development, analysis of activities, benchmarking against other owners with similar or competitive facilities, and other steps involved in setting appropriate goals and objectives for the project to be a success.

Identifying facility options

The first opportunity to consider sustainability in the pre-design process is during the fundamental identification of facility needs. This process can be improved by embedding it in the context of sustainability needs analysis, where facility needs are forecast based on long-term sustainability of the organization and its physical resources as a whole. Future users and community members may also be involved (see *Case study: Bardessono Hotel*). For large, institutional owners, systematic identification of likely long-term facility needs often takes place on a periodic basis to support long-term financial planning. However, it may not always consider the potential impacts of possible future events such as more stringent legislation and regulations; resource shortages; rising energy, water and other infrastructure costs along with potentially diminished capacity; and external development encroachment and/or synergies.

Such long-term thinking can result in capital investments that benefit the organization both financially and in terms of its ability to conduct business, and prevent costly disruption of operations or service failures. The level of risk associated with the ability to conduct sustained operations for some programmes can be extremely high, with severe penalties being imposed if operations cannot continue on a sustained basis. Investing in resilient facilities that can survive likely future threats, even if they are not abrupt and catastrophic but instead gradual, is part of a long-term sustainable capital projects programme to support the organization's

Case study: Bardessono Hotel, Yountville, CA

The Bardessono Hotel is a boutique luxury hotel located in the heart of Napa Valley (Figure 5.10). The hotel includes 62 luxury rooms, a spa, four treatment rooms, a 75-ft-long rooftop infinity pool, a fine dining restaurant and a meeting space. Bardessono was developed by MTM Luxury Lodging of Kirkland, Washington and opened in February 2009. Recognizing the value of sustainability and environmental issues as well as the importance of providing a luxurious guest experience, the MTM development team was guided by the follow-ing mission statement: 'A hotel can provide a fully luxurious guest experience and be very sustain-able at the same time, and environmental initiatives can be implemented in a manner that is practical, economic and aesthetic.' To achieve those goals, Bardessono has implemented sustainable practices not only during the design and construction phase of the development but also at the operation stage of the hotel. The hotel was awarded the LEED Platinum certification by USGBC in January 2010.

Bardessono Hotel

- Project size: 55,159 sq ft, with 62 rooms and a restaurant
- Project cost: $46 million
- Sustainable features: second LEED Platinum hotel in the USA
- Developer: MTM Luxury Lodging
- Architect: WATG
- Contractor: Cello & Maudru Construction

Figure 5.10 The Bardessono Hotel

continued

Pre-design in the Bardessono Hotel

During the project delivery of the Bardessono Hotel, the project team fully recognized the value of an integrated design process including a collaborative multidisciplinary approach among all project team members. The major team members consisted of MTM Lodging (Developer), WATG (architect), O'Brien & Company (sustainable consultant) and Cello & Maudru Construction Company (general contractor). These organizations committed to share specialized expertise and coordinate their individual design efforts to achieve a well-functioning, sustainable hotel.

One of the first integrated design processes was the development of sustainable hotel guidelines that set both general goals for the project and specific parameters for hotel design, products, systems and siting. To guide this process, the integrated project team established a 'Project mission statement'. Based on this mission statement, the integrated project team established sustainable hotel goals, defined the process to achieve those established goals, and developed a clear understanding of the expected results from sustainable practices at the pre-design phase. One of the first processes was to have a 4-hour sustainable practice charrette with members of the project team (Table 5.1) to identify and evaluate the project's sustainable design features using the LEED for New Construction (LEED NC) rating system.

Project mission statement

- To make environmental responsibility one of the key design criteria for the hotel's design, development and sustainability.
- To commit, in as many steps, to minimize the environmental cost of the hotel's construction operation.
- To recommend all practical environmental actions in regards to this project and research further methods to reduce the impact of the hotel on the environment.

At the design charrette, the developer demonstrated the vision of the Bardessono Hotel, 'Luxurious Environmental Hospitality', to enhance the guest experience and satisfaction and conserve energy and resources through integrating sustainable practices. Based on the developer's vision and mission statement, the project team members identified the following sustainable strategies:

- **Air** – Noticeably fresh air in rooms; ceiling fans; operable windows; hard surfaces and minimal carpets; vacuums with highly effective filters to minimize dust and allergens; bedding and linens with no toxins or allergens; low or no-VOC interior flooring, furniture, finishes, etc.
- **Food** – Emphasize local and organic food and wine on hotel's menu.
- **Lighting** – Energy efficiency and beautiful aesthetics.

Table 5.1 Bardessono Hotel Charrette Participants

Role	Company
Client/developer	MTM
Principal resort management	MTM Management
Architect	WATG
Interior designer	Inside Out Design
Landscape architect	George Girvin Associates
Electrical engineering/CAD	Travis Fitzmaurice
Lighting designer	Luminae Souter
Mechanical engineer	Ecotope
Civil engineer	Bartelt Engineering
LEED project manager	O'Brien & Co.
Contractor	Cello & Maudru Construction

continued

- **Connection to the outdoors** – Maintain lushness in the landscape design; compost kitchen waste; look for opportunities for dual-purpose systems in landscaping.
- **Aesthetics** – Demonstrate sustainability in an elegant, not overt, way.
- **Stewardship in operation** – Natural cleaning products; kitchen waste composting; organic landscaping; high-tech systems controls; dispensers vs. bottles for shampoos; Earth-friendly linens such as Beechwood linens.
- **Education** – See things in action; practice 'apparent' sustainability; provide LEED features for others to incorporate in their designs; be a leader in the hospitality industry for sustainable design.

In addition, the project team also discussed additional considerations for LEED certification including prerequisites, LEED implementation, USGBC fees and a disclaimer.

After the first charrette, O'Brien & Co., a sustainability consultant, performed a detailed LEED analysis of the project based on conversations with team members and independent research. A report was developed to deliver the results of the LEED analysis, including an updated LEED scorecard, a detailed explanation of the project's approach and required actions for each credit, and next steps to ensure success in pursuing LEED certification. This first charrette and the report created an integrated collaborative environment among project members to establish sustainable design and construction criteria and guidelines, and also to set priorities for the project design criteria. This integrative design process at the pre-design phase actually enhanced the project team in its quest to successfully move to the design and construction phases of the project.

Sustainability features in the Bardessono Hotel

During construction and operations, the Bardessono Hotel project implemented and continued to use sustainable building practices to not only minimize negative environmental impacts, but also to enhance guests' satisfaction and comfort. In addition, the hotel also wanted to achieve cost savings by using less energy, water and natural resources over the building life. Toward this end, the project team of the Bardessono Hotel addressed the following criteria: site sustainability, efficiency with water, energy, atmosphere, materials, resources, indoor environmental quality, design innovation and operation. The Bardessono Hotel started with adopting an integrative design approach and an integrative team process among all project participants at the design phase to achieve the goals of sustainability and luxury while eliminating or minimizing the first cost premiums.

Sustainable site practices included construction activity pollution prevention strategies (Figure 5.11), public transportation access, bicycle storage and changing rooms, low-emitting and fuel efficient vehicles, stormwater management and measures to reduce the urban heat island effect. The Bardessono also adopted multiple strategies to reduce water consumption, including dual-flush toilets and low-flow fixtures, thereby reducing 34 per cent of projected water consumption, from 603,618 gallons/year to 398,400 gallons/year. In addition, drought-resistant landscaping using native California species and underground emitters for irrigation were projected to reduce 64 per cent of potable water use for landscaping, from 1,463,452 gallons/year to 526,876 gallons/year. Planting native trees and flowers at the Bardessono is beneficial, as it not only enhances the authentic experience, but also helps save water because native plants will be most suited to the climate.

The Bardessono also incorporated multiple energy-saving strategies, including passive solar design, low-e glass, sensor technologies, geothermal heat pumps combined with seventy-two 300-ft-deep geothermal wells, LED and fluorescent lamps, and 940 solar panels (Figure 5.12). Passive design strategies in the Bardessono included building orientation, daylighting using natural light,

continued

Figure 5.11 Erosion control practices at the job site

Figure 5.12 Solar panels on the roof to generate electricity

Figure 5.13 Construction waste management

Figure 5.14 Indoor environment of the Bardessono Hotel

and shading devices to mitigate solar heat gain and increase the overall energy efficiency. With these strategies, it is possible to reduce energy consumption by 31.5 per cent (2980 MBtu/year) along with solar panels that generate 889 MBtu/year. The Bardessono also purchased Green-e accredited Tradable Renewable Certificates for 70 per cent of its power requirements to encourage the development and use of renewable energy technologies.

The Bardessono salvaged multiple building materials including tufa, a type of local limestone, and various kinds of trees including walnut, cypress and redwood to reuse in the exterior and interior of the hotel. The project team also implemented a construction waste management plan to recycle construction waste (Figure 5.13), resulting in a diversion rate of 92 per cent or 1053 tons.

Since the Bardessono is a high-end boutique hotel, the hotel implemented multiple strategies related to improving the indoor environment, including daylighting with shading devices, LED and fluorescent lamps, and low-VOC building materials including all glues, adhesives, finishes, paints, carpets and fabrics used in the project. The high quality of the indoor environment (Figure 5.14)

continued

provides benefits to occupants' health, comfort and well-being. Daylight and views also provide hotel occupants and guests with a connection between indoor spaces and the outdoors.

The Bardessono also uses green practices as part of hotel operations, including:

- planted areas that are managed organically
- vegetable waste composted in an 'Earth Tub' and reused in planted areas as fertilizer
- organic and locally produced food, including fruit, vegetables and meats
- two culinary gardens to supply food to its own dining facilities, including one on-site and one off-site
- organic bath and cleaning products
- electronic and bio-diesel vehicles used by the hotel to minimize air pollution.

By implementing green practices over the building lifecycle, the Bardessono can not only achieve its goals for sustainability, but also improve guests' satisfaction with the recognition as being the most sustainable luxury hotel in the world.

Source: All photos courtesy of Cello & Maudru Construction

mission. *Passive survivability* is an important feature of sustainable build-ings and communities in contexts where threats may interrupt the supply of water or power from centralized infrastructure. These buildings and communities can continue to provide habitable shelter and other important functions even in the absence of infrastructure services. Projects such as the Visionaire building in New York City (see *Case study: The Visionaire*), designed to include on-site water collection and treat-ment systems, can offer advantages in terms of passive survivability compared other buildings that rely only on public services.

Portfolio management and resource allocation

A second opportunity occurs when developing a sustainable business case for a project competing for resources with other projects in the owner's portfolio. The quality of an investment in terms of its proactive contribution to the mission of the organization is likely to improve when links are made to formal investment criteria and the strategic plan of the organization. Project planners must understand how the project will contribute to the larger mission of the organization as well as improv-ing sustainability parameters in making a successful case to invest funds (see *Considerations for prioritizing project sustainability improvement opportunities*).

Projects often exist within a larger organizational context where a finite set of resources must be allocated across multiple opportunities relevant to the organization's mission. During project prioritization, formal consideration of sustainability criteria can help to ensure that projects with sustainability benefits are given proper weighting in the prioritization process. Sustainability prioritization criteria should all be considered from a long-term organizational perspective. Examples of such criteria include project impact on the environment, human health

Case study: The Visionaire, New York, United States

Completed in 2008, the Visionaire is a green mixed-use, high-rise residential condominium and headquarters for the Battery Park City Parks Conservancy located in Battery Park in New York. The Visionaire was developed as a public–private partnership between the Battery Park City Authority and Albaness Organization, Inc. and incorporates many green building strategies and technologies for enhancing energy efficiency, indoor air quality and water conservation such as its own waste-water treatment plant (Figure 5.15). Due to incorporating green building strategies and technologies, the Visionaire has been certified LEED Platinum for New Construction in 2009 by the US Green Building Council.

Internal environment

- A high efficiency air filtration system continually replenishes and cleanses the air in all residences.
- Twice-filtered fresh air is heated and humidified in dry winter months, and cooled and dehumidified in summer months, for optimal resident comfort.
- Programmable digital thermostats provide year-round climate control.
- Eco-friendly paints, adhesives and sealants emit no or low VOCs, helping to maintain the integrity of indoor air quality.
- The kitchen exhaust system allows residents to increase air exhaust rates on demand for improved air quality and energy efficiency.

Figure 5.15 A 25,000-gallon per day wastewater treatment plant located in the basement recycles water to resupply to toilets and provide make-up water for the HVAC system cooling tower

continued

Building systems

- A 24-hour indoor air quality (IAQ) monitoring system ensures optimal filtration.
- The building's central heating and cooling system is powered by natural gas and uses minimal electrical energy compared to a typical New York City building.
- Environmentally responsible operating and maintenance practices are used.
- Carbon monoxide monitoring and ventilation is managed by a control system in the parking garage.

Interior finishes and fit-out

- Warm-toned natural materials were selected for their intrinsic beauty, texture and environmental integrity:
  - Floors are quarter sawn/rift cut oak floors sustainably harvested from Forest Stewardship Council (FSC)-certified forests.
  - Kitchen cabinetry is made from rapidly renewable bamboo.
  - Kitchen countertops are of river-washed absolute black granite.
  - Backsplashes consist of bricks of art glass.
  - Bathroom floors and walls are made from rich limestone with glass mosaic accents.
  - Bathroom cabinetry is made from teak.

Landscaping

- Rooftop gardens provide beautifully landscaped vistas and open-air entertainment patios.
- Rapidly renewable plantings, such as bamboo trees in the indoor pool area, are carefully nurtured with pesticide-free nutrients.
- Rooftop landscaping provides an extra layer of natural insulation for the building, reducing the urban heat island effect.

Recycling

- Over 85 per cent of site construction waste materials were collected and processed for recycling.
- Construction materials contain a minimum of 20 per cent recycled content.

Residences

- To minimize energy consumption, all residences feature Energy Star® appliances.
- A single master switch at the entrance of each residence allows bedroom, living room and hallway lights to be turned off simultaneously when exiting.
- A four-pipe fan coil heating and air conditioning system provides optimal comfort and energy savings.
- Residences are prewired for automated internal solar window treatments by each resident.

Energy sources

- A 48 kW photovoltaic solar power system integrated into the building façade generates electricity for the building.
- Geothermal systems provide heating and cooling for the Battery Park City Parks Conservancy section of the building.
- 35 per cent of the building's base electricity load is provided through Green-e certified renewable energy sources.
- Natural gas fuels the heating and cooling systems for the building, contributing to a substantially lower peak demand on New York City's electric grid.

continued

Building systems

- High-efficiency natural gas-fired heaters and microturbine simultaneously generate electricity and hot water.
- The ventilation systems utilize energy recovery technology for substantial energy savings.
- The heating and air conditioning systems utilize high-efficiency pumps and motors.
- The humidification systems utilize high-efficiency natural gas fired equipment.
- Occupancy light sensors minimize electric use in common areas including corridors, stairs, garage and mechanical control rooms.

Building envelope

- High-performance exterior terracotta and glass curtain wall system provides daylighting.
- High-performance radiant low-E insulated glazing provides superior performance in solar energy control and reducing energy costs.

Materials transport

- A 500-mile resource boundary for 50 per cent of building materials reduced transportation energy consumption and pollution during construction.

Living environment

- Light is utilized as a central feature of the Visionaire to enhance the living environment for residents at every level (Figure 5.16).

Figure 5.16 Daylighting and fresh filtered air delivered to each living space provide an optimal indoor environment to residents

continued

- Generous, open floor plans integrate a flood of natural light into every residence.
- Floor-to-ceiling windows in living rooms provide an abundance of natural light.
- A sky-lit indoor swimming pool and hot tub overlook two beautifully landscaped roof gardens offering panoramic views of the city and New York Harbor.

Building design

- The building's streamlined curved façade optimizes natural light and allows generous river and city views from all exposures.
- The building's radiant low-E glass delivers superior performance in maximizing natural light transmittance and reducing harmful UV rays.
- PV panels integrated into the building's façade harvest solar power to generate a portion of the building's electric load, while reducing greenhouse gases and the use of fossil fuels.
- Public areas feature occupancy sensor controls that automatically raise and lower lights to conserve electricity when areas are not in use.

Filtered water

- Centrally filtered water is supplied to all residence baths, showers, taps and icemakers.

Conservation

- Plumbing fixtures and appliances, including front-loading clothes washers, minimize water consumption.
- Toilets by Toto provide a dual flush feature for added water savings.
- A 25,000 gallon per day wastewater treatment plant located in the basement recycles water to resupply toilets and provide make-up water for the HVAC system cooling tower (Figure 5.15).
- A roof garden catchment system harvests up to 12,000 gallons of rainwater for irrigation, helping to reduce the effects of storm surges that can overwhelm municipal water treatment facilities and flush sewage into local water supplies.

and welfare, occupant satisfaction and productivity, resource requirements and associated costs and risks over the lifecycle. In light of criteria such as these that are often considered externalities to the process, the true benefits of green or sustainable projects can be examined accurately in comparison with traditional alternatives and their advantages adequately given credit in the prioritization process.

Elements of a sustainable business case that should be developed by project advocates include considerations beyond the basic economic investment criteria that are traditionally considered. For instance, what are the environmental liabilities and risks associated with the project as opposed to not doing the project? What are human health and welfare risks and benefits? How will the project be adaptable for presently unforeseen future needs on a 25–30-year time frame? How can investing in higher efficiency or more adaptable technologies now reduce long term risk for the organization? A growing body of evidence exists to support the business case for sustainable facilities, and arguments can easily be made based on precedents set for other projects of many types (see Chapter 9).

Considerations for prioritizing project sustainability improvement opportunities

- Contributions to the organizational mission.
- Impact on the environment.
- Human health and welfare.
- Occupant satisfaction and productivity.
- Resource requirements and costs.
- Risks over the lifecycle.

Establishing functional requirements and sustainability goals

During the development of the programme of requirements for a facility, opportunities exist to incorporate sustainability considerations and goals to guide the project. As part of requirements definition, a clear set of project sustainability goals should be developed that will guide the thinking of the project team throughout the project. At the requirements definition stage, sustainability goals should address major issues such as resource consumption, stakeholder involvement and outcomes, and ecological impacts that will be affected by the project, but they should not propose specific solutions or approaches. Instead, sustainability goals should reflect measurable outcomes toward which the design and construction teams will strive in developing specific project solutions. These goals should also fit within the larger sustainability goals of the institution for which the project is being built.

Formal operationalization of project sustainability objectives is also necessary as part of the final programme of requirements for the project. This process involves specific articulation of how the project sustainability goals are relevant for each component of the programme, including implications by functional space areas, any known constraints that must be observed, and photos, descriptions, sketches or other information from comparable or similar facilities to show desired sustainability-related outcomes. The overall project sustainability goals should also be included as part of the programme of requirements.

Planning project delivery

The project procurement strategy significantly influences the outcome of the project, because it establishes the framework of responsibilities and risk allocation for the execution of all tasks required by the project (see *Key elements of a sustainable project procurement strategy*). The project procurement strategy has four elements. First, the phasing strategy defines how the project may be split into parts that are constructed sequentially or concurrently to account for constraints such as owner budget, resource availability or other factors. The next element is the delivery system to be used on the project, which defines the contractual relationships between or among project team members, and also establishes the relationships and sequences among the design, procurement and construction phases of a project. Types of project delivery systems include design–bid–build, design–build, fast-track, integrated project delivery (IPD) and others. The third element is the construction contract type, which defines the primary compensation approach within a specific contractual relationship between two parties and allocates risk among them. Types of construction contracts include firm fixed price, cost plus fixed fee and many others. Finally, the fourth element refers to the method by which members of the project team are selected. Typical selection methods include low bid selection, best value selection and qualifications-based selection.

Key elements of a sustainable project procurement strategy

- Phasing strategy.
- Delivery system.
- Contract type.
- Selection mechanisms.

A sustainable project procurement strategy should address several key issues. First, based on scheduling constraints, fundraising progress and other considerations, a phasing strategy for a capital project should seek to promote an integrative design and construction process while maximizing flexibility of delivery. Although a project may need to be phased to deliver functionality while funds are still being raised, special attention must be given to coordinating the phases of the project in terms of functionality and design, and to coordinating stakeholders to ensure that all information is properly transferred between parties. All parties must also be involved in project alignment process to ensure that everyone understands and can deliver on the sustainability goals and objectives for the project.

Second, the delivery system for the project should be selected based on maximizing effective communication among parties while minimizing temptations to cut costs on design. While many advocates of design–build delivery systems argue that design–build promotes sustainable construction due to greater integration and better communication among disciplines and reduction of adversarial relationships, others point out that this delivery system can incentivize teams to cut costs on design to maximize profit. When properly executed, design–build can enhance the constructability and sustainability of the design, but careful measures should be taken to ensure that the design meets all objectives as the process proceeds. Other possible project delivery systems such as IPD can also offer sustainability advantages (see *Integrative processes for project design and delivery*, page 232).

Third, the project contract type can also be an entry point for sustainability based on the nature of the project itself. For instance, contractual incentives such as risk and revenue sharing can be built into the agreement between parties to motivate sustainable outcomes such as recycling construction waste. Techniques such as relational contracting are another example, where the trust between parties in a contractual relationship is maintained across multiple projects and affords implicit terms and understandings that may not be laid out specifically in the contract but are inherent in how the parties relate to one another in the interest of maintaining long-term relationships. For all contracts, the party playing the construction management role should be an educated advocate with the authority to enforce sustainability objectives and ensure coordination among other team members.

Finally, all selection mechanisms employed on the project, from selection of programming and design agents, to contractor and subcontractor selection, to selection of vendors and products for the project, should be aligned with sustainability goals and objectives wherever possible. While owners do not traditionally have the authority to specify specific vendors or subcontractors inside the overall construction contract, they can include performance criteria and documentation requirements as part of all contracts to ensure that these decisions made by their agents fall within larger sustainability goals. For instance, the US federal government has a broad variety of contracting requirements that must be met by all vendors, ranging from use of recycled content and bio-based products, to use of socially disadvantaged or underrepresented subcontractors for a portion of the work, to required demonstration of experience on green building projects. These requirements can serve as precedents for other organizations seeking to develop green procurement programmes.

With these tactics in mind, the remainder of this chapter introduces a case study of the LEED Platinum-rated Trees Atlanta Kendeda Center, located in Atlanta, Georgia, USA. In this chapter, the case is presented and the primary players are introduced. The chapter concludes with a look ahead at possibilities for pre-design in the year 2020.

Case study: Trees Atlanta Kendeda Center, Atlanta, GA

Trees Atlanta is a non-profit organization dedicated to the protection and improvement of the urban environment through the conserving and planting of trees (see Figure 5.17). After more than 20 years of operating, Trees Atlanta constructed a new office space in 2008 by renovating an existing facility donated by Chip Robert & Company. Using resources generated from a 3-year capital campaign, Trees Atlanta developed a new office, the Trees Atlanta Kendeda Center (here called the Trees Atlanta building), with tremendous support from the local community. Since Trees Atlanta's mission is to 'protect and improve Atlanta's urban forest by planting, conserving and educating', the Trees Atlanta building has implemented many sustainable design and construction features along with providing comfortable spaces including staff offices, support spaces, a working exterior yard for tree planting and maintenance work, and an education centre.

Figure 5.17 Trees Atlanta mission
Source: Trees Atlanta

continued

Project site and description

The project site is located in the Reynoldstown neighbourhood of Atlanta (Figure 5.18) – just north of Interstate Route 20 and a few miles east of I-75/85. The Trees Atlanta project is on the periphery of the Cabbagetown and Reynoldstown neighbourhoods, and adds to the exciting changes that are occurring in the adjacent Memorial Drive corridor. This urban location allows the organization to reside in the heart of the city, adjacent to the future beltline development and as part of an emerging mixed-use zone. The existing site is just under an acre and includes an existing vacant manufacturing building built in the 1940s with some additional miscellaneous structures (Figure 5.19). Since one of the first goals for sustainability is reuse, the existing vacant building has been renovated and updated to provide a state-of-the-art sustainable building housing both the non-profit's headquarters and meeting and event spaces used by the community. In addition, the Trees Atlanta building also includes the Bartlett Tree Experts Urban Forestry Demonstration Site and the Home Depot Program Operations Center, where volunteers meet each Saturday before planting and maintaining trees throughout the community.

Building overview

The aim of the Trees Atlanta project was to remodel and upgrade the existing warehouse building into a state-of-the-art office and education building incorporating many sustainable design and construction strategies in order to achieve the goals of sustainability. The key parameters of the Trees Atlanta project are summarized in Table 5.2. Table 5.3 shows the key stakeholders involved in the project. Figure 5.20 shows the site plan and floor plan for the Trees Atlanta building.

Pre-design of the Trees Atlanta building

As the first step in the building development process, pre-design includes the development of the facility's functional and operational requirements as well as sustainable design and construction goals. Since decisions made during pre-design not only set the project direction and sustainability of the project, but also have to prove cost-effective over the life of the

Figure 5.18 Existing location of the Trees Atlanta building within a previously developed industrial district. *Source: Trees Atlanta*

Figure 5.19a, b Front view of the existing building prior to construction and a rendering of the same perspective for the final Trees Atlanta building. *Source: Trees Atlanta*

continued

Table 5.2 Trees Atlanta building overview

Project name	Trees Atlanta Building	Owner type	Non-profit organization
Project location	Atlanta, GA., USA	Occupant type	Non-profit corporation
Building type	Office and community centre	Date of occupancy	January 2008
Project cost	$2.5 million	Construction type	New construction: 18% Renovation: 82%
Site conditions	Previously developed	Project scope	Single building
Total site area	44,321 sq ft	Total building footprint	13,442 sq ft

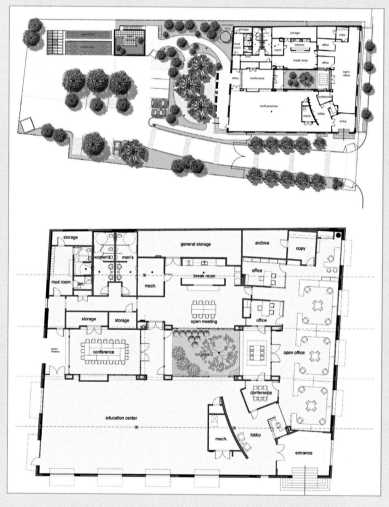

Figure 5.20 Site plan and floor plan of the Trees Atlanta building, *Source: Trees Atlanta*

continued

Table 5.3 Project participants

Stakeholder	Name of firm
Owner	Trees Atlanta
Architect	Smith Dalia Architects
Civil engineer	Eberly and Associates
Structural engineer	Palmer Engineering
Mechanical / electrical / plumbing engineering	Moinar Jordan
General contractor	Gay Construction

project, the project team for the Trees Atlanta building faced a significant challenge to establish sustainable project goals, define the process with which to achieve these goals, and develop a clear understanding of the expected results. Integrative design is also a cornerstone for developing a sustainable building, and results in efficiently combined systems for coordinated and environmentally sound products, systems and design elements. The project team, including Smith Dalia Architects and Gay Construction, intended to use an integrative building design approach, which was launched during the pre-design phase. The project also included a team approach to achieve the goals of sustainability in the project. As a key part of these processes, the Trees Atlanta building project implemented the following practices at the pre-design stage:

- Have a kick-off meeting.
- Establish a vision statement that embraces sustainable principles and an integrative design team approach.
- Establish the project's sustainable design and construction goals.
- Establish sustainable design and construction criteria.
- Set priorities for the project design criteria.

The first step at the pre-design stage was to have a kick-off meeting among three key parties including owner Trees Atlanta, Smith Dalia Architects and Gay Construction which would serve as the general contractor, in order to develop functional requirements and sustainability goals for the building. At the first meeting, the project team developed the vision statement for the Trees Atlanta building.

Vision statement for the Trees Atlanta building

Since Trees Atlanta is a nationally recognized citizen group that protects and improves Atlanta's urban forest by planting, conserving and education, the Trees Atlanta building has to not only provide comfortable spaces but also must achieve the goals of sustainability through the implementation of sustainable design and construction strategies.

Based on the vision statement, the project team established the project's sustainable design and construction goals. The sustainable design and construction goals were organized into five areas: sustainable sites, water efficiency, energy efficiency, materials and resources, and indoor environmental quality.

continued

After setting sustainable design and construction goals at the kick-off meeting, the project team planned two different charrettes for the project. The first charrette was to focus on developing detailed sustainable design and construction criteria and setting priorities for the project's design. The second charrette was specifically related to native and adaptive planting, since Trees Atlanta is dedicated to planting trees and securing a green belt in the City of Atlanta. From the sustainable design charrette, the project team was able to achieve the following goals:

- Develop early consensus on project design priority.
- Generate early expectations for final energy and environmental outcomes.
- Provide early understanding of the potential impact of various sustainable design strategies.
- Develop a checklist of sustainable strategies (LEED credit checklist).
- Initiate a design process to reduce project costs and schedules, and obtain the best energy and environmental performance.
- Identify project strategies for exploration with their associated costs, time constraints and the needed expertise to eliminate costly 'surprises' later in the design and construction processes.
- Identify partners, available grants and potential collaborations that can provide expertise, funding, credibility and support to the project.
- Set a project schedule and budget with which all team members feel comfortable.

Sustainable design and construction goals: Trees Atlanta building

Sustainable sites

- Minimize stormwater run-off.
- Use native/adaptive planting.
- Achieve site light pollution reduction.
- Install green roof for demonstration purposes.

Water efficiency

- Collect rainwater for onsite (and possible off site) uses.
- Minimize water consumption in the building by waterless fixtures.

Energy efficiency

- Achieve 50 per cent reduction in energy consumption.

- Utilize solar energy (photovoltaic and thermal).
- Keep the systems simple.

Materials and resources

- Use 100 per cent FSC certified wood products.
- Install recycled/refurbished furniture.
- Use locally harvested building materials.
- Install recycled content materials.

Indoor environmental quality

- Provide access to daylight and public amenities.
- Use non-volatile organic compound materials.

The first charrette was comprised of two sessions (morning and afternoon – see *Structure of a one-day charrette*). At the end of the first charrette, the participants developed a charrette report that included objectives, sustainable strategies to achieve the objectives and strategies for measurement and concerns. Figure 5.21 shows the sustainable development strategies for energy, and Table 5.4 describes the part of the charrette result report pertaining to energy and sustainable sites.

continued

Structure of a one-day charrette

In the morning session, tasks include:

- Introduction, ground rules, goals of the day, overview of the charrette process
- Project description
- Introduction to sustainable design and construction
- Why sustainable design and construction for Trees Atlanta building?
- Integrative and sustainable design process

The afternoon session includes:

- Small group meetings
- Brainstorming: develop strategies to reach the goals
- Facilitated session: what are the objectives, strategies, measurement needed?
- Outcome: cost premium and schedule effect
- Whole group meeting – Summarize the small group meetings
- Closing the charrette – Where will the project go from today?

The second charrette was specifically related to landscaping and planting trees and vegetation at the project site. At the second charrette, the Trees Atlanta project team set up the following objectives based on the first charrette:

- types of native and adaptive plants and trees
- innovative site details and products
- innovative planting technologies
- viable site technologies
- irrigation system including the rainwater harvesting system
- schedule of Trees Atlanta building
- interactive plan for education at the site.

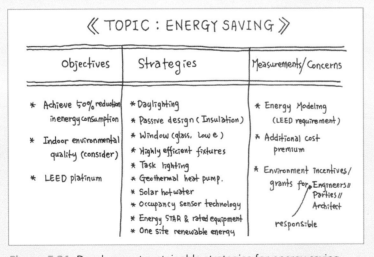

Figure 5.21 Development sustainable strategies for energy saving

continued

Table 5.4 Charrette result report from the Trees Atlanta building

Priority	LEED credit	Objectives	Strategies	Measures	Concerns
1	EA	• Energy efficiency (reduce 50% energy consumption in the building)	• Daylighting • Passive design (Insulation) • Glazing (low-e glass) • Highly efficient fixtures (T-5 lighting) • Task lighting • Geothermal heat pump • Solar hot water heater • Occupancy sensor technology • Energy Star rated equipment and appliance • Carbon offsets through purchase of green power credits	• Energy modelling by the design team • Energy Star rated equipment and appliance	• Additional cost premiums • Government incentives and grants investigation
2	SS	• Restore native and adapted vegetation • Plant trees • Provide habitat for vegetation and wildlife	• Plant various types of native and adapted vegetation and trees • Possibly adopt Silva Cell technology • Green roof technologies • Rainwater harvesting system for irrigation • Courtyard at the middle of the building • Permeable concrete for parking spaces • Bioswales (vegetated water retention)	• Stormwater modeling of site features • Monitoring of water levels in rainwater storage tank	• Second charrette for planting and landscaping by Trees Atlanta volunteers • LEED points (integrative approach)

Figure 5.22 The second charrette for the Trees Atlanta project. *Source: Trees Atlanta*

continued

Sustainability opportunities and challenges

Opportunities

- soil condition
- paved and open ratio
- budget for landscaping and planting trees and vegetation.

Challenges

- power lines on the site
- building schedule
- approval process.

In addition, the second charrette (Figure 5.22) defined potential opportunities and challenges associated with planting trees at the project site. From the charrette, the project team identified the following opportunities and challenges related to landscaping and planting trees and vegetation. Through the in-depth charrette, the project team achieved the goals set forth for the event and solved challenges associated with the project.

After completing two charrettes in the Trees Atlanta building project, the design team led by Smith Dalia Architects moved forward into the design phase, incorporating as many of the objectives and strategies (mainly sustainable design and construction strategies) presented at the charrettes as could be incorporated in an integrative manner. In addition, the design team also began to investigate the potential grants and incentives related to sustainable design and strategies that could be integrated into the project. With the completion of these background tasks, the pre-design phase concluded and the design phase began.

Pre-design in the future

What can we expect of the pre-design process in the coming years? Almost certainly new types of project delivery approaches will evolve based on new ways of thinking about project teams and risk sharing. However, the most significant changes are likely to be driven by new ways of collecting, analysing and using data.

Improved understanding of the lifecycle behaviour of buildings constructed using sustainable design and construction strategies will lead to a greater confidence in planning for the costs associated with these projects. Especially for owners of large capital facility portfolios, better tracking of building performance will enable real-time management to optimize investment in improving existing buildings and balancing facility needs with resources. This information will play a key role in pre-design because it will enable owners to better predict costs as they define facility needs and set sustainability goals. New models of building lifecycle performance that incorporate human and environmental factors will improve the accuracy of long-term forecasting.

With regard to integrative design and delivery processes that include collaborative opportunities such as charrettes, innovations in social networking, virtual collaboration and virtual reality will enable new stakeholders to be more actively involved in the pre-design process in new ways. At present, much of the creativity resulting from charrettes comes from the energy and enthusiasm of participants interacting together during intense work sessions. Quality facilitation is also key to making the most of these sessions (Roberts 2016). Emerging information technology will make this type of interaction easier without the requirement for everyone to be located centrally. Rapid retrieval of information from information systems such as Google Earth about the project site will be immediately available to answer questions that arise, and new easy-to-use, open-source tools will continue to evolve that can support rapid visualization and prototyping of ideas (see *Case study: Freedom Park, Naples, FL*). Indeed, the skeleton of a building information model may be developed as part of the charrette process instead of later in the process as it occurs today.

Case study: Freedom Park, Naples, FL

Collier County Stormwater Management Division

Completed in 2009, Freedom Park is 50-acre natural stormwater management facility constructed by Collier County in Florida on the Gordon River (Figure 5.23). Flood control, water quality improvement, habitat restoration, public recreational use and environmental education are all achieved in a beautiful park setting designed as a community asset. The project is the culmination of years of planning to conserve, restore and protect the adjacent Naples Bay and its tributary ecosystem from adverse effects of urbanization. The objectives of the park were to:

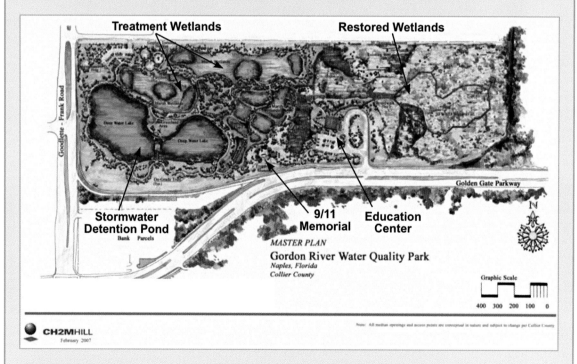

Figure 5.23 Freedom Park serves multiple purposes, including stormwater management and treatment, education and recreation

continued

- Detain stormwater before it is discharged to the Gordon River and lessen chronic flooding concerns in the Gordon River Extension Basin.
- Improve river water quality by wetland treatment of stormwater and base flow.
- Restore and rehydrate the rare subtropical bald cypress floodplain swamp wetlands.
- Create an attractive facility that is well-suited for a range of passive and active recreational uses (Figure 5.24), including a public centre for environmental education and nature study.
- Conserve upland and wetland habitats for public open space in a developed urban area.
- Commemorate the events occurring during the 9/11 terrorist attack in the United States.

The total project cost was US$30.2 million, including land acquisition, design costs and facility construction. Value engineering was used to identify US $600,000 in cost savings that could be used as a buffer for contractor needs during construction, resulting in no change orders during the project.

Figure 5.24 Boardwalks and walking trails provide access to the many ecologically restored habitats throughout the park

The project combines ponds, wetlands, habitat restoration, trails, boardwalks, educational facilities and natural landscaping within a passive park setting designed to build and sustain public use and interest. The treatment system consists of a 5-acre pond for stormwater storage, followed by constructed marshes designed to enhance stormwater polishing by submerged aquatic vegetation and native herbaceous marshes that remove harmful pollutants from the stormwater and river water prior to discharge to the on-site natural wetlands. A passive periphyton marsh treatment system included in the park is modelled after natural chemical processes from the nearby Florida Everglades and removes phosphorus from stormwater in the park.

Figure 5.25a, b During design, visualization techniques such as Google Sketchup models (a) and renderings (b) were used during community meetings to obtain public input and increase understanding and acceptance of the project

continued

More than 10,000 new native plants and trees were planted during construction. Shallow marshes are populated with native emergent marsh species, including pickerelweed, spikerush, sawgrass, duck-potato and fireflag. Deep marshes in the park include white water lily as well as native species of submerged aquatic vegetation. Extensive infestations of non-native vegetation were removed and the area was replanted with native subtropical cypress floodplain species during construction. Approximately 20 acres of native upland habitat was restored from use as citrus groves and is preserved as an upland preserve area designed for gopher tortoise conservation.

The sustainably designed 2500-sq-ft educational centre provides a centre of activity to the park and both an origin and destination to site visitors. It includes restrooms, six lookout pavilions, water fountains and walking trails. Educational and informational signage is available throughout the park. There are 3,800 feet of boardwalks throughout the park and over two miles of walking trails. In its first year of use, over 18,000 people visited the park for recreation as well as educational programmes such as summer camps, natural history lectures, site walks, silent guided tours and home schooling support.

Workshops were conducted with the public during the design process to capture community input and build support. A variety of visualization techniques including Google Sketchup and traditional renderings (Figure 5.25) were used during the process, resulting in better public understanding and acceptance of the proposed project.

Sources: All images courtesy of Collier County Growth Management Division and CH2MHill

A larger body of historical data from existing high-performance buildings, coupled with interactive pre-design models, will enable quick trade-off analysis of options without intensive effort. This body of knowledge will enable quick visualization of alternatives, increasing the ability of laypeople and non-experts to substantially contribute their reactions and ideas as part of the process. Ultimately, better access to historical data, increased participation of a broader range of stakeholders and better ability to predict future performance will be the foundations of pre-design in the future.

Discussion questions and exercises

5.1   Consider the building in which you are presently located. When was it built? What was its design life or projected service life? What do you expect will be its actual service life given development trends in your area?

5.2   Who are the external and internal stakeholders for the building in which you are presently located? Identify the major stakeholders for earlier phases of the building lifecycle. For example, what firm(s) designed the building? Who built it? What developers were involved? What public agencies played a role?

5.3   Contact an owner organization or design firm and obtain a copy of a programme of requirements for one of their projects. Public agencies such as local schools or governments are likely to have such documents and may be willing to make them available. What elements are included in the programme of requirements? Is sustainability specifically addressed? If not, how could it be incorporated as part of the programme?

5.4   Obtain a copy of the construction documents for a local project. If possible, obtain the documents for the same project for which you reviewed the initial programme of requirements. What are the major sections of the construction documents? Is sustainability specifically addressed? If not, how could it be incorporated as part of the construction documents? How does the final set of construction documents measure up in terms of the original programme of requirements?

5.5   Identify a local public sector project in the planning or design phases of its lifecycle. Contact project representatives and determine the schedule for any upcoming public review meetings. If no meetings are presently scheduled, ask to be contacted when a date has been determined. Attend the public meeting if possible. What opportunities are there for input to the project? How could sustainability issues be raised and addressed in these meetings?

5.6   How will major trends such as more stringent legislation and regulations, resource shortages, rising energy and water costs, competition for infrastructure capacity, and projected future development influence projects in your community? What is likely to be the most significant influence in the next 5 to 10 years?

5.7   For the project identified earlier, contact the owner to determine what project procurement strategy has been selected for the project, and why. What delivery system, contract type and selection methods will be used? Is the project typical in this regard? Is a phasing strategy planned for the project? If so, what major phases are planned? Has sustainability been implicitly or explicitly taken into account in developing the project delivery strategy? If so, how?

5.8   Contact the local chapter of professional design associations (such as the American Institute of Architects in the United States) and ask about charrettes that are planned for any upcoming projects. Search the Internet to identify any organizations in your area that offer charrette facilitation services.

5.9   Plan to attend a charrette for a local project as an observer if your schedule allows. How is the charrette organized? What techniques and technologies are used to document participant input? How is sustainability taken into account? What is the dynamic among charrette participants? What is the outcome of the event?

References and resources

Allen, E. (1980). *How Buildings Work: The natural order of architecture*. Oxford University Press, New York, NY.

American Hospital Association. (2015). "Integrating Sustainable Principles into the Project Delivery Process." *Sustainability Roadmap for Hospitals*. www.sustainabilityroadmap.org/topics/sustprinciples.shtml (accessed 14 January 2017).

Churchman, C. W. (1979). *The Systems Approach and Its Enemies*. Basic Books, New York, NY.

Halliday, S. P. (1994). *Environmental Code of Practice for Buildings and their Services*. Building Services Research and Information Association, Bracknell, Berkshire, UK.

Hendrickson, C. T. and Au, T. (1989). *Project Management for Construction*. Prentice-Hall, Englewood Cliffs, NJ.

Korkmaz, S., Swarup, L., Horman, M., Riley, D., Molenaar, K., Sobin, N., and Gransberg, D. (2010). *Influence of Project Delivery Methods on Achieving Sustainable High Performance Buildings: Report on Case Studies*. Charles Pankow Foundation. www.dbia.org/resource-center/Documents/CPF_ThrustII_05212010_Final.pdf (accessed 14 January 2017).

Lindsey, G., Todd, J.A., Hayter, S.J., and Ellis, P.G. (2009). *Handbook for planning and conducting charrettes for high-performance projects*, 2nd ed. National Renewable Energy Laboratory (NREL), Golden, CO. www.nrel.gov/docs/fy09osti/44051.pdf (accessed 14 January 2017).

Pearce, A. R. (1999). *Sustainability and the Built Environment: A process and metric for prioritizing improvement opportunities*. PhD dissertation, School of Civil & Environmental Engineering, Georgia Institute of Technology, Atlanta, GA.

Roberts, T. (2016). "How to Run a Great Workshop: 37 Tips and Ideas." *Environmental Building News,* 25(5), May 3. www.buildinggreen.com/feature/how-run-great-workshop-37-tips-and-ideas (accessed 29 December 2016).

Todd, J.A., and Lindsey, G. (2016). "Planning and Conducting Integrated Design (ID) Charrettes," *Whole Building Design Guide.* www.wbdg.org/resources/planning-and-conducting-integrated-design-id-charrettes (accessed 14 January 2017).

Toronto Artscape, Inc. (2017). "Planning a Visioning Charrette." *DIY Creative Placemaking Toolbox.* www.artscapediy.org/Creative-Placemaking-Toolbox/Who-Are-My-Stakeholders-and-How-Do-I-Engage-Them/Planning-a-Visioning-Charrette.aspx (accessed 14 January 2017).

Vanegas, J. A. (1987). *A Model for Design/Construction Integration During the Initial Phases of Design for Building Construction Processes.* PhD dissertation, Civil & Environmental Engineering, Stanford University, Palo Alto, CA.

Vanegas, J. A., Hastak, M., Pearce, A. R., and Maldonado, F. (1998). *A Framework and Practices for Cost-Effective Engineering in Capital Projects in the A/E/C Industry.* RT 112, Construction Industry Institute, Austin, TX.

Von Bertalanffy, L. (1968). *General System Theory.* George Braziller, New York, NY.

Willis, D. (2010). "Are Charrettes Old School?" *Harvard Design Magazine,* 33/Design Practices Now, Vol. II. www.harvarddesignmagazine.org/issues/33/are-charrettes-old-school (accessed 14 January 2017).

Yeang, K. P. (1995). *Designing With Nature.* McGraw Hill, New York, NY.

Chapter 6

Sustainable design opportunities and best practices

In the design phase of project delivery, many opportunities exist to influence the sustainability of the resulting facility project. This chapter describes the integrative design process that is key to developing sustainable projects, and presents a range of strategies that can be incorporated in design to improve project sustainability. It concludes with two case studies to illustrate these strategies in practice: the design phase of the Trees Atlanta project, and the Bank of America Tower in New York City.[1]

The integrative design process

Integrative design of all facets of the project with each other is a key element of capital project sustainability. Figure 6.1 shows a schematic of a sustainable design process for buildings with details of key sustainability-related processes that can be undertaken to ensure a sustainable design outcome. Selected steps in each of three tracks (resource analysis and modelling, materials analysis and ecological site planning) are associated with each of the four phases of the design process. Crosscutting all of these processes are integration milestones, indicated by the green bars, involving activities such as interdisciplinary charrettes, facilitated design reviews and project alignment sessions to ensure that all sustainability strategies are being developed in complementary ways throughout the process. A similar process can also be applied for integrative design of infrastructure systems, with a focus on the types of resources employed for each specific type of project.

As with the Integrative Process used in pre-design, integrative strategies can also be employed in design to anticipate and prevent problems in later lifecycle phases. The Prevention through Design (PtD) process is one such strategy employed during the design phase that can be incorporated to prevent future health, safety and other sustainability problems for building occupants, construction workers and other key stakeholders (see *Prevention through Design*).

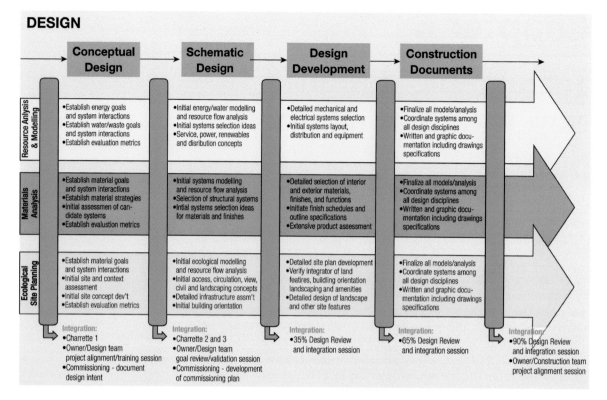

Figure 6.1
Integrative sustainable design process for buildings

From a health and safety standpoint, this process is rewarded under the LEED v4 rating system due to its positive impacts for health, safety and well-being of building stakeholders. However, the concept can also be applied from a broader sustainability standpoint to address a wider range of sustainability 'hazards'. Using key concepts from Chapter 2, a hazard from a sustainability standpoint can be defined as:

- anything that reduces the carrying capacity of a natural ecosystem to provide services (i.e. environmental damage);
- anything that decreases the sustainable yield of a resource base or reduces functionality of resources currently in circulation (i.e. resource depletion);
- anything that threatens to decrease human health, safety or well-being.

Using hazard assessment and control methods from occupational safety and health, Prevention through Design measures that benefit both humans and the environment can be identified and considered at the earliest stages of project design, while changes can be easily made without significant cost to the project.

Prevention through Design (PtD): Creating safer, more sustainable projects

Prevention through Design (PtD) is a philosophy and approach to projects that involves identifying potential hazards during the earliest stages of a project, and changing the design itself to avoid those hazards during the construction or occupancy phase of the building's lifecycle. While not all hazards can be effectively mitigated or eliminated through design changes, PtD provides a framework to identify and consider improvements based on lifecycle consideration of possible hazards.

In the context of safety and health, a *control* is a measure or intervention undertaken to mitigate, reduce or eliminate a hazard that could lead to injury or sickness. Familiar types of health and safety controls include personal protective equipment worn by workers to protect them from injury on the job site, ranging from the common hard hat, gloves and safety glasses to more complex full body suits and breathing devices to protect from chemical hazards. While these measures can be effective in preventing injuries, they can also have other unwanted impacts such as discomfort for employees, additional cost for contractors and reduced productivity. They also require knowledge and appropriate behaviour to work effectively.

Other types of controls also exist that can mitigate potential safety and health hazards, such as safety training and work policies, equipment guards or warning systems on hazardous equipment.

At the highest level of effectiveness is the elimination or prevention of a hazard by changing the design of the project (i.e. Prevention through Design). Personal Protective Equipment (PPE) is at the lowest end of the spectrum, since it assumes that the hazard cannot be otherwise controlled and puts the burden on individual workers to know what to do, then change their behaviours to achieve the desired outcome. Controls can be organized into different categories based on their relative effectiveness and reliability in preventing hazards to workers. Known as the Hierarchy of Controls (Figure 6.2), this framework can be used to identify and prioritize possible health and safety interventions, as well as interventions to address other kinds of sustainability problems encountered during the project lifecycle.

A similar approach can be used to address other kinds of sustainability hazards or threats, such as resource depletion, damage to or destruction of natural ecosystems and others. Figure 6.3 shows an adapted Hierarchy of Controls for Sustainability Hazards. As with occupational safety and health, the most effective solutions eliminate potential problems by designing them out of solutions, while lower-level solutions include fixes for sustainability problems that are allowed to happen.

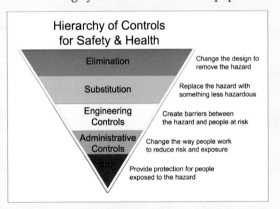

Figure 6.2 Hierarchy of Controls for occupational safety and health

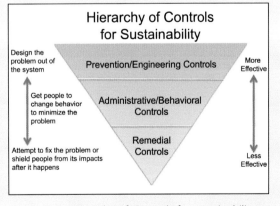

Figure 6.3 Hierarchy of Controls for sustainability

continued

Table 6.1 lists some examples of potential sustainability challenges along with controls at different levels of the hierarchy. Not all problems can necessarily be solved using higher levels of controls within a project's design parameters, and generally there are tradeoffs among solutions that will influence the feasibility and attractiveness of options. Ultimately, the integrative project team must consider control opportunities from a variety of perspectives to optimize project outcomes. However, considering alternatives using this approach is a worthwhile exercise at the early planning and design stages where significant changes can be made at very low cost.

Table 6.1 Sustainability challenges and potential controls

Challenge	Remedial controls	Administrative/ behavioral controls	Prevention through design controls
Air pollution resulting from lawns that require regular mowing	Use a more efficient gas-powered mower	Mow less frequently; adjust expectations regarding lawn appearance	Choose a ground cover species that does not require mowing
Energy used by building lighting	Install on-site renewable energy to power lights	Encourage users to turn off lights when not using space	Design building to incorporate natural daylight for lighting needs
Destruction of site ecosystems during construction	Restore ecosystems after construction is complete	Limit disturbance area during construction; purchase offsets	Reuse an existing building or already developed site where no ecosystems exist

Best practices for sustainable design

In conjunction with a more integrative design process, many opportunities exist to adapt the design elements of a built facility to make its lifecycle more sustainable. The following subsections describe best practices in six major categories:

- **Sustainable sites** – this category includes choosing a good site for a facility, efficiently placing the facility on the site avoiding damage to the site, and restoring the quality of ecosystems on site. It applies to both buildings and infrastructure systems.
- **Energy optimization** – this category includes eliminating unnecessary use of energy, using energy more efficiently, balancing energy demands and seeking alternative sources for energy. It applies primarily to buildings.
- **Water and wastewater performance** – this category includes eliminating unnecessary use of water, using water more efficiently and seeking alternative sources and sinks for water and wastewater. It applies primarily to buildings.
- **Materials optimization** – this category includes eliminating unnecessary use of materials, using abundant renewable materials, using multifunctional materials and seeking alternative sources of

materials. It also includes eliminating or preventing waste, reusing waste within the facility system, sharing it with other systems and storing it for future use. It applies to both buildings and infrastructure systems.

- **Indoor environmental quality** – this category includes preventing problems at the source, segregating polluters, taking advantage of natural forces and giving users control over their environment. It applies primarily to buildings.
- **Integrative strategies** – this category includes practices that result in multiple benefits from one action. It includes capitalizing on construction means and methods, making technologies do more than one thing and exploiting relationships between systems. It applies to both buildings and infrastructure systems.

Together, these practices illustrate the broad spectrum of actions that can be taken to reduce negative impacts of the built environment on the natural environment during the design phase.

Sustainable sites

The first area of best practice involves facility site and landscape. After a need has been established for a new facility, the most important decision for a project is the selection of a good site. After the best site has been chosen, the next most important decision is where to locate structures and developed areas on the site. These two decisions will affect a building's ongoing performance over its lifecycle. Next, the development of the site itself should use low-impact principles and features. Lastly, restoring ecosystems on the site to their best quality will help to keep the site functioning well.

Quite possibly the most important decision when constructing a facility is to choose a good site. The choice of site will not only affect the performance of the facility itself in terms of energy and environmental impact, it will also govern how much and via what mode people travel to and from the facility. For infrastructure projects especially, choosing the best site can strongly influence the ability of the project to perform efficiently and effectively over its lifecycle. For example, locating power and water/wastewater plants centrally within a service area can reduce pumping or transmission requirements and line losses. The terrain, existing development and other considerations will also strongly influence the best choice of site for an infrastructure project. In all cases, the lifecycle performance of the project should be taken into account when siting infrastructure projects, as should social equity and other important considerations. Siting to avoid damaging high-quality ecosystems and to reduce the threat of damage from natural forces such as flooding is also key. When siting infrastructure projects with very long service lives, possible impacts of climate change and population growth and migration should also be taken into account, including sea-level rise, expansion and evolution of neighbourhoods and changes in temperature.

> **Priorities for sustainable sites**
>
> - Choose the best site for the project.
> - Choose the best location on the site for the project.
> - Develop the site using low-impact principles and features.
> - Restore ecosystems on and off the site.

For buildings, choosing a site in an already developed area will provide building occupants with access to existing amenities, and might allow them to take advantage of walking, cycling or using public transport to access the building instead of having to drive a car. Greyfields, which are previously developed sites that are economically obsolete or underutilized such as vacant stores, offer the potential to contribute to social and economic sustainability for the community while leveraging resources such as parking and infrastructure that are already in place. Redeveloping such sites can provide a positive contribution to an existing neighbourhood. Choosing a site that is a brownfield (with real or perceived environmental contamination) and cleaning it up not only represents a positive step for the community, it might also offer a tax credit or development incentive for the developer.

Considerations for facility siting

- Location with respect to demand.
- Terrain.
- Potential damage to high-quality ecosystems.
- Vulnerability to natural disasters.
- Proximity to amenities/amenity sharing.
- Opportunities to remediate environmental damage.

Avoiding sites that either have valuable ecological resources or higher levels of risk from environmental damage is also wise. For instance, avoiding sites with wetlands or habitats of threatened or endangered species not only preserves these valuable ecological resources, but also avoids the need to mitigate any impacts to these habitats, saving considerable expense. Finally, the choice should consider the long-term risks of resource depletion that might make development difficult in the long run. Water supply in some areas, for example, is becoming more scarce and may mean restrictions on future facility development and use.

Wherever possible, seek sites that will allow amenity sharing with other sites. Siting projects next to facilities that have complementary hours allows them to share common resources such as parking lots. For instance, a bank that is only open during the day might share parking with a nightclub or restaurant district that operates at night, thereby reducing the need to have two separate parking lots. Choosing sites in already developed areas provide the project with access to existing infrastructure such as streets, water, power and sewers, and can save project time and budget while minimizing impact in the long term. Some sites such as areas under elevated freeways offer overlooked areas with significant possibility (see Figure 6.4).

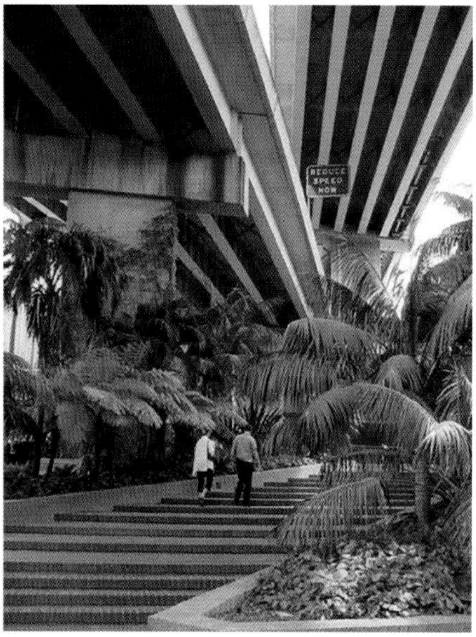

Figure 6.4 This park area in Sydney, Australia creates a connection between neighbourhoods in an otherwise unusable space

After selecting the best site possible for the project, the next step is to determine where on the site to place buildings or built-up areas. Wisely locating a building and orienting it with respect to the sun can save considerable development costs and 30–40 per cent in heating and cooling costs over the lifecycle of the building. Passive solar design takes advantage of the sun's energy to provide for a building's lighting and space conditioning requirements without using electrical or combustion energy. For instance, orienting a building with its long axis running from east to west and most of its windows on the south side of the building in the northern hemisphere or the north side in the southern hemisphere is a passive solar design strategy. In the summer when the sun passes higher in the sky, overhangs shade the windows from receiving too much sunlight

and overheating the building. In the winter when the sun passes lower in the sky, sunlight can enter the windows under the overhangs, providing free heat to the building during the seasons when it is most needed (see Figure 6.5).

A complementary action is to preserve existing trees that may surround the building site. Deciduous trees on the sunny side of a building can provide valuable shade in the summer, reducing cooling costs (see Figure 6.6). In the winter, they lose their leaves and allow solar energy to reach the building, providing warmth and heat gain. Trees on the sides of the building facing prevailing winds can also provide shelter to reduce the heating and cooling loads of the building. Trees can also have positive effects on water retention, air quality and property values.

After arranging buildings and developed areas on the site, another opportunity applicable both to infrastructure and building projects is to develop low-impact site amenities. These include softscape – the vegetated parts of the site – and hardscape – the parts of the site that are paved or otherwise developed. From a softscape standpoint, native species require little or no irrigation, fertilizer or pesticides once established, since they are well adapted to the climate and can survive typical conditions. Avoid exotic species from other parts of the world, since they often out-compete natives and can become invasive in the landscape. Group plants with like needs

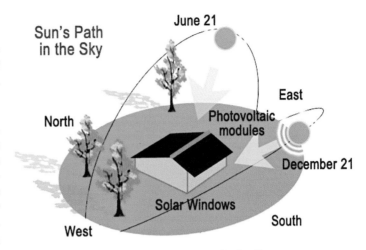

Figure 6.5 Careful orientation can provide significant energy to a building, in this case, in the northern hemisphere. Solar hot water heating can also be provided on the southern exposure. Windows should be minimized on the east and west sides of the building

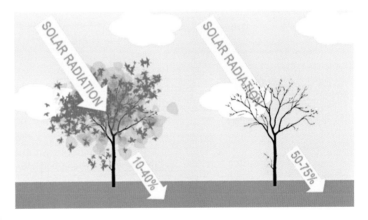

Figure 6.6 Deciduous trees provide shade in summer and allow sunlight through in winter; only use on the east and west sides of buildings to avoid shading photovoltaic systems on the south side

together in the landscape. This 'zoning' of landscape areas permits focusing irrigation, fertilizer and pesticides only on the plants that require it. Zoned landscaping can also help reduce the risk of wildfire around buildings by keeping areas near buildings clear of flammable vegetation and debris. Use mulch, organic plant matter such as bark or pine straw, to suppress weeds and help retain moisture around landscape plantings. Not only does mulch improve the appearance of many landscapes, it also helps to reduce the need for irrigation and pesticides, and helps to stabilize new plantings until they can become established. Xeriscaping is a series of strategies involving design of landscaped areas that require no irrigation, including use of native plants that can survive without additional water after they have been established.

Bioswales and rain gardens are vegetated areas that absorb and treat stormwater runoff from paved areas and allow it to percolate slowly back into the soil (see Figure 6.7). Plants within a bioswale also help to remove contaminants from runoff that would otherwise have to be treated at a wastewater treatment plant or contaminate local rivers and streams. They are often used in conjunction with alternative pavement configurations to direct runoff to the appropriate areas.

Alternative pavement materials such as pervious concrete, stabilized soil or grid systems are also an option. These systems allow stormwater to infiltrate the site instead of being collected and requiring special treatment (see Figure 6.8). Stormwater that comes into contact with pavement, particularly areas where vehicles drive or park, becomes contaminated with engine drippings, rubber and asbestos particles, and other automotive residuals. These contaminants, when concentrated in runoff, can pollute local streams and rivers. Stormwater runoff also causes stream damage because there is an increase in water temperature when it contacts the pavement, leading to thermal pollution of local streams. Permeable pavements capture stormwater and allow it to percolate back into the soil rather than running off. High albedo pavement that is light in colour and reflective helps to minimize the urban heat island effect. Urban heat islands are areas with higher temperature than surrounding areas because of dark pavements and buildings absorbing more heat from the sun. Using light-coloured concrete, for instance, can reduce overall temperatures during summer months and lower air-conditioning bills in surrounding buildings.

Amenities to promote the use of alternative transportation, such as sheltered public transport stops, bicycle storage facilities and bike paths, sidewalks or pedestrian trails, and alternative fuel vehicle refuelling stations, are ways to improve project sustainability by encouraging low-

Figure 6.7 This bioswale collects and treats stormwater from surrounding paved areas

Figure 6.8 This pervious pavement consists of decorative gravel bonded with a liquid bonding agent; which can be poured around site features and allows water to percolate to the subsoil

or no-carbon mobility. Designing to encourage physical activity among building users can also contribute to social sustainability by improving occupant health and well-being. Setting aside special parking for carpools or alternative fuel vehicles also rewards drivers who use these greener transportation options.

After developing the site using green amenities, look for opportunities to restore natural ecosystems on the site to the extent possible. Native species require little or no maintenance and provide landscape beauty as well as habitat for birds, butterflies and other local creatures. Set aside areas of the site to remain undeveloped, or designate areas as open space that can serve the needs of both humans and local fauna. Work with adjoining sites to create common undeveloped areas to increase the size of habitat available to local fauna. Collaborate with owners in the area to restore local ecosystems that benefit everyone, such as local streams and waterways.

Energy optimization

The energy used in the building sector is significant, and has considerable impact on the natural environment. The best way to reduce impact is by avoiding the use of energy where functional requirements can be achieved using other means. Increasing the efficiency with which energy is used is a second way to reduce system demands, followed by balancing electrical loads, since centralized power systems are dynamic and cannot efficiently store electrical energy. The final way to increase the sustainability of energy use in buildings is to seek alternative energy sources based on renewable energy technologies.

One approach to reducing demand for energy in buildings is using timers or occupancy sensors to reduce operating hours for equipment not needed during hours when a facility is not occupied, such as water heaters, or by switching off lights or reducing ventilation in spaces that are not in use. These and other strategies can reduce demand with no changes required in occupant behaviour. Passive heating, cooling and daylighting are ways to provide desirable conditions for building occupants while reducing or eliminating the use of electrical power to provide these services. Energy recovery ventilation systems also help to reduce demand by capturing waste heat from exhaust air in the winter or incoming air in the summer, and using it to preheat incoming air in the winter or exhausting it with outgoing air in the summer. Some systems also equalize humidity in the same way by using a desiccant wheel to transfer humidity to its desired location.

Another simple yet under-utilized technology is high-albedo surfaces. These reflective roof or pavement materials and coatings reduce heat gain by reflecting solar energy rather than absorbing it (see Figure 6.9). These surfaces need not be white and shiny – new coatings are available that reduce heat gain even with darker-coloured roofs. In many cases, the albedo (or reflectivity) of a roof can be increased at no cost simply by choosing a different colour for roof finishes during design. High-performance building envelopes can also eliminate unnecessary uses of

Priorities for energy optimization
• Avoid unnecessary energy use.
• Increase energy efficiency.
• Balance electrical loads.
• Seek alternative energy sources.

Demand reduction strategies
• Timers/occupancy sensors.
• Passive heating and cooling.
• Daylighting.
• Energy recovery ventilation.
• High-albedo surfaces.
• High-performance building envelopes.
• Dual switching/labelling.
• Thermal mass.

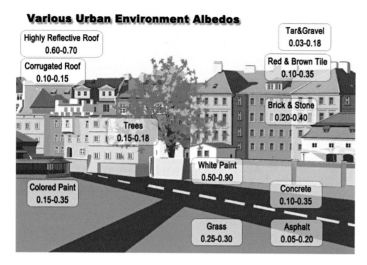

Figure 6.10 This dual-switched light fixture allows the outer lamps to be used alone, the inner lamp to be used alone or all three to be used at once

Figure 6.9 Albedo of various materials in the urban environment (0 = perfectly absorptive; 1 = perfectly reflective)

energy for heating and cooling. Ways to improve the building's envelope include increasing insulation in wall cavities and attics, sealing cracks and using air/vapour barriers to reduce infiltration, installing high-efficiency windows and doors, and using low-emissivity paint or radiant barriers in attics to reduce heat gain. All of these envelope enhancements can reduce the demands on facility space conditioning equipment and increase occupant comfort during all seasons of facility operation.

Effective control systems for lighting can help to eliminate the use of energy during hours when daylight provides enough light. Dual switching – wiring light switches so that combinations of lamps can be switched on or off depending on how much light is needed – helps users to adjust light levels to appropriate levels (see Figure 6.10). Labelling switches also helps users to make more effective decisions about how to operate lighting systems.

A second set of strategies focuses on optimizing the efficiency of energy use within a facility. One way to save considerable energy is by paying attention to proper sizing of building heating, cooling and ventilation equipment (see Figure 6.11). Proper sizing can greatly enhance the efficiency with which mechanical systems use electrical power to meet occupant needs. With mechanical systems, oversizing equipment means these systems operate at peak efficiency only a few days of the year. To achieve peak performance, it is far better to use two parallel systems where both systems are used concurrently to meet peak demands. Under normal conditions, only one system is required, and it operates at peak capacity. The investment in additional capital equipment is offset by operational savings in energy and also provides redundancy when needed for repairs or in emergencies.

Water heating is responsible for a significant share of total energy use (nearly 10 per cent) in the United States, for example, and represents a significant share of the operating cost of many buildings. Careful design of potable water systems and water conservation measures is an important way to reduce overall energy use, both at the individual building scale for water heating and at the infrastructure scale for pumping and treatment. Replacing tank-type hot water heaters in buildings with tankless models is another way to save energy in some applications. Although tankless models often cost more initially, they may last longer than tank-type models if properly maintained. They also eliminate standing heat loss since water is only heated as needed.

High-performance heating, ventilation and air conditioning (HVAC) systems also save energy. The use of variable speed fans, pumps and motors enables HVAC to operate at optimum levels rather than frequently cycling on and off. This increases user comfort as well as saving energy. Optimized distribution systems also help to reduce heating, cooling and ventilation costs. Sealing and insulating ducts and routing them within conditioned space is an important way to ensure that the energy expended to condition air actually benefits the users of the building. Technologies such as chilled beams (see *Case study: Sydney Water HQ, New South Wales, Australia*) offer a more efficient approach to air conditioning than conventional technologies and can be integrated as part of the structure of the building as well.

Figure 6.11 Especially with systems such as ground source heat pumps, proper sizing is critical to keep installation costs low. Oversized systems mean longer trenches for burying geothermal loops, resulting in significant additional installation costs

Case study: Sydney Water HQ, New South Wales, Australia

Sydney Water's head office is located at One Smith Street, in the historical city of Parramatta, west of Sydney's central business district. It has received a Five Star Green Star Office as Built v2 rating, achieving a score of 72 from the Green Building Council Australia. The building, designed by Denton Corker Marshall and owned by Brookfield Multiplex, is a 17-storey, 23,000 sq m office tower, and has incorporated many sustainable strategies and technologies including water and energy efficiency and recycling. Sydney Water's HQ includes green features such as:

- brownfield remediation to clean up asbestos contamination at the site
- an onsite water recycling plant to provide recycled water for toilet flushing, cooling towers, fire testing and irrigation
- rainwater harvesting to provide additional water for toilets and cooling towers
- solar heating panels to supplement hot water requirements
- a high-performance glass façade with shading that can control heat entering the building without limiting natural light
- construction materials made from renewable sources and/or with high recycled content
- a rooftop garden to reduce the urban heat island effect and manage stormwater
- chilled beam cooling instead of conventional air conditioning
- improved air quality and office environment through using products with low volatile organic compounds (VOCs) and continuous fresh air supply.

continued

Figure 6.12 A high-performance glass façade has shading that can control heat entering the building without limiting natural light

By implementing these sustainability features, the Sydney Water HQ has cut greenhouse gas emissions by 30 per cent and reduced drinking water use by 75 per cent over a typical office building. In addition, the reduction in water used at the site has reduced the flow of wastewater to the sewerage system by 90 per cent. The building also utilizes a convective chilled beam system to cool the building and its occupants, which requires fewer refrigerants and is more energy-efficient than conventional cooling systems. Chilled beams work by allowing air to circulate around pipes placed at ceiling level filled with cool fluid. As the air circulates, it becomes cooler and sinks to the occupied area of the room. The system aims to improve air quality within the building while its heat absorption from lighting and equipment reduces energy demand and therefore greenhouse gas emissions. Since the Sydney Water HQ is located next to a major transport interchange, the building contains efficient showers, bicycle racks and other facilities to encourage staff to commute using healthier and more environmentally sustainable methods.

Load balancing is a way to limit the loads placed on the power grid so they occur when there is capacity, limiting the need for new power plants. The electrical grid is a complex system of power plants, distribution networks and end users that consume the power being generated. To keep the system functioning, energy supply must remain exactly balanced with energy demand, or the system becomes unstable. When demand exceeds supply, additional generators are brought online to meet the additional need. If these peak generators cannot meet demand, the system may experience brownouts because there is insufficient power to meet everyone's needs. In the worst case, the system becomes so unstable that safety mechanisms activate to take parts of the grid offline. This is known as a blackout.

The highest demand for energy is typically during mid-afternoon in the hottest part of summer, when industry is operating at maximum output and commercial buildings are operating air conditioning at maximum capacity. Shutting down unnecessary equipment during these periods is one way to reduce the need to bring peak generators online. Peak shaving means reducing demand during peak times of the day, and shifting that demand to off-peak times. Utilities often charge lower rates during off-peak times, providing an incentive for users to shift the timing of their demands to these periods. Thermal storage systems can exploit this opportunity very effectively: energy is used at night to super-cool a thermal storage medium, which then absorbs heat to provide cooling during the day.

Building energy management systems use electronic controls to balance loads during peak times by reducing the amount of power sent to equipment that is tolerant of voltage variability (like some air conditioners, pumps, fans and motors) while maintaining a steady stream of power to equipment that cannot tolerate this variability (like plug loads from computers). Depending on the local or regional climate for power production, the price for electricity during peak hours can be quite high. In many areas, investments in equipment and controls to balance loads can pay back rapidly. Often, utility companies will finance or invest in these systems for built facilities as part of a smart grid system. They may also provide technical assistance to support implementation.

After reducing demand, optimizing efficiency and balancing loads, a final strategy is to explore alternative sources of energy. Two primary ways exist to obtain alternative energy: buying energy from a green power provider, or generating power on site. The first approach, buying from a green power provider, depends on whether a provider exists in the area. Green power providers generate electricity from renewable energy sources such as wind power and solar power, while avoiding non-renewable power sources such as fossil fuels. Even if local power companies do not sell green power, offsets can be purchased from firms that specialize in brokering 'green tags' that represent carbon saved by other organizations.

An alternative is to explore the possibility of on-site energy generation such as photovoltaics, wind turbines (see Figure 6.13), microturbines, fuel cells or micro-hydro generation. Photovoltaics, wind and micro-hydro systems provide power from renewable sources, while fuel cells and microturbines typically draw upon fossil fuel sources but can also be powered by bio-based fuels. Fuel cells and microturbines are often

Energy-efficiency strategies

- optimal equipment sizing
- tankless water heaters
- optimized distribution systems for heating, ventilation and air conditioning (HVAC) and hot water
- high-performance lighting.

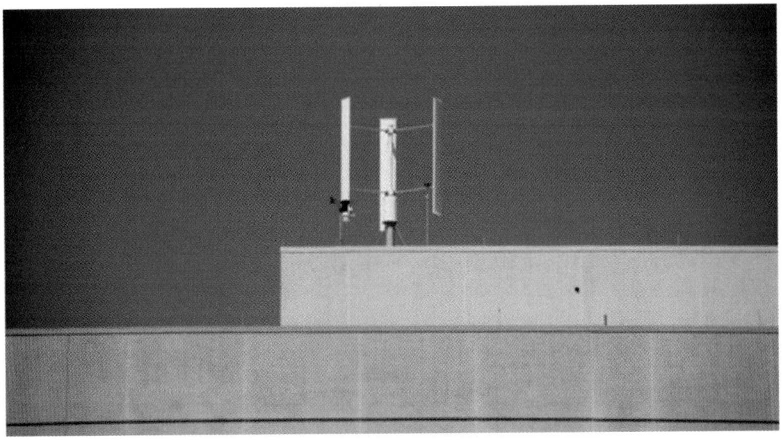

Figure 6.13
Small-scale, vertical-axis wind turbines can be used to provide renewable energy

used as co-generation systems which provide heat for water and space heating as well as electrical power. Such systems can be costly, so reducing, optimizing and balancing energy demands of the building before investing in on-site energy is key to manage the cost of this investment.

Given the difficulty of storing electrical energy, the most cost-effective way to use on-site alternative power systems is with a grid inter-tie, which allows the owner to return excess power back to the electrical utility when a surplus is generated. Power is purchased from the utility during times when the on-site system does not meet demand, such as at night in the case of photovoltaic systems. This strategy avoids the need for costly and high-maintenance battery storage systems as part of the on-site generation system. With photovoltaic systems in particular, peak capacity falls at the same time of day as peak demand (during the afternoon), thus providing a way to better meet peak demand without additional power plants. In parts of the country where there is not much excess generating capacity, utilities may provide financing and technical assistance to help pay for alternative energy systems for interested customers.

With demand for energy only likely to grow in the future, designers need to be aware of opportunities to manage consumption of electrical power in built facilities. Energy investments save money over the life-cycle of a facility. They can also increase the ability of the facility to withstand fluctuations in power supply. This reduces vulnerability to natural disasters and other threats. As the world's need for and dependence on power grows, the vulnerability of power systems will increase as well. Facilities with well-managed, minimal, energy requirements that include passive systems for lighting, heating and cooling and on-site power generation capabilities will be the least vulnerable to fluctuations in power (and prices) from utility suppliers.

Water and wastewater performance

The next set of best practices deals with sources and uses of water for a facility and sinks for its wastewater. Nearly all facilities require a source of water to meet the needs of occupants such as drinking, washing and waste disposal. Sometimes this water must be potable, i.e. drinkable, but other times potable water is used for purposes where it is not really needed. While water appears to be abundant on our planet, less than three per cent of all water on Earth is fresh water. The remaining 97 per cent is salt water, which is mostly unusable for human purposes. Of the small percentage of fresh water available, 69 per cent is trapped as ice and snow cover, 30 per cent is stored as groundwater, and less than 1 per cent is available on the surface as fresh water lakes and rivers (UNEP 2002). While it is true that the hydrologic cycle continues to recharge sources of fresh water supply through precipitation, runoff and groundwater infiltration, in many areas the rate of use exceeds the recharge rate, leading to aquifer depletion.

Alternative energy sources

Off-site:
- green power provider contracts
- green tags/offsets

On-site:
- photovoltaics
- fuel cells
- wind turbines
- microturbines
- micro-hydro

Priorities for water and wastewater

- Eliminate unnecessary uses of water.
- Increase efficiency of water use.
- Develop alternative water sources.
- Develop alternative wastewater sinks.

To conserve this precious resource, the most important tactic is to eliminate unnecessary uses of water, followed by increasing efficiency or using less water to accomplish what is required. For existing facilities, a careful water audit is critical to identify opportunities for improvement and find and repair leaks that waste water. For new facilities, proper commissioning of water systems will ensure that all systems are installed properly and function according to their design intent.

Another way to eliminate water use is to install water-free toilets and urinals. Waterless urinals exploit the fact that urine is a liquid itself and does not require water to be conveyed through wastewater pipes. Waterless urinals use a special trap with a low-density fluid or gel to provide a seal that allows urine to pass through but prevents sewer gases from escaping (see Figure 6.14).

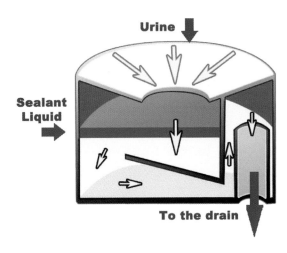

Figure 6.14 The design of a waterless urinal trap uses gravity to convey liquid waste and a viscous fluid instead of water to provide a trap for sewer gases

Alternative toilets are also available that require no water to flush. Composting and incinerating toilets function without the use of water, and collect and dispose of human waste. Composting toilets function most effectively with either a passive or powered ventilation fan, depending on use. These toilets convert human waste into harmless compost that can be used as a soil amendment on-site. Incinerating toilets use electrical energy or other fuel to literally burn away human waste, leaving an ash residue that can be disposed of more easily. The energy required to power an incinerating toilet is considerable, though, and should be taken into account as a tradeoff against the water savings offered by these toilets.

The second tactic is to increase the efficiency of water uses that cannot be eliminated. Many different fixtures are now available on the market to increase the efficiency of water use in showers and sinks. Aeration, the introduction of air into the water flow, is a common technique to reduce the actual flow of water in taps and shower heads while maintaining the perception of high volume flow. While this is a useful tactic for washing, it can be frustrating when the aim is to fill a container with water for cooking, cleaning or other purposes. Some aerated kitchen faucets provide a lever that allows the user to control the degree of aeration and restore full flow when needed. Other types of alternative controls include foot-operated sinks, which allow precise control of flow timing while leaving hands free. More precise automated flow controls are also becoming available for sinks and toilets. With all of these systems, overall system design is important to consider to ensure that the water delivered at each fixture is safe and clean (see *Unanticipated consequences of water conservation*).

Toilet technology has improved considerably over the last decade, with multiple options available to improve the efficiency of water use. In addition to conventional low-flow toilets, multiple manufacturers offer dual-flush toilets that allow a full-volume flush for solid waste and a half-

Water efficiency tactics

- water-efficient fixtures
- high-efficiency toilets/ alternative flush mechanisms
- water-efficient landscaping
- high-efficiency appliances

Unanticipated consequences of water conservation

William Rhoads, Amy Pruden-Bagchi and Marc Edwards

Adoption of green/sustainable plumbing designs is driving a paradigm shift in the water industry. Consumers have typically relied on drinking water utilities to produce and deliver safe drinking water, and modern drinking water treatment, including filtration and disinfection of drinking water supplies, is considered one of the greatest engineering and public health achievements of the 20th century (NRC 2006). However, new green plumbing designs are using much less water than conventional designs, and as a result, unprecedented volumes of water are being saved by using a broad selection of conservation strategies and technologies. While such systems can play a role in advancing water and energy savings in the future, it is important to work towards also advancing understanding of their potential impacts on water quality and public health.

Water age, defined as the residence time in a system measured from the point of entry to the point of use, is a critical factor in all water systems to maintain safe potable water. When green buildings save large amounts of water, water age can increase by up to a factor of ten, potentially initiating water quality changes that can affect public health (Edwards et al. 2014; Rhoads et al. 2015). Water age is increased in potable water systems when water demand is reduced without also reducing total system volume. For instance, this occurs when conventional end-point devices (e.g. water faucets, shower heads) are simply replaced with low-flow devices. High water age is also inherent to some green building designs. For instance, using alternative sources of water such as rainwater for landscaping, toilet flushing or potable water decreases overall water utility water demand. Similarly, in some solar-powered hot water systems, hot water storage is increased to maintain adequate supply during periods with no sun.

Water age has been documented as a key cause of water main distribution system problems such as corrosion, development of taste and odours, and regrowth of microorganisms. The characteristics of premise plumbing (i.e. plumbing to the building from water mains and inside the building) relative to main distribution systems, such as higher surface area to volume ratios, variable use patterns, water temperature fluctuations and more reactive plumbing materials, adds complexity to issues encountered in premise plumbing compared to those encountered by drinking water utilities in main distribution systems. A variety of potential water quality issues have been identified that tie specifically to green building practices and strategies (see below *Unintended consequences of green plumbing system design*; synthesized from Rhoads et al. 2014 and 2015).

Unintended consequences of green plumbing system design

- **Rapid loss of disinfectants** – in systems with high water age, chemical, physical and biological reactions can consume disinfectant residual at a much faster rate than in system with lower water age, increasing growth of bacteria. These changes have also been associated with pinhole leaks and staining in copper piping, as well as taste and odour problems.
- **Unsatisfactory water temperatures** – water in low-use water lines with high water age tends to warm or cool to room temperature, which can lead to consumer dissatisfaction and increased levels of water flushed to obtain desired temperatures. Temperature settings in storage water heating tanks may also be reduced to save energy, which can lead to the proliferation of opportunistic pathogens.
- **Lead leaching** – pre-2014 brass plumbing components contain significant amounts of lead in the United States, which are at risk to leach into water supplies particularly under

continued

conditions of high acidity (pH < 8) or low alkalinity (< 30 mg/L as $CaCO_3$). These conditions are often observed in harvested rainwater systems.

- **Microbial regrowth** – in systems with high water age, pathogenic microbes have a higher likelihood to regrow as disinfectant residual diminishes. Some water conserving devices such as low-flow metered and electronic faucets have also been measured to have high levels of increased pathogen growth.
- **Low flow rate** – More laminar flow in water supply piping may facilitate pathogen growth in piping biofilms. Downstream of fixtures, low or reduced flow rates may reduce the volume of water available for waste conveyance, causing potential clogging, unwanted chemical reactions in conveyance piping, and potential odour and sanitation problems.

An example of one cause for concern regarding public health risks is potential exposure to opportunistic pathogens in premise plumbing (OPPPs). These microorganisms are native to many potable water systems. Low-flow rate (≤0.5 gallons/minute) electronic faucets, which are specifically designed to use less water than manually operated conventional faucets (typically ≥1.5 gallons/minute) and avoid the spread of germs, sometimes support growth of much higher concentrations of waterborne opportunistic pathogens (Moore and Walker 2014; Sydnor et al 2012; Yapicioglu et al. 2012; Zingg and Pittet 2012). Despite substantial field evidence documenting these problems, plumbing codes across the United States are adopting the requirement to have low-flow hands-free faucets installed in public lavatories. When problems arise (typically in hospital settings where monitoring for pathogens regularly occurs), faucets have been removed and replaced with conventional faucets with no follow-up investigations. Thus, in addition to the unintended public health consequences sometimes created by these faucets, the solution involves both relinquishing water savings from the original fixtures and doubling the environmental impact due to replacing the fittings needed for each tap.

Harvesting rainwater for indoor potable use is another water sustainability practice that is becoming more popular; however, the quality and safety of rainwater used for potable water uses is relatively unstudied in a modern context. Natural rainwater typically has a pH of about 5.2 and negligible alkalinity, properties that make it naturally corrosive to metals used in plumbing systems. It can also be contaminated by local industry and land use with atmospheric lead, airborne herbicides/pesticides, or other contaminants from the roof collection/conveyance system, including faecal contamination from wildlife. While treatment such as filtration and ultraviolet light are commonly used to treat the water, monitoring of potable rainwater systems is very limited. A survey of approximately 2700 rainwater systems reported that only 12 per cent of homeowners and 3 per cent of businesses monitored for faecal derived waterborne pathogens on a quarterly basis, and did not report any other sort of contamination monitoring at all (Thomas and Greene 1993). In addition, rainwater storage tanks require enough water storage to maintain adequate water supplies, with water ages that easily reach the order of months by design.

The movement toward net zero water may result in more on-site treatment systems, which can trigger costly requirements to comply with all the national primary drinking water standards. Drinking water utilities are also experiencing additional financial stress due to decreases in overall water demand by the communities they serve, termed a 'conservation conundrum'. Many communities are using less water than a decade ago, which translates to less revenue from water sales for water utilities at a time when the physical state of treatment facilities and distribution networks is in urgent need of upgrades. At the same time, utilities are expected to maintain service connections for seasonal customers, such as 'snowbirds' and buildings that require only emergency

continued

connections to utility water supplies. Although there are emerging financial solutions for these very low-end and seasonal users as explored elsewhere (Hughes et al. 2014), the amount paid for water will have to increase to cover the widening funding gap.

A secondary impact of the conservation conundrum that has not yet been explored is the potential for water quality to be degraded across entire distribution systems due to conservation. For example, as a result of selling 40 per cent less water in 2012 than projected when designing and sizing their distribution system, the water age at a Virginia utility roughly doubled (Ramaley 2013). When demand in distribution systems that are oversized to meet predicted future demand actually decreases with time, distribution system water age is higher than expected. Thus, total water age is increased even in buildings that do not adopt water conservation.

Flushing water to lower water age and introduce fresh water from the drinking water utility (when a building is connected to a municipal system) is currently considered an adequate temporary solution for reducing problems associated with water age. The volume needed to reduce issues is not well-studied, but in one case, <1 per cent of the total daily water demand was successful (Nguyen et al. 2012). This implies that even with the 'wasted' water that is flushed, the building likely still reduces its overall water footprint and the water flushed could be put to other uses if it is considered during the design of the building plumbing system and not installed retroactively to address issues. The perception of wasting water can be a significant barrier to the use of this strategy. Unfortunately, it is not feasible for off-grid systems with storage tanks and limited water resources. For these buildings, alternate means of treating and monitoring water to ensure public health will have to be explored.

Making progress towards sustainable water systems is necessary to ensure adequate water resources for future generations. Moving forward requires a 'shared responsibility' model for provision of safe and good-tasting water to consumers. It requires all stakeholders, including water utilities, consumers, green building owners and organizations, building designers, plumbers, code setting organizations and device manufacturers to play a role in preventing and solving water quality problems in buildings. As this brief overview illustrates, green plumbing systems present a wide range of challenges. We are at the earliest stages of a green building revolution, where achieving extreme reductions in water demand are becoming more commonplace while scientific understanding is lagging far behind. There is sufficient evidence to cause legitimate concern, and there are specific areas in need of research, including how we can successfully avoid elevated water age in buildings while still conserving water or, if high water age is unavoidable, what materials, treatments or other strategies are needed to reduce potential water quality problems. We must explore how drinking water utilities can continue to provide high quality water to the public while also overcoming the funding challenges they are facing. Finding answers to these key issues will lead to green building plumbing systems that truly meet the needs of their occupants while maintaining resource efficiency, having minimal ecological impact and providing good and safe drinking water.

Acknowledgements

This information is based on work supported by the Water Research Foundation, as part of Project 4383 Green Building Design: Water Quality Considerations, and financially supported by the Alfred P. Sloan Foundation Microbiology of the Built Environment program. Any opinions, findings, conclusions or recommendations expressed here are those of the authors and do not necessarily reflect the views of the sponsors.

References

Edwards, M., Rhoads, W., Pruden, A., Pearce, A.R., and Falkinham, J.O., III. (2014). "Green water systems and opportunistic premise plumbing pathogens." *Plumbing Engineer*, November, 63–65.

continued

Hughes, J., Tiger, M., Eskaf, S., Berahzer, S.I, Royster, S., Boyle, C., Batten, D., Brandt, P., and Noyes, C. (2014). *Defining a resilient business model for water utilities.* Final Project Report, Water Research Foundation Project #4366, Denver, CO.

Moore, G., and Walker, J. (2014). "Presence and control of Legionella pneumophila and Pseudomonas aeruginosa biofilms in hospital water systems." *Biofilms in Infection Prevention and Control: A Healthcare Handbook,* 311.

Nguyen, C., Elfland, C, and Edwards, M. (2012). "Impact of advanced water conservation features and new copper pipe on rapid chloramine decay and microbial regrowth." *Water research* 46(3), 611–621.

NRC (National Research Council). (2006). *Drinking water distributions systems: Assessing and reducing risks.* National Academy of Sciences, Washington, DC.

Ramaley, B. (2013). "The changing water utility." *Environmental and Water Resources Seminar,* Virginia Tech, Blacksburg, VA.

Rhoads, W., Edwards, M., Chambers, B.D., and Pearce, A.R. (2014). *Green building design: Water quality and utility management considerations.* Final Project Report, Water Research Foundation Project #4383, Denver, CO.

Rhoads, W.J., Pearce, A.R., Pruden, A., and Edwards, M.A. (2015). "Anticipating the effects of green buildings on water quality and infrastructure." *Journal of the American Water Works Association* 107(3), 2ff.

Sydnor, E.R., Bova, G., Gimburg, A., Cosgrove, S.E., Perl, T.M., and Maragakis, L.L. (2012). "Electronic-eye faucets: Legionella species contamination in healthcare settings." *Infection Control And Hospital Epidemiology* 33(3), 235–240.

Thomas, P., and Greene, G. (1993). "Rainwater quality from different roof catchments." *Water, Science and Technology* 28(3), 291.

Yapicioglu, H., Gokmen, T.G., Yildizdas, D., Koksal, F., Ozlu, F., Kale-Cekinmez, E., and Candevir, A. (2012). "Pseudomonas aeruginosa infections due to electronic faucets in a neonatal intensive care unit." *Journal of Paediatrics And Child Health* 48(5), 430–434.

Zingg, W., and Pittet, D. (2012). "Electronic-eye faucets – Curse or blessing?" *Infection Control and Hospital Epidemiology* 33(3), 241–242.

volume flush for liquids. On the commercial level, dual-flush flushometer valves are indicated by a green plunger on the unit. Moving the plunger up results in a half-volume flush, and pushing it down produces a full-volume flush (see Figure 6.15). One manufacturer of this technology has also released an automated dual-flush control unit which dispenses either a regular or low-volume flush based on how long the user is within range of the flush sensor. Manual controls are also included with this unit to allow users to control the flush level if desired. An interesting addition to tank-type toilets is the tank-top hand washing unit. This unit directs potable water through a tap fixture incorporated on the tank lid. Users can wash their hands following use of the toilet, then the wastewater from hand washing is collected in the tank and used for the next toilet flush. These units are available both as retrofits and as new installations.

For landscaping uses of water, drip irrigation provides a slow release of water directly to the root zone of the plant, avoiding much of the evaporative losses from more common sprinkler systems. Moisture sensors and timers control the operating periods of irrigation systems, ensuring that water is only dispensed when needed and during the early morning when it has time to reach plant roots before evaporating. With both systems, it is critical to properly locate and maintain the distribution elements to prevent or repair leaks. Including meters as part of these systems can help to quickly identify problems that may not be otherwise apparent.

Figure 6.15 This flushometer valve can provide either a half-volume flush or full-volume flush, depending on what is needed

---

Alternative water sources

- Rainwater harvesting.
- Greywater recycling.
- Water reuse.

Water-efficient appliances also present an opportunity to increase the efficiency of water use. High-efficiency dishwashers and washing machines are available that reduce water use by up to 50 per cent over conventional units. Since these units use less hot water, they also save energy.

A third option for increasing sustainability of water use is to seek alternative sources of water, particularly for uses that do not require water to be treated to drinking-water standards. Much of the water used in homes and businesses does not require this high level of water quality, offering the potential to save considerable amounts of energy presently used for water treatment and conveyance.

With water scarcity becoming a more serious issue worldwide, rainwater harvesting systems are becoming increasingly common. These systems require a collection surface, typically a roof, that is relatively debris-free and made from a chemically stable substance. Older metal roofs often used lead-based solder at the joints, and are not suitable candidates for rainwater harvesting because of the potential for lead contamination of the water source. Asphalt shingles are also not recommended since they tend to shed particles of sand and other residue as they age. Metal, slate and synthetic roofs are all good candidates as collection surfaces. In addition to the collection surface, rainwater systems also require filtration, storage, overflow and piping to connect components (see Figure 6.16). Most existing roof drainage systems can be easily retrofitted to accommodate rainwater harvesting, provided the collection surface is suitable.

Another alternative source of water is greywater – the water from sinks, showers and laundry facilities that has been used but does not contain human pathogens. Water from toilets and dishwashers is considered to be blackwater, since it contains pathogens either from human waste or from food waste associated with dishwashing. Unlike greywater, blackwater requires additional levels of treatment before it is suitable for reuse. Full-building greywater systems can be designed to capture greywater for use in irrigation or for toilet flushing. Separate waste piping is required for greywater and blackwater, in addition to storage receptacles, filtration and piping. The design of greywater systems to meet code requirements often requires special attention, since greywater is an excellent harbour for bacterial growth if it is stored for any length of time. Provisions must be made to balance supply with demand and to ensure that greywater does not remain in the system for a significant length of time before it is reused or purged. Small-scale greywater systems are also available that can capture water from one fixture such as a sink or shower and divert it for use in an adjacent fixture such as a toilet. These small systems can often be designed to fit under a bathroom sink, and work well for retrofit situations where the opportunity to install dual piping is limited.

A third option to be explored at the infrastructure scale is water reuse – the direct recovery of water following wastewater treatment and repurposing for landscaping or non-potable uses. In some cultures, water reuse faces a significant psychological barrier, even though most treated potable water is ultimately derived from the wastewater of upstream communities. However, water reuse is receiving increased attention in areas such as the southwestern United States, where existing reserves of water are dwindling and climate change threatens ongoing supply. Reused water is recovered directly from municipal or local treatment facilities and provided through separate piping for non-potable use in buildings. It is essentially the same approach as greywater recycling, but on a municipal scale.

Figure 6.16 This residential rainwater collection system has three underground storage tanks

A final opportunity to increase the sustainability of a building's water systems involves finding alternative uses or treatments for the wastewater generated. In addition to greywater and blackwater, a building's wastewater stream also typically includes stormwater, which may be contaminated with particulates or drips from parking lots and roofs, and is likely to also be at a much higher temperature than local waterways into which it might run off. Alternative treatment systems are available to treat all of these types of wastewater streams. For instance, greywater systems can capture water that has not been heavily contaminated and redirect it for certain uses without treatment as permitted by code. These systems can also be the first step in a more comprehensive system, where used water is treated on site and then reused for various purposes.

One type of alternative wastewater technology that can handle all levels of contamination is plant-based constructed wetlands (Figure 6.17). Constructed wetlands have been used in a variety of climates and for a variety of applications, both small and large. The series of ponds is filled with plants that provide increasing levels of treatment to water as it cycles through. The system is preceded by a primary treatment step that removes solids from the water stream. Following treatment, wastewater is discharged to a local stream. Smaller scale plant-based wastewater treatment systems are also available, and can be sized to treat the wastewater from a single building or multiple buildings. Living Machines such as the one included in Emory University's WaterHub include plants, bacteria and other organisms as part of a simulated ecosystem that uses wastewater as a nutrient source (see *Case study: the WaterHub at Emory*).

Figure 6.17 Constructed wetlands can handle multiple types of wastewater

Case study: WaterHub at Emory, Atlanta, GA

Emory University, USA

The WaterHub is an ecological wastewater recycling plant located at Emory University in Atlanta, GA. Brought online in March 2016, the WaterHub treats wastewater from campus buildings using a plant-based treatment process to a level at which it can be used as process makeup water in Emory's steam and chiller plants, as well as for toilet flushing in several residence halls. By replacing these potable water uses with recycled water, the WaterHub is able to supply nearly 40 per cent of the university's water needs in a city where water is becoming an increasingly vulnerable resource.

The WaterHub can treat up to 400,000 gallons per day at full capacity, both reducing loads on local wastewater treatment facilities and displacing demand for treated water provided by Dekalb County. The plant has both an above-ground glass and concrete structure (the Glasshouse – Figure 6.18) in which ecological processing of the wastewater is done, and a series of underground concrete processing and storage tanks (Figure 6.19) that provide a reserve of treated water to support campus operations as well as a series of process reservoirs to support the treatment process.

The WaterHub at Emory

- **Treatment capacity**: 400,000 gallons per day
- **Owner**: Emory University, Atlanta, GA
- **Designer**: Sustainable Water, Charlottesville, VA
- **General contractor**: Reeves Young, Atlanta, GA
- **Mechanical engineer**: McKim and Creed

The treatment process is based on hydroponic technology and reciprocating wetlands that mimic the ebb and flow of tidal marshes. In the hydroponic part of the system, beneficial bacteria and microorganisms live in a series of bioreactors in a habitat of free moving plastic pellets, fixed engineered textiles and plant roots. Together, 2000 to 3000 different microorganisms inhabit this engineered ecosystem and break down the waste, consuming the nutrients in the wastewater and producing high-quality reclaimed water. The reciprocating wetlands provide a redundant treatment path included as a demonstration and educational/research opportunity for students and faculty at the university's Rollins School of Public Health. This highly energy-efficient technology is comparatively land-intensive and is well-suited for rural application. The performance of both systems is continuously monitored as part of ongoing studies.

Figure 6.18 The above-ground Glasshouse contains lush vegetation growing from wastewater nutrients and a diagram of the redistribution network embedded into the floor

Figure 6.19 Attractive planters are the only visible part of underground processing tanks

continued

Before entering the treatment stream, wastewater diverted from the normal sewer system passes through a rotary screen to remove solids and debris. An anoxic moving bed bioreactor (Figure 6.20) metabolizes nitrogen and carbon in a low-oxygen environment, followed by an aerobic moving bed bioreactor that uses different microbes in a high-oxygen environment. Plant roots then clarify the water in a hydroponic reactor that is followed by a clarifying step where particles settle out. Treatment concludes by passing the clean water through a felt filter to remove any remaining particulates and an ultraviolet disinfection system that kills any remaining bacteria. The final treated water enters a storage tank before distribution through a purple pipe system for use on campus (Figures 6.21a and 6.21b).

As the first facility of its kind in the United States, this infrastructure retrofit offered several significant lessons for others considering similar systems. First, understanding the characteristics of available wastewater as well as chemical requirements for recycled water are absolutely essential to keep the costs of the system down. Emory needed to develop a comprehensive understanding of its wastewater stream, including timing of discharges from laboratories that might be harmful to the plants and microbes in the treatment system. Also critical was the volume and timing of both available wastewater and reclaimed water needs, both from a mass balance standpoint and an economic standpoint. Reclaimed water used for process makeup water demanded a very specific water chemistry, which required additional pre-use treatment steps not anticipated in the original design.

Special approvals were required by the local municipality, which had to consider both changes to the hydraulics of the existing sewer network due to large quantities of diverted wastewater as well as potential loss of revenue. Presently, the price of both wastewater and potable water are lumped together as one fee by the county, since in most cases the amount of wastewater produced by a user is roughly

Figure 6.20 Honeycomb plates inside the anoxic moving bed bioreactor host microorganisms that metabolize carbon and nitrogen in a low-oxygen environment

Figure 6.21a, b Purple pipes are used to distribute reclaimed water for reuse, to distinguish it from the conventional municipal treated water supply

continued

Figure 6.22a, b The aesthetics of the finished system blend well in the urban/suburban environment

proportional to the amount of water purchased. However, on-site treatment and reuse of water changes this balance. Emory University is presently negotiating with its utility provider to receive credit for the wastewater it treats, instead of automatically assuming it is proportional to potable water purchased. If successful, this credit would further improve the economic performance of the WaterHub.

The project was financed by a Design-Build-Operate water services agreement that provides treated water for less than the municipal supply at a stable rate over a 20-year period, based on projected water and sewer pricing for the Atlanta area. The vendor, Sustainable Water, leases the land on which the plant is located from Emory University and owns and operates the plant over the contract period, selling water at a reduced rate back to the University. The vendor retains liability for water quality as well as meeting all regulations applicable to wastewater treatment plants.

The visible parts of the design feature lush plantings and no odours, with much of the treatment works located underground. This allows it to be incorporated seamlessly in dense urban or suburban areas, unlike conventional municipal treatment plants (Figures 6.22a and 6.22b). Along with the aesthetic benefits, systems such as this one can provide valuable local capacity to overtaxed treatment systems in already-developed areas. Particularly for institutional owners who can directly use reclaimed water to offset other uses of potable water, the economics of this technology make sense in areas where water rates are high or unstable due to water scarcity or rapid population growth.

References

Zern, B. (2016a). "From waste to resource: WaterHub at Emory University." *High Performing Buildings*, Summer.

Zern, B. (2016b). *Personal interview*. Assistant Director of Facilities for Operational Compliance, Emory University. June 14.

Source: All photos courtesy of Brent Zern, Emory University

Materials optimization

The materials used to construct facilities and to operate and maintain them over their lifecycle are among the biggest contributors to their impact on the natural environment. For infrastructure projects such as highways and bridges, materials represent the largest impact these facilities have over their lifecycle. In addition to the raw materials used to build, operate and maintain a facility, solid waste also contributes to a facility's impact on the natural environment. Increasing the sustainability of materials use and solid waste generation starts by eliminating the unnecessary use of new materials. This includes using materials that come from waste, salvage or recycled sources. It also includes reusing and adapting existing buildings instead of building new, or possibly even deciding that a new building is unnecessary.

Pollution prevention is the careful design of products and processes to eliminate the use or waste of materials in the first place. One example in building construction is to eliminate finishes and expose structural materials instead. For example, concrete floors can be stained, polished or textured to create beautiful surfaces that do not require additional finishes. Exposed ceilings and structure eliminate the need for extensive dropped ceiling systems and can provide visual interest and even educational opportunities while saving costs (see Figure 6.23).

Salvaged materials can also be a very environmentally friendly contribution to projects. If an existing building is being demolished, there may be ways to reuse materials in the new building. Often, concrete and masonry rubble can be reused for fill, pavement subbase or drainage. Sometimes, materials from demolished buildings are higher quality than those that are available new. Timber in good condition can be reused for structures, or remilled for finish flooring or siding. Masonry units may be reused as well. Materials that cannot be reused on new projects may be valuable for salvage in other projects.

New materials with recycled content also reduce the use of virgin materials. Recycled content materials range from steel and concrete to finishes, structural materials and landscaping materials (see Figure 6.24). Recycled materials like steel and concrete may be impossible to distinguish from virgin materials. Other products may include recycled content as composites. Recycled plastic lumber (RPL) is one such product. RPL combines post-consumer or post-industrial plastic waste with wood fibres or other materials to produce an extremely durable wood substitute.

Recycled content can come from a variety of sources. Some material such as fly ash is produced as a byproduct of manufacturing processes. This material is called post-industrial or pre-consumer recycled content.

> **Priorities for materials optimization**
>
> - Eliminate unnecessary use of materials.
> - Increase the efficiency of materials use.
> - Seek better sources for material supplies.
> - Find better sinks for material waste.

> **Green materials strategies**
>
> - Reuse existing facilities and materials.
> - Employ pollution prevention.
> - Use salvaged materials.
> - Use materials with recycled content.

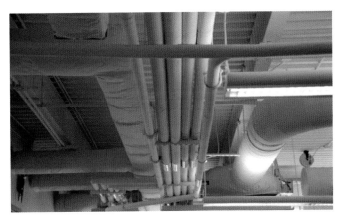

Figure 6.23 Exposed ceilings eliminate the need for materials in a dropped ceiling

Figure 6.24
Recycled content
construction materials

A. Recycled plastic lumber
B. Insulated concrete form material with
   recycled polystyrene pellets
C. Recycled rubber turf stabilizer
D. Recycled paper fibre board
E. Recycled paper fibre countertop
F. Recycled rubber shingles

G. Recycled content carpet (face
   fibre and backing)
H. Salvaged lumber
I. Recycled paper and glass cement tiles
J. Recycled glass countertops
K. Recycled rubber flooring

It can be recovered and reused as feedstock for other manufacturing processes. Other recycled content comes from products that have been produced and then used by consumers, then recovered for recycling at the end of their useful life. This material is called post-consumer recycled content, and it is preferable to post-industrial recycled content because it truly closes the material loop. Post-industrial recycled content can sometimes represent waste or inefficiency in manufacturing, as in the example of trimmings or cutoffs of materials from a finished product. Rather than recycling this material, it would be better to improve the manufacturing process to eliminate the waste in the first place.

A second tactic for increasing material sustainability is to make more efficient use of materials. This means getting more benefit from the materials that are used, or using fewer materials to achieve the same benefit. One way to do this is to use multi-function materials. Multi-function materials play more than one role as part of a building. They can speed up construction considerably along with saving building materials and reducing waste. Every system and technology used in a project comes with overhead waste – the packaging, transportation and other costs that do not directly add value to the product itself. When one product serves multiple purposes, the overhead for the system is only a fraction of the overhead associated with the multiple systems it replaces. Many multi-function materials serve as the primary structure of the building, such as insulating concrete forms (ICFs), structural insulated panels (SIPs) and aerated autoclaved concrete (AAC) (Figure 6.25). Each of these structural systems also provides the building enclosure, insulation system and mounting surface for interior and exterior finishes. When these products are used, the construction of one system achieves as much as three or four conventional building systems combined. This saves time, labour, packaging, transportation and ultimately money.

An energy-related example of multi-function materials is building-integrated photovoltaics (BIPVs). These solar cells are built into components that play other roles in a building. For instance, some are incorporated as coatings on windows and can generate electricity whenever the sun shines on the window. Other BIPVs are manufactured as solar shingles – they generate power while acting as the roof itself. Still others are used as shading devices for windows or parking lots.

Vegetated roofs and walls are a second example (see *Case study: Living wall – Hotel InterContinental, Santiago, Chile*). Vegetated roofs have been used in Europe for many years, and are becoming more widely adopted in the United States due to their durability and thermal performance benefits as well as their contributions to stormwater management. Vegetated roofs are typically installed on top of a roof membrane, and can be of varying complexity (see Figure 6.29). They help to ballast the roof itself. They also protect the membrane from solar radiation, which causes the material to break down. They stabilize temperatures on the roof, which also reduces stress on the roof membrane and keeps the building cooler. They help to absorb rainfall, reducing the rate at which stormwater runs off the site. Finally, they are a microhabitat for birds, insects and plants, they help to clean the air. This integrated roof system, although it may cost more than a traditional roof, provides significant benefits that can make it a good investment. All benefits and costs must be considered together when deciding what kind of system to use.

Another approach is to use smart materials, which work by changing in response to environmental conditions. For instance, smart windows may automatically increase their opacity in response to higher levels of sunlight. Smart window blinds or mounts for solar panels may automatically adjust to follow the sun's path. Other materials can change colour or other properties as well. For instance, phase change materials

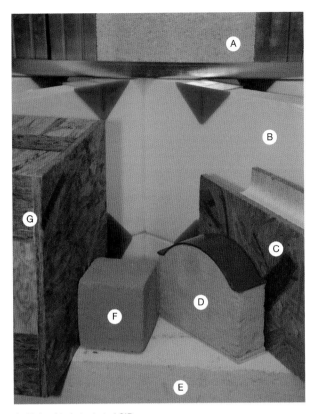

A. Embedded steel stud SIP
B. Polystyrene insulating concrete form
C. Polystyrene stress skin SIP
D. Aerated glass stress skin SIP
E. Aerated autoclaved concrete
F. Fibre-reinforced aerated concrete
G. Straw core stress skin SIP

Figure 6.25 Multi-function construction materials

absorb or shed significant quantities of heat at a specific temperature as they change from one phase to another (e.g. solid to liquid or liquid to solid). With this behaviour, such materials can be used as thermal mass to stabilize the temperature inside a building and make HVAC systems perform more efficiently. They can absorb heat during the day when it is warm, and release that heat during the night when it is cool, all while maintaining a consistent temperature.

Case study: Living wall, Hotel InterContinental, Santiago, Chile

Located in Santiago's Financial Center, the InterContinental Hotel contains one of the largest living wall systems in the world as part of a 16-storey addition to the existing hotel. The 17,000 sq ft modular green wall was installed on the southern and western elevations by local contractors, and features four different types of plants: ophiopogon, ajuga reptans and ceratostigma mixed with musgo. Automated irrigation was used to facilitate the health of the plantings, and water sensors control the amount of irrigation to minimize wasted water.

The exterior wall system is anticipated to provide a variety of benefits, including:

- savings of between 40 and 60 per cent on cooling costs due to shading and evapotranspiration
- sound dampening of noise from adjacent streets
- removal of pollutants from surrounding air.

An interior living wall is also installed in an atrium off the lobby of the club tower, providing a welcome quiet space for guests to relax and enjoy the ambience.

Source: Greenroof Projects Database (2011). 'Hotel InterContinental, Santiago, Chile.' www.greenroofs.com/projects/pview.php?id = 1205 (accessed 28 September 2011).

Figure 6.26 An interior atrium features a living wall that provides a quiet and peaceful venue for hotel guests

continued

Figure 6.27 Living wall modules cover the western and southern façade of the club tower

Figure 6.28 Automated irrigation with water sensors provides water to the variety of plants in the interior and exterior living wall systems

Lightweight modularized construction systems are available that can do more with less. Prefabricated, factory-assembled building components reduce waste and use materials more efficiently since they are manufactured in a controlled environment. Production of multiple runs allows manufacturers to optimize the use of materials and produce the most efficient components possible. Modular construction components include carpet tiles, raised floor systems and demountable furniture systems. These systems allow rapid reconfiguration to meet changing user needs. They also permit replacement on a unit-by-unit basis in the case of damage, avoiding the need to replace entire rooms of carpet when there is one stained or worn area (Figure 6.30). To make the best use of modular materials, the overall dimensions of spaces should be selected to best match available module sizes.

Figure 6.29 Vegetated roofs provide a number of functions to serve the building, including thermal buffering, stormwater management, habitat and aesthetics

Figure 6.30
Carpet tile is a modular material that reduces waste and speeds installation

Properties of green materials

- Abundant.
- Renewable/rapidly renewable.
- Sustainably harvested.
- Local/regional.
- Bio-based.
- Recyclable.
- Recycled-content/waste-based.
- Non-toxic.

Seeking better sources for materials can help to ensure an ongoing supply of products to meet the needs of future generations. Substituting abundant, renewable materials helps to preserve the limited supply of non-renewable resources. Ensuring that materials are sustainably harvested means that the supply of those materials can continue indefinitely. Using local materials also helps to reduce the impacts of transportation while supporting local economies.

Rapidly renewable materials are a special class of materials that can be regrown quickly (see Figure 6.31). According to the US Green Building Council, a rapidly renewable material is any material that can be sustainably harvested on a less than 10-year cycle. Examples of rapidly renewable materials are bamboo, cellulose fibre, wool, cotton insulation, corn-based carpet, blown soy insulation, agrifibre, linoleum, wheatboard, strawboard and cork. Each of these materials is bio-based – they are made from plant products.

Other materials can be considered green because they are abundant. Buildings made from soil – rammed earth, adobe, earthbag or cob construction – fall into this category. Some earth-based construction methods include small quantities of cement or lime to provide additional strength. An innovative material in this category is Papercrete, a fibre cement product that uses waste paper for fibre.

Strawbale construction uses abundant bio-based materials. Straw bales used for this purpose are typically three to five times more densely packed than agricultural bales. Strawbale construction can be either structural or used as infill with a wood frame or other structure. Bales are stabilized by hammering reinforcing steel or bamboo spikes through the layers. Bond beams are used to ensure a load-bearing surface along the tops of walls. Proper detailing to prevent moisture intrusion is critical for success. Deep overhangs and moisture barriers between bales and foundations are common. Straw bales can have very high insulating value and a two-hour or greater fire rating depending on the exterior and interior finishes.

Sustainably harvested materials offer another more sustainable option. The Forest Stewardship Council (FSC) is an organization that evaluates wood products to determine whether their harvest is sustainable. Sustainable harvest means that wood is harvested in such a way that it could continue to be harvested indefinitely without degrading the forest source.

Wherever possible, local sources of materials should be sought out for projects. Using locally produced materials reduces one of the biggest impacts of building materials – the energy required to transport them. Often, locally harvested materials can be obtained at lower cost than imported materials. Using local materials can save time for procurement, and also helps support the local economy.

Finally, finding better sinks for waste materials from the construction, operation and end-of-lifecycle of built facilities reduces environmental

Figure 6.31
Rapidly renewable
construction materials

A. Soy-based foam insulation
B. Wool carpet
C. Corn-based carpet fibre
D. Coir/straw soil stabilization mats
E. Corn-based carpet backing
F. Coconut wood flooring
G. Wheat straw fibreboard

H. Sorghum board
I. Bamboo flooring/plywood
J. Linoleum
K. Hemp fiber fabric
L. Cork flooring
M. Cotton fibre batt insulation

impact. On- and off-site recycling and using biodegradable or reusable packaging are some of the ways to improve performance in this area.

As the world population continues to grow and people demand a higher quality of life, demand for raw materials will only increase over time. Smart use of materials for construction can help to ensure an ongoing supply of materials to meet human needs both now and in the future.

Indoor environmental quality

Contemporary humans spend over 90 per cent of their time indoors in some countries, often in climate-controlled buildings that are sealed to prevent air infiltration with the goal of maximizing energy efficiency. At the same time, the materials used to build those structures have become largely composite and/or synthetic. From carpets to engineered wood products, the indoor environment now consists of products that are made from and emit chemicals over their lifespan.

Together, these factors have led to a significant increase in building-related symptoms of ill-health. Sick building syndrome is a disease associated with symptoms that occur whenever people occupy a particular building. These symptoms include headaches, fatigue and other more severe problems that increase with continued exposure. With

Priorities for indoor air quality

- Prevent problems at their source – avoid toxins in buildings.
- Segregate contaminants from occupied areas.
- Increase ventilation to dilute contaminants.
- Ensure proper building maintenance.

increased concerns for energy conservation, building operators often reduce ventilation rates in unoccupied spaces or during nights and weekends. Reduced ventilation rates plus more airtight buildings plus more chemical emissions from building products inevitably leads to problems. On top of all this, many products in modern buildings such as drywall, carpet and ceiling tiles serve as excellent food sources for mould. Inadequate ventilation, moisture and availability of food sources have led to explosions of mould growth in some construction projects, with disastrous results for building occupants. Humans also contribute to the problem by virtue of their activities. The basic act of breathing increases carbon dioxide levels in a room. Higher concentrations of carbon dioxide can lead to drowsiness, reduced productivity, or even headaches and discomfort. Activities such as smoking, cooking or operating printers and photocopiers contribute pollutants to the indoor air.

The indoor environment is critical for a building to do what it was designed to do. Keeping occupants happy, healthy and productive is essential for good business and stakeholder satisfaction. The first tactic is to prevent potential problems at their source, before they have a chance to create a problem. Often, decisions made during the design of a building can have a huge impact on indoor environmental quality. For instance, locating buildings upwind of major pollution sources such as power plants and away from noise sources such as highways can eliminate the need to try to mitigate these problems later. Careful placement of air intakes for ventilation systems can also prevent problems. All vents and exhaust areas should be located downwind from air intakes, and the features of neighbouring sites and buildings should be considered when deciding where to locate air intakes and exhausts.

The materials used to construct the facility have the potential to create problems as well if they are not carefully selected. Many modern finishes such as paints, sealants, carpets and composites contain solvents and adhesives that release VOCs as they age (Figure 6.32). If occupants can smell an odour, the product that caused it is already on its way to their lungs. Products should be specified that contain minimum or no VOCs to avoid this problem. Most major paint producers have lines of low-VOC paints that perform just as well and cost about the same as traditional paints. Carpets can also be certified for indoor air quality.

Proper drainage, exterior detailing and landscaping can help to prevent water intrusion into the building envelope. Whenever water or moisture exists at moderate or warm temperatures with an available food source, mould or mildew is likely to grow. Careful detailing and maintenance can help to prevent this problem, along with proper and adequate ventilation. In particular, the areas surrounding the facility must be graded to properly drain water away from the building to prevent moisture problems in basements and crawl spaces.

Careful selection of landscape plants can also help to prevent problems. Many ornamental trees and shrubs produce considerable pollen, as do species such as olive, acacia, oaks, maples and pines. Pollen produced by these trees can aggravate allergies as well as contaminating air intakes and rainwater harvesting systems.

Figure 6.32 Composite products can offgas VOCs

Figure 6.33 Non-absorptive finishes such as this recycled glass tile floor are easy to clean and do not trap particulates and vapours

Indoor air quality can be improved by specifying finishes that are non-absorptive such as stained/polished concrete, tiled floors and painted walls (Figure 6.33). Not only are these finishes easier to clean, but also they do not trap particulates and vapours that can later be released to cause occupant discomfort. Choosing finishes with biocides or mould-resistant coatings can also help to prevent problems. Natural linoleum, for instance, has biocidal properties and is a good choice for flooring for this reason.

A second category of tactics to improve indoor environmental quality is to segregate activities and materials that create pollution from other parts of the building. This also applies to segregating the building itself from polluters in surrounding areas. Activities often occur within buildings that have the potential to introduce contaminants into the indoor environment. These include but are not limited to kitchens, office equipment rooms, housekeeping areas, and chemical mixing and storage areas. Research has shown that the primary sources of indoor chemical pollutants are ordinary consumer products such as paint, cleaning compounds, personal care products and building materials. Everyday tasks such as bathing, laundering, cooking and heating can all contribute to poor indoor air quality.

Separate ventilation should be provided for areas where these activities take place to minimize the risk of pollutants spreading to other parts of the building. In addition, these areas can be isolated from other spaces in the building through full deck-to-deck partitions and sealed entryways, depending on the risk level of contamination. Indoor smoking areas are special cases that require careful design and construction to ensure that non-smokers do not receive unwanted exposure to environmental tobacco smoke.

Entrance control is important to keep outside dirt and pollutants from getting inside the building. Air lock entryways and vestibules are one approach that can also save energy by keeping conditioned air inside the building and unconditioned air outside (Figure 6.34). These and

Figure 6.34
Vestibules both save energy and improve indoor air quality

other entry areas can benefit from the use of walk-off mats and dirt collectors, particularly in areas where inclement weather may lead to snow and ice being tracked into the building. Walk-off mats are also available for use outside construction areas to help prevent contamination of other areas of the building.

Additional strategies for indoor air quality include increased ventilation rates and proper maintenance. Increased ventilation rates can help to ensure that any contaminants within occupied areas are exhausted and replaced with fresh air. The American Society for Heating, Refrigeration and Air Conditioning Engineers (ASHRAE) specifies recommended ventilation rates by type of occupied space that should be considered minimums for design. Increasing ventilation in mechanically conditioned spaces also can increase energy use and introduce problems with humidity if not carefully controlled, so these issues must also be considered in developing an effective ventilation strategy. Proper maintenance is also critical for indoor air quality, and green housekeeping plans are discussed further in Chapter 8 as a means to achieve this requirement. Maintenance requirements should be considered during design, with a system selected that can be properly maintained by available personnel.

Other aspects of the indoor environment extend beyond air quality. A variety of natural forces, from psychology to physics, can affect the indoor environment. Paint colours influence the user's experience of a space. They can be employed to create different perceptions in different ways. The laws of physical science can also be used to provide natural lighting, ventilation, heating and cooling in a space. This not only saves energy, but also can provide better indoor environmental quality for occupants.

Natural daylight can be used to enhance the indoor environment in many ways. If not managed well, it can also cause problems, such as excessive heat gain, fading of finishes and glare. Light shelves are one way to incorporate daylight into a space without overheating spaces along the building perimeter. Light shelves bounce sunlight onto the ceiling of a room and reflect it further back into the building space, while shading areas adjacent to windows from excessive glare and heat gain. In combination with reflective or light-coloured ceilings, light shelves spread the benefits of natural light throughout a space.

Natural ventilation, when carefully designed, can also create a comfortable indoor environment with minimal energy, but its success is often a function of ambient humidity levels. Even in humid climates, however, there may be seasons in which operable openings in the building envelope allow users to control comfort levels without using mechanical heating and cooling. Giving users control over their environment is an effective way to improve indoor environmental quality from the standpoint of building occupants. A variety of technologies old and new are available to provide this function, and the effects on building users can be significant. Studies have suggested that occupants who can control their spaces are happier, have greater job satisfaction, take fewer sick days and are more productive.

Operable windows and doors (Figure 6.35a and b) have gone in and out of vogue in architectural design over the past decades. Traditional buildings without mechanical heating and cooling had operable windows to allow users to control the indoor climate. Many contemporary buildings, however, have mechanical systems whose operation can be impeded with uncontrolled introduction of outside air. For this reason, many commercial and institutional buildings do not include operable windows, which can create a threat to passive survivability if the power supply to run mechanical equipment becomes unavailable. In many climates, operable windows and doors can be used in lieu of mechanical ventilation during the swing seasons (fall and spring) when

a)

b)

c)

Figure 6.35a–c Operable windows (a and b) and UFADs (c) give building occupants control over their environment

outdoor temperatures are mild. This can save considerable energy for operations. Operable windows and doors can also contribute to user satisfaction with the space and enhances flexibility of use.

Lighting, temperature, humidity and ventilation controls are also becoming more sophisticated in modern buildings. Many commercial buildings rely on elaborate sensor and control systems to control these variables. However, each individual user of a building is different. Not all users are comfortable with the same environmental conditions. Providing individual controls at each workstation or office can help users adjust conditions to meet their individual needs and increase their satisfaction with the space.

Under floor air distribution systems (UFADs – Figure 6.35c) increase the level of control users have over their individual workspaces. UFADs provide the ability to control ventilation rates and perceived temperatures at each individual workspace. These systems also increase ventilation effectiveness and promote a more uniform flow of conditioned air through the workspace. UFADs used in office environments can be combined with modular office furniture to allow rapid reconfiguration of space to meet changing user needs. They maximize the ability to make adjustments to space while maintaining proper ventilation and space conditioning in each work area. Modular control units can be moved around the space to provide airflow wherever users are located. This level of flexibility and control means that users are happier in their workspace and less likely to need supplemental space heaters or fans to be comfortable.

Integrative strategies

Many of the tactics and technologies described in the previous subsections offer multiple benefits for the natural environment. The challenge for the designer is to find Integrative green solutions that meet the owner's needs for a facility, do not damage natural ecosystems or deplete resource bases, and do not exceed the project budget. Building systems are inherently related to each other, and the design of one system affects the design of another. For instance, increasing the weight of the structure means that the foundations have to be increased as well. Sometimes these relationships can be exploited to pay for investments in one system through savings in another.

For instance, Integrative design means that larger areas of high-performance windows might be included as part of the building envelope design to provide for daylighting (raising total project cost), but the benefits of better envelope performance and reduced heat load from light fixtures are recouped by reducing the capacity of the building cooling system (lowering total project cost). Additionally, a smaller HVAC system might mean smaller pumps, fans and motors, reduced duct sizes, smaller plenums, and reduced floor-to-floor height, also reducing the cost of the facility. Reduced floor-to-floor height means less surface area of the building envelope, which means lower material costs for the system. It also means that the overall weight of the building

---

Integrated strategies for sustainable buildings and infrastructure systems

- Integrated systems design.
- Dematerialization.

is reduced, meaning that foundations can be smaller and more efficient as well.

In the end, the overall increase in the total first cost of the project may be negligible if the benefits of improving one system are captured in the design of related systems. More importantly, lifecycle cost savings can be even bigger with these more efficiently designed systems. HVAC systems in particular will be much more efficient if they are right-sized for the facility, allowing them to operate at maximum efficiency over the lifecycle of the facility.

Another overarching strategy being used to improve the sustainability of capital projects is dematerialization or using services instead of products to meet user needs. Office managers have successfully used this concept for years when leasing rather than purchasing photocopiers for use in their businesses. Instead of buying the machine, the user pays a per-copy charge to the company that owns and maintains the machine. Rather than take ownership responsibility (and associated liability) for equipment, the user pays another company to provide the functional benefits of that equipment. That company, often the manufacturer of the equipment, has an incentive to provide the most efficient equipment possible, since its profits are based on a revenue per unit of service provided. It also has incentive to design products that can be easily repaired, upgraded, or disassembled, since it retains responsibility for the ongoing maintenance and eventual disposition of the equipment.

Advances in information technology such as cloud computing and distributed computing are also examples of this approach. Cloud computing moves key functions such as storage and processing to specialized central facilities that can be operated more efficiently using economies of scale. Cloud data centres use enormous amounts of energy to operate, and maintaining reliability requires that a significant amount of energy be used for functional redundancy. However, they can be located in areas more easily served by renewable energy (such as the midwestern US with its vast wind energy resources, or Iceland with its geothermal and hydroelectric power generation capabilities), thereby reducing the overall carbon footprint of operations.

A variety of building systems and components ranging from roof systems to flooring to mechanical and electrical systems such as fuel cells (Figure 6.36) are available as services from companies that will install appropriate systems to provide a level of performance defined in the contract on a fee or pay-for-performance basis. Some companies also incorporate maintenance services to optimize the performance of their systems. Manufacturers typically know more about maintaining their products than anyone else. Often owners can afford higher performance on a lease basis than they can buy outright. In addition, leasing is typically part of operational budgets, whereas purchase of building elements may be part of initial capital costs. Depending on the nature of organizational funding for built facilities, leasing may allow costs to be shifted from one kind of budget to another while enabling more frequent upgrades to take advantage of technology advances.

Figure 6.36
A leased fuel cell is an example of a dematerialization strategy

When considering what actions to take on a project, making smart choices is critical. Every action taken comes with a cost. Recognizing those costs and considering the benefits of those actions can result in actions that achieve their intended results. An important issue to consider is whether users will have to change their behaviour to achieve the desired effects. Some changes are completely transparent to users, whereas others require significant change in habits or procedures. Changes are more likely to have the desired outcomes if they do not require people to change their behaviour. When possible, choose solutions that can get the job done without requiring users to change their habits.

The degree to which the change is compatible with existing infrastructure is also important. Some changes, such as lighting retrofits, can be as simple as changing a light bulb. Other changes may not be compatible at all with existing buildings, or may not be possible given the skills, resources and equipment of the people responsible for the change. Look for easy changes that can be undertaken with existing skills and tools wherever possible.

To illustrate sustainable design opportunities and strategies used in the design phase of the facility lifecycle, the next sections of the chapter present sustainable design best practices using two in-depth case studies: the Trees Atlanta building located in Atlanta, Georgia and the Bank of America Tower in New York City, both in the USA.

Case study: Trees Atlanta Kendeda Center, Atlanta, GA

The Trees Atlanta building incorporated a wide range of sustainable design strategies to improve its performance. Best sustainable practices for these buildings are described in five categories, as follows:

- sustainable site
- energy optimization
- water and wastewater performance
- materials optimization
- indoor environmental quality.

The first set of strategies involved the building's site, as follows.

Sustainable site

Trees Atlanta is a non-profit organization, dedicated to protecting and improving the urban environment by planting and conserving trees. In congruence with its organizational values, some of the most important design considerations and decisions for the Trees Atlanta building were made regarding the choice of site and its design.

continued

Sustainable site considerations in the Trees Atlanta building

Site selection

- using a previously developed site and building
- locating in an urban setting with existing public infrastructure
- locating in an urban setting with existing public transportation.

Alternative transportation

- locating the building near public transport routes
- providing secure bicycle racks and storage
- providing preferred parking for low-emitting and fuel-efficient vehicles.

Low-impact site amenities

- installing porous surfaces to allow for stormwater retention

- installing bioswales (vegetated water retention)
- installing a green roof on the shed building.

Site development

- conserving existing natural areas
- restoring damaged areas
- maximizing open spaces
- installing advanced exterior lighting fixtures.

Heat island effect

- installing a highly reflective roof on the main building
- planting Georgia native and adapted trees and plants
- installing a green roof
- using an open-grid pavement system.

Site selection

For its new headquarters, Trees Atlanta found a site that had previously been developed in the city of Atlanta (Figure 6.37). In addition, it made a decision to renovate an old and existing warehouse to become a state-of-the-art facility with many sustainable features because this caused less environmental impact than a new building project. Due to the renovation, the Trees Atlanta building was able to save previously undeveloped land and did not affect land that is prime farmland, is subject to flooding, provides a habitat for threatened or endangered species, is near or includes bodies of water or wetlands, or is available as a public park or open space. In addition, since the project site had been previously developed, it also meant that there was no need for expanded transportation and utility infrastructure. The site also affords the building occupants more access to alternative transportation, further reducing the overall environmental impact of the development project.

As the Trees Atlanta building involved converting an existing warehouse to a multi-purpose facility with access to existing infrastructure and public transportation, it was very easy to connect the development with the existing urban community and residential areas and neighbourhoods. In addition, the building is close to basic services such as schools, restaurants, libraries and medical services with good pedestrian amenities, which provides occupants easy access without cars. This community connectivity also helps to reduce urban sprawl, a growing problem that affects quality of life and requires commuters to spend an increasing amount of time in automobiles. The redevelopment of the Trees Atlanta building in an urban area helped to restore, invigorate and sustain established urban living patterns, and to create a more stable and interactive community. This choice

continued

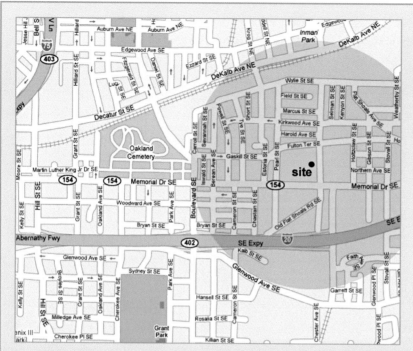

Figure 6.37
The existing facility location near public transport and amenities. Public transport (heavy rail) runs along the train line north of the site, and bus lines run along Memorial Drive just south of the site. Interstate access is available to the west (I-75) and south (I-20). Surface streets provide bicycle and pedestrian access

Credit: *Smith Dalia Architects*

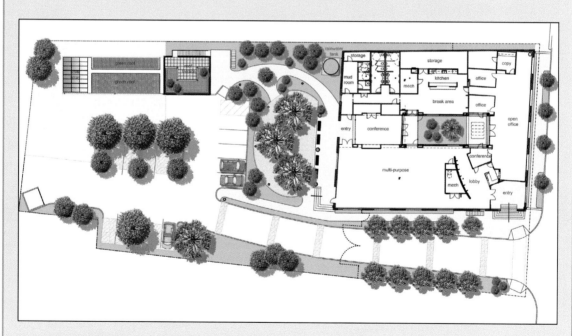

Figure 6.38 Site plan of the Trees Atlanta building
Credit: *Smith Dalia Architects*

continued

reduced the environmental impact from automobile use, minimized the need for installation of new infrastructure and transportation, and also improved the quality of life of building occupants.

Building location and parking

Even though the building was pre-existing at the site, there were many opportunities to incorporate sustainable design strategies. One of the major sustainability strategies was to create a courtyard to provide daylight in the building interior spaces (Figure 6.38). The project team also developed a landscape plan to plant trees that surround the building site with the purpose of providing valuable shade in the summer that would eventually reduce cooling loads in the summer. In the winter, the trees lose their leaves and will allow solar energy to reach the building, providing warmth and heat gain.

The Trees Atlanta building also used multiple approaches to reduce pollution and land development impacts from automobile use, encouraging occupants to use alternative transportation with a lower environmental impact. The building is located in an urban area with existing public transportation, and is within a quarter-mile of one or more stops for two or more public bus lines. There are secure bicycle racks/storage with shower/changing facilities inside the building (Figure 6.39). Trees Atlanta also maintains several bicycles with safety helmets to lend occupants for local trips during the work day.

The Trees Atlanta building provides preferred parking for low-emitting and fuel-efficient vehicles (Figure 6.40) as well as carpools and vanpools. Parking was sized to meet but not exceed minimum local zoning requirements. These approaches to minimize private automobile use can save energy, specifically petroleum, and can avoid the associated environmental problems such as vehicle emissions that contribute to smog, air pollution and greenhouse gas emissions, as well as environmental impact associated with oil extraction and petroleum refining. The restriction of the size of the parking lots also reduces negative impacts on the environment from stormwater runoff and urban heat island effects from the asphalt surface.

Low-impact site amenities

The project incorporated multiple low-impact amenities as part of its site design. Native trees and plants were installed throughout the site that require little or no irrigation, fertilizer or pesticides once they are established (Figure 6.41). Mulch made from ground wood pieces was laid not only to suppress

Figure 6.39a, b Secure bicycle racks and the shower/changing facility

Figure 6.40 Preferred parking for low-emitting and fuel-efficient vehicles is reserved by the sign shown in the photo

continued

Figure 6.41 Planting native trees and plants that require little or no irrigation

weeds and help retain moisture around landscape planting, but also to improve the appearance of the landscape and help reduce the need for irrigation and pesticides. These strategies also reduced the challenges associated with stormwater runoff from the site.

Stormwater has been identified as one of the major sources of pollution for all types of water bodies, because large volumes of stormwater runoff can discharge both pollutant loading and excess heat into receiving water, seriously damaging water quality and harming aquatic life (USEPA 2007). Minimizing stormwater runoff reduces the need for stormwater pipes and infrastructure to convey and treat runoff. The Trees Atlanta building developed a stormwater management plan to reduce the amount of impervious area and increase infiltration using pervious paving materials, harvesting stormwater for reuse in irrigation and indoor non-potable water applications, designing infiltration swales and retention ponds, and installing a vegetated roof.

To reduce the impervious areas in the parking lots, pervious pavers made of precast brick (Figure 6.42) were installed to filter pollutants and reduce stormwater runoff flow rate, volume and temperature. This type of pervious paver is one of the less expensive strategies from a construction and maintenance standpoint. It also helps recharge groundwater supplies.

Runoff storage capabilities were created by installing rain barrels and cisterns to collect water from the membrane roof. This collected water is used for irrigation and toilet flushing to reduce potable water consumption.

Vegetated bioswales (Figure 6.43) were also installed to improve water quality of runoff, attenuate flooding potential and convey stormwater away from critical infrastructure. Bioswales use a combination of plantings, geofabrics, soil and aggregate to strategically divert and filter stormwater using the roots of the plants installed in the swale. By installing bioswales at the building site, the Trees Atlanta building achieved the following benefits:

- Improve water quality using soil, vegetation and microbes.
- Reduce total volume of stormwater runoff.

Figure 6.42a, b The parking area incorporates pervious pavers to capture stormwater for tree roots and recharge groundwater supplies

continued

- Increase infiltration and groundwater recharge.
- Provide a multifunctional conveyance system.
- Create an aesthetic part of the landscape that improves bio-diversity on site.

A vegetated roof was installed in some areas to capture stormwater, which then slowly evaporates from the otherwise impervious roof area. This vegetated roof system can reduce peak stormwater runoff rates and volumes as well as filtering runoff to produce a clear effluent.

By incorporating these sustainable strategies for low-impact site amenities in the Trees Atlanta building, it was possible to limit disruption and pollution of natural water flows and to provide native trees and species that require little or no irrigation, fertilizer or pesticides.

Figure 6.43 Installing bioswales to manage stormwater runoff

Site development

In congruence with its role as a citizens group that protects and improves Atlanta's urban forests by planting, conserving and edu-cating, Trees Atlanta actively sought to restore native and adapted vegetation and other ecologically appropriate features to the site. Damaged areas were restored using native and or adapted plantings for over 50 species, including three different types of green roofs (Figure 6.44). This also provided a significant open space as habitat for vegetation and wildlife, promoting natural biodiversity. Trees Atlanta installed a subsurface tree protection and stormwater infiltration system that supports traffic loads while providing uncompacted soil volumes for large tree growth and on-site stormwater management. The installed modular framework (Figures 6.45 and 6.46) provides unlimited access to healthy soil, a critical component of tree growth in urban environments. This subsurface tree protection and stormwater infiltration system also provides a reservoir to manage stormwater, reducing overheated stormwater surges to local watercourses. The site's open space with many trees and native and adapted vegetation also reduces the urban heat island effect, increases stormwater infiltration and provides the users with a connection to the outdoors.

Illumination of the building, site and supporting facilities such as pavements, parking lots and roadways was designed to limit the light emitted, because excess and obstructive artificial light can affect a site's nocturnal ecosystem and negatively impact fauna. Light pollution also limits night sky observations by humans and represents an inefficient use of energy. To mitigate those concerns, the Trees Atlanta building implemented several features including:

Figure 6.44a, b Vegetated roofs for the Trees Atlanta building

- choosing interior building light systems that only provide the necessary light for the use of a space.

continued

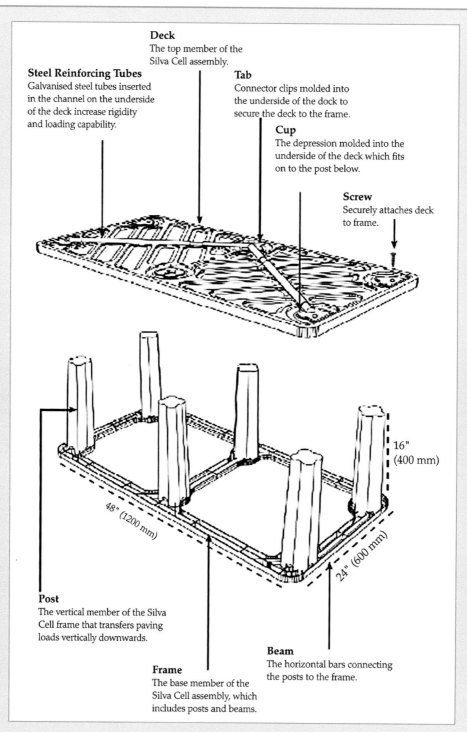

Deck
The top member of the Silva Cell assembly.

Steel Reinforcing Tubes
Galvanised steel tubes inserted in the channel on the underside of the deck increase rigidity and loading capability.

Tab
Connector clips molded into the underside of the dock to secure the deck to the frame.

Cup
The depression molded into the underside of the deck which fits on to the post below.

Screw
Securely attaches deck to frame.

16"
(400 mm)

48" (1200 mm)

24" (600 mm)

Post
The vertical member of the Silva Cell frame that transfers paving loads vertically downwards.

Frame
The base member of the Silva Cell assembly, which includes posts and beams.

Beam
The horizontal bars connecting the posts to the frame.

Figure 6.45 Subsurface tree protection and stormwater infiltration system
Source: Deep Root Green Infrastructure, LLC

continued

Figure 6.46 The subsurface tree protection and stormwater infiltration system being placed
Source: Trees Atlanta

Figure 6.47 Lighting controls
Source: Trees Atlanta

- installing an automatic dimming system that automatically controls the luminance of light and turns off light fixtures when a space is not being used (Figure 6.47).
- using the minimum amount of light necessary for safety and comfort.
- installing covers on site lights that do not allow light to project upwards or outside the building boundary (Figure 6.48).

Heat island effect

The installation of dark, non-reflective surfaces for parking, roofs, pavements and other hardscape contributes to the urban heat island effect, which elevates temperature during the summer. As a result the annual mean air temperature of a city with one million people or more can be

Figure 6.48 Reduction of exterior light pollution by using shields that direct light downward where it is needed, rather than upward into the night sky

continued

1–3°C (1.8–5.4°F) warmer than its surroundings. In the evening, the difference can be as high as 12°C (22°F). The urban heat island effect affects communities by increasing summertime peak energy demand, air conditioning costs, air pollution and greenhouse gas emissions, heat-related illness and mortality, and water quality. In addition, it can affect site habitat, wildlife and animal migration corridors.

To mitigate this effect, the Trees Atlanta building used materials with higher solar reflectance properties (with a solar reflectance index (SRI) of at least 29) in the site design, provided shaded areas using trees and vegetation, and reduced hardscape surfaces in the large open spaces surrounding the building. The SRI is a measure

Figure 6.49 The high-albedo roof surface reflects sunlight instead of absorbing it

of how much radiant energy is reflected from a surface and absorbed from it, with 100 being a perfectly reflective surface and 0 being a perfectly absorptive surface. The Trees Atlanta building also included a high-albedo roof surface shown in Figure 6.49, with an SRI of 78. There is also a vegetated roof, a layered system that consists of vegetation, growing medium, filter fabric, drainage and a waterproof membrane set on top of a conventional roof (Figure 6.44).

Energy optimization

The building sector accounts for approximately 40 per cent of the energy and 74 per cent of the electricity consumed annually in the United States (EIA 2010), so it is very important to increase energy efficiency and generate energy using on-site renewable energy systems wherever possible. This reduces the environmental and economic impacts associated with excessive energy use. Multiple energy-related best practices were employed in the Trees Atlanta building, resulting in an estimated 70 per cent energy saving over ASHRAE 90.1–2004 (the applicable minimum standard). Table 6.2 provides an energy analysis.

Sustainable energy features of the Trees Atlanta building

- Daylighting
- Geothermal heat pumps
- Solar hot water heater
- Highly efficient fluorescent fixtures and ballasts
- Lighting occupancy sensor controls

- Task lighting
- Energy Star rated equipment
- Onsite renewable energy (solar panels on shed building)
- Carbon offsets through purchase of green power credits.

Since the energy performance of a building depends significantly on its design, the Trees Atlanta building considered both passive design strategies, including the building's massing and orientation, materials, construction methods, and building envelope, and active strategies including HVAC and

continued

Table 6.2 Energy analysis compared with a conventional building

Building element/item	Unit of measure	Baseline building	Proposed building
BUILDING			
	Visible Light Transmission	0.75	0.58
Type 3	SHGC	0.76	0.23
	U-Value	1.16	0.28
	Visible Light Transmission	0.75	0.38
Type 4	SHGC	Code minimum	0.21
Small side windows	U-Value	Code minimum	0.28
	Visible Light Transmission	Code minimum	0.34
Atrium	Glass (New)	Code minimum	Types, 1, 2 and 3
Main Entrance Side	Glass (New)	Code minimum	Types, 1, 2 and 3
Entrance	Glass (New)	Code minimum	Types, 1, 2 and 3
Window/Wall Ratio Zone	Orientation	WWR	WWR
Multipurpose	West	38%	Same
	South	38%	Same
	East	50%	Same
	North	50%	Same
	North (Atrium)	37%	Same
Office	Southwest	16%	Same
	South	35%	Same
	East	28%	Same
	North	0%	Same
	West (Atrium)	66%	Same
Backspace	North West	0%	Same
	South (Atrium)	8%	Same
		37%	Same
Conference	West	66%	Same
	East (Atrium)	66%	Same
Light Shelves			
East	Depth (Feet)	None	5'-0"
South	Depth (Feet)	None	5'-0"
West	Depth (Feet)	None	5'-0"
Skylights	None	n/a	n/a
HVAC			
Design temp-cool	deg F	75	Same
Design temp-heat	deg F	70	Same
Supply temp-cool	deg F	55	Same
Supply temp-heat	deg F	95	Same

continued

Table 6.2 *continued*

Building element/item	Unit of measure	Baseline building	Proposed building
Occupied	Fan operation	Continuous	Same
	Fan operation	On/Off	Same
Dining	W/sq ft	1.40	0.98
Locker	W/sq ft	0.60	0.42
EXTERIOR LIGHTING POWER			
Connected load Operation	Watts Schedule	7.850 On when dark (Photocell)	1.141 Same
INTERNAL LOADS			
Office equipment	W/sq ft	1.10	Same
Office task lighting	W/sq ft	0.20	Same
Conference rooms	W/sq ft	0.25	Same
Multi-purpose	W/sq ft	0.25	Same
OCCUPANCY			
Occupancy	Full time FTE	11	Same
Hours of Operation	Transient	50 (2hr stay)	Same
	Mon–Fri	8:00 am to 5:00 pm	Same
	Sat, Sun	Closed	Same
	Holidays	Closed	Same
SPACE VENTILATION RATES			
Entire building	Cfm	ASHRAE plus 30% (min)	Same

lighting systems. The building also includes on-site renewable energy via solar PV panels to reduce the environmental and economic impacts associated with fossil fuel energy use. Finally, the building has adopted an Integrative, whole-building approach to optimize energy efficiency. The following subsections describe these strategies in more detail.

Daylighting

A significant portion of all the lighting energy used by a building can be saved through daylighting, and daylight also can provide comfort for the indoor environment. The Trees Atlanta building includes several daylighting options. The first architectural decision related to daylighting was to open up the centre of the existing industrial building, creating a new interior courtyard to bring daylight into the space (see Figure 6.50). This new interior courtyard provides an outdoor meeting place and has become a place to plant the symbol of the entire mission of the organization: trees. In addition to the new interior courtyard, the daylighting components include the use of window overhangs for shading and interior light shelves for bouncing light into the space (Figure 6.51), light tubes in interior rooms (Figure 6.52) and detailed selection of fenestration characteristics to optimize daylighting performance.

continued

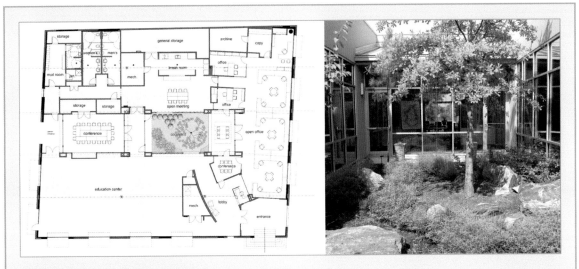

Figure 6.50 The courtyard provides enough daylight and a warm environment. *Source: Smith Dalia Architects*

Figure 6.51a, b
Window overhangs for shading and interior light shelves

Figure 6.52a, b
Light tubes to bring daylight into the interior

continued

The HVAC system

HVAC systems can be the largest energy consumer in a modern building, requiring energy to provide heating, cooling, humidity control, filtration, fresh air makeup, building pressure control and comfort control. In the Trees Atlanta building, the HVAC system utilizes a ground source heat pump system (Figure 6.53) made up of four indoor units, each serving a unique zone, and a heat recovery unit for the office space unit (Figure 6.54). Using pipes to exchange heat (Figure 6.55), the system retrieves heat from underground during cool months and returns heat to the ground in hot summer months. The units are controlled by programmable thermostats (Figure 6.56) and are strategically located to ensure that the ductwork runs simple, straight and with minimum bends. The air conditioning design load was established to accurately reflect conditions and prevent oversizing the HVAC systems. The initial budget load estimate was over 35 tons of cooling, but when the high performance features of the building were incorporated, the final cooling load was about 24 tons.

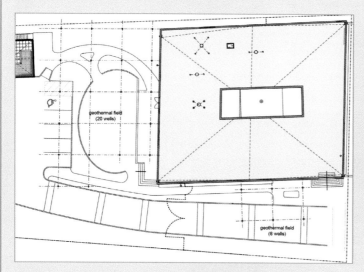

Figure 6.53 Geothermal site plan. *Source: Smith Dalia Architects*

Figure 6.54 The heat recovery system to recover heat from air

Figure 6.55a, b The geothermal trench and underground piping. *Source: Trees Atlanta*

Figure 6.56a, b The geothermal system and programmable thermostats

continued

Figure 6.57 Trees shading the west exposure

Building envelope

The building envelope of the Trees Atlanta building was optimized through energy modelling analysis early in the design phase, as shown in Table 6.2. The original building was an existing concrete masonry unit (CMU) structure with a simple built-up roof and no insulation. The final design included furred-out walls with R-13 insulation and additional roof insulation (R-20) as well as the selection of high-performance fenestration products. The new roof is white to reflect summer sun. Since the western exposure of the building is often a difficult exposure in the southern US climate, it features a planting of trees that provide shading for this exposure (Figure 6.57).

Figure 6.58a, b Solar hot water system elements

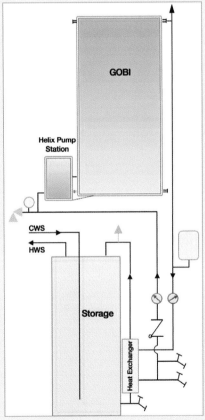

Figure 6.59 Solar hot water system schematic

continued

Domestic hot water system

Domestic hot water is provided by a solar thermal hot water system (Figure 6.58) consisting of roof-top panels, a solar PV powered circulation pump, and a well-insulated storage tank of 80 gallons. The system is an indirect type in that a glycol (environmentally friendly) solution is circulated through a heat exchanger located in a tank connected to the city water supply (Figure 6.59). A backup electric element is provided for periods when solar heating is limited. The solar fraction for this installed system is over 70 per cent and the system's estimated energy output is 2119 kWh per year with system rating at 3.9 kW.

High efficiency lighting fixtures and ballasts with occupancy sensors

Lighting is a major building energy user and also is very important for peoples' daily life and health. To reduce energy demands while preserving indoor environmental quality, the Trees Atlanta building has maximized the use of daylight as a primary lighting source for the building. In addition, high-efficiency lighting fixtures and ballasts were installed (Figure 6.60a) that can reduce lighting density in watts per square foot and use 30 per cent less energy than the minimum ASHRAE 90.1 2004 lighting power density. Lighting occupancy sensors (Figure 6.60b) were installed to control indoor light in designated spaces including rest rooms, conference rooms, bicycle storage and the like, which can turn lights on automatically when someone enters a space. These installed occupancy sensors can reduce lighting energy use by turning lights off soon after the last occupant has left the space. In addition, supplemental task lights also have been installed to provide additional light at each workspace for user comfort.

To minimize energy consumed through plug loads, Energy Star rated equipment has also been installed.

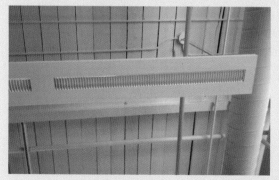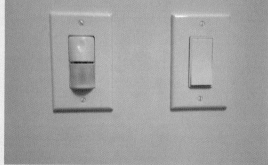

Figure 6.60a, b High-efficiency lighting fixtures and occupancy sensors

Solar photovoltaic (PV) systems

The Trees Atlanta building has installed a 5-kilowatt solar electric PV system comprising 24 SunPower SPR-210 (210 watt) modules mounted on the roof (Figure 6.61). The installed PV system can annually generate 6635 kWh based on a DC to AC derate factor of 0.80 and tilt of 8 degrees. The fixed panels face due south to receive full solar exposure. In addition, the system has incorporated a SunPower 5000 m inverter and net metre for a direct grid intertie system with the local power company (Georgia Power).

continued

Water and wastewater performance

Since the consumption of public water supply in facilities in the United States increased 12 per cent between 1990 and 2000, to 43.3 billion gallons per day (11 per cent of total withdrawals and slightly less than 40 per cent of groundwater withdrawals), it is vital to increase water efficiency to not only reduce water consumption but also reduce the amount of waste water and energy consumption. As described in Chapter 1, a significant portion of US energy demand goes to treating, pumping and heating water. The Trees Atlanta building has several sustainable features that increase water efficiency to minimize water consumption and wastewater generation. Sustainable water features in the facility also have lowered the lifecycle cost for building operations and the costs for the municipal supply and treatment facilities.

Landscaping irrigation practices consume a significant amount of potable water in the United States. Outdoor uses, mainly irrigation for lawns and plants, account for 30 per cent of the 26 billion gallons of water consumed daily (USEPA 2010). Planting native and adapted plants on the project site can reduce water consumption for landscaping, and irrigating using captured rainwater also eliminates the need for potable water. The Trees Atlanta building has not installed a permanent irrigation system, and has native and adapted plantings for Georgia's climate. The building also captures rainwater in storage tanks, thereby eliminating potable water use for landscaping (Figure 6.62).

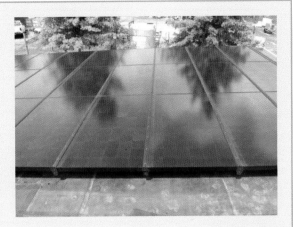

Figure 6.61 Roof-mounted solar photovoltaic panels

Sustainable water strategies in the Trees Atlanta project

- No permanent irrigation system installed
- Native and adapted plants for the Georgia climate
- Rainwater collection for toilet flushing and site irrigation
- Low-flow fixtures, including toilets, urinals and showerheads
- Self-generating, aerating taps.

Figure 6.62a, b Native plants and rainwater harvesting tanks

continued

It is very important to increase water efficiency in buildings to reduce the burden on municipal water supply and wastewater systems. The Trees Atlanta building includes several green features for this purpose: low-flow fixtures (1.28 gallons per flush (gpf) toilets, 0.5 gpf urinals, aerated self-generating taps and low-flow shower heads, some of which are shown in Figure 6.63), and using rainwater for the flushing toilets and landscaping. There is a 40ft × 40ft catchment area comprising a quarter of the roof area (Figure 6.64). The annual rainfall in Atlanta is 50.1 inches per year. The captured water is collected into two 4,500-gallon aboveground rainwater collection tanks (Figure 6.64) that preserve the water for use in toilet flushing and landscaping. In using these sustainable features for water efficiency, the building is estimated to save 83.3 per cent of typical potable water usage (Table 6.3).

Figure 6.63 Low-flow fixtures including aerated self-generating taps and low-flow urinals

Table 6.3 Water saving in the building

Baseline case: annual water consumption (gal):	34,590 gallons/year
Design case: annual water consumption (gal):	20,425 gallons/year
Total annual non-potable water consumption (gal):	14,660 gallons/year
Total water savings	83.3%

Sustainable material strategies

A large amount of waste is generated during construction-related activities, including construction, operation and demolition. The construction sector is a major consumer of materials and resources. Sustainable design and construction best practices related to materials and resources include:

- selecting sustainable materials that have minimal environmental, social and health impacts during extraction, processing, transportation, use and disposal.

Figure 6.64 Rainwater catchment and two rainwater collection tanks

continued

- practising waste reduction by maintaining occupancy rates in existing buildings to reduce redundant development and the associated environmental impact of producing and delivering new materials.
- minimizing waste at its source, including reducing the overall demand for products.
- reusing and recycling materials, resources and existing facilities.

In the Trees Atlanta building, the majority of the existing building's envelope (75 per cent: 31,453 sq ft/41,932 sq ft) was reused to conserve resources, reduce waste and reduce the environmental impact of the materials, manufacturing and transport involved in a new building (Figure 6.65).

The Trees Atlanta building also achieved its goals for the project with a construction waste management plan to recycle or salvage for reuse 90 per cent of the waste generated on-site. For example, in an effort to minimize the material waste from the construction effort, all of the demolished concrete was crushed and reused on site. With this and other strategies such as recycling and salvage (Figure 6.66), the project diverted 90 per cent of construction and demolition debris from disposal in landfills and incineration facilities.

Sustainable materials strategies in the Trees Atlanta project

- building envelope reuse
- construction waste management planning, including reuse of waste concrete on site

- use of local or regional materials
- use of recycled content materials
- use of rapidly renewable materials
- use of fsc certified wood products

Over 10 per cent of the total material cost was for materials that were extracted, processed and manufactured regionally or locally. This strategy reduced transportation impacts as well as stimulating local economies. To increase demand for building products that incorporate recycled content materials, the Trees Atlanta project employed a high level of recycled-content building materials – 13 per cent by total material cost – including structural steel components, metal studs, drywall, carpet, window frames, wood doors, bike racks, toilet partitions, batt insulation materials and cast concrete countertops (Figure 6.67).

The design team also planned to use as many rapidly renewable materials as possible, including bamboo flooring materials and medium-density fibreboard (MDF), to reduce the use and depletion of finite raw materials and long-cycle renewable materials (Figure 6.68).

Figure 6.65 Existing CMU walls were reused in the new structure. *Source: Smith Dalia Architects*

Figure 6.66 Recycling skips helped to divert construction waste. *Source: Smith Dalia Architects*

continued

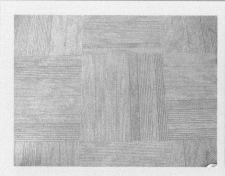

Figure 6.67a, b Recycled content materials included carpet tiles and cast concrete countertops

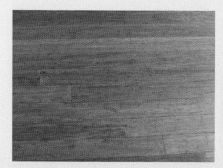
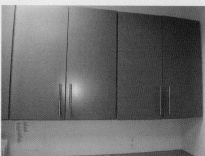

Figure 6.68a, b Bamboo flooring and MDF cabinets

Since Trees Atlanta is a non-profit organization dedicated to planting and preserving trees, a major goal for the project was to use sustainably harvested wood products whenever possible. The building project team ensured that 63 per cent of all wood-based materials and products used in the building were certified in accordance with the Forest Stewardship Council's (FSC's) sustainable harvesting criteria (see Chapter 4, p. 172). The main purpose of using FSC wood-based materials and products was to encourage environmentally responsible forest management.

Indoor environmental quality strategies

People spend a significant amount of their time indoors, on average more than 90 per cent in the United States, so the quality of the indoor environment has a significant influence on human well-being, productivity and quality of life (USEPA 2001). Trees Atlanta understands the importance of indoor air quality in creating a good working environment in the building, and their building incorporated many features to ensure that it occurs.

Sustainable materials strategies in the Trees Atlanta project

- Improving air ventilation
- Managing air contaminants
- Using low emitting materials to prevent indoor environmental quality problems
- Allowing occupants to control desired settings
- Providing daylighting and views.

---

continued

Indoor air ventilation

To improve occupant comfort, well-being and productivity, the building manages indoor air quality (IAQ) by mechanical ventilation systems exceeding the minimum requirements of the ASHRAE Standard 62.1 2004, Ventilation for Acceptable Indoor Air Quality. The building was calculated to need a minimum total of 1994 cubic feet per minute (cfm) of air exchange to meet minimum air quality requirements. To improve upon this, an outside air energy recovery ventilator supplies one of the ground source heat pumps units with up to 2400 cfm of outside air while exchanging heat with exhaust air for energy efficiency. The other three ground source heat pump units serving the facility have 430 cfm each ducted into the facility. This is a total of 3690 cfm of outside air introduced into the building compared with the minimum ASHRAE Standard 62.1–2004 requirement of 1994 cfm.

Low-emitting materials

One of the strategies to improve indoor air quality is to reduce the quantity of indoor air contaminants including adhesives, sealants, sealant primer products, paints, coatings and carpets that are odourous, irritating and/or harmful to the comfort and well-being of installers and occupants. Therefore, all materials that emit contaminants including VOCs were avoided to the maximum extent possible. Instead, low-emitting materials were used for all adhesives, sealant, sealant primer products, paints and coatings (Green Seal products). All carpets met the testing and product requirements of the Carpet and Rug Institute Green Label Plus Program, which identifies carpets with very low VOC emissions (Figure 6.67).

Occupant control over systems

One of the significant factors related to occupant satisfaction is providing a high level of lighting system control by individual occupants or groups in multi-occupant spaces. There are 15 individual workstations including private offices and cubicles in the building, and each one has an individual lighting control. Table 6.4 lists the lighting controls that have been installed in the shared multi-occupant spaces within the building.

Every user of the space also has direct access to natural ventilation: each bay of workstations has access to its own operable windows to provide natural air (Figure 6.69). The mechanical system was designed for four separate work zones that can be controlled independently, saving energy by not

Table 6.4 Multi-occupant space lighting control

Multi-occupant space	Description of installed lighting controls
Small conference room	Manual light switch with an occupancy sensor
Break-out room	Manual light switch
	Mostly daylit space – manual shades available at individual window sections for controllability of natural light
Large conference room	Eight-scene controller for specific room uses
	Electronic shading devices for daylight control
Multi-purpose space	Eight-scene controller for specific room uses
	Manual shading devices for daylight control

continued

ventilating the zones that are not in use while also providing independent user controllability for each zone. Thermostats were provided in the office area, break-out/kitchen area, large conference room, small conference room and multipurpose space. The majority of the multi-occupant spaces have access to either exterior or interior courtyards.

The HVAC systems provide typical conditioning for an office-type facility, with temperature and humidity design criteria as shown in Table 6.5.

Table 6.5 Trees Atlanta temperature and humidity design criteria

Season	Maximum indoor space design temperature (deg F)	Minimum indoor space design temperature (deg F)	Maximum indoor space design humidity (%)
Spring	78.0	68.0	70.0
Summer	78.0	68.0	70.0
Fall	78.0	68.0	70.0
Winter	72.0	68.0	70.0

Figure 6.69 Operable windows provide natural air into occupied spaces

Figure 6.70 Daylight in the building and open views across spaces

Figures 6.71a, b Shading devices and interior light shelves

continued

The multiple daylighting strategies provide occupants with a connection between indoor spaces and the outdoor spaces through the introduction of daylight and views into regularly occupied areas of the building (Figure 6.70). The first approach used was to expand existing window openings at the exterior walls. Second, the large interior courtyard provides ample natural light to virtually all interior spaces, while providing open views through and across the entire building. The glass used for the exterior windows was carefully considered. High-performance glass was used in the areas that receive the most severe solar gains (east/west) and near regular building occupants. Shading devices included canopies and manual blinds to shade extra light (Figure 6.71a). Daylighting shelves (Figure 6.71b) also were provided at the southern window openings. Due to the importance of daylighting for the Trees Atlanta building, virtually all of the regularly and non-regularly occupied spaces (with the exception of only the storage areas and restrooms) have direct views to an exterior space and to the outside (Figure 6.72). These exceptional views were achieved through the creation of large openings at the exterior walls, and the courtyard, which was planted with native trees and ground cover to afford a serene exterior setting for the occupants of the facility.

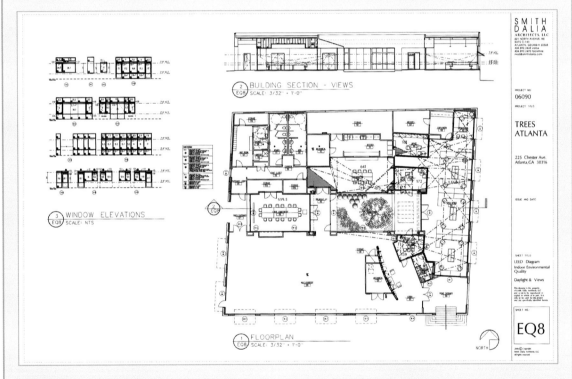

Figure 6.72 Floor plan for daylight and views. *Source: Smith Dalia Architects*

The Trees Atlanta Kendeda Center incorporates a variety of sustainable design strategies to improve its lifecycle performance that are appropriate for low-rise commercial construction and that support Trees Atlanta's primary mission and goals. The next case study illustrates very different types of sustainability strategies used by the Bank of America Tower in New York City, all of which are appropriate for a high-performance tall building in a dense urban environment.

Case study: Bank of America Tower

The Bank of America Tower is second only to the Empire State Building in New York City in square footage. It is also the second tallest building (336 m with 55 storeys) in New York, again after the Empire State Building (381 m with 102 storeys), with a mast rising about 46 metres taller than the Chrysler Building's spire (319 m with 77 storeys). The tower is located in the heart of Manhattan, across from Bryant Park at 42nd Street and 6th Avenue, and is a landmark not only for sustainable and energy-saving design, but also for its crystalline form and sheer size. The tower project was conceived in 2000, the ground-breaking occurred in 2004, and the tower officially opened following construction in May 2010 (Figure 6.73).

Bank of America, a major international corporation, initiated the Bank of America Tower project early in 2000 so that it could be one of the world's largest financial institutions to provide financial services in New York. Since Bank of America is genuinely interested in corporate sustainability and environmental initiatives including climate change, one of the company's long-standing efforts is to sustainably operate and develop its buildings in ways that can:

- reduce greenhouse gas emissions
- reduce energy consumption
- reduce water consumption
- reduce materials consumption
- provide healthier indoor environments.

The Bank of America Tower is a prime example of this comprehensive commitment to sustainability. At the programming and design stage, the project team included Bank of America, Durst Organization and Cook + Fox Architects. The aim was to create a sustainable world landmark building that would also be the highest-performing building, integrating many sustainable design and construction strategies that would enhance the health and productivity of its tenants, minimize water and energy consumption, reduce waste and promote environmental sustainability.

Bank of America Tower

- Location: New York City, NY, USA
- Building type: Office
- Date of groundbreaking: 2004
- Date of completion: May 2010
- Total area: 195,000 sq m (2,100,000 sq ft)
- Total height: 336 m (1200 ft)/55 floors
- Cost: $1 billion
- Green Building Certification: LEED Platinum

- Owner: Bank of America
- Developer: Durst Organization
- Architect: Cook + Fox Architects LLP
- Structural Engineer: Severud Associates
- Contractor: Tishman Construction Corporation

Sustainable sites

Since the tower is located in the heart of Manhattan, occupants can enjoy the New York public transport systems with zero parking spaces, reducing automobile use and greenhouse gas emissions. The tower also includes a mid-block pedestrian passage known as Anita's Way and Urban Grade Room (3,500 sq ft, 330 sq m) at the corner of Sixth Avenue and 43rd Street (Figure 6.75), which acts as a front

continued

porch by providing public space and a sheltered extension of Bryant Park (Figure 6.74a). A carbonized bamboo ceiling extends through the curtain wall and hovers 25 ft (7.6 m) out over the sidewalk to provide a natural habitat for pedestrians (Figure 6.74b). To reduce the urban heat island effect in the city, the tower incorporates a green roof and highly reflective pavers. Finally, the project also includes the newly opened Stephen Sondheim Theatre to provide social services to the public. The theatre, originally built as Henry Miller's Theatre in 1918, is now the first 'sustainable' theatre on Broadway.

Energy optimization

One of the main objectives of sustainability for Bank of America was to reduce energy consumption in its buildings. To achieve this goal, the Bank of America Tower incorporated multiple energy saving strategies. The tower employs high-performance curtain walls to reduce solar heat gain through low-emissivity glass and heat-reflecting ceramic frit as well as minimal air infiltration (Figures 6.76 and 6.81). The Tower also includes daylighting concepts along with occupancy sensor and automated daylight dimming technologies to reduce energy for lighting.

Figure 6.73 Bank of America Tower in New York

Source: Courtesy of Cook + Fox Architects

Sustainable site features of the Bank of America Tower

Public transportation and parking

- Include zero parking spaces.
- Locate near public transport.

Urban garden room

- Provide an urban garden room to provide green public space, reinforcing the building's street-level interactions as well as its connection to Bryant Park.

Public circulation spaces

- Provide three times more public circulation space than is mandated by as-of-right zoning.

Heat island effect reduction

- Install a green roof.
- Install highly reflective pavers.

continued

Figure 6.74a, b Sustainable site features in the Bank of America Tower – a dense urban location

Sustainable energy features of the Bank of America Tower

High-performance curtain wall
- Reduce solar heat gain through low-emissivity glass and heat-reflecting ceramic frit.
- Reduce air infiltration.

Daylighting system
- Reduce artificial lighting requirements through higher ceilings and highly transparent glass.
- Install an occupancy sensor system.
- Install an automated daylight dimming system.

On-site co-generation plant
- Install a 4.6 mW cogeneration system.
- Provide 65 per cent of building's annual electricity requirements.
- Reduce daytime peak electricity demand by 30 per cent.
- Generate most of the heating energy for the building.

Absorption chiller
- Provide heat in winter and cooling in summer (see Figure 6.79).

Glycol (ice storage) system:
- Provide approximately 25 per cent of building's annual cooling requirements
- Reduce daytime peak loads on city's electricity grid at night, excess electricity from the co-generation system is used to produce ice, which is melted during the day to supplement the cooling system.

Figure 6.75 Anita's Way and Urban Grade Room at the corner of Sixth Avenue and 43rd Street

continued

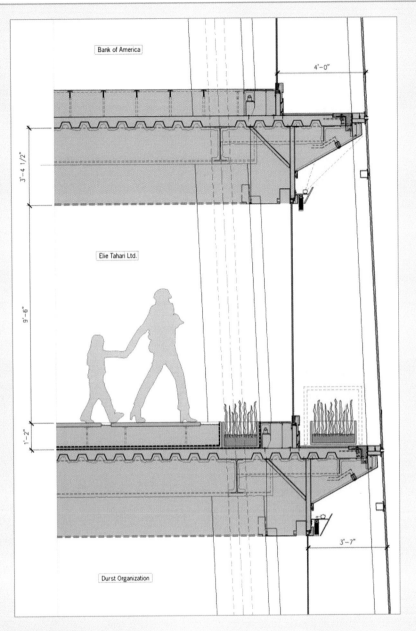

Figure 6.76 Sustainable envelope design

Source: Courtesy of Cook + Fox Architects

A 4.6 mW cogeneration plant provides 65 per cent of the building's electrical energy, and the heat generated by the plant is recovered to run other building systems (Figure 6.77a). Efficient generation of the building's own electricity through the natural-gas-fuelled cogeneration plant is cheaper than purchasing electricity from local utilities. The plant also significantly reduces the need for additional electricity grid development in the city.

continued

Air conditioning (AC) is supplied by chillers in a variety of sizes, ranging from 850 to 1,200 tons (770 to 1,100 metric tons), which provide optimal efficiency. The staggered sizes make it possible to monitor AC demands and adjust chiller usage to get the most efficient use out of each one. The electrical demand also is reduced through the use of a glycol system (Figure 6.77b) that creates ice at night, when demand is lower and energy is cheaper, which is then used to help cool the building during the day. Forty-four tanks, each with a capacity of 625 cu ft (17.7 cu m) of glycol, are frozen at night and then melted the next day to shave the building's electrical energy demand for air-conditioning by about 5 per cent.

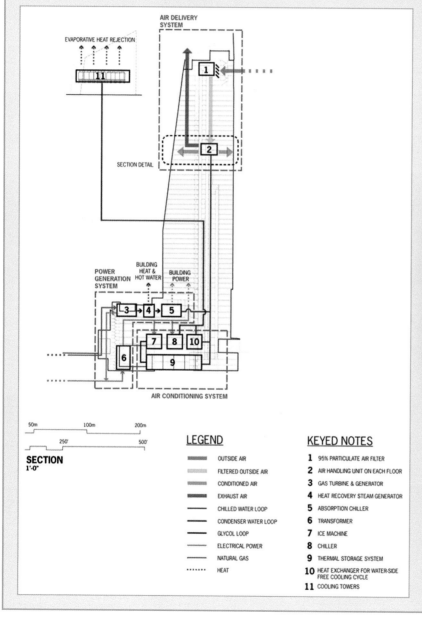

LEGEND

	OUTSIDE AIR
	FILTERED OUTSIDE AIR
	CONDITIONED AIR
	EXHAUST AIR
	CHILLED WATER LOOP
	CONDENSER WATER LOOP
	GLYCOL LOOP
	ELECTRICAL POWER
	NATURAL GAS
....	HEAT

KEYED NOTES

1 95% PARTICULATE AIR FILTER
2 AIR HANDLING UNIT ON EACH FLOOR
3 GAS TURBINE & GENERATOR
4 HEAT RECOVERY STEAM GENERATOR
5 ABSORPTION CHILLER
6 TRANSFORMER
7 ICE MACHINE
8 CHILLER
9 THERMAL STORAGE SYSTEM
10 HEAT EXCHANGER FOR WATER-SIDE FREE COOLING CYCLE
11 COOLING TOWERS

Figure 6.77a, b
Diagram of Tower's HVAC Systems (a) and glycol ice storage (b) for energy optimization

Source: Courtesy of Cook + Fox Architects.

continued

These energy-saving strategies make it possible not only to save significant energy but also to reduce greenhouse gas emissions.

Water and wastewater performance

Water conservation was a major consideration in the Tower's design. The project team included a rainwater harvesting system as well as a greywater harvesting system not only to reduce water consumption but also to minimize the generation of wastewater. The concept was to recycle all water in the building except toilet water by treating all collected rainwater and greywater.

Sustainable water features of the Bank of America Tower

Rainwater harvesting and greywater system (5,000 gal/19,000 l per day):

- Capture and reuse 48 inches of annual precipitation.
- Capture and reuse cooling coil condensate and greywater generated on-site.
- Collect groundwater in the basement of the building (not necessarily a more sustainable water source, but reducing the use of potable water and its resource-intensive treatment).
- Install four large holding tanks (total about 60,000 gallons) located at four different floors heights to feed toilets and for other greywater uses.

Water/Wastewater reduction:

- Install waterless urinals and low-flow plumbing fixtures.
- Save 7.7 million gallons per year (45 per cent reduction in the use of city water).
- Reduce 95 per cent of wastewater to the city's sewer system.

The Tower can capture about 5,000 gallons (19,000 litres) of rainwater, cooling-coil condensate water and groundwater[2] per day in order to treat and reuse it (Figure 6.78). There are four large holding tanks – totalling about 60,000 gallons (230,000 litres) located on different floors in order to feed toilets and for other grey-water uses, such as the cooling towers on the roof. These integrative water strategies, coupled with water conserving plumbing fixtures, can reduce water consumption from the conventional level by about 45 per cent. It was possible to cut water sent to the sewers by about 95 per cent, which substantially reduces demands on the city's sewer system.

Material optimization

The Bank of America Tower used substantial quantities of local or regional building materials, including 1.6 million sq ft of glass in the curtain wall that was manufactured within 500 miles of New York City (Figure 6.79). In addition, 91 per cent of all construction and demolition debris was recycled or otherwise diverted from disposal in landfills. The Bank of America Tower used as many building materials as possible that contain recycled content, including structural steel (87 per cent recycled content) and concrete (45 per cent recycled content). All building materials were carefully sourced and reviewed before installation to confirm that they were low-VOC, sustainably harvested, manufactured locally, and/or contained recycled content wherever possible. Finally, salvaged artefacts from Henry Miller's Theater were reused in the new Tower (see Figure 6.80).

continued

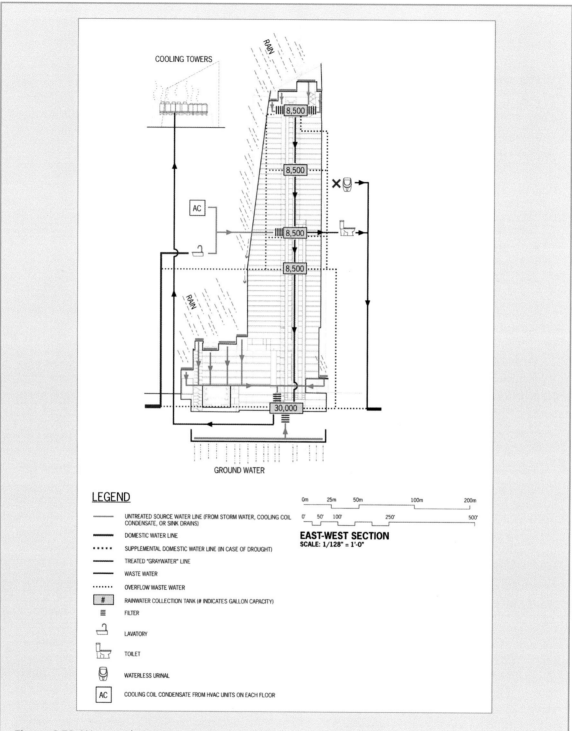

Figure 6.78 Water and waste water saving strategies

Source: Courtesy of Cook + Fox Architects

continued

Examples of regionally sourced materials

Red: Manufacturing location
Green: Harvesting location

1 Structural steel (Columbia, SC)
2 Curtain wall (Montreal, CA)
3 Concrete (Port Chester, NY)
4 Bathroom countertops (Brooklyn, NY)
5 Quarried stone (various locations in Vermont)

6 Stone fabrication (Patterson, NJ and Bronx, NY)
7 Millwork (Jamaica, NY)
8 Access flooring (Red Lion, PA)
9 Gypsum wallboard (Shippingport, PA)

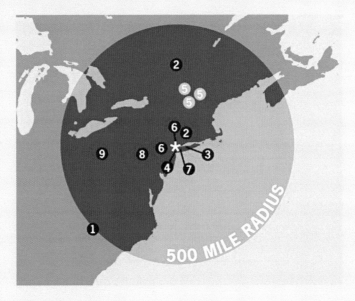

Figure 6.79
Regionally harvested and manufactured materials. *Source: Courtesy of Cook + Fox Architects*

Sustainable materials features of the Bank of America Tower

Recycling

- Recycle or divert 91 per cent of all construction and demolition waste from landfill.

Recycled materials

- Use structural steel materials that contain 87 per cent recycled content.
- Use concrete that contains 45 per cent recycled content (blast furnace slag).

Regional materials

- Use 1.6 million sq ft of glass in the curtain wall that was assembled within 500 miles of the site.
- Use sustainably harvested wood crating for glass transit.

Reuse materials

- Reuse salvaged artefacts from Henry Miller's Theatre.

Low-VOC materials

- Use low-VOC building materials.

continued

Figure 6.80 Sustainable features for material optimization – recycled content steel and reuse of materials from the old Henry Miller's Theater. The building façade was reused in its original, intact form.
Courtesy of Cook + Fox Architects

Indoor environmental quality features of the Bank of America Tower

Air ventilation

- Remove 95 per cent of particulates by air filtration.
- Intake outside air at 800 ft.

Daylight and views

- Access natural daylight and views through higher 9 ft 6 in ceilings.

Under-floor ventilation system

- Allow for individual climate control through in-floor air diffusers.

- Provide more efficient and healthy cooling and fresh air through floor-by-floor air handling units.

Air monitoring system

- Track carbon dioxide, carbon monoxide, VOCs and small particulates.
- Ensure consistently high quality of ventilation air.

continued

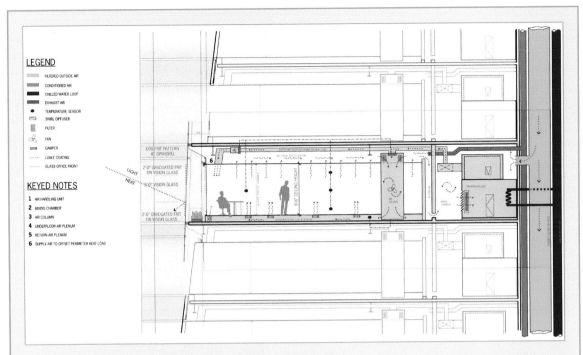

Figure 6.81 Diagram of floor section and air intake and circulation system

Source: Courtesy of Cook + Fox Architects

Indoor environmental quality

The Tower was designed and constructed to provide a high-quality indoor environment to building occupants, which can improve occupants' health and productivity. An under-floor ventilation system was installed throughout the building to allow for individual climate control through in-floor air diffusers. This raised floor system makes the routing of cables and computer infrastructure much easier (Figure 6.81).

The makeup air inside the building is pulled in through intakes at 800 ft (300 m) above the street, cleaned through an air filter that captures about 95 per cent of particulates, and then cleaned again before it is distributed throughout the building (Figure 6.81). These systems result in providing very clean hospital-grade air to all building occupants. In addition, there is an air monitoring system that measures air quality throughout the building. This system monitors carbon dioxide, carbon monoxide, VOCs and small particulates. Finally, occupants in the tower enjoy natural daylight and views through higher 9 ft 6 in ceilings and floor-to-ceiling windows glazed with extremely transparent, low-iron glass (Figure 6.81). With these design strategies, the Bank of America Tower can provide an exceptional indoor environment to improve occupants' health and productivity.

Altogether, the sustainable design strategies employed in the Bank of America Tower have resulted in a world-class landmark facility exemplifying the best in sustainable design for high-rise facilities. With the anticipated increase in world population and increasing numbers of people living in cities, there is a real need to develop better ways to design and build structures like this, both now and in the future.

Sustainable design in the future

How will we design built facilities and infrastructure systems in the future? What will they look like, and what expectations will their owners have for them? Not only will the technologies and materials we use be different from those of today, but also the processes we use to determine human needs and develop design solutions will be different. Indeed, the very nature of the built environment is likely to shift in response to drivers such as higher population density in urban centres, climate change and increased resource scarcity.

We can certainly expect an increase in the numbers of tall buildings and the challenges associated with them as we look for better ways to house ourselves in dense urban settings. New technologies will be developed to take better advantage of resources in place, such as solar and wind energy, natural rainfall, and waste heat, water and other resources. The buildings of the future will, by necessity, be far more self-sufficient than their counterparts of today, with systems in place to passively light, ventilate and condition their spaces. Water will be cycled and treated extensively, refreshed primarily by rainfall instead of dependence on local water supplies. Future buildings will also be much more mixed-use than buildings of today, integrating nature as part of vast urban gardens that provide food and services to their residents and occupants. They will support a greater degree of virtual enterprise and commerce based on information technology, with occupants working from home or travelling locally to satellite offices instead of commuting long distances to a central workplace.

The very process of design will continue to evolve and change. New generations of building and urban information models will be established very early in design and will carry through continuously to operations and the end of the lifecycle of capital projects. These models will be interoperable between and among all stakeholders involved in project delivery, unlike today where they must be awkwardly translated over time and between stakeholders. Information models from existing projects will provide valuable performance data and lessons learned that can be easily tapped to make better design decisions and better understand the integrated performance of building systems. Likewise, the body of scientific research along with experience-based knowledge will be increasingly incorporated as part of an evidence-based design process to achieve desired outcomes (see *Evidence-based design for sustainability*).

Evidence-based design for sustainability

Sheila J. Bosch, PhD, EDAC

Without evidence, design is guess work. Those who practice sustainable design rely heavily on evidence from research and professional practice to inform design decisions that result in more sustainable built environments. As such, sustainable design can be considered a subset of evidence-based design (EBD), which is defined by the Center for Health Design as 'the process of basing

continued

decisions about the built environment on credible research to achieve the best possible outcomes'. 'Outcomes' within EBD are numerous and may include, for example, patient satisfaction, staff retention, or family presence in healthcare settings; standardized test scores or incidences of aggressive behaviour in learning environments; or even sustainability-related goals, such as lowering energy use or increasing access to nature.

Evidence, when referred to by most EBD practitioners, is not limited to scientific research; it may also include documented expert knowledge and client values (Stichler 2009). However, it is appropriate to give more weight to high quality research evidence. Chong et al. (2010) describe an evidence-based design practice as 'a model of rigorously seeking or conducting research to predict how well specific design proposals will support desired performance outcomes or conversely, inadvertently cause harm'.

Research evidence is fundamental for improving both human and ecological outcomes through design. As an example of a typical EBD-inspired research study, Bosch et al. (2016) investigated the impact of a decision to relocate patient beds on several outcomes, including noise, patient satisfaction and staff presence in patient rooms. In the pre-move condition, the radial floor layout of the nursing unit afforded visibility of all 10 patients in the unit from the care team work station located in the centre of unit, unless a privacy curtain was drawn. Staff liked being able to see the patients from the work station and believed that good visibility helped reduce patient falls, since they could see a patient trying to get out of bed and provide assistance. When the administration decided to relocate patient beds to the opposite wall during a build out to reduce noise and increase patient privacy, the nursing staff feared that patient falls in this orthopaedic unit would occur more frequently. Indeed, when comparing nine months of data pre-move with nine months of data post-move there was a

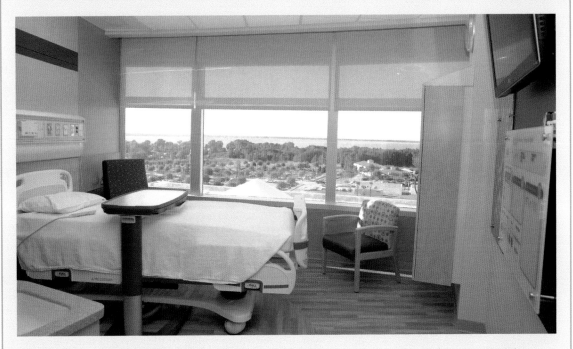

Figure 6.82 An example of evidence-based design is providing daylight and views of nature, which have been associated with a range of benefits for patients, families and staff, including stress reduction, pain reduction and increased satisfaction

continued

slight increase in the patient fall rate, from 2.5 to 3.7 falls per 1000 patient days, although this increase was not statistically significant. In addition, however, investigators demonstrated through behaviour mapping that staff members were more frequently observed in patient rooms ($p < 0.05$) post-move. Documented patient satisfaction was higher, based on Hospital Consumer Assessment of Healthcare Providers and Systems (HCAHPS) scores, both in terms of rating the area around patient rooms as 'always' quiet at night (55% post-move vs. 46% pre-move) and the percentage of patients rating the hospital a 9 or 10 (top box category), which increased from 66% to 69%. Furthermore, a third-party acoustician found that the noise reaching the patients' heads in the new configuration was on average 5 dB lower than it would have been if patient beds were directly visible from the care team work station. These findings will likely influence future design decisions, and they demonstrate the importance for designers to find the right balance between competing goals such as visibility and privacy, as well as other tradeoffs important to sustainability.

As sustainability models evolve, there is a greater emphasis on the integral relationships between social and ecological systems (Du Plessis and Brandon 2015), and EBD is an important approach for achieving design that meets goals in both domains. Many outcome variables of interest to researchers and practitioners of both sustainability and traditional EBD overlap, such as providing views to nature and daylight, enhancing beauty (i.e., creating positive distractions), and improving human health and well-being. Using the lessons of EBD, design firms are increasingly interested in both applying existing research and in conducting their own original, empirical investigations to ensure that the facilities they design enhance both human and ecological well-being, both of which are integral for the design of sustainable solutions.

References

Bosch, S.J., Apple, M., Hiltonen, E, Worden, E., Lu, Y., Nanda, U., and Kim, D. (2016). "To see or not to see: Investigating the links between patient visibility and potential moderators affecting the patient experience." *Journal of Environmental Psychology* 47, 33–43.

Chong, G.H., Brandt, R., and Martin, W.M. (2010). *Design informed: Driving innovation with evidence-based design.* John Wiley & Sons, Hoboken, NJ.

Du Plessis, C., and Brandon, P. (2015). "An ecological worldview as basis for a regenerative sustainability paradigm for the built environment." *Journal of Cleaner Production* 109, 53–61.

Stichler, J. (2009). "Balancing three essential components of evidence." *Health Environments Research and Design* 2(3), 3–5.

Source: Photo by Mark S. Gall; courtesy of Gresham, Smith and Partners

Stakeholders, both laypeople and professionals, will have greater access to the design process through social networks and project information systems through which they can contribute their ideas, requirements and aspirations for the project. The interests of stakeholders traditionally outside the project such as members of the surrounding community and construction workers will also be actively included as part of design considerations through methods such as Prevention through Design. Just as today's space-conditioning systems are becoming more easily controlled by individual building occupants, so too will the design process of tomorrow be more responsive to the concerns of the whole set of stakeholders, resulting in buildings that better meet the needs of all stakeholders over time.

Discussion questions and exercises

6.1   Identify a local project in the design phase of development. Contact the design team and learn more about the project, including its sustainability goals and objectives. If possible, attend one or more design phase meetings as an observer. How is sustainability taken into account in these meetings?

6.2   Consider the building in which you are presently located. Given its design, how well did it address the major priorities for sustainable sites? What sustainable site features were included? How might it have been better designed to take site sustainability into account?

6.3   Where is your building located with respect to major infrastructure support systems such as power plants, water and wastewater treatment facilities, solid waste facilities and transport networks? Obtain a local or regional map and determine how close your building is located to these critical support facilities. How is your building connected to this critical infrastructure? What options are available?

6.4   Where is your building located with respect to major natural and cultural assets? In what watershed or drainage basin is your building sited? Where does stormwater go when it leaves your site? Where is the nearest functioning natural ecosystem? How do your development and surrounding neighbourhoods interact with that ecosystem? What possibilities exist for synergies with surrounding development? What environmental vulnerabilities exist?

6.5   How is your building oriented on its site? Obtain a site plan and evaluate the building's solar orientation. What could be improved about the building's orientation and geometry to improve its solar performance? How does the current landscape influence the building's performance, and how might it be improved?

6.6   Inventory the amenities and landscape features on your site. How could the site design and amenities be improved to increase the sustainability of your building?

6.7   Consider the building in which you are presently located. Given its design, how well did it address the major priorities for energy optimization? Inventory the energy-optimizing features that were included in the design. How might the building have been better designed to take energy into account? What features or technologies could have been included? Which would have the greatest potential impact?

6.8   Contact the staff responsible for operating your facility, and map the operating schedule for your building. What are the set points and operating hours for your building? How could the operating schedule be adjusted to increase energy efficiency? What, if any, trade-offs would there be for building occupants?

6.9   Obtain a schematic of the mechanical systems of your facility if one exists. What technologies are included to improve the energy and water performance and overall resource efficiency of those systems?

6.10 Contact the organization that provides electrical power to your facility. What is the rate structure and pricing scheme for power? Is any portion of the energy portfolio provided by renewable sources? Are green tags available? Does your facility's design take into account the energy supply scheme and pricing incentives available to it?

6.11 Identify the nearest sources of renewable energy to your location. Some nearby sources may be distributed on local sites rather than produced centrally. Visit one or more of these sources and interview the operators. What are the major challenges to their effective operation?

6.12 Consider the building in which you are presently located. Given its design, how well did it address the major priorities for water efficiency and wastewater performance? Inventory the water-optimizing features that were included in the design. How might the building have been better designed to take water into account? What features or technologies could have been included? Which would have the greatest potential impact?

6.13   Locate a local facility that uses alternative water or wastewater technologies. Interview users of those technologies to determine how well they perceive the technologies to work. Interview operations or maintenance staff and document their perceptions. What, if any, unexpected maintenance issues have arisen? Have any technologies been discontinued or abandoned?

6.14   Consider the building in which you are presently located. Given its design, how well did it address the major priorities for materials optimization? Inventory the material-optimizing features that were included in the design. How might the building have been better designed to take materials into account? What features or technologies could have been included? Which would have the greatest potential impact?

6.15   Locate a local facility that used green or high-performance materials in its construction. Interview users of those technologies to determine how well they perceive the technologies to work. Interview operations or maintenance staff and document their perceptions. What, if any, unexpected maintenance issues have arisen? Have any materials been discontinued or replaced?

6.16   Consider the building in which you are presently located. Given its design, how well did it address the major priorities for indoor environmental quality? Inventory the IEQ-optimizing features that were included in the design. How might the building have been better designed to take IEQ into account? What features or technologies could have been included? Which would have the greatest potential impact?

6.17   Locate a local facility that included high-performance indoor environmental quality in its design. Interview building occupants to determine how well they perceive the technologies to work. Interview operations or maintenance staff and document their perceptions. What, if any, unexpected maintenance issues have arisen? Have any products or systems been discontinued or replaced?

6.18   Perform an Internet search to identify building products and systems that can be purchased with manufacturer service, maintenance or leasing agreements. What are the terms of these agreements? How might they increase the sustainability of a facility on which they are employed?

Notes

1   The images of the case study of the Bank of America Tower are from Cook+Fox Architects and are used with permission.
2   The Bank of America Tower collects groundwater from the basement of its building site to eliminate groundwater discharge to the public sewer system.

References and recommended reading

7group, Reed, W., and Fedrizzi, S.R. (2009). *The integrative design guide to green building: Redefining the practice of sustainability.* Wiley, New York, NY.

ASLA – American Society of Landscape Architects. (2017). "Sustainable Design Guides." Professional Practice website. https://asla.org/guidesandtoolkit.aspx (accessed 14 January 2017).

Keeler, M., and Burke, W. (2009). *Fundamentals of integrated design for sustainable building.* Wiley, New York, NY.

Kibert, C.J. (2007). *Sustainable construction: Green building design and delivery.* Wiley & Sons, New York, NY.

Kwok, A., and Grondzik, W. (2006). *The green studio handbook: Environmental strategies for schematic design.* Architectural Press, London, UK.

Lechner, N. (2008). *Heating, cooling, lighting: Sustainable design methods for architects.* 2nd ed. Wiley & Sons, New York, NY.

Mendler, S.F., and Odell, W. (2005). *The HOK guidebook to sustainable design.* 2nd ed. Wiley & Sons, New York, NY.

Mumovic, D., and Santamouris, M. (eds). (2009). *A handbook of sustainable building design and engineering: An integrated approach to energy, health, and operational performance.* Earthscan, London, UK.

Pearce, A.R., Bosch, S.J., DuBose, J.R., Carpenter, A.M., Black, G.L., and Harbert, J.A. (2005). *The Kresge Foundation and GTRI: The far-reaching impacts of green facility planning.* Final Project Report to the Kresge Foundation, Troy, MI, 30 June.

Means, R.S., Inc. (2010). *Green building: Project planning and cost estimating.* 3rd ed. Reed Construction Data, Kingston, MA.

Sustainable Sources, Inc. (2017). "Sustainable Sources." Online directory and guidelines for sustainable design. http://sustainablesources.com (accessed 14 January 2017).

UNEP – United Nations Environment Programme. (2002). *UNEP in 2020,* United Nations, www.unep.org/pdf/annualreport/UNEP_Annual_Report_2002.pdf (accessed 30 December 2016).

USEIA – US Energy Information Administration. (2010). *Annual Energy Outlook,* www.eia.doe.gov/forecasts/aeo/index.cfm (originally accessed 10 November 2010; 2016 version accessed on 30 December 2016).

USEPA – US Environmental Protection Agency. (2001). *An Introduction to Indoor Air Quality,* www.epa.gov/iaq/iaintro.html (accessed 30 December 2016).

USEPA – US Environmental Protection Agency. (2007). *Developing Your Stormwater Pollution Prevention Plan: A guide for construction sites.* USEPA https://epa.gov/npdes/pubs/sw_swppp_guide.pdf (accessed 30 December 2016).

USEPA – US Environmental Protection Agency. (2010). *Conserving Water.* USEPA, www.epa.gov/greenhomes/ConserveWater.htm (accessed 10 June 2010).

US General Services Administration. (2017). *Sustainable Facilities Tool.* http://SFTool.gov (accessed 14 January 2017).

Whole Building Design Guide. (2017). "Design Recommendations". https://wbdg.org/design/design-recommendations (accessed 14 January 2017).

Chapter 7

Sustainable construction opportunities and best practices

During the construction phase of project delivery, careful attention must be paid to the aims established in design as well as the specific needs and opportunities of the project site and context to ensure that the project is delivered in a sustainable fashion and that the outcome meets the design intent. This chapter describes opportunities in the construction phase of project delivery that apply to infrastructure, building and industrial projects alike. The chapter begins with an overview of the changing nature of construction services, followed by a range of strategies that can be incorporated to improve project sustainability, from pre-construction through turnover of the project at the end of construction. Detailed case studies of the Trees Atlanta Kendeda Center and the Proximity Hotel and Print Works Bistro illustrate these strategies in practice.

Construction opportunities

Managing a project to increase its sustainability requires careful attention to construction operations as well as the way in which those operations are managed. Opportunities also exist for construction stakeholders to contribute to the project before construction begins. Many practices can be employed in each of these areas to improve the sustainability of a construction project, including:

Preconstruction services – these are activities undertaken by construction firms to support owner project development during earlier planning and design phases, including contributing to project sustainability goal setting, tracking project team decisions and preparing variance reports, preparing bid lists, providing bidders with scopes of work, pre-qualifying bidders and holding pre-bid conferences.

Construction engineering best practices – these are practices that pertain to the means and methods used on site to construct a building, including site development; earthworks and underground construction; temporary construction; prefabrication and modularization; green materials and specifications; materials handling and utilization; commissioning, testing and balancing; and resource recovery.

Project management best practices – these are practices that involve managing people, equipment, information and other resources to achieve the goals of a specific project related to schedule, budget, level of quality and sustainability, including green procurement; logistics and transportation; job site operations; education and training; management plans; and project surroundings and public outreach.

As with design, the breadth of opportunities to incorporate sustainability tactics during the construction portion of the delivery process is extensive. Based on the nature of the owner, its procurement constraints, and the project and its specific needs and requirements, applicable tactics can vary considerably from project to project, so it is difficult to predict at the early stages of a project what specific tactics will apply. Preconstruction services are one opportunity a contractor has to better understand the opportunities offered by a particular project. The range of sustainability opportunities in this part of a project is discussed next.

Preconstruction services

Integrative project design and delivery offer a much greater role for contractors in project planning and design. Historically, the role of contractors prior to construction has been limited to providing cost estimates in the form of bids for projects, even if they are not ultimately selected for project delivery through conventional design-bid-build. As project complexity increases and the costs of construction continue to escalate, owners have begun to recognize the incredible value that can be brought to early planning and design by those who will ultimately have to construct the designed facility.

The project delivery methods of construction management–agency, construction management at risk, and design–build are seeing increased use, especially in the public sector, to draw upon the additional value construction knowledge can bring to the facility delivery process while delivering a completed project faster than ever before. These delivery methods involve construction personnel, especially estimators and pre-construction managers, for a much longer time than the traditional bid preparation process. In addition to conventional tasks such as quantity takeoffs, pricing and subcontractor/vendor bid management, these stakeholders are now involved with tasks such as helping to define the scope of work for a project; assessing alternative systems, materials, or methods; and assisting the design team and owner team in defining the design intent for the project.

In projects with sustainability goals, these roles are further enhanced. Contractors are involved in the integrative design process from the very beginning, and may be hired only to perform preconstruction services. Or, they may also be permitted to bid on the project, depending on the owner's limitations and requirements. In addition to information on the cost and schedule impacts of design decisions, construction managers must assist in providing information on the sustainability impacts of those decisions, including the extent to which materials and systems will contribute to a project's potential rating under one of the many green rating systems (see Chapter 4), or the degree to which selected systems will impact the project's site and surroundings during construction. Providing this level of service requires an increased focus of construction companies on the task of knowledge management, and effective capture of lessons learned from earlier experiences with green projects.

The role of preconstruction manager has emerged among construction management firms as a distinct function separate from the traditional construction project manager. This construction stakeholder shepherds a project through the planning and design phase and assists in evaluating the implications of potential decisions at every point along the way. One of the most important functions of this individual is to

Table 7.1 Selected preconstruction practices and their role in project sustainability

Practice	Description	Sustainability role
Contribute to project sustainability goal setting	Contribute to the development, evaluation and selection of project sustainability goals based on construction knowledge.	Include knowledge of means and methods associated with choices of materials and systems; identify impacts of those means and methods that may not be apparent to other team members.
Track project team decisions	Maintain tracking log of all decisions made by the project team during design, including current status.	Tracking log should include decisions that may be cost-neutral but would otherwise affect sustainability goals for the project.
Prepare variance reports	Develop estimates that will allow easy tracking of changes from previous and future estimates to explain what has changed, and why.	Variance reports should capture not only impact on project cost and schedule, but also impacts on sustainability goals. For example, how might the decision affect the project's score on a green rating system? What credits would be affected?
Prepare bid lists	Working with the owner and design team, develop a list of bidders who should be invited or encouraged to bid on the project.	Assist owner in developing evaluation criteria that include relevant experience with sustainability projects and key qualifications.
Provide scopes of work to bidders	Provide a clear scope of work for all bidders, including a bid manual that incorporates key requirements such as insurance, safety and general provisions.	Include sustainability goals and implications for project delivery as part of the bid manual; highlight key sustainability requirements that may impact the cost, schedule, means or methods of the scope of work.
Pre-qualify bidders	Confirm that all bidders have the resources to perform the work, a history of successful performance, ability to complete the work given current workloads and can meet insurance and bonding requirements.	Evaluate sustainability-related capabilities such as the number of accredited or certified personnel for relevant rating systems; identify experience with comparable sustainability projects.
Hold pre-bid conference	Conduct a comprehensive pre-bid conference to encourage successful bidders.	Include explicit discussion of project sustainability goals and requirements, and clarify how these may impact project delivery; discuss qualification criteria.

provide a complete picture of the project budget and schedule at any point during this process, including all appropriate contingencies and qualifications at each stage. As the project nears completion of the design phase, the preconstruction manager will systematically work with the design and owner teams to reduce the need for contingencies and qualifications by identifying areas of risk and aiding in decision making to reduce and manage those risks.

The preconstruction manager also plays the critical role of packaging work and communicating its scope to bidders, which can have a significant impact on the ultimate project cost. This pivotal person is also often in charge of constructability review and value engineering or value analysis, since he or she possesses the most comprehensive perspective on the project and how it will ultimately need to be implemented. Table 7.1 shows some of the key practices that are generally the responsibility of the preconstruction manager, and how they play an important role in green projects.

After the preconstruction phase of the project is complete and a construction provider has been selected, the next important steps in project delivery involve deciding how the project site and operations will be managed to meet sustainability objectives, through good construction engineering and management.

Construction engineering best practices

The next category of best practices pertains to the means and methods by which construction occurs on a site. This may include optimizing processes to minimize damage to ecosystems or consumption of resources, using different types of materials or equipment, or handling materials in different ways. This section covers best practices for site development; earthworks and underground construction; temporary construction; prefabrication and modularization; green materials and specifications; materials handling and utilization; commissioning, testing and balancing; and resource recovery.

Site development and protection

In general, the starting point of any project is the selection of a project site, which will have a tremendous impact on the project regardless of whether it is a green building or a conventional building. In the ideal case, the construction process will leave the site in a better ecological condition than it was originally. This section explores measures to achieve this goal, including ecosystem protection and restoration, sedimentation and erosion control, stormwater management and control of fugitive emissions before, during and following the construction process.

Ecosystem protection and restoration

Ideally, a sustainable site development plan would prevent ecological damage in the first place by inventorying and protecting valuable ecosystems and avoiding ecological damage during construction. Measures such as tree protection fencing and optimized site layout help to prevent damage, and vegetation salvage helps to recover valuable plants from a site that would otherwise be displaced by construction. After construction, measures can be taken to undo some of the damage to soils and plants resulting from construction activities, but this should be considered as a last resort, with prevention a much more preferable option. Finally, for ecological damage that cannot be avoided, habitat or biodiversity offsets can be developed or purchased to compensate for damage to the site. The **Bosques de Escazú** (see *Case study*), a green condominium project in Costa Rica, is an example of reforestation included as part of development.

Case study: Bosques de Escazú, Costa Rica

Roberto Meza and Rodolfo Valdes-Vasquez

As of 2016, the Bosques de Escazú development is Costa Rica's largest green condominium project. The project was built on a mountainous area with 100-year-old trees, trails and an interdependent connection with surrounding nature. Within this land, there are four towers with a total of 110

Figure 7.1 The Bosques de Escazú development

continued

Figure 7.2 On-site odourless wastewater treatment plant

Figure 7.3 Solar water heaters, helping with 35 per cent reduction in the energy bill

Figure 7.4 Swimming pool without chlorine or chemicals (cleaning by ionized salts)

Figure 7.5 Permeable Interlocking concrete pavers to manage rainwater runoff

condominiums (Figure 7.1). The flats have one to three dorms and offer panoramic views towards the central valley of San Jose. One of the most important aspects of the project is the reforestation of nearly two hectares of green areas that surround the condominiums, as only 25 per cent of the land was occupied during construction. Specifically, the development includes the reforestation of 250 trees, which makes this the largest condo/reforestation project in Costa Rica.

continued

Under the motto 'Live Green', this multi-family residential development is a demonstration of green technologies in Costa Rica. The innovative materials used and construction techniques make this an example project for green developments throughout Latin America. For instance, the green technologies utilized allowed a 90 per cent savings in electricity consumption for lighting and 35 per cent overall savings in electricity bills by using solar water heaters. Other green technologies include an odorless wastewater plant, low water consumption toilets and faucets, and an architectural design which supports natural lighting, among other technologies. Some of the environmental elements included in each apartment include:

- LED lights allowing up to 90 per cent reduction of electricity consumption for lighting.
- cutting edge technology in wastewater treatment, where treated water is reused for irrigation of green areas.
- taps and toilets with low flow rates, saving water.
- cross natural ventilation to minimize the use of air conditioners or fans.
- natural lighting to avoid the use of electric lighting during the day.
- thermal glass.
- non-toxic glues and paints.
- more than 8000 m^2 of forest trails, designed by the Costa Rica National University to attract butterflies, birds and small mammals.
- certified timber.

SPHERA was the local sustainability consulting firm that has helped to design and manage construction projects with a socially responsible, profitable and environmental mindset. This firm works closely with owners, designers and construction firms in Latin America. Their work is interdisciplinary and offers sustainable consultancy during the lifecycle of the project. Bosques de Escazú is another showcase project for Costa Rica, which is a country rich with renewable energy. The country gets more than 90 per cent of all its electrical energy from clean sources and is aiming to be the first country to become carbon neutral in Latin America. Some of Costa Rica's energy sources include geothermal energy; the burning of sugarcane waste, biofuels and other biomass; solar energy; and wind energy. For more information about the project and the consulting firm, please visit the following websites:

http://bosquesdeescazu.com/

www.spherasostenible.com

Sources: All photos courtesy of Engineer Roberto Meza and Dr. Rudolfo Valdes-Vasquez

Figure 7.6 Residential tower with several windows providing natural lighting (to avoid use of electric lighting during the day)

Optimized site layout, equipment selection and other measures to avoid ecological damage can be achieved by planning all construction activities and location of material storage/staging, equipment movement paths, washdown areas, temporary site offices and other construction assets on site to minimize disturbance to soils and existing ecosystems and features. It may also take into account minimizing disturbance to human occupants of neighbouring sites. While all projects can benefit from this strategy, it is particularly applicable to sites with undisturbed or mature ecosystems in place, disturbed sites with the potential to create problems for other neighbouring sites, sites with significant historical or archaeological features that should be protected, or sites with sensitive land uses on surrounding sites.

Disturbing the site creates potential soil erosion and instability problems. It can also damage tree roots, even if the tree trunk and branches appear to be undisturbed. Careful attention must be paid to limiting or prohibiting vehicle traffic in areas with tree roots. This is especially true for heavy equipment. The weight of the vehicles and equipment compacts the soil and damages the ability of the tree roots to absorb nutrients from the soil. This eventually can kill the tree.

The most common means for limiting site disturbance is to erect construction fencing as a boundary for operations. Chain-link or security fencing is often used around the perimeter of the site for security purposes. Lightweight plastic fence is more often used within the site boundaries to mark areas that should be left undisturbed. The fence is placed at least as far from the trunk as the drip line of the widest branch (Figure 7.7). This ensures that heavy equipment will stay off the area in which tree roots exist.

A second important factor in limiting site disturbance is proper selection of construction equipment. Equipment is often selected based on cost or productivity factors, but ecological impact is an important consideration as well. Since skilled labour has historically been the most expensive aspect of construction projects, equipment is often selected to maximize human productivity and minimize the length of time a task requires.

Figure 7.7a, b (a) Fencing can be used to limit disturbed areas, (b) tree protection fencing should be placed at or beyond the tree's drip line

Figure 7.8 Signs remind equipment operators of contract penalties for damaging vegetation

Figure 7.9 Rubber-tracked equipment does less damage than metal tracks on soft ground beyond the tree's drip line

Contractual penalties and incentives can also be used to motivate more sustainable behaviour on site. For instance, projects for which preservation of mature trees is important may impose fines for damaging trees targeted to be saved on site. This fine is often a function of tree diameter. In one case, a project manager actually posted large price tags on the trees to be preserved on his project site (Figure 7.8). This visual reminder made the contractual penalties for damaging trees very clear to the excavation subcontractor and its equipment operators. As a result, no trees were damaged during the project.

Selecting equipment strictly based on productivity does not take into account the ecological damage heavy equipment can cause. For example, metal-tracked equipment can cause much more serious damage to site soil than rubber-tracked equipment (Figure 7.9), and tyre-based equipment may cause even less damage depending on soil conditions. While ecological damage may not be the only criterion for equipment selection, it should at least be considered as a factor in making more sustainable choices.

Optimizing a site and construction operations to avoid damage to ecological resources can improve the likelihood of those resources remaining viable after construction. However, protecting site ecosystems may require additional movement or travel distances by equipment that can increase the overall carbon footprint of the project, and may be resisted by equipment operators driven by productivity concerns. Careful attention should be given to balancing the *preservation* of existing resources vs. the *restoration* of those resources after construction is complete. Some damage caused by construction is essentially irreversible in any meaningful time frame, such as damage or removal of mature trees, loss of native topsoil and soil compaction. Site resources should be carefully evaluated during the project planning phase and considered in the design of all site operations. Ideally, initial site selection should avoid sites with significant ecological assets in favour of already developed sites.

Vegetation salvage involves identifying and moving transplantable vegetation on a site to other sites before construction begins. This practice applies to project sites with healthy vegetation, generally at the scale of very young trees, bushes or smaller plants. Benefits include salvage of ecological assets that can be reused elsewhere, saving both removal costs for the site developer and purchase costs for the organization salvaging the vegetation. Costs depend on the arrangement used to implement the salvage. Some areas have non-profit organizations such as Master Gardener programs dedicated to vegetation salvage who will come to a site upon request, inventory vegetation, and remove viable candidates for transplant. Projects in other areas may need to find interested third parties to remove salvageable vegetation at their own cost, or may be required by local ordinances to pay for professional transplant. The likelihood of a successful transplant depends on many factors including the initial health of the vegetation, its age and degree of establishment, conditions at the site receiving the transplant, season or time of year and others. Even in cases where salvage occurs at no cost to the site developer, through working with interested third parties, additional time may be required in the schedule to accommodate the salvage process.

Post-construction site restoration involves reversing damage to soils and intact vegetation following construction on a project site. It can be complemented with additional plantings and landscaping. Approaches include:

- soil decompaction and enrichment via scarification, vertical mulching, fertilization, radial trenching and air tilling
- pruning of dead or diseased tree branches and dead tree removal
- root collar excavation to remove soil that may have accumulated around the base of a tree during construction
- tree irrigation and fertilization
- soil replacement in eroded areas
- transplanting new native vegetation to complement existing vegetation.

Specific approaches for a particular site depend on site conditions. Benefits include not only enhancing the value and ecological health of the site, but also increasing the health of the site's vegetation and preventing future tree death and replacement costs.

It is important to note that soil compaction is considered by many to be essentially irreversible in less than two generations (40 years or more). As such, protection of key areas is a much more effective approach than damaging and planning to restore the site later. The cost of this practice depends on the measures used. Additional time may be required in the schedule to accommodate the restoration process. The success of site restoration activities depends on many factors, including the degree of damage caused by construction, season or time of year. In some cases, the success or failure of the effort will not be apparent for several years, as in the case of trees that eventually die after they have depleted their resource reserves over a period of years. A professional arborist or horticulturist should be involved in all restoration activities to provide expert advice.

Habitat/biodiversity offsets require remediating or creating habitat elsewhere to 'offset' the damage to habitat on the project site during construction on a site. It can be a condition of development or can be done voluntarily. Land can also be purchased elsewhere and placed into a conservation easement to preclude future disturbance and preserve that site in perpetuity. Mitigation, Conservation and Biodiversity banks have been developed where offsets can be purchased by a developer to compensate for habitat destruction on a project. Specific approaches for a particular project depend on the type of habitat present before construction and objectives of the project or regulatory requirements. Plans that require creating new ecosystems on sites where they did not naturally evolve have a low likelihood of success and are likely to require heavy investments both in construction and operation. The location and context of the offset ecosystem is also critical, since fragmented eco-systems are less likely to survive and prosper than well-integrated ecosystems. Some studies suggest that habitat offsetting is not effective from an overall biodiversity standpoint. Restoration of degraded ecosystems is more likely to be successful.

Stormwater, erosion and sediment control

As a construction site is cleared of natural vegetation and graded to develop an appropriate surface for the construction phase, the natural vegetation and topsoil is often either severely damaged or removed entirely and the natural contours of the land altered, both of which are sensitive issues for the environment. Erosion and sediment control is typically regulated extensively in developed countries, with mandatory requirements for dealing with erosion and sediment to reduce the envir-onmental damage caused by construction activity on the surrounding environment. Some projects also go beyond minimum requirements, for example, by reducing the limits of disturbance on site to protect existing ecosystems. The following subsections describe best practices currently in use to manage stormwater, erosion and sediment control during construction, including minimal sequenced exposure of soil, managed runoff paths and infiltration measures, sediment trapping and filtration, topsoil protection, soil stabilization and construction entrance management. Table 7.2 lists some of the more common measures for erosion and sedimentation control.

Sequencing to minimize soil exposure (Figure 7.10) involves carefully planning and managing the exposure of bare soil during construction through limits to disturbance, avoidance of sensitive areas and delib-erate sequencing and stabilization of exposed areas as soon as possible. Sequencing the disturbance of the construction site involves completing development in stages, where disturbed parts of the site are permanently stabilized before adjacent areas are disturbed. This helps to preserve

Figure 7.2 Common erosion and sedimentation control measures

Control measure	Description
SOIL STABILIZATION	
Temporary seeding	Plant fast-growing grasses or 'green manure' species to temporarily stabilize soil in areas where the soil will again be disturbed in the future.
Permanent seeding	After soil has been placed in its final location on site, plant grass, trees, shrubs and other plantings to permanently stabilize soil.
Mulching	Place straw, cut grass, wood chips or gravel on the surface of disturbed soil to hold it in place. Choose an organic mulch that will quickly decompose if the soil will later be moved to another location.
STRUCTURAL CONTROL OF SOILS	
Earth dike	Construct a dam of stabilized soil to divert runoff from disturbed areas into sediment basins or sediment traps. Use seeding or other measures to stabilize the earth dike.
Wattle	Install bundles of straw or straw bales continuously across areas of erosion to filter sediment from stormwater and slow down rate of flow. Wattles should be staked in place to prevent movement.
Silt fence	Install 2 inch x inch stakes no greater than 6 feet apart to support filter fabric that removes sediment from stormwater. The base of the filter fabric should be placed in a trench at least 4 inch x 4 inch and backfilled with native soil or gravel to prevent blow-out.
Sediment trap	Excavate a pond area or construct earth embankments to contain stormwater. This trap will retain stormwater and allow sediment to settle while the water percolates into the soil.
Sediment basin	Construct a pond with a controlled water release structure to allow sediment to settle out of stormwater before it is discharged into the wastewater system.

soil resources and water quality while remaining in compliance with regulations. Overall, avoiding damage by limiting disturbance can cost less than remediating problems later, and can avoid fines imposed for violating regulations. As with other site protection measures, sequenced soil disturbance can impact construction operations and may make it more difficult to effectively modify terrain as needed for development. Limiting areas of disturbance may result in a need to store/ stage materials off-site, or to use different types of construction equipment that can function in constrained spaces. Transportation requirements may increase, causing a commensurate increase in carbon footprint.

Managed runoff paths and infiltration measures include devices and terrain modifications designed to direct, distribute and slow the path of water flow on site during storm events, to provide opportunities for stormwater to infiltrate rather than run off, and to provide opportunities for soil to remain in place on site rather than be mobilized as sediment that must then be controlled at the site boundaries. Specific measures employed during construction include:

Figure 7.10 Phased clearing of a site involves waiting until work is ready to begin before removing vegetation

- **Check dams** – small dams built from logs, stones or sandbags across minor drainage channels , which slow the flow of water during storm events, allow sediment to settle and prevent gullying.
- **Earth dikes** – earth structures built up from ground level to block or redirect the flow of stormwater.
- **Level spreaders** – linear structures that convert high-velocity concentrated stormwater flow to sheet flow across large areas, thus slowing its speed, reducing erosion and facilitating infiltration.)
- **Perimeter dikes/swales** – terrain modifications around the perimeter of a site that are either built up (dikes) or dug down (swales) to inhibit the flow of stormwater off site.
- **Pipe slope drains** – pipe drains running from the top to the bottom of a slope to route stormwater in a controlled fashion that does not allow erosion of the slope.
- **Temporary storm drain diversion** – rerouting storm drains from their normal outlets to other structures designed to trap sediment.
- **Temporary swales** – depressions created below ground level along contour lines to intercept gravity flow of stormwater and encourage infiltration.
- **Water bars** – a system of trenches and adjacent mounds placed perpendicular to downhill water flow paths to slow and divert stormwater.

Some or all of these measures can also be integrated with permanent site measures such as infiltration swales or basins, bioretention system, and percolation trenches, provided sedimentation during construction does not compromise the future function of these measures. This practice applies to all construction projects where soil will be disturbed during construction. As with other site protection tactics, these measures can have consequences for construction operations.

Sediment trapping and filtration involves devices designed to intercept stormwater runoff and remove sediment from water that escapes into stormwater collection systems or runs off site. Specific measures employed during construction include:

- **Compost filter berm** – a dike made of compost or compost product placed perpendicular to sheet flow runoff to retain sediment and other pollutants; can be vegetated and left permanently in place.
- **Portable sediment tank** – a tank connected to a temporary stormwater diversion path to remove sediment from incoming water.
- **Rock dams** – a basin or depression in the soil with sides constructed of rock and gravel, where collected stormwater is released gradually through the spaces between the rocks and sediment is trapped within the basin.
- **Sediment basin** – a basin or depression in the soil with sides constructed of compressed earth, where collected stormwater is detained until sediment can settle out, and water is released through a single riser.
- **Sediment traps** – small impoundments, generally used at the outlets of stormwater diversion structures and runoff conveyances to allow sediment to settle out of construction runoff.
- **Silt fence** – a temporary sediment barrier made of porous fabric held in place by posts or metal mesh fencing, generally placed around the perimeter of disturbed areas.
- **Storm drain inlet protection** – temporary barriers placed around storm drains to filter sediment from stormwater passing through the barriers into the storm drains, including sandbags, porous fabrics and blocks/gravel, or excavation around the perimeter of the drop inlet.
- **Straw bale dike** – temporary barriers constructed of straw or hay bales to slow the flow of stormwater and filter sediment (not preferred).
- **Vegetative buffers** – areas of natural or established vegetation maintained around disturbed areas to protect the water quality of neighbouring areas.

These measures should be applied on all construction projects where soil will be disturbed during construction. To the extent that construction operations can be planned to accommodate future site features, additional costs can be reduced by combining construction-phase measures with permanent site features.

Topsoil protection and management involves activities designed to preserve, supplement or enhance existing topsoil on construction sites during and following construction activities. Topsoil imported from other sites can cause significant damage to those sites and should be avoided in favour of amending or improving a site's own topsoil when possible. Specific measures recommended during construction include:

- **Cover crops** – planting short-lived vegetation that improves the soil by fixing nitrogen or accumulating other beneficial nutrients while stabilizing the soil on which it is planted.

- **Importing soil from sustainable sources** – choosing sources for imported topsoil that do not harm ecosystems on other sites; avoiding sources of soil such as greenfields or prime farmland.
- **Stockpiling topsoil** – carefully separating topsoil from subsoil during initial site clearing and stockpiling it separately for future use.

Tradeoffs include potential effects on construction operations, but use of these practices will preserve and enhance local soil resources and avoid the cost of future soil amendment following construction.

Soil stabilization measures are intended to stabilize exposed or unstable soil on construction sites, both during and following the construction phase. Specific measures employed during construction include:

- **Brush matting** – the use of cut brush to stabilize slopes while revegetation occurs. Often employed in riparian areas during stream restoration.
- **Mulching** – the application of organic matter such as wood chips, often in conjunction with sheet products for weed suppression, to protect the soil from direct exposure to rainwater.
- **Temporary seeding** – the application of fast-establishing seeds and nutrients to exposed soil that is expected to be disturbed again in future.
- **Permanent seeding** – the application of seeds and nutrients to exposed soil that is in its final location and is not expected to be disturbed in future.
- **Soil blankets** – the use of biodegradable meshes, fabrics or composite sheets to prevent direct exposure of soil to wind and rain and hold soil in place until plantings become established.
- **Soil roughening** – the scarification of the surface of compacted soil using plows to provide pathways for plant roots and water infiltration.
- **Wattling** – using small branches or fibres either bundled together or woven around stakes to hold back soil.

When using seeding strategies, use native or non-invasive plants to minimize the potential for future landscaping problems. Some plant species are known as 'green manure' and become established quickly after planting, fixing nitrogen in the soil in which they are grown that then acts as a nutrient for other plants. Recommended species of green manure include legumes (members of the pea family) and various types of clover. Consult a local agricultural school or appropriate government agency for recommended species for your local climate and conditions.

As with other erosion control measures, benefits of soil stabilization include preservation of local soil resources and water quality and avoiding the cost of future soil amendment. Chemical products should be carefully evaluated for potential environmental impacts prior to use, and appropriate measures should be selected in the context of likely storm events and other soil stressors.

Priorities for erosion and sedimentation control during construction

- Avoid landscape disturbance.
- Isolate disturbed areas from the rest of the site using earth dikes, wattles or silt fences.
- Divert stormwater from disturbed areas to sediment traps or basins.
- Stabilize disturbed soil and control airborne dust in travel pathways during construction.
- Use permanent landscape features where possible, and temporary features elsewhere, to manage stormwater:
  - Encourage sheet flow of water.
  - Lengthen flow paths.
  - Reduce gradients to reduce velocity.
- Use temporary seeding or mulching in infrequently disturbed areas:
  - Native plantings.
  - Green manure/nitrogen-fixing species.
- Maximize the restoration potential of the site.

Construction entrance management and path stabilization helps to control potential problems at the entrance to the site and on paths used for movement of equipment and vehicles around the site during construction. Specific measures employed for this purpose include:

- **Stabilized construction entrance** – using a stabilized pad of aggregate underlaid with filter cloth at any point where traffic will be entering or leaving a construction site to or from other public areas.
- **Construction road stabilization** – placement of gravel road base, sloping/grading and filter cloth to provide a surface for moving equipment and vehicles during construction (may include removal of aggregate following construction). May also be accomplished under certain conditions with chemical soil binders.
- **Roadside drainage controls** – the use of ditches, grading and runoff control measures to divert water away from temporary roadways.
- **Vehicle wash racks/pads** – washing stations to remove dirt and sediment from the wheels and undercarriage of vehicles leaving the site and reduce trackout of sediment onto roadways. Ideally these systems will be coupled with water recycling and solids separation tanks to allow reuse of water.
- **Rumble grates** – ridged metal plates (an alternative to oversize rock) at the entrance to the site to dislodge heavy solids from vehicle wheels.

Costs depend on the specific approaches used, including removal of measures at the conclusion of the job, and/or remediation of soils to allow for desired landscaping. Benefits include preservation of water quality, avoiding problems with site neighbours due to dirt in local roadways and more efficient construction operations with increased on-site mobility.

Control of fugitive emissions through pollution prevention

After construction begins, pollution of many types can threaten workers on the site and people and ecosystems on neighbouring sites. Pollution can be defined as any harmful substance or effect introduced into the environment as a byproduct of another activity. Also known as fugitive emissions in construction, pollution includes unwanted spillovers of noise, light, vibration, dust, heat and other unintentional discharges resulting from construction activities to areas surrounding the work site. Table 7.3 lists sources of these types of pollutants along with examples of reactive and proactive measures for their control. As with other measures, some measures are more appropriate than others, beginning with prevention, followed by suppression, with containment as a last resort. Pollution prevention can also include creative approaches such as redefining security perimeters to avoid unnecessary vehicle idling (see *Case study: Air Force Weather Agency (AFWA) Headquarters Building*).

Dust control measures reduce the creation of dust wherever possible, and suppress or otherwise stabilize it if it does occur, including measures such as:

- **Soil stabilization** – using revegetation, mulching or stabilizing chemicals mentioned in previous sections to prevent dust occurring due to soil disturbance on site.
- **Dust suppression** – including wetting down of haul routes over disturbed soil, wet cleaning of paved areas and use of wet methods for cutting, grinding, sanding, sand-blasting, welding/soldering and other dust-producing activities.
- **Minimizing drop heights** – for all earth moving or excavation processes.
- **Avoiding burning** – debris on site, including site clearing debris and construction waste.
- **Enclosing dust producing areas/isolation of sensitive areas** – involves physical boundaries using materials such as plastic sheeting and duct tape; should be accompanied with ventilation equipment that ensures negative pressure in the dust-producing area relative to surrounding areas.
- **Personal Protective Equipment (PPE)** – respiratory protection worn by individual workers to prevent occupational exposure to dust, particularly in cases where dust poses specific known health risks such as silica dust, lead-based dust or asbestos.

Special measures may be required for sites that include mobile crushing plants, concrete batching plants, or other high-risk activities. Local permits will specify what measures are appropriate. Containment or enclosure of these activities is preferred to control inevitable dust, along with filtered exhaust ventilation.

Noise and vibration control measures help to reduce the auditory impacts of construction on workers and neighbours, and are particularly

Table 7.3 Examples of pollution prevention and mitigation measures

Pollutant	Source(s)	Reactive (mitigation) measures	Proactive (prevention) measures
Light	• Night operations. • Welding or cutting operations.	• Shielding or redirecting light fixtures to focus only on the work site.	• Revising construction schedules to avoid night operations. • Using smaller lights focused directly on task areas.
Noise and vibration	• Equipment operation.	• PPE for workers. • Arranging work shifts to allow worker breaks. • Perimeter fencing for a noise barrier.	• Revising construction schedule to avoid operations during sensitive times. • Choosing equipment with lower noise production.
Dust and airborne particulates	• Equipment operation. • Wind erosion of exposed soils.	• PPE for workers. • Surface treatment of exposed soils with water or dust suppression chemicals.	• Limiting site disturbance. • Leaving existing vegetation intact. • Covering exposed soil with temporary or permanent seeding.
Airborne chemical emissions	• VOCs from the offgassing of new synthetic materials.	• PPE for workers. • Increased ventilation rates during product installation.	• Using low- or no-VOC products. • Design using exposed surfaces instead of finishes. • Using pre-finished materials.
Soil and groundwater pollution	• Engine drippings. • Refuelling. • Accidental spills. • Improper disposal.	• Spill countermeasures such as berms, absorbent mats and barriers. • Contained storage for chemicals and hazardous materials. • Spill cleanup plans/ equipment.	• Using non-hazardous materials where possible. • Spill prevention training for employees. • Proper equipment maintenance. • Centralized refuelling.
Surface water pollution (heat and contaminants)	• Engine drippings. • Accidental spills. • Exposed soil without erosion control measures. • Paved surfaces.	• Spill countermeasures. • Perimeter silt fences. • Spill cleanup plans/equipment. • Contained storage. • Stormwater detention basins.	• Proper equipment maintenance. • Pervious surfaces. • Seeding exposed soil. • Limiting construction disturbance. • Infiltration basins.
'Tracked' soil on neighbouring streets	• Vehicle wheels.	• Vehicle wash stations.	• Limiting construction disturbance. • Off-site materials staging. • Just-in-time delivery.

important for sites located next to schools, medical facilities, daycares or sensitive ecological sites. Measures include:

- **Noise barriers** – erected around the perimeter of the site or noise-producing activity, including temporary barriers such as stockpiled soils, walled enclosures or loaded vinyl curtains.
- **Equipment selection and controls** – the use of engine mufflers or specialized equipment to reduce noise resulting from normal operation of equipment. Also includes rerouting vehicle traffic and noise-producing operations away from sensitive areas to reduce noise.
- **Process substitution** – substitution of alternate methods when possible such as helical piles, cast piles or vibratory pile driving instead of impact driven piles, or cutting bridge decks into sections for removal rather than impact or explosive demolition.
- **Work scheduling** – avoiding sensitive times such as night construction, or to reduce the number of concurrent vibration producing activities occurring at once on site.
- **Off-site fabrication** – displacing noise-producing activities to less sensitive or more controlled facilities, thus reducing the duration and intensity of noise on site.

Control of light pollution is important to avoid disturbance of neighbouring buildings or sensitive ecological areas, and may influence energy use and construction operations as well. Measures include:

- **Light barriers** – erected around the perimeter of the site or light-producing activity, including curtains, temporary barriers such as stockpiled soils and others.
- **Shielded fixtures/cutoff luminaires** – fixtures containing shields to contain and direct light to desirable areas. Should be coupled with proper lighting orientation to minimize lighting spillover.
- **Shadow-reduced lighting** – fixtures designed to simulate daylight conditions on construction sites to improve visibility and increase safety.
- **Work scheduling** – avoiding night construction and thus reduce or eliminate the need for supplemental lighting.
- **Off-site fabrication** – displacing noise-producing activities to less sensitive or more controlled facilities, thus reducing the duration and intensity of noise on site.

Fugitive emission management practices apply to all construction projects with outdoor activities, with special measures being required during development near sensitive human populations such as schools or sensitive ecological areas. Some measures may have to be removed at the end of the job, whereas others may be incorporated as part of the finished design. Some measures such as shadow reduced lighting may involve higher initial cost but pay back in terms of productivity and safety gains. While these practices attempt to mitigate social and ecological disturbance, some may extend the overall duration of construction due to tactics such as daylight or off-hours work scheduling. Careful attention should be paid to ensure that the net effect of fugitive emissions control measures is positive for key stakeholders and ecosystems.

Case study: Air Force Weather Agency (AFWA) Headquarters Building, Offutt Air Force Base, NE

US Air Force Air Combat Command and Kenneth Hahn Architects

This award-winning LEED Gold project was one of the first buildings certified under a new green building policy implemented by the US Air Force (USAF). Despite being a data centre, this building was designed to consume less than half the energy of a typical office building. It sets the standard for future Air Force buildings, both at Offutt Air Force Base and throughout the US Department of Defense.

Sustainable sites

Construction of parking lots was avoided on the project by repurposing an adjacent abandoned runway for parking. Storm water detention ponds keep the storm water discharge rate at pre-development levels. The detention ponds also contribute to a goal of maintaining more than twice the area of the building footprint as permanent open space adjacent to the facility. Preferred parking areas are provided for eco-friendly vehicles, and bike racks and shower facilities encourage building users to bike to work. Light pollution is reduced through the use of low-cut-off parking lot light fixtures, which are carefully located to prevent light spill onto adjacent properties. Roofing is a highly reflective white membrane, which reflects the majority of the sun's radiation instead of absorbing it.

Water efficiency

Outdoor irrigation was eliminated on the project by selecting native plants and plants that are well adapted to the growing conditions of the region. Once established, these plants can survive under normal rainfall conditions. Indoor water use was reduced by 30 per cent through the use of waterless urinals, low-flow shower heads and ultra-low-flow lavatory taps with automatic sensors set for 12-second duration.

Energy and atmosphere

Data centres are one of the most energy-intensive types of facilities constructed today. As the agency responsible for collecting and storing extensive amounts of weather-related data for the Air Force, the AFWA Headquarters building houses significant data processing and storage facilities. Despite this fact, AFWA has achieved an overall reduction in energy usage of over 50 per cent compared to a typical office building. Careful siting of the building to optimize sun angle, the use of sunshades and light shelves inside the windows, and use of highly efficient window glazing are part of the overall strategy to keep energy usage to a minimum. Energy efficiency is also achieved through the use of a HVAC system designed for flexibility and individual control. Key to this efficiency is the use of an under floor air distribution (UFAD) system, which delivers tempered air to individual floor diffusers located at each work area. Using this system, occupants have the ability to adjust their own air flow easily at each workstation. Use of this system allows lower speed fans and lower velocity air to provide greater comfort with less energy use than conventional systems. Light fixtures use low-wattage lamps

continued

Figure 7.11 The AFWA Headquarters Building received a LEED Gold certification based on innovations including reusing an old runway as a parking lot

Figure 7.12 Concrete work was an extensive part of the building's construction. Delays due to security inspection of ready-mix trucks were eliminated by relocating the base's perimeter fence to exclude the building during construction.

Source: Courtesy of U.S. Air Force Air Combat Command Headquarters

with energy efficient electronic ballasts, and are controlled by occupancy sensors. When daylight is adequate for illumination, the lights are turned off for energy savings.

Materials and resources

Special areas on each floor are set aside for the collection of recyclables, and there are also storage areas for recyclables outdoors. During construction, more than 99 per cent of the waste generated on-site was diverted from landfills, through either recycling efforts or reuse of products on-site. The reuse of the existing 24-inch thick concrete runway for parking contributed greatly to achieving such a high rate of diversion. Products specified for this project contain on average over 10 per cent recycled content. Specific products containing high recycled content include structural steel, concrete and carpet. Over 20 per cent of the products specified were manufactured within 500 miles of the project, including structural steel, metal decking, concrete and face brick.

Indoor environmental quality

During construction, special attention to keeping the site clean and eliminating smoking indoors helped to prevent contaminants from accumulating in the building. Before occupancy, the building was flushed with outside air to remove lingering contaminants from the construction process. Low-emitting (low-VOC) materials used included adhesives, paints, carpets, and wood products throughout the building, in order to minimize offgassing into the indoor environment. Walk-off mats in exterior vestibules keep contaminants from entering the building. Increased ventilation at copy centres prevents noxious odors from migrating to the work areas. Throughout the facility, carbon dioxide is monitored and ventilation rates are increased to ensure that sufficient air exchange occurs to maintain indoor air quality. High-efficiency filters are installed on all mechanical systems.

continued

Project innovations

Security is of paramount concern for all Department of Defense installations, and Offutt AFB is no exception. Limited access controlled through security gates is a primary means for ensuring base security, but controlled access also can result in long queues of idling vehicles awaiting inspection. To reduce the carbon impacts resulting from additional idling vehicles during construction, AFWA HQ was constructed 'off-base', as if the project were not located on a military base, by temporarily relocating the base perimeter security fence to exclude the project site during construction. All construction traffic and deliveries came on-site without having to go through any security checks. This saved waiting in line, a vehicle search and a nearly 3-mile-long escorted trip from the security gate on the opposite side of Offutt AFB. This innovation greatly simplified the construction process, saving over 400,000 miles of travel and thousands of labour-hours that would have been required if the facility was built on-base.

A second major innovation was employed to reduce future costs and solid waste during the building's lifecycle. The majority of the work spaces in AFWA are enclosed with modular wall panels that are completely movable and reusable for many future reconfigurations. This will save a great deal of drywall, metal studs, insulation and paint, every time a reorganization creates a need for a different floor plan layout. Use of these demountable walls will keep materials out of the landfill and will prevent construction noise and dust from making an unpleasant work area for the building occupants.

Earthworks and underground construction

Earthwork and underground construction has seen the development of new techniques with sustainability-related benefits, including methods that allow for underground utility improvements without trenching and measures to increase the structural properties of soil to reduce the amount of material and fuel use needed to place foundations. Of particular interest are measures that can be used to reduce the amount of concrete used, since concrete has high embodied energy and also is responsible for chemical release of carbon dioxide during cement production.

Ground improvement

Ground improvement methods can be used to improve the structural properties of soil on a site to reduce the requirements for structural foundations. The impact on project sustainability is very site-dependent, and methods such as lifecycle assessment should be used to evaluate appropriate engineering options based on specific project requirements (see *Selecting more sustainable solutions in geotechnical engineering and ground improvement*). Techniques include:

* **Deep dynamic compaction** – reduces voids and densifies soil through a process of compaction by dropping a heavy weight in a grid pattern on weak soil.

- **Vibro stone columns** – improves the bearing capacity of soil and reduces settlement by using a vibrating probe to place pockets of aggregates into granular soil, thereby removing soil voids and creating a solid bearing surface.
- **Compaction grouting** – involves injecting a grout mixture under pressure under the soil surface to create a bulb of grout that densifies surrounding soil.
- **Vertical (wick) drains** – accelerates settlement of disturbed soil through a series of drains that remove pore water from the soil matrix, reducing settlement time from years to within months so that it can be mostly complete by the end of construction.

Selecting more sustainable solutions in geotechnical engineering and ground improvement

C. Shillaber

All buildings and other civil infrastructure are ultimately supported by the ground, and require a foundation capable of supporting the weight of the structure and the imposed forces of nature without undesirable deformation. Geotechnical engineering is the sub-discipline of civil engineering that involves the application of engineering technology and principles to earth materials such as soil and rock (Holtz et al. 2011). Geotechnical engineers provide the input needed to develop structural solutions to support buildings and infrastructure on the natural soil and rock underlying a site using either shallow foundations such as spread footings, or deep foundations such as piling (Das 2016). As the number of sites with good foundation soils available for development has decreased, geotechnical engineers and contractors have developed a variety of ground improvement methods and technologies to improve the engineering properties of soil and rock at a site. Ground improvement methods vary widely in function and construction technique. They can be simple, such as dropping a heavy weight on the ground to densify soils at depth, or complex, such as injecting or mixing different types of cementing agents into the ground, or facilitating cementation using active biological or biologically inspired methods. A substantial catalogue of ground improvement technologies and information is available at www.geotechtools.org (Transportation Research Board 2014).

Ground improvement methods are important to consider when seeking more sustainable geotechnical solutions, as they may result in lower costs and environmental impacts compared to conventional deep foundation designs (Egan and Slocombe 2010; Spaulding et al. 2008). Figure 7.13 shows photographs of a ground improvement technology being used to provide support for a six-storey building.

Often, more than one viable geotechnical solution will accomplish the purposes needed for a project. Therefore, it is important to consider how different geotechnical alternatives meet sustainable development goals. Through the project planning process, requirements are set for social, economic and environmental factors. These requirements are then implemented in the contract documents, which guide the subsequent engineering and construction of the facility (Shillaber et al. 2016a). Geotechnical engineers have historically focused on developing and selecting foundation and ground improvement solutions that meet established performance requirements (e.g., load carrying capacity, tolerable settlements) while minimizing monetary costs (Holt et al. 2010). While this

continued

Figure 7.13 Rammed aggregate piers are a ground improvement technology where holes bored in soil are backfilled with compacted aggregate. In this case, the piers were constructed to strengthen the foundation soil and support shallow foundations for a six-storey building, thus eliminating the need to install deep foundation elements composed of concrete and/or steel. Where feasible, ground improvement technologies like this have the potential to reduce costs and adverse environmental impacts compared to deep foundation solutions. These photographs show aspects of the construction of rammed aggregate piers: (a) drilling; (b) placing aggregate in the bored hole for compaction; (c) the tamping/compaction device, which densifies the backfill through vibration; (d) the tamper extended to the bottom of the hole showing the hydraulic vibratory drive at the top of the tamper rod

continued

approach leads to functional geotechnical works, it fails to address their environmental consequences beyond meeting environmental guidelines for construction established in the project master plan and specifications. In fact, Shillaber et al. (2016a) have suggested that since all viable geotechnical solutions must meet project performance criteria (i.e. they are assumed to be functionally equivalent), costs and environmental impacts are key considerations for geotechnical engineers and contractors in seeking to meet sustainable development goals.

Well established methods exist within the geotechnical professional community for quantifying the performance and monetary costs of geotechnical solutions. Design guidance and cost information for many ground improvement technologies is available from www.geotechtools.org (Transportation Research Board 2014); however, the profession lacks standard methods for quantifying environmental impact. To that end, Shillaber et al. (2016b) developed the Streamlined Energy and Emissions Assessment Model (SEEAM) for geotechnical works. The SEEAM is a streamlined Lifecycle Analysis (LCA) methodology for quantifying embodied energy (EE) and carbon dioxide (CO_2) emissions, where EE is all energy required to bring something into its present state (Chau et al. 2008). LCA is a quantitative process of evaluating environmental impacts over the whole lifetime of a product (EPA 2006) and may be streamlined by limiting the assessment to narrowed boundary conditions or specific impact factors (Todd and Curran 1999). In the case of the SEEAM, the specific impact factors are EE and CO_2 emissions.

The SEEAM relies on a database of industry average unit EE and CO_2 emissions coefficients, which represent the amount of energy or CO_2 emissions per unit of a construction input (e.g. kg, L) (Shillaber et al. 2016b). The basic calculation methodology in the SEEAM is straightforward; EE and CO_2 emissions are determined by summing the product of the quantity of each construction input (Q_i) and its corresponding unit energy or CO_2 emissions coefficient (Ci) for all n construction inputs, as shown in Eq. 1 (Shillaber et al. 2016b):

$$\text{Embodied energy or } CO_2 \text{ emissions} = \sum_{i=1}^{n} Q_i C_i \tag{1}$$

A challenge that arises for SEEAM analyses is uncertainty, which is present in both the industry average unit coefficients and the subsurface conditions (Shillaber 2016; Shillaber et al. 2016c). These authors suggest that uncertainty in the coefficients may be accounted for by assuming they are lognormally distributed and implementing a Monte Carlo simulation. In the Monte Carlo simulation, each coefficient is randomly generated from its lognormal distribution for the calculation of EE and CO_2 emissions; the calculations are then repeated multiple times with different randomly generated coefficients to produce simulated data sets for total EE and CO_2 emissions. The mean and standard deviation for total EE and CO_2 emissions may then be determined directly from the simulated data sets. Shillaber (2016) suggests that 1000 values in the simulated data sets are usually sufficient for the error in the mean to be less than ±2.5 per cent. The simulated data sets may also be used to compare the EE and CO_2 emissions for different geotechnical alternatives through statistical inference, which is useful for the decision process.

Accounting for variability in the subsurface conditions is more complex, but it is especially important where the quantities of construction inputs are highly dependent upon the subsurface conditions encountered in the field during construction. Subsurface conditions include factors such as depth to bedrock, groundwater table location, stratigraphy and the engineering properties of the soils and rocks. To account for subsurface variability in the analysis, Shillaber (2016) has proposed a framework that involves identifying the geotechnical parameter(s) that control material quantities, determining the distribution of values of the parameter(s) based on observations from the

continued

geotechnical investigation, then randomly generating a value of the parameter(s) for determining the quantities of each construction input in the Monte Carlo simulation.

The results from streamlined LCA analyses for the EE and CO_2 emissions associated with geotechnical works can be improved with additional data informing the unit coefficients and through better knowledge of subsurface conditions. Research directed at improving how subsurface conditions are characterized and interpreted is particularly important for sustainable development because geotechnical engineers tend to use excess conservatism in their designs in order to account for variable conditions. This conservatism can lead to greater resource consumption than may be necessary.

Armed with information regarding environmental impacts such as EE and CO_2 emissions, engineers are better equipped to select geotechnical solutions that meet sustainable development goals through comparing quantitative information regarding performance, cost and environmental impacts. The comprehensive decision-making process should address all three factors; however, further research is needed to provide guidance on how to weight these factors in selecting a geotechnical solution, particularly given that no two solutions are truly functionally equivalent.

References

Chau, C., Soga, K., Nicholson, D., O'Riordan, N., and Inui, T. (2008). "Embodied energy as an environmental impact indicator for basement wall construction." *Proc., GeoCongress 2008*, ASCE, Reston, VA, 867–874.

Das, B. M. (2016). *An Introduction to Foundation Engineering*, 8th ed., Cengage Learning, Boston, MA, USA.

Egan, D., and Slocombe, B.C. (2010). "Demonstrating environmental benefits of ground improvement." *Proceedings of the Institution of Civil Engineers – Ground Improvement*, 163(GI1), 63–69.

EPA (2006). *Lifecycle assessment: Principles and practice*. Report No. EPA/600/R-06/060. Cincinatti, OH, National Risk Management Research Laboratory, U.S. Environmental Protection Agency.

Holt, D.G.A., Jefferson, I., Braithwaite, P.A., and Chapman, D.N. (2010). "Sustainable geotechnical design." *Proc., GeoFlorida 2010*, ASCE, Reston, VA, 2925–2932.

Holtz, R.D., Kovacs, W.D., and Sheahan, T.C. (2011). *An introduction to geotechnical engineering*, 2nd ed., Pearson Education, Inc., Upper Saddle River, NJ, USA.

Shillaber, C.M. (2016). "Toward sustainable development: Quantifying environmental impact via embodied energy and CO_2 emissions for geotechnical construction." Ph.D. Dissertation, Virginia Polytechnic Institute and State University, Blacksburg, VA.

Shillaber, C.M., Mitchell, J.K., and Dove, J.E. (2016a). "Energy and carbon assessment of ground improvement works. I: Definitions and background." *Journal of Geotechnical and Geoenvironmental Engineering*, 142(3).

Shillaber, C.M., Mitchell, J.K., and Dove, J.E. (2016b). "Energy and carbon assessment of ground improvement works. II: Working model and example." *Journal of Geotechnical and Geoenvironmental Engineering*, 142(3).

Shillaber, C. M., Mitchell, J. K., Dove, J. E., and Hamilton, M. (2016c). "Streamlined energy and emissions assessment model (SEEAM) v. 1.0 spreadsheet calculator user manual." CGPR #85, Center for Geotechnical Practice and Research, Blacksburg, VA.

Spaulding, C., Masse, F., and Labrozzi, J. (2008). "Ground improvement technologies for a sustainable world." *Proc., GeoCongress 2008*, ASCE, Reston, VA, 891–898.

Todd, J.A., and Curran, M.A. (1999). *Streamlined lifecycle assessment: A final report from the SETAC North America streamlined LCA workgroup*, Streamlined LCA Workgroup, Society of Environmental Toxicology and Chemistry.

Transportation Research Board (2014). "Geotech Tools, Geo-construction information & technology selection guidance for geotechnical, structural, & pavement engineers." Release 1.10. www.geotechtools.org (9 June 2016).

Photographs included in Figure 7.13 were taken by C. M. Shillaber in March of 2014 during construction of Pearson Hall on the Virginia Tech campus.

Underground construction

Underground construction can be employed to rehabilitate existing infrastructure or place new infrastructure with minimal disruption to site ecosystems and human uses of the site. Also known as trenchless technology, it involves techniques to conduct work on underground utilities without disturbing the surface of the site, including:

- **Cured-in-place pipe lining** – the installation of an invertible resin-saturated felt tube, generally made of polyester, inside a damaged pipe. The liner is cured using hot water, UV light or steam, or cured under ambient conditions to repair damaged pipes without excavation and soil disturbance.
- **Directional drilling for infrastructure placement** – the use of horizontal boring or drilling techniques to install utility lines such as telecommunication or electrical lines without surface disturbance.
- **Microtunneling** – an operation using a remotely operated small tunnel boring machine to create underground tunnels through which casing is inserted to create a stable opening.

These techniques can be applied on projects with poor soil or with significant underground utility requirements. Some practices also apply to existing utility repair or renovation projects. Benefits include avoidance of the significant damage to site ecosystems and extensive embodied energy associated with foundation improvement and underground utility work. Costs depend on the specific measures used. Costs may initially seem greater than conventional cut and cover techniques or deep foundation techniques, but remediation costs and also productivity and safety costs must also be taken into account when comparing these low-disturbance techniques. Ground improvement techniques can be used to reduce the raw materials required for structural foundations, but they may also encourage building in areas of higher risk. Underground construction methods have limitations to applicability but are generally preferred when feasible. Cured-in-place pipe linings have many benefits but ultimately reduce the diameter of the pipe, which may reduce flow rates in some conditions.

Temporary construction

Temporary construction includes a variety of measures used during construction to facilitate the construction process that are subsequently removed before occupancy. Multiple innovations in temporary construction methods improve sustainability from a resource or environmental standpoint, many of which have to do specifically with concrete formwork. Specific measures include:

- **Bio-based form release agents and curing products** – made from soy or other biological materials, these products are used to release formwork from cast products such as concrete, or to create a seal on top of newly poured concrete to aid in curing.

- **Protective barriers** – products in this category help to protect installed systems and finishes after installation as construction continues. Examples include coverings for finished floors and membranes to block air intakes on mechanical equipment. These products may contain recycled content, biobased content or other green materials.
- **Construction barriers** – these products block access to unsafe areas of the site or areas where work is actively going on. They can be manufactured with recycled content or using certified wood.
- **Certified wood** – can be used both for the constructed building and also for temporary structures such as formwork and construction barriers. Certified wood has been determined by an independent evaluation agent such as the Forest Stewardship Council (FSC) to be certified sustainably harvested, with adequate replanting to offset harvest rates.
- **Reusable formwork** – designed to be used repeatedly rather than discarded after a single use, reusable formwork may be made from metal or plastics to reduce degradation from concrete pours. Metals typically have high recycled content. This method reduces the waste stream from the construction phase of the project.
- **Removable fasteners/mechanical connections** – incorporated in the construction of temporary structures such as formwork and construction barriers to allow for easy disassembly following use, and for future reuse on new projects. Includes screws or double-headed nails instead of conventional nails or adhesives. Reduces the waste stream from the construction phase of the project.

These tactics are generally applicable on all projects. Reusable formwork in particular is applicable on projects with standardized dimensions that fit reusable forms. Benefits include reduction of the solid waste stream from the construction phase of a project as well as closing the material source loop by using recycled content materials.

Careful attention to design for reuse can spread the costs of temporary construction materials over multiple projects. Additional labour costs may be required to recover or disassemble temporary construction carefully enough to ensure reusability. Marginal costs of materials that are biobased, certified sustainably harvested or of recycled content depend on the specific product but are often competitive with conventional options where available. While these measures reduce waste and use of materials, they may also require additional labour and storage of materials between projects. Costs of temporary construction materials must be amortized over multiple projects, which increases accounting complexity. Rental of reusable forms and construction barriers can simplify accounting.

Prefabrication and modularization

Prefabrication and modularization refer to techniques designed to increase construction efficiency through taking advantage of repetitive elements. These practices move processes from field construction into controlled factory conditions and/or standardize construction to minimize waste and improve efficiency of resource use.

Prefabrication refers to constructing portions of a structure in a factory setting, then transporting components to the construction site and assembling them in place. Examples include:

- **Prefabricated rooms** such as bathroom pods (Figure 7.14) or even entire hotel or hospital rooms that can be installed complete with finishes inside field-constructed structures and connected to central utilities.
- **Prefabricated insulated balcony connections**, which improve consistency of detailing at a critical interface for better building performance as well as reduce the time spent working at height to construct these details in the field.
- **Prefabricated structural elements** such as prefabricated concrete beams, planks, or columns that increase speed of construction and consistency of product.
- **Hybrid sheathing/air barrier products** that integrate air and moisture barriers as part of sheathing for structures, eliminating the need for a separate barrier layer and allowing airtight construction to be achieved by simply taping joints.
- **Prefabricated door and window units** that include integrated frames, flashing, sheathing and other elements to aid better integration with the building envelope.
- **Prefabricated Mechanical/Electrical/Plumbing** (MEP) solutions for equipment rooms and other applications.

Modularization involves designing buildings to take advantage of standardized dimensions of construction materials, such as the typical 2 ft incremental sizes typical of lumber and sheet good products in the U.S. Modularization may also involve changing standard dimensions of

Figure 7.14
This polymer concrete slab will be part of a prefabricated bathroom module for a hospital

materials to accommodate typical construction practices or to facilitate maintenance, such as:

- In the United States, 93-inch framing studs which, when coupled with a single top and bottom plate, allow for a finished 8-ft wall without cutting.
- Carpet tiles in uniform sizes to allow for incremental replacement of tiles instead of needing to replace entire sheets of broadloom carpet.
- Sheathing with slightly larger panel sizes and tongue and groove edges to result in uniform 4 ft spacing.

Modular construction has also been used to refer to buildings constructed of prefabricated modules, often designed to be relocatable at the end of their function on a particular site. These buildings can be used for temporary space or in contexts where speed and flexibility are required.

All construction projects involving repetitive elements, such as hotels, hospitals, bridges and schools, can benefit from these techniques, as can groups of buildings with consistent parts such as housing sub-divisions. The benefits of prefabrication extend mostly to structures where repetition enables economies of scale in construction, or in situations where the construction site itself is especially sensitive or difficult. Prefabrication can also occur in reverse, for instance, removal of segments of a building for more careful deconstruction in a factory setting.

Potential benefits include increased efficiency and consistency of production under controlled conditions and reduction of time and risk spent constructing in uncertain field environments (Figures 7.15a and 7.15b). The potential for damage to materials due to exposure on the construction site is significantly reduced with prefabrication methods. Significant on-site schedule reductions can also be achieved. Additional measures can be used to avoid or recover waste in a factory setting, and modularization also helps reduce waste by maximizing the use of

Figure 7.15a, b Custom-fabricated Structural Insulated Panels (SIPs) can be easily moved on site, greatly speed construction and organized and labelled in the factory to facilitate proper placement

standard dimensions. Prefabricated parts can be designed to optimize transportation and assembly on site, and the most effective solutions consider both transport logistics as well as minimizing handling and staging requirements on site prior to installation.

Costs depend on the specific measures used. With panelized or structural prefabricated members, additional attention must be paid to interfaces between units with regard to potential separation and cracking. Different types of equipment are required for transportation and erection of materials, given the much larger weights and sizes of building components. Safety risk may be greater due to hoisting and placement of large elements in spaces with tight clearances, although overall time on site is reduced along with attendant safety risks of other types of construction activities.

Prefabrication may sometimes encourage the use of standardized dimensions in situations where a customized approach would be more sustainable, although this is uncommon. Modularization may also increase waste in situations where economies of scale involve procurement of a single size of material that must later be adapted on site, such as structural insulated panels or insulated concrete forms. Significant additional front-end design and planning is required, but substantial time savings for on-site activities is the payoff. There is often an increase in equipment costs due to the need for heavy equipment to transport and install modules, but a reduction in labour costs due to reduced installation time in the field compared to conventional construction.

Green materials and specifications

Products employed in the construction process offer a variety of sustainability benefits based on their origin, manufacturing and transport processes, and material components. While many materials are explicitly specified during design, some degree of freedom is available in procuring green products during construction. Key issues to consider include the degree to which a product will produce emissions that contaminate occupied spaces, the use of materials to improve the lifecycle performance of concrete, incorporation of products for effective moisture management to extend building longevity and reduce the potential for indoor air quality problems, and the use of materials with environmental benefits over their conventional counterparts.

Low emitting materials

Low emitting materials have zero or reduced emissions of volatile organic compounds that can cause negative human health effects both during installation and subsequent occupancy. Available low-emitting products include:

- paints, stains and other coatings
- adhesives, caulks and sealants
- Floor coverings such as resilient tile, carpet, padding and engineered planks

- fire caulking and fire-resistant materials
- composite wood and agrifibre products, including particleboard, plywood wheatboard, strawboard, panel substrates and door cores

Additional products such as low-emitting systems furnishings and cabinetry products are also available but are often determined explicitly during design and not under the control of construction stakeholders. These products are especially relevant for projects involving sensitive human populations, such as schools and hospitals. The primary role of the contractor is to procure materials that meet or exceed specified standards and to monitor installation to ensure that unauthorized product substitutions do not occur. Many low emitting products are cost neutral compared to conventional products. Additional costs may be associated with monitoring compliance and documenting materials for projects seeking third party certification such as LEED.

Low emitting products typically result in a more positive indoor environment and can reduce downtime and productivity loss during construction in occupied buildings. However, care should be taken to use products with equivalent functional performance to the conventional product to ensure that more frequent replacement is not needed.

Green concrete

As a high-energy material frequently employed in construction, concrete is a significant contributor to negative environmental impacts from construction such as embodied energy and carbon emissions. CO_2 is produced not only from consumed energy, but also from decarbonation of limestone, kiln fuel combustion and vehicles involved in the manufacture, transport and placement of concrete. Green concrete reduces the environmental impacts and lifecycle performance of concrete products in various ways. Specific measures employed for green concrete include:

- **Insulated concrete forms (ICFs)** – these stay-in-place concrete forms are typically constructed of extruded or expanded polystyrene connected with plastic form ties and can be quickly assembled and filled with concrete to result in an energy-efficient building envelope (Figure 7.16).
- **Green curing compounds and admixtures** – a variety of curing compounds and admixtures are available, including bio-based curing agents and admixtures, with environmental benefits over conventional products. The use of curing compounds in general also has environmental benefits due to the reduced need for water to ensure hydration.
- **Hydration control admixtures (HCAs)** – can be used to slow or stop hydration and preserve concrete in a pre-cured state to extend placement time and reduce waste.
- **Supplementary cementitious materials (SCMs)** – substitutions to conventional cement can be made using waste products such as flyash (a byproduct of coal burning) or ground granulated blast furnace slag (a byproduct of steelmaking).

- **Alternative aggregates** – including recycled aggregate from demolition of concrete and other materials such as tyre rubber or glass cullet.
- **Recycled content reinforcing** – involves the use of high-recycled content reinforcing steel.
- **Return concrete recycling** – involves the recovery of unused concrete and re-integration into the production of virgin concrete or into concrete masonry units or precast concrete.

a)

b)

c)   d)

Figure 7.16a–d Insulated concrete forms provide a stay-in-place, modular formwork system that reduces thermal bridging and increases wall performance. Forms are collapsed to save space during shipping (a), then assembled and adjusted on site using integrated bracing (b). The webbing connecting the foam sides supports rebar (c) and concrete is placed in lifts using a pump (d)

- **Concrete reclaiming** – involves applying a washing process to unhardened concrete to separate aggregate and a cement/water slurry that can be processed back into raw materials to make new concrete.

Additional measures are also associated with concrete production and installation, including use of concrete washout systems, on-site water management for concrete production process water, and other practices such as managing fugitive emissions through filtration during sandblasting of equipment and using dust suppression to non-paved areas of the plant. Concrete can also be used for other environmentally friendly applications such as aerated autoclaved concrete for structural purposes, or pervious concrete for site development purposes.

Detailing for durability

This practice involves measures dedicated to reducing potential damage to buildings and their components during and following construction. In particular, products and practices exist to prevent moisture from entering buildings and causing future problems. Various products can be used to manage the flow of water and water vapour within or outside the building envelope to keep it from entering a building, including:

- **Subsurface drainage plane products** – provide a vertical barrier between subsurface walls and surrounding soils to direct water to perimeter drains rather than toward the building (Figure 7.17).
- **Structural drainage plane products** – provide a vertical barrier between exterior wall sheathing and exterior finish to allow moisture penetrating the outermost wall layer to drain.
- **Sub-roof ventilation layers** – provide an airspace between the outermost layer of roofing and roofing underlayment. These products are used frequently with standing seam metal roofs to provide a means for water vapour leaking from the interiort to escape rather than condense on the bottom of the metal roof pan.

Many products of this type are available that contain recycled content. Their use, especially in roof systems, may also affect the thermal performance of the building due to introducing a layer of air beneath the top roof surface where circulation can occur. Correct use of these products requires proper detailing to ensure that diverted water has a way to escape. Creating continuous drainage paths is essential, and may include details such as:

- **Weep holes** – small holes at the bottom of drainage planes to allow water to escape from wall cavities.
- **Capillary breaks** – air gaps between material surfaces to prevent water from travelling between surfaces through capillary action, created using furring strips or other products

Figure 7.17
Roller-applied bituminous sealant covered by dimpled drainboard provides a subsurface drainage plane on the up-gradient face of this building

.• **Paths for stack effect ventilation** – continuous air gaps vertically behind exterior wall finishes that allow air flow to ventilate the wall and allow the wall to dry if moisture has intruded.
• **Flashing** – the use of impermeable membranes or sheet products, typically metals, to direct water away from building openings or interfaces between surfaces.

These practices reduce the risk of damage to the structure during its lifecycle by preventing moisture intrusion and allowing ways for it to dry if it does occur. Although these practices may increase the first cost of a project, their use may avoid catastrophic moisture damage later that cannot be effectively remediated without substantial reconstruction of the building.

Environmentally preferable materials

As discussed in earlier chapters, materials used in green projects are often selected because they offer environmental benefits compared to conventional materials. Construction managers must be familiar with the properties that make materials more environmentally preferable so that they can ensure that the right products are procured and installed (see *Types of environmentally preferable materials*). Specific properties of a building material that make it environmentally friendly include its recycled content, recyclability, compostability, biodegradability, use of bio-based components, regionality, rapid renewability and use of sustainably harvested components (Figure 7.18).

Figure 7.18 This Structural Insulated Panel is made from oriented strand board that is certified by the Sustainable Forestry Institute

Materials handling and utilization

Good materials handling practices optimize the use of construction materials and reduce damage to and waste of those materials during construction. Deliberate attention to these practices by construction planners and managers can reduce the overall environmental impact of a project.

Cutting optimization

Optimizing cuts allows workers to get the most useful product out of raw materials that must be cut to size on the job site, and minimizes the amount of material wasted through cutoffs. Centralized cutting areas (Figure 7.19) for raw materials facilitate the use of smaller scraps when needed, rather than cutting from full size stock. Smart saws and planned

Figure 7.19 This centralized cutting area keeps cutoffs organized and easily available for use when smaller size pieces are needed

cutting using optimization algorithms can also help to optimize the process, and are most efficient in situations where multiple instances of the same cuts must be made or where complex cuts in expensive materials such as stone must be made. Computer numerical control (CNC) cutting can help with implementing cutting plans designed using optimization algorithms.

Types of environmentally preferable materials

Recycled content materials generally require less energy and raw materials to produce than virgin content materials. The recycled content in a building product can come from two different sources: pre-consumer and post-consumer. Pre-consumer recycled content comes from waste generated during the manufacture of the material. It is better to use this waste in the manufacture of new materials. However, ultimately it represents inefficiency in the manufacturing process that should be engineered out of the process. Post-consumer recycled content comes from products *after* they have been used by the consumer and returned for recycling. The amount of recycled content in a product is often stated on the product's label. Recycled content can be certified by a third party organization so that users know the manufacturer's claim is authentic.

Recyclable materials are able to be recovered for remanufacture into the same product or a derivative product following use. Even better are materials that can be reused or reconditioned for use in their original form.

Compostable materials can be placed into a composting process where they will naturally degrade into a biologically useful product at the end of their lifecycle.

Biodegradable products will also eventually degrade, but not necessarily at a reasonable rate.

Bio-based materials are made only with renewable content, although some materials such as timber may require many years to regenerate in the natural environment.

Regionality is the degree to which a product has been manufactured near the point of use. Using regional or locally-produced materials is important because it reduces the energy required for transport and minimizes the pollution resulting from that transport. It also increases the health of the local economy by supporting local or regional businesses.

Rapid renewability is the degree to which a product's raw materials or components can be regrown quickly without damage to the environment, typically in less than a 10-year renewal cycle. Many building materials are becoming available that are made from rapidly renewable components, including flooring, finishes, structural materials and others.

Certified sustainably harvested materials have been harvested from the natural environment in a way that could be continued indefinitely without harm. Measures that must be undertaken to receive this certification include managed replanting of plant-based raw materials. These materials are generally labelled by a third-party certification organization that reviews the harvest practices to ensure they are sustainable.

Just-in-time delivery

Scheduling materials to arrive on the job site at the last possible moment before installation reduces the need to allocate space on the job site for material storage and staging. It also reduces the possibility of damage to products before they are installed. Off-site staging in protected areas can also provide these benefits, although at additional cost for storage. Just-in-time delivery is fairly common for projects involving large components such as precast concrete or steel members for which on-site storage would be difficult. Just-in-time delivery can also be used for situations where materials are more readily available and scheduling constraints are looser. Commodities such as drywall, drop ceiling panels, carpet, carpet pad and insulation are best delivered as close to installation as possible.

Optimized material storage and staging

Depending on how predictable and reliable the supply chain is, having products arrive on site well before they are needed may be necessary to avoid construction delays. When products arrive before they can be installed, adequate product storage and staging must be provided to ensure their safety and minimize threats of theft, moisture damage, ultraviolet light exposure, physical damage (crushing, puncture, etc.), chemical spills or absorption of contaminants from the surrounding environment. Stock rotation systems can help to ensure that older products are used first, and optimized stacking and protection of materials is essential.

> **Threats to sustainable product storage**
>
> - material security
> - moisture/humidity
> - temperature changes
> - physical damage
> - chemical spills
> - absorption of environmental contaminants
> - ultraviolet radiation
> - failure to offgas synthetic materials

Off-site offgassing of materials

Materials containing synthetic components or adhesives may need to be unwrapped and allowed to offgas before installation. This prevents potential indoor air quality problems after the building is complete. Products containing fabrics, foam, composite wood, adhesives and finish materials all may need to offgas. During this time, they must also be carefully protected from other types of damage. For this reason, off-site storage areas may be used instead of storing materials on site where other activities are occurring.

Hazardous materials management and storage

Chemicals, solvents and other synthetic products pose chemical hazards on site and should be carefully stored, labelled and isolated from each other. Separate ventilation should be provided for storage and mixing areas to prevent build-up of dangerous vapours and reduce the possibility of fires. A management plan should be developed to address spills and accidents with hazardous materials. Where possible, hazardous materials should be replaced with less hazardous products to avoid exposures.

Excess materials

After products have arrived on site and been installed in a building, any remaining excess materials must be addressed. Some materials may remain with the building as attic stock for future repairs or maintenance. Others may not be suitable for subsequent use and must be disposed. Leftover or excess materials can sometimes be returned to the vendor, supplier or manufacturer for reuse. Protection of excess materials from potential damage is critical to enable product takeback, especially moisture and UV damage (Figure 7.20). In other cases, usable leftover materials can be donated to charities such as Habitat for Humanity. Local universities, colleges or trade schools may also be interested in leftovers to use as part of education and training, particularly for innovative new products.

Product tracking and procurement optimization

Electronic materials management systems can help to optimize procurement and management of materials for a project, and may include electronic tracking using technologies such as RFID chips or bar codes. Combined with Building Information Modeling (BIM) and multi-dimensional scheduling during construction, this information can be

Uses for excess materials

- Provide to owner as attic stock for future repairs.
- Avoid excess via careful procurement.
- Return to vendor, supplier or manufacturer.
- Donate to charity such as Habitat for Humanity ReStores.
- Donate to universities, colleges or trade schools for use in education or training.
- Donate certain materials to schools for use in art or science projects.

Figure 7.20 This job site could do a better job in storing excess materials to prevent damage and increase job site safety

used to quickly visualize what has arrived and been installed on site, vs. what was planned in the project schedule.

Whole building commissioning

Commissioning is a systematic process for quality control of building systems. This process covers the quality from inception through project delivery. The aim of commissioning is to ensure that building systems as installed perform according to the design intent and owner's requirements. Whole Building Commissioning involves measures dedicated to clearly articulating and understanding the owner's intent and monitoring the project delivery process to ensure that intent is met. Since contractors are involved in delivering and installing building systems to create a complete facility, they represent a critical link in the commissioning chain.

Commissioning is typically conducted by a third party not otherwise involved in the project to ensure the objective review and evaluation of the building systems' condition and performance. Results are then reported directly to the owner. In some cases, the commissioning agent can be an employee of one of the companies involved in project delivery. This may be the architect, engineering firms or contractors. In more complex projects, the agent must be an independent consultant or a qualified employee of the owner. In all cases, the commissioning agent should act independently of the project's design and construction management.

The following systems are typically included as part of the scope of all commissioning efforts:

- Heating, ventilation, air condition and refrigeration (HVAC&R) systems, both mechanical and passive, plus associated controls.
- Lighting and daylighting controls.
- Domestic hot water systems.
- On-site renewable energy systems such as photovoltaics or wind turbines.

More extensive commissioning efforts may also include systems such as the building envelope, which behaves dynamically and affects energy performance. Fire suppression, stormwater management systems, water treatment systems, data/IT systems and other specialty systems are also often part of the commissioning scope. The commissioning process involves, at a minimum, the following steps:

- Designating an individual as the commissioning authority to lead, review and oversee the completion of commissioning process activities.
- Reviewing owner project requirements (developed by the owner) and basis of design (developed by the design team) to ensure clarity and completeness.

- Developing commissioning requirements and incorporating them into construction documents.
- Developing and implementing a commissioning plan.
- Verifying the installation and performance of systems within the scope of the plan.
- Completing a summary commissioning report to the owner.

A more detailed level of commissioning, Enhanced Commissioning, involves additional requirements:

- Conducting a commissioning design review prior to the development of construction documents.
- Reviewing contractor submittals applicable to systems under the scope of commissioning.
- Developing systems manuals for commissioned systems.
- Verifying that requirements for training on commissioned systems are met.
- Reviewing building operation within 10 months after substantial completion.

Many of these requirements should be implemented in conjunction with members of the project team. The commissioning authority is ultimately responsible for reviewing and approving the outcomes developed by the team. Commissioning requirements are most often described in a specific section of the general conditions of the construction specifications. They may also be referenced on the drawings, in bid forms, or be a part of specification sections related to the systems being commissioned.

The commissioning process and Intent

Depending on the project delivery method being used, the construction team may be involved in commissioning activities throughout the project delivery process. For instance, in design-build projects, the construction team may be invited to participate as part of pre-construction activities such as project planning or design. In these cases, the construction team can contribute know-ledge and expertise on how construction means and methods will affect system delivery and performance, which may contribute to the owner's project goals and requirements. The commissioning authority (CxA) may also be a member of the contractor's organization, particularly if the project is being delivered by a construction management 'at risk' delivery method. The overall commissioning process parallels the project delivery process as shown in Figure 7.21.

Construction-related commissioning activities

The primary involvement of the contractor with commissioning activities will occur during planning and implementing the construction process. Specific roles played by the contractor during this time may include:

COMMISSIONING ACTIVITIES

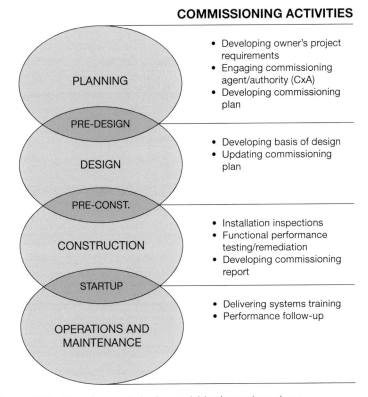

PLANNING

- Developing owner's project requirements
- Engaging commissioning agent/authority (CxA)
- Developing commissioning plan

PRE-DESIGN

DESIGN

- Developing basis of design
- Updating commissioning plan

PRE-CONST.

CONSTRUCTION

- Installation inspections
- Functional performance testing/remediation
- Developing commissioning report

STARTUP

OPERATIONS AND MAINTENANCE

- Delivering systems training
- Performance follow-up

Figure 7.21 Typical commissioning activities by project phase

- Reviewing and understanding commissioning requirements in the construction documents and commissioning plan.
- Preparing a commissioning milestone schedule.
- Undertaking any responsibilities specifically assigned to the construction supervisor in the commissioning plan.
- Informing and coordinating responses of subcontractors with commissioning requirements in the construction documents or commissioning plan.
- Coordinating the presence of the commissioning authority with scheduled construction activities such as system installation and performance testing.
- Notifying the commissioning authority of any relevant changes in the project scope or schedule.
- Verifying that the commissioning authority has met the requirements of the commissioning plan.
- Providing copies of submittals and documentation pertaining to commissioned systems to the commissioning authority for review.
- Coordinating training on commissioned systems as part of project closeout.

Working with the contractor, the commissioning agent will also be involved with installation inspection and performance testing. The purpose of these activities is to ensure that equipment is properly installed and able to perform as desired. Installation inspection of commissioned systems often occurs at startup of individual system components. The commissioning agent will follow prefunctional checklists or startup and checkout forms to document the startup process. Depending on the requirements of the commissioning plan, the installing contractor might complete the forms instead of the commissioning agent. The purpose of these inspections is to discover any improperly installed components prior to performance testing.

System performance testing (also called functional performance testing) occurs after all system components have been installed. All components within each system to be commissioned should be fully ready for operation under part- or full-load conditions. Performance testing then involves testing each process in the sequence of operations, including startup, shutdown, capacity modulation, emergency and failure modes, alarms and interlocks to other equipment. The purpose of the test is to simulate conditions under all modes of operation to be sure the equipment performs as expected. The outcomes of the test will identify performance problems that need to be remedied before the system will be acceptable to the owner. The contractor will work with other members of the project team to develop and implement a plan to fix any identified problems.

Resource recovery

The goal of resource recovery is to minimize the quantity of solid waste leaving the project site and maximize the recovery of useful materials that would otherwise be disposed. Demolition debris and construction cutoffs are the most obvious sources of waste on the construction site. However, significant quantities of waste come from other sources that could be avoided, including:

- **On-site fabrication of building components** – fabrication on site increases overall waste rates. The likelihood of reusing scrap materials is much smaller on the job site than in a centralized fabricating facility where similar products are regularly made.
- **Inefficient procurement and transport of products to site** – purchasing materials in bulk can often avoid significant packaging waste. However, it may also result in excess product on site that is easier to dispose than return to the manufacturer or reuse.
- **Careless materials management** – poor storage and staging practices result in materials that are wasted due to physical damage, exposure to moisture or weather, or other damage.
- **Poor coordination of subcontractors** – failure to coordinate the work of subcontractors can result in damage of installed systems or a need to rework finishes.

- **Poor site layout** – inefficient location of scrap storage, cutting areas or disposal areas make it easier to pull a new piece of material from supply instead of searching for a scrap piece that will work.
- **Poor protection of installed work** – a lack of effort to protect installed systems, especially finishes, can require that damaged systems be torn out and reinstalled. This not only generates waste, but also increases project cost and extends the completion schedule.

Each of these practices increases the quantity of solid waste generated on a project. Some occur due to a desire to save time or labour costs. Others can actually *increase* installation time, such as poor protection of installed work.

Many opportunities exist for reducing the amount of solid waste from a project. Additional opportunities exist for diverting unavoidable waste to better destinations. When preparing for construction and procuring products and materials, significant waste can be avoided in the first place by:

- **Prefabricating components in the shop instead of on site** – this can optimize raw material use and encourage greater use of scrap material, since it is located where similar products will be made in the future.
- **Designing for modular construction** – this reduces waste by sizing building elements to fit standard material sizes, thus reducing the need to cut and customize parts. This approach can also save significant construction time.
- **Ordering materials cut to size** – as with shop fabrication, this encourages centralized cutting, which increases reuse of scrap for future projects and reduces mistakes.
- **Working with manufacturers to reduce packaging waste** – this includes manufacturer takeback of reusable packaging as well as the use of compostable or biodegradable materials.

During construction, process waste can be reduced by:

- **Managing procurement from vendors** – requesting that products be shipped with returnable packaging eliminates the need to dispose of the packaging as waste. Ask suppliers to take back or buy substandard, rejected or unused items, especially if they are responsible for the problem in the first place.
- **Careful excavation planning** – balancing project cut and fill requirements reduces the need to dispose of soil spoils off site. Stockpiling topsoil for reuse also reduces the need to import new soil at the end of the project.
- **Limiting site disturbance** – this reduces the amount of land-clearing debris generated during the project and also minimizes the cost of re-landscaping at the end of the project.
- **Effective materials management** – this ensures that materials are not damaged while on site *before* they are installed. A staging plan that reduces material movement also reduces the chance for damage.

- **Careful coordination of subcontractors** – properly sequencing installation of products, especially finishes, can reduce the chance of material damage *after* installation.
- **Protection of installed work** – taking the time to mask off completed areas or install coverings over installed work reduces the likelihood of damage and need for rework.

In projects requiring removal of existing buildings or systems, waste can be reduced by:

- **Engaging salvage organizations** – this provides an opportunity for companies with established markets to recover useful products from the building prior to demolition activities.
- **Deconstruction instead of demolition** – this involves carefully taking apart building systems so that they can be reused.
- **Protection of installed work** – sealing off areas where demolition is occurring reduces the likelihood of damage to materials and systems that are not part of the project.

For unavoidable waste generated during demolition, renovation or construction, waste can be diverted from landfills or incinerators by:

- **Capturing recyclable materials** – this involves diverting recyclable materials from the solid waste stream either through on-site separation or commingling for off-site separation. Recycling opportunities differ by location based on availability of facilities, market value and other factors (Figure 7.22).
- **Composting** – this involves segregating biodegradable wastes such as certain types of packaging or land-clearing debris for diversion to off-site composting facilities. The effectiveness of this option depends on the types of waste generated and the availability of local facilities.
- **Using off-site separation services for co-mingled construction waste** – this service is available in some locations instead of on-site waste separation and generally results in higher levels of diversion at lower cost.

On-site disposal options may also be available for certain types of waste, including:

- **Biodegradable materials** – certain types of packaging, limited amounts of wood waste, and site-clearing and landscaping debris can be processed on site for use as landscaping mulch.
- **Drywall** – in limited quantities, drywall can also be applied on site as a soil amendment after being processed through a grinder or chipper. Whether this is a good idea depends on soil characteristics.
- **Waste concrete** – can be used as subbase for paved areas or foundations, or also as riprap for stormwater management or soil retention. It can also be processed on site using crushing equipment for reuse as aggregate.

Figure 7.22 Separate skips or dumpsters can be used for on-site waste sorting

Project management best practices

The next category of best practices pertains to managing people, equipment, information and other resources to achieve the goals of a specific project related to schedule, budget, level of quality and sustainability. This may include optimizing processes for managing the supply chain and obtaining materials, managing construction crews in new ways, collecting new types of documentation or doing new types of analysis. At a larger scale, construction ethics (see *Ethics in sustainable construction*) and occupational safety and health are also key considerations. This section covers best practices for green procurement; logistics and transportation; job site operations; management plans; and project surroundings and public outreach.

Ethics in sustainable construction

Dr. Kenneth Sands

The construction industry is unparalleled in its ability to create complex structures from pure imagination. As an industry, we are able to manipulate natural resources to produce products that contribute to our physical and social environment both aesthetically and functionally. At times, however, the ethical integrity of the construction industry can go overlooked. Over its history, construction has experienced an inundation of poor ethical business practices (Table 7.4), aligning with the contentious assertion of Friedman (2007) that the 'ethic' of business is to maximize profits while acknowledging that legal constraints apply. Unethical practices of the construction industry are so prevalent that according to Transparency International (2005), the construction industry is 'more corrupt than any other sector of the international economy'. This should not be the case.

continued

Table 7.4 Unethical practices of the construction industry (Sands 2012)

Source	Unethical issues/practice(s)
Ray (1999)	**Ethical issues in tendering (bidding)**: Issues include bid withdrawal, bid cutting, cover pricing (bid appears competitive but is submitted with the intention to lose and remain on client bid list), compensation of tendering and collusion.
Van Gorp and van de Poel (2001)	**Ethical issues in design**: Emerge at formulation of construction requirements, specifications writing and creating design criteria and during decisions of acceptable trade-offs (quality, economics and standards).
Fellows et al. (2004)	**Ethical issues in briefing (pre-design and design)**: Issues include shielding responsibility (placing liability for design on the client) and client manipulation through specialized knowledge over the client.
Rahman et al. (2007)	**Ethical issues during construction**: Issues include poor quality management and poor quality material substitution (cost savings).
King et al. (2008)	**Ethical issues of engineers**: Issues include negligence, pollution, specifications based on interest, collusion, bid modifications and job responsibility disputes.
Fewings (2009)	**Ethical issues of designers**: Issues include consideration of professional reputation over social responsibility (timing of design error identification), and use of proprietary design vs. system diversity and functionality requirements. **Ethical issues of constructors**: Issues arise as contractors choose not to identify weakness of design prior to receipt of contract to claim for extra work, bid shopping, use of cartels (collusion) and cover pricing.
Tow and Loosemore (2009)	**General ethical issues of construction industry**: Issues include bribery, intimidation, insensitivity to minority groups, poor quality of service, poor quality material selection and use, poor safety records, broken promises, bid rigging, price fixing, cover pricing, use of cartels and poor employment practices.

A significant change is required in how we address ethical issues as an integral part of the Triple Bottom Line. Ethics is defined as 'the moral principles by which any particular person is guided and the rules of conduct recognized in a particular profession or area of human life' (Fellows et al. 2004). In the construction industry, ethics is instilled by relating a practitioner's actions to their due diligence to maintain their contract with the profession by adhering to that profession's code of ethics. The American Society of Civil Engineers (ASCE) provides an exemplary code of professional ethical practice that considers sustainability as an integral part of that practice. Canon 1 of this code (see box) highlights the relationship between sustainability and ethics.

Subparts e and f of this canon state that engineers should seek opportunities to engage in the protection of the environment and safety, health and well-being of their communities through practice of sustainable development, with the commitment to '[adhere] to principles of sustainable development so as to enhance the quality of life of the general public'. Maintenance of this contract requires consideration of *all* stakeholders, whether they are able to be vocal

> Engineers shall hold paramount the safety, health, and welfare of the public and shall strive to comply with the principles of sustainable development in the performance of their professional duties.
>
> (ASCE Code of Ethics, Canon 1, 23 July, 2006)

continued

Table 7.5 Societal issues relevant for ethical practice (Sands and Pearce 2014)

Module	Topics
Social Responsibility	Benevolence; Consideration of Public Welfare; Waste of Public Resources; Deterioration of Public Economy
Social Environment	Minority, Race, & Gender Discrimination; Harassment (in general); Derogatory Name Calling; Disrespectful Behaviour; Racist Graffiti; Racist Jokes; Rumours; Cultural Norms & Divides
Sustainability	Carbon Footprint; Global Warming; Degradation of Urban, Suburban, and Rural, Environments; Ecological Sustainability; Energy Efficiency; Water Use; Environmental Protection; Pollution; Global Warming; Ethical Construction Products; Recycling; Toxic Waste Dumping; Triple Bottom Line of Business
Safety & Health	Occupational Health; Injury and Fatalities (direct/indirect result of practice); Dangerous Working Conditions; Safe Products
Human Resources	Child Labour; Employee Substance Abuse; Employee Use of Company Resources for Personal Gain; Illegal Migrant Work; Internal Fraud by Employees; Unfair Labour Practices

with their displeasure of our actions or not. In alignment with this broader view, construction ethical practice must go beyond professional considerations to include a broad variety of societal issues as well (Table 7.5).

Sustainable construction must be practiced with an ethical mindset to relay its importance for society (El-Zein et al. 2008). Sustainable construction is more than simply adhering to the requirements for green rating systems. It is about doing what is right. We need to step up as an industry for superior practice to achieve Triple Bottom Line success. We must be steadfast in understanding and practicing sustainability as part of ethical construction practice, always taking into account the importance of society and the environment in all of our dealings (Fewings 2009).

References

ASCE – American Society of Civil Engineers. (2006). *Code of Ethics.* 23 July update. www.asce.org/code-of-ethics/ (accessed 1 January 2017).

El-Zein, A., Airey, D., Bowden, P., and Clarkeburn, H. (2008). "Sustainability and ethics as decision-making paradigms in engineering curricula". *Int J of Sus in Higher Ed,* 9(2), 170–182.

Fellows, R., Liu, A., and Storey, C. (2004). "Ethics in construction project briefing." *Science and Engineering Ethics,* 10(2), 289–301.

Fewings, P. (2009). *Ethics for the built environment.* Taylor & Francis, Abingdon, Oxon, UK.

Friedman, M. (2007). "The social responsibility of business is to increase its profits." *Corporate ethics and corporate governance,* Springer-Verlag, Heidelberg, Berlin, Germany, 173–178.

King, W.S., Duan, L., Chen, W., and Pan, C. (2008). "Education improvement in construction ethics." *Journal of Professional Issues in Engineering Education and Practice,* 134(1), 12–19.

Rahman, H.A., Karim, S.B.A., Danuri, M.S.M., Berawi, M.A., and Wen, Y.X. (2007). "Does professional ethic affects construction quality." *Proc., Quantity Surveying International Conference,* Kuala Lumpur, Malaysia.

Ray R.S., Hornibrook, J., Skitmore, M., and Zarkada-Fraser, A. (1999). "Ethics in tendering: A survey of Australian opinion and practice." *Construction Management & Economics,* 17(2), 139–153.

Sands, K.D. (2012). "Toward Sustainability-Ethics Synergy for Construction Higher Education." Department of Building Construction, Myers-Lawson School of Construction, Virginia Tech, Blacksburg, VA.

continued

Sands, K.D. and Pearce, A.R. (2014). "Toward a Framework for Construction Ethics Education: A Meta-Framework of Construction Ethics Education Topics." *Proceedings, construction research Congress*, May, Atlanta, GA.

Tow, D. and Loosemore, M. (2009). "Corporate ethics in the Construction and Engineering Industry." *Journal of Legal Affairs and Dispute Resolution in Engineering and Construction*, 1(3), 122–129.

Transparency International. (2005). "Global corruption report 2005." *Rep. Prepared for Transparency International*, Berlin, Germany, 31–155.

Van Gorp, A. and van de Poel, I. (2001). "Ethical considerations in engineering design processes." *Technology and Society Magazine*, IEEE, 20(3), 15–22.

Green procurement, logistics and transportation

Not only is it important to choose green products for use in a sustainable construction project, but also it is important *how* those products are brought to the site. Key elements of this process include both product procurement and product delivery. Product procurement involves identifying and selecting a source for products and communicating product requirements and delivery expectations to the supply chain. Green product delivery involves the means and methods of transportation used to deliver the product to the site, and the packaging that is used to transport and protect the product during delivery. It is followed up by ensuring that what is delivered meets those requirements and working with the vendor or supplier to correct any problems.

Supply chain management for green procurement

Procuring green products may require using different vendors and suppliers than is customary for a company. New relationships, accounts and lines of communication may need to be established. These relationships are a good investment for companies working on green projects. However, they represent a risk of the unknown. Lead times for products may be unfamiliar, and more time may need to be built into the schedule. New requirements for product documentation may also require additional attention. Manufacturer training and certification may also be required for some types of products. Building these relationships proactively will help companies be more competitive as green projects become the norm. Obtaining necessary training and building a product documentation file for green products will reduce the learning curve and uncertainty for future projects. Systematic attention to product procurement is worth the effort in the long run.

Packaging optimization and takeback

Adequate packaging is essential to prevent damage to products during and following delivery. However, that packaging need not be an environmental liability. Some manufacturers can provide their products with returnable or reusable packaging. This prevents the packaging from turning into waste, provided that it is properly returned to the source. For instance, delivery of small sizes of sand and aggregate can be arranged using returnable heavy-duty bags that are removed from the

---

Green product procurement and delivery strategies

- Establish relationships and lines of communication early with new vendors/suppliers.
- Allow for lead time uncertainty in product delivery.
- Review and understand new requirements for product documentation.
- Establish a product documentation file.
- Obtain necessary manufacturer training and certification.
- Specify reusable/returnable packaging and return it to the manufacturer or supplier.
- Use compostable or recyclable packaging and appropriately compost or recycle it.
- Use just-in-time delivery where possible.

delivery truck by crane. These bags allow multiple types of materials to be delivered at once. This has lower transportation impacts than bringing each material separately loose in a truck bed. It also keeps these materials from being contaminated on site or spreading to unwanted areas, and it speeds clean-up.

Other types of packaging may be compostable or biodegradable. For instance, many types of plastic are now being made from plant products such as corn and soybeans. Depending on environmental conditions, these plastics may be compostable, especially in municipal composting facilities where higher temperatures are maintained than in home composting systems. Be sure to check with local composting facilities to verify acceptability of these types of wastes. Still other construction material packaging can be recycled. Corrugated cardboard is a fairly high value waste for recycling and can be recycled in most locations. Storing it in a protected location helps keep the material dry and makes it easier to handle for recycling. Wood pallets that are not reusable can be recycled along with clean wood waste or chipped and used as mulch.

Molded polystyrene is commonly used with cardboard to secure products within packaging, but it is difficult to recycle in all but a few locations. Sheet polystyrene or packing peanuts may be reusable; check local shipping outlets to see if they accept such materials for reuse. Waste paper can be used, either shredded or crumpled, as packaging material that can subsequently be recycled or composted. Packing peanuts may also be made of rapidly renewable, starch-based biodegradable materials. Be sure to verify that this is the case before mixing these items with compostable waste. Sometimes popcorn is even used as an environmentally friendly packing material.

Optimized shipping and reverse logistics

Optimized shipping and reverse logistics involve receiving the maximum amount of transportation benefit per unit of fuel expended. Optimized shipping can occur with prefabricated building elements that are designed to be carefully packaged onto one or more trucks where offloading matches the order of assembly. Use of distribution warehouses to ensure full shipments and planning of deliveries is also essential to minimize trips per unit material. Reverse logistics and back-hauling can be used to ensure that trucks are full both coming to and leaving the site. Arrangements can sometimes be made to back-haul waste materials or packaging as new deliveries come in.

Job site operations

Considering sustainability as part of general job site operations is an important way in which goals can be met during the construction phase. Specific measures and best practices exist to promote indoor environmental quality control, use green job trailers and facilities, employ green vehicles and equipment, optimize the use of temporary utilities, and ensure occupational health and safety.

Indoor environmental quality control

This practice involves measures dedicated to protecting future building occupants from problems in the indoor environment that can threaten their health. Specific measures employed for this purpose during construction include:

- use of low-emitting products and materials for pollutant source control.
- maintaining a tobacco-free workplace during construction.
- proper housekeeping practices during construction.
- isolating polluting activities from HVAC and occupied building areas through pathway interruption.
- sequencing construction activities to avoid introducing contaminants into building finishes or finished areas.
- proper material storage to protect vulnerable materials from moisture damage that can lead to mold).
- protection of installed work to prevent damage and contamination.
- ventilation with temporary equipment during construction instead of using permanent building equipment that can become contaminated.
- control of the interior environment during later stages of construction to prevent excess humidity and mold growth.
- conducting a building flush-out before occupancy.

The benefits of controlling indoor environmental quality include reduction of health risks for both construction workers and later building occupants. Costs depend on specific measures used. Some tasks involving procedural changes can be completed with no additional cost.

Temporary utilities

Temporary utilities are used during construction operations before permanent utilities are installed and typically include temporary power and water supply. They may also include temporary lighting, heating, cooling or ventilation for spaces in the building after it has been enclosed. Temporary utilities and green services are also used to provide suitable working conditions and amenities for workers in the building and to maintain the necessary conditions for installation of interior materials. They may be used to manage moisture, cure concrete, create proper conditions to install interior finishes or ventilate contaminants from an indoor space. Best practices include:

- locating temporary services as centrally as possible to avoid long power cords or hoses.
- locating temporary services where they are unlikely to be damaged or to interfere with construction operations, including areas near truck traffic or where excavation will occur.
- protecting temporary water lines to avoid freezing temperatures.
- using temporary frame structures to enclose areas requiring heating while temporary heat sources are being provided.

Workers should be instructed regarding the proper, efficient and safe use of these systems, and should be sure to turn off all temporary equipment when it is not needed. However, they must be instructed on when equipment should be left on to maintain necessary environmental conditions in the building. Regular inspections and maintenance should occur for all temporary water lines to ensure that leaks are identified and promptly repaired. This not only reduces wasted water but also reduces the likelihood of moisture or water damage to installed work and stored materials.

Green job practices and facilities

Green job practices and temporary facilities on site offer not only the opportunity to conserve resources, but also present a chance to educate workers and stakeholders visiting the site on principles of sustainability. This involves actions dedicated to reducing resource use and waste generation, closing the materials loop and generating new resources on site.

The same practices that will be employed in the final building should also be considered for the site's temporary construction office. For instance, recycling facilities for paper, beverage containers and other types of recyclable waste should be available for workers on site. Other containers may be appropriate to collect rechargeable batteries, compostable food waste or other types of waste. Closing the recycling loop is also important. Using recycled content paper and office products helps to provide a market for recyclable materials. Refurbished office furniture also saves raw materials and reduces waste. Paper goods such as toilet paper can be obtained with recycled content. Recycled content paper towels can be collected and composted if facilities are available.

It is possible to rent environmentally friendly portable toilets in some markets. These toilets use non-toxic chemicals that are formaldehyde- and alcohol-free. During servicing, they may be cleaned with bio-degradable, non-toxic cleaning products. Paper products may be chlorine-free and made from recycled content. Water used in flushable stalls can also be dispensed through water-conserving fixtures.

With the advent of Building Information Modeling (BIM), job sites may become increasingly paperless in the future. Plans, specifications and other project documents may be shared among the project team electronically rather than through paper copies. The LEED Online system for managing LEED project submittals is one example of this type of approach. It reduces the need for raw materials on which to print documents and the energy needed to print them. It also reduces the need for physical storage space at the end of the project. Virtual project meetings are now possible using widely available, low cost infrastructure available on many projects. Conference calls and video or computer-based conferencing can replace the need to physically travel to the job site. Webcams, remote sensing and other digital technology can provide documentation of project conditions that reduce the need for physical transportation to the site.

Green job practices

- Green job trailers.
- Office waste recycling.
- Food waste composting.
- Hazardous waste management.
- Use of recycled content products.
- Electronic documentation/ paperless operations.
- Virtual meetings.
- Job site gardens.
- Green job site toilets.

Green job site operation practices can be useful for public relations and conveying a positive corporate image, although they may require behavioural changes and additional measures to sustain. Rental costs for specialized equipment such as eco-friendly toilets may be greater in some areas. Virtual meetings and electronic documentation can save substantial time and resources but may not be perceived by some firms to be as effective as conventional methods.

Green vehicles and equipment

Increased attention has focused on the impacts of construction equipment on air quality. Heavy construction equipment and other tools such as portable generators have significant impacts on local air quality. The burning of diesel fuel for heavy equipment in particular has been linked with human health problems such as asthma and bronchitis. Diesel exhaust contains nitrogen oxides and extremely fine particulates that can lodge in human lungs. These can cause both short- and long-term health problems. Diesel exhaust has been linked to cancer in several studies. Children and adults with respiratory problems are especially vulnerable to health problems from diesel emissions. Better practice involves the use of equipment that employs clean combustion technology or alternative fuels. Specific measures employed for this purpose include:

- **Diesel exhaust control** – the use of diesel equipment with emissions control to meet EPA standards. Primarily applies to new equipment, since retrofits may not be widely available for older equipment.
- **Reduced idling time** – careful operation of equipment where engines are turned off when not in use for extended periods. This measure can also be implemented through improved planning of operating schedules and patterns.
- **Spill prevention during refuelling** – this includes the use of spill containment systems as well as centralized refuelling and other measures.
- **Biodiesel** – the use of diesel made from renewable resources to operate diesel-powered equipment on site. This requires a separate fuel storage and refuelling process and also reduces or eliminates the ability of conventional diesel to be used in designated vehicles and equipment due to different properties of the fuels.
- **Alternative fuel and hybrid vehicles** – the use of equipment that operates with alternative combustion fuels such as propane or hybrid combustion-electric engines and motors.
- **Electric vehicles and cord/battery operated tools** – this involves substituting electricity as a power source in lieu of combustion engines. Requires a power source on site and may generate a waste stream of batteries that must be disposed. Not all tools are available at a reasonable cost with this power option.
- **Solar-powered generators** – can be used in lieu of fossil fuel-powered generators to produce electrical energy for use on site.

- **Alternative heating systems** – these include systems fuelled by electricity instead of combustion appliances used as temporary utilities on site.

Replacing equipment is generally done incrementally as older equipment goes out of service, so these strategies may only realize benefits over an extended time. Operating characteristics are different between combustion-fuelled tools and equipment and electrically powered equipment, and must be taken into account in planning construction operations. During transitions, dual infrastructure may need to be maintained (e.g. separate fuel storage for plain diesel and biodiesel) to support both types of equipment.

Occupational health and safety on green projects

Green building and infrastructure projects have been recognized in the literature as having significant characteristics that distinguish them from conventional projects, including:

- tightly coupled designs and multifunction materials and systems (Riley et al. 2003; Rohracher 2001)
- procurement of unusual products with limited sources (Klotz et al. 2007; Pulaski et al. 2003; Syphers et al. 2003)
- existence of incentives and resources not available to other projects (Grosskopf and Kibert 2006; Pearce 2008; Rohracher 2001)
- requirements for additional information and documentation (Lapinski et al. 2005, 2006; Pulaski et al. 2003)
- greater involvement of later stakeholders in earlier project phases along with greater integration of their input (Cole 2000; Gil et al. 2000a; Matthews et al. 1996; Pulaski and Horman 2005; Pulaski et al. 2006; Reed and Gordon 2000; Rohracher 2001).

In the literature on occupational safety and health, some of these qualities are identified as posing greater safety risks for workers on green projects, including (e.g. Dewlaney et al. 2012; EASHW 2013a; NFPA 2010; Oregon Solar 2006):

- product unfamiliarity to installers, operators and emergency responders
- unfamiliarity with work context such as work at height for landscapers on green roofs, or electricians installing renewable energy systems (Figure 7.23)
- additional exposure to potentially hazardous work conditions while assembling documentation for certification
- installing and maintaining performance monitoring equipment not used on conventional projects
- inexperienced companies entering the construction market to take advantage of subsidies and incentives
- inexperienced workers entering the workforce based on green jobs incentives or workforce shortages

Figure 7.23 These workers have been trained to work safely at height and are using personal protective equipment for fall protection

- compressed project schedules and processes to meet incentive deadlines
- new types of potentially hazardous systems such as wind turbines for which workforce experience is relatively scarce.

To address these potential hazards on green projects, a different approach to construction safety is required that can take advantage of the integrative process used on green projects. Under Prevention through Design, methods such as the CHAIR process and other approaches for Job Hazard Assessment (JHA) can be used to identify and address potential health and safety hazards as early as possible (see *Managing hazards in green projects*). Identified hazards can be prioritized in terms of their probability of occurrence per exposure, likely severity of impact, and frequency of exposure of workers to each hazard. The Hierarchy of Controls, discussed earlier in Chapter 6 (page 271), should be used to find appropriate interventions to mitigate or eliminate these hazards, starting in the design phase. Hazards for work during construction as well as operations and maintenance should be considered and addressed.

Managing hazards in green projects

In addition to the hazards inherent in any construction project, the technologies and strategies used in green projects have the potential to introduce new hazards to workers during the construction and operation of the facility. Rajendran et al. (2009) found suggestive evidence that green building projects experienced higher injury rates than comparable non-green building projects. A second study by Fortunato et al. (2012) found that workers on LEED projects are exposed to work at height, electrical current, unstable soils and heavy equipment for a greater period of time than workers on traditional projects as well as being exposed to new high-risk tasks such as constructing atria, installing green roofs and installing photovoltaic (PV) panels. In a study of both incident/fatality reports and 31 international green rating systems, Pearce and Kleiner (2014; 2015) identified common characteristics of green projects that tend to increase worker risk on green projects, including:

- increased material handling requirements throughout the lifecycle;
- greater use of electrical sensors and controls to monitor building conditions and make operational adjustments;
- greater use of on-site renewable energy and alternative fuel technologies;
- use of district heating and cooling requiring greater temperature ranges in occupational contexts;
- potential exposures to pathogens resulting from alternative water technologies such as on-site water treatment and stormwater management facilities;
- increased use of building envelope components such as walls and roofs as platforms for other functions; and
- preference for previously developed sites, including existing buildings and brownfields, in dense urban areas.

These studies and others (e.g. Dewlaney et al. 2012; Dewlaney and Hallowell 2012; Gambatese et al. 2007; Gambatese and Tymvios 2012) also highlight ways in which green projects promote increased occupational safety and health, including reducing or eliminating the use of building products with hazardous chemicals and using products with increased service life such as LED lighting that reduce the need for work at height. However, given the increasing evidence suggesting that green projects may pose unique safety and health hazards compared to conventional projects, a need exists to proactively address these challenges. Although green projects offer the potential for improved environmental performance, they are ultimately unsustainable if they compromise the health, safety or quality of life of their constructors, their occupants or the workers who operate and maintain them.

As discussed in Chapter 6, Prevention through Design (PtD) is a proactive approach to hazards that threaten human or ecological health and well-being. Some PtD strategies such as product substitution are already embraced by green rating systems, such as the reduction or elimination of red list materials. Others may not be as obvious. Table 7.6 identifies specific safety and health hazards associated with the noted risks, along with examples of possible PtD measures that could be used to address them.

Integrative design and delivery processes on green projects offer a means to identify PtD opportunities. For instance, the Construction Hazard Assessment Implication Review (CHAIR) process developed in Australia is a systematic PtD technique that involves multiple project stakeholders coming together in facilitated sessions to identify and reduce safety risks over the whole project lifecycle through design improvements (Workcover NSW 2001). The CHAIR process is divided into discrete steps occurring at conceptual design review and detailed design review.

continued

Table 7.6 Green project risks, hazards and PtD measures

Risk	Potential hazard(s)	Possible PtD measures
Increased material handling	• Stress/strain injuries from lifting components such as high performance windows or construction waste • Punctures/toxin exposure due to demolition from reuse of existing sites and buildings	• Integrated attachment points to allow lift by equipment • Modular construction that minimizes on-site assembly of building components, particularly at height • Encapsulation of hazards
Greater use of sensors/controls	• Shock/electrocution due to increased exposure to live electrical elements	• Electrical lockout or de-energizing lock required to access spaces • Insulated conductors and shrouded terminals
Greater use of on-site renewable energy/alternative fuels	• Shock/electrocution due to increased exposure to live electrical elements • Toxin exposure due to batteries and alternative fuels	• Modular package units that can be removed for off-site maintenance or replaced • Isolating circuit design and switch disconnectors
District heating and cooling	• Burns due to increased exposure to high temperatures	• Guards to isolate high temperature elements • Provision of adequate space for maintenance to allow proper clearance
Alternative water and wastewater/stormwater systems	• Pathogen exposure from untreated or partially treated water and treatment residues	• Incorporating processes to significantly reduce pathogens (PSRPs) as part of treatment systems
Use of building envelope as platform for other functions	• Slips, trips and falls due to work at height, especially by trades that do not typically work at height • Glare from reflective roofs and building elements	• High parapet walls on roof areas • Permanent tie-off points • Roughened surface treatments and permanent walkways for roofs • Access openings from inside building
Preference for previously developed sites	• Toxin exposure due to contaminants • Struck-by or crush hazards due to heavy equipment use in congested areas	• Sequencing demolition activities based on structural properties of existing structures to minimize collapse risk

The detailed design review considers both construction safety as well as operations and maintenance safety. The CHAIR process is partially modelled after the Hazard and Operability (HAZOP) process widely used in the petrochemical industry, and involves the use of guidewords to stimulate discussion and prompt review for particular hazards within the design. Examples of guidewords include heights/depths, position/location, movement/direction, load/force, energy, environmental conditions, toxicity and others. Participants use the guidewords to identify specific risks, causes and consequences, then develop a set of safeguards, actions and responsible parties to address each risk.

continued

Table 7.7 Hazard criticality assessment – safety and health

Factor	1	2	3
Severity	First Aid, Medical Aid and/or Minor Property Damage ($5,000 or less)	Lost time, Injury and/or Significant Property Damage ($5,000–$10,000)	Permanent Disability, Fatality and/or Major Property Damage (more than $10,000)
Probability of incident	Unlikely to occur	Could occur	Will occur if not attended to
Frequency of exposure	Rarely (less than once per month)	Often (multiple times per month)	Frequently (every day)

Source: University of Calgary (2011)

After hazards have been identified, they are typically prioritized in terms of their severity, probability of occurrence and frequency with which workers are exposed to situations where they could potentially occur (Table 7.7), so that the most critical hazards can be addressed first.

Each hazard is rated using this rubric for each of the three factors, and the total score is determined by summing across factors, as follows:

Total risk = severity + probability + frequency

- High Risk (7–9): Requires immediate attention
- Moderate Risk (5–6): Requires attention
- Low Risk (3–4): Requires monitoring
- No Risk (1–2): No action needed

The same approach can be applied to different kinds of sustainability hazards by changing the rubric to account for different measures of severity for relevant variables such as site disturbance, materials and waste, or environmental quality. Table 7.8 provides examples of these rubrics that can be used to evaluate and prioritize potential sustainability hazards. Probability and Frequency rubrics remain the same for each factor. Total risk can be determined for each factor by evaluating the factor's severity, probability and frequency.

Table 7.8 Severity measures for key sustainability factors

Factor	1	2	3
Site disturbance	Damage can be self-remediated by site ecosystems	Damage can be remediated by human interventions on a project time scale	Damage cannot be remediated on a project time scale with standard measures
Materials and waste	Materials can be reused for the same purpose, or repurposed on site	Materials can be recycled, downcycled or repurposed off site	Significant portion of materials cannot be recovered and must be disposed through landfill or incineration
Environmental quality	Pollutants can be contained or avoided at the individual task scale	Pollutants will require special cleanup measures after the task is complete	Pollutants are likely to persist into building occupancy and affect occupants

continued

Additional resources

ANSI/ASSE Z590.3-2011 Prevention Through Design: Guidelines for Addressing Occupational Hazards and Risks in Design and Redesign Processes
www.asse.org/publications/standards/z590/docs/Z590.3TechBrief9-2011.pdf
Provides guidance for implementing PtD and safety design reviews.

CHAIR (Construction Hazard Assessment Implication Review)
www.workcover.nsw.gov.au/formspublications/publications/Documents/chair_safety_in_design_tool_0976.pdf
Provides an overall process and guidewords to facilitate PtD reviews at conceptual design and prior to construction when the full detailed design is known, focusing on construction safety as well as maintenance and repair issues.

Control of Pathogens and Vector Attraction in Sewage Sludge
www.epa.gov/sites/production/files/2015-04/documents/control_of_pathogens_and_vector_attraction_in_sewage_sludge_july_2003.pdf
Highlights environmental regulations and available technology to control or reduce pathogens in wastewater.

Design Best Practice – Promoting Safety through Design
www.dbp.org.uk/welcome.htm
Database of best practices for PtD in architecture, civil, structural, mechanical, electrical and other disciplines. Developed by practitioners.

Hazard Assessment and Control Procedure
www.snyder.ucalgary.ca/files/snyder/hazard_assessment_and_control_procedure_2011_11_28_final_1.pdf
Process and tools for identifying and addressing occupational safety and health hazards in capital projects developed by University of Calgary.

Prevention through Design – Design for Construction Safety website
www.designforconstructionsafety.org/index.shtml
Describes various aspects of PtD and provides links to international PtD regulations and resources. Provides presentations, case studies and links.

Sustainable Construction Safety and Health (SCSH)
http://sustainablesafetyandhealth.org/
A web-based rating system tool developed by researchers at the Oregon State University to use a lifecycle approach to evaluate safety and health on construction projects.

Washington State Construction Center for Excellence Green Building Safety Curriculum
www.constructioncoejobs.com/small-business-incubator/sustainability-resources-and-trainings/green-building-safety-curriculum
High quality slides reviewing potential safety and health issues associated with a wide variety of green features can help facilitate design and constructability discussions.

Wastewater Treatment for Pathogen Removal
http://publications.iwmi.org/pdf/H042608.pdf
Describes design features that can be used at different scales and in different contexts to remove or eliminate pathogens from wastewater.

References

Dewlaney, K.S., Hallowell, M.R., and Fortunato, III, B.R. (2012). "Safety Risk Quantification for High Performance Sustainable Building Construction." *Journal of Construction Engineering and Management*, 138(8), 964–971.

Dewlaney, K.S., and Hallowell, M.R. (2012). "Prevention through design and construction safety management strategies for high performance sustainable building construction." *Construction Management and Economics*, 30, 165–177.

Fortunato, III., B.R., Hallowell, M.R., Behm, M., and Dewlaney, K. (2012). "Identification of Safety Risks for High-Performance Sustainable Construction Projects." *Journal of Construction Engineering and Management*, 138(4), 499–508.

continued

Gambatese, J.A., Rajendran, S., and Behm, M.G. (2007). "Green Design and Construction: Understanding the Effects on Construction Worker Safety and Health." *Professional Safety*, 52(5), 28–35.

Gambatese, J., and Tymvios, N. (2012). "LEED Credits: How They Affect Construction Worker Safety." *Professional Safety*, 42–52.

Pearce, A.R., and Kleiner, B. (2014). *Construction in the International Green Economy: A Systematic Review.* Final Project Report, National Institutes for Occupational Safety and Health, Washington, DC, 28 June.

Pearce, A.R. and Kleiner, B.M. (2015). "Safety and Health in Green Buildings: An Occupational Case Analysis and Evaluation of Rating Systems." *Proceedings, Engineering Sustainability Conference*, Pittsburgh, PA, 19–21 April.

Rajendran, S., Gambatese, J.A., and Behm, M.G. (2009). "Impact of Green Building Design and Construction on Worker Safety and Health." *Journal of Construction Engineering and Management*, 135(10), 1058–1066.

University of Calgary. (2011). "Hazard Assessment and Control Procedure." *Occupational Health and Safety Management System, Hazard Identification and Assessment.* Dept of Environment, Health, and Safety. www.snyder.ucalgary.ca/files/snyder/hazard_assessment_and_control_procedure_2011_11_28_final_1.pdf (accessed 2 January 2017).

Workcover New South Wales. (2001). CHAIR: Safety in Design Tool. www.designforconstructionsafety.org/Documents/Chair%20Safety%20in%20Design%20Tool.pdf (accessed 28 December 2016).

Management plans

Management plans play a valuable role in explicitly laying out objectives and means to achieve them during the project. Key sustainability-related management plans include plans for indoor air quality management, construction waste management and materials management, discussed in the following subsections.

Indoor Air Quality (IAQ) management plan

Multiple options exist for improving indoor air quality. The Sheet Metal and Air Conditioning Contractors National Association (SMACNA) has developed a standard for managing indoor air quality during construction, demolition or renovation of occupied spaces. Chapter 3 of the standard focuses on control measures and guidelines to be used during construction. These include:

HVAC protection: The contractor should protect all HVAC equipment, both temporary and permanent, from dust and odours during construction (Figure 7.24). Before installation, all duct and equipment openings should be sealed with plastic. Equipment operated during construction should be protected with filters at all return air openings or on the negative pressure side of the system. Unducted plenums in the construction work zone should leave ceiling tiles in place and seal off return air grilles. The contractor should check for leaks in return ducts and air handlers and seal them promptly to prevent introduction of contaminants. Mechanical rooms should not be used for storage during construction. All filtration media should be replaced before the building is occupied.

Source control: Where possible, low-emitting materials should be used. Where materials containing VOCs must be used, proper control measures should be employed. For instance, areas where such materials

Figure 7.24
HVAC protection

are stored or installed should be isolated from other areas and properly ventilated. Equipment that generates exhaust fumes should also be isolated or avoided if possible. This includes exhaust from idling vehicles near air intakes for the building. Dust collection systems should be used on all equipment used for cutting or sanding.

Pathway interruption: Construction activities should be physically isolated from clean or occupied areas. This can be accomplished with temporary barriers such as plastic sheeting and duct tape. Isolated areas should be kept under negative pressure to keep contaminants within the space. Ventilation with outside air should be used, especially during construction tasks that generate air pollution. All absorptive materials should be protected both before and after installation to prevent contamination.

Housekeeping: Proper maintenance and cleaning should be regularly undertaken during construction, especially after spaces have been enclosed. This provides a safer work environment and also prevents problems for future occupants. Porous materials should be protected or cleaned with dry methods instead of wet methods to prevent exposure to moisture. High efficiency vacuum cleaners can be used to clean dusty areas without spreading dust further.

Scheduling: Sequencing of construction activities can be used to prevent dirty activities from contaminating previously installed materials. If buildings are occupied during regular business hours, dirty or odorous tasks should be scheduled during off hours to reduce impacts to occupants. Adequate time should be included in the schedule for building flush-out and/or air quality testing after construction is complete. The schedule should also include a reminder to replace all filtration media prior to occupancy.

The process of developing an IAQ Management Plan is similar to planning for materials management and pollution prevention. In fact, these plans have a number of elements in common and should be reviewed and coordinated before the project begins. When developing an IAQ Management Plan, the first step is to identify potential threats to IAQ during the construction process. These are typically associated with specific construction tasks. A template can be used to assess the risk associated with installation of specific products, materials and systems on the project. Specific construction *processes* should also be reviewed to identify potential threats. IAQ management generally focuses in more detail on construction processes that happen after the building has been enclosed.

The next step for an IAQ Management Plan is to identify options to reduce or eliminate threats. In general, options that prevent pollution are preferable to options that try to fix it after the fact. Specific measures recommended by SMACNA should be reviewed as potential options and incorporated into the plan.

Finally, the last part of IAQ management planning is to evaluate solutions in terms of their costs and benefits for the project. In addition to the hard costs of each option, nonmonetary costs and benefits should also be included. Reduced callbacks, owner satisfaction and contractor reputation are all worth considering. Reduced environmental liability is also important. Contractors can be involved in litigation for sick building syndrome long after the project is complete. Careful attention to IAQ and documentation of efforts can be invaluable in reducing liability if problems occur later in the building's lifecycle.

All subcontractors on the project should be made aware of IAQ goals for the project and should be given copies of the IAQ Management Plan. Periodic inspection by the project manager can help subcontractors comply.

While IAQ management offers benefits such as reduced exposure to toxins and irritants, reduced cleanup costs and subsequent improvements in productivity and satisfaction, these practices may extend the overall duration of construction due to additional time needed to protect installed work and other systems. Careful attention should be paid to ensure a net benefit to all stakeholders.

Construction waste management plan

Managing solid waste on the construction site creates tangible, visible benefits. Psychologically, dumpster contents form a first impression of how green a project is. Although there are more impactful ways to increase project sustainability, effective waste management is a good place to start. Three fundamental activities are part of waste management planning, as follows.

IAQ control measures during construction

- HVAC protection.
- Source control.
- Pathway interruption.
- Housekeeping.
- Scheduling.

Elements of a waste management plan

Waste stream assessment:
- Types and quantities of waste by task.
- Timing of waste generation in construction schedule.
- Possible disposal mechanisms.

Evaluation of local markets:
- Availability of potential destinations for each waste stream.
- Requirements for on-site collection/separation/storage.
- Value for each type of waste.

Cost/benefit analysis of options:
- Pros and cons for individual project.
- Feasibility of options.

Waste stream assessment: The first step involves reviewing the project to estimate the likely types and quantities of waste associated with each project task, including items to be salvaged, recycled or composted. Identifying the period in the schedule when these wastes will occur is also necessary. Knowing the timing of each waste stream will help to plan site layout and ensure that no more dumpsters are on site than are needed at a particular point in the project. For instance, projects that recycle drywall waste do not need a drywall dumpster on site until interior construction begins. Likewise, a concrete waste dumpster is not likely to be needed after demolition is finished. The project manager will be responsible for coordinating waste management activities on site. Table 7.9 provides a template for organizing this information.

Local market evaluation: The second step is to investigate the local and regional market for recycling and salvage. This will determine potential destinations for each waste stream. It will also establish a value for each type of waste. This evaluation should include waste haulers, recyclers, on-site waste subcontractors and other market options for handling waste streams such as Habitat for Humanity or nonprofit organizations.

Cost/benefit analysis of options: The third step is to evaluate the costs and benefits of options for each waste stream. The nuances of the project will also determine what is feasible in terms of space constraints for dumpsters. The cost/benefit analysis will suggest the best approach for the individual project conditions. Both direct costs, such as landfill and hauling fees, and indirect costs, such as increased management time and labour costs, should be taken into account.

Other considerations, such as likely diversion rates, project waste diversion goals and potential liabilities and schedule constraints, are also important to consider. Four major categories of options include:

- **On-site segregation**: This option includes all scenarios where waste is sorted into multiple dumpsters located on site. Most commonly, subcontractors are required to sort their own wastes as part of their scope of work. Or, it may include the use of a housekeeping subcontractor that takes care of sorting and housekeeping on site. Bins or dumpsters may be kept in a centralized location, or may involve multiple smaller bins located near work areas. The latter option is likely to result in a higher diversion rate, but also requires additional labour to consolidate and empty the bins.

- **Commingling for off-site separation**: This option is available in areas where centralized construction and demolition waste sorting facilities exist. All on-site wastes are deposited by subcontractors into a common bin or dumpster, which is periodically removed and sorted in an off-site facility for subsequent recycling. A specialty waste management subcontractor generally provides this arrangement. It often results in higher diversion rates. The subcontractor provides

---

Considerations for waste management strategy selection

- Direct costs (disposal/tipping, hauling, etc.).
- Indirect costs (management time, labour costs, etc.).
- Project diversion goals.
- Likely diversion rates.
- Health and safety implications of waste handling by employees.
- Potential liability.
- Schedule constraints.
- Space constraints.

Table 7.9 Waste stream assessment template

Project:	Expected quantity (unit)	Generated by (task)	Time period	Diversion Method				Comments
Material type				Recycle to facility	Recycle to salvage	Recycle on-site	Dispose to facility	
MIXED MATERIALS								
Mixed C&D waste								
SEPARATED MATERIALS								
Asphalt								
Brick/Masonry								
Concrete								
Lumber								
Composite wood								
Drywall								
Roofing materials								
Cardboard								
Paper								
Polystyrene foam								
Plastic								
Metals								
Gravel								
Soil								
Compostables								
OTHER								

required documentation for LEED credits or other purposes. The cost of this option depends on the types and quantities of wastes to be generated as well as market conditions. Waste management companies offering this service have well-established market connections. They are able to consolidate waste streams to achieve higher value of recycled materials.

- **On-site processing**: This option may be appropriate for some waste streams such as land-clearing debris or large-scale concrete waste. It involves contracting specialized equipment to come on site to deal with a single type of waste that will then be re-used on site, such as crushed concrete for aggregate or wood waste for mulch.

- **Salvage/deconstruction**: If the project involves work with existing facilities on site, salvage or deconstruction may be economically desirable to recover useful materials. Specialized firms can be employed to salvage reusable products, often in exchange for the value of those products. Demolition subcontractors often provide this service prior to demolishing the building. Capturing documentation of the fate of these materials is important for LEED certification.

> **Major waste management options**
>
> - On-site segregation.
> - Commingling for off-site separation.
> - On-site processing.
> - Salvage/deconstruction.

Often, projects will involve a hybrid of these approaches based on local market conditions and the specific nature of waste streams involved. Table 7.10 provides a list of key questions to ask salvage or recycling organizations when investigating the local market for recycling and salvage.

After inventorying the likely waste streams from the project and checking the local market, the last step in developing a waste management plan is to evaluate the costs and benefits of each option. Spreadsheet tools available online can help evaluate the costs of disposing vs. recycling different materials for the project. Analysis may also include non-monetary costs and benefits such as likelihood of meeting diversion goals associated with certification and corporate environmental image.

Environmental and economic benefits can result from resource recovery, but tradeoffs may exist with regard to additional time requirements and planning needs. Firms are entering the market to provide construction waste management recovery as a service. Managing waste materials on site may require additional site disturbance or take up additional space. Benefits include possible reductions in disposal costs or even generation of new revenue streams for contractors.

Materials management plan

A materials management plan requires an inventory of materials management needs, identification of potential solutions and a cost/benefit analysis to determine the best course of action. Needs for materials management will correlate closely with the construction schedule. Therefore, the construction schedule is a good starting point for identifying needs and requirements.

Table 7.10 Questions to ask salvage or recycling organizations (adapted from Kingcounty.gov)

Accepted materials	• What materials do you accept? • Are there minimum or maximum quantities that are acceptable?
Material transport	• Will you pick up the materials at the job site? • Is there a charge for pick-up service or hauling? • Who do I call to arrange for pick-up? • How much lead time is required? • If I must make separate hauling arrangements, where and how should materials be delivered?
Containers	• Will you provide containers? • Do you charge a container rental fee? • Must containers be protected from moisture?
Material value	• Will you pay for the materials? • What potential contaminants are problematic? • What conditions will cause a load to be rejected?
Documentation	• Can you provide proof of waste diversion by weight or volume? • What signage or information can you provide to ensure proper separation?

A materials management needs inventory should begin by identifying materials associated with each construction task and evaluating the procurement and storage/staging requirements for each type of material. It should be prepared in consultation with subcontractors to ensure that their material storage/staging needs are also met.

After all material storage needs have been inventoried, the next step is to map these needs over time, based on the level of completion of the building and other site needs. This helps to determine the amount of space required, special material protection requirements, and total storage time. Arrangements may be possible for items with long lead times to not take possession of the item until shortly before it is needed on site, thereby minimizing the risk of damage. Access to materials within the storage/staging area should also be considered. Heavy items requiring equipment to move should be placed where they can be accessed without damaging other items.

After the materials inventory is complete, each item in the inventory should be assessed in terms of potential threats that should be guarded against. Table 7.11 lists potential threats to be considered for each material in the material inventory, along with mitigation measures. Provisions should also be made in the schedule to ensure that offgassing materials have adequate time to release chemicals before inspection begins.

Excess materials at the conclusion of each task should be considered in the materials management plan and also addressed specifically in the project waste management plan.

Elements of a materials management plan

• Inventory of needs:
  – Materials by task.
  – Procurement requirements.
  – Storage/staging requirements.
  – Special subcontractor requirements.
• Storage requirements vs. schedule:
  – Total space required.
  – Material protection requirements.
  – Schedule dependencies.
  – Access within storage area.
  – Inventory of threats.
• Identification of potential solutions.
• Cost/benefit analysis of options.

Table 7.11 Threats and mitigation measures for material storage

Threats to material integrity	Mitigation measures
Moisture • Exposure to precipitation • Excess humidity • Absorption from ground contact	• Indoor product storage • Placement to allow ventilation • Preventing ground contact • Adequate covering • Active ventilation/heating
Photodegradation • Exposure to UV radiation	• Indoor product storage • Adequate covering
Material security	• Indoor product storage • Protected/locked storage
Temperature fluctuation	• Indoor product storage • Active ventilation/heating
Physical damage • By equipment during moving • By equipment while stored • Improper orientation/support	• Indoor product storage • Adequate support • Following manufacturer stacking/protection recommendations
Contamination • Exposure to spills • Exposure to dust • Absorption of contaminants from surrounding materials	• Separate storage of absorptive items from potential contaminants • Adequate covering • Sealed openings • Active ventilation • Cleaning before installation

Implementing the materials management plan requires attention throughout the project. Key considerations include:

- **Coordinating product delivery to manage congestion** – delivery times should be arranged when possible to avoid conflicts with other site traffic. Products that require equipment for offloading from delivery vehicles must also have adequate space to reduce the potential for damage to other products in storage.
- **Minimizing the need to move products on site** – the risk of physical damage is increased every time a product is moved from one place to another. Ideally, products would be installed right off the truck through just-in-time delivery. For most products, however, this is impractical. Advance planning can help to ensure that products are not stored in areas where they will interfere with concurrent tasks.
- **Minimizing risk of schedule disruption** – managing this risk is critical for products with long lead times or from unfamiliar sources. Additional time may need to be built into the procurement schedule to ensure that construction is not delayed. Arrangements should be discussed with suppliers to delay taking control of products until close to their installation time.
- **Storage of excess materials to preserve value** – careful storage of excess materials at the end of installation is important to preserve

their value. Products to be taken back by vendors or suppliers are especially important. If donation to charity or freecycling will be used, it makes sense to accumulate excess materials and dispose of them together rather than individually. This requires setting aside a space for excess materials where their value can be preserved.

Improving materials management on the job site can help to reduce costs of replacing damaged materials, as well as the need to remediate problems later in the lifecycle due to moisture damage and other threats. Overall, these practices will increase the efficiency of resource use and may reduce risk of delays due to supply chain issues. However, they generally will require additional planning up front.

Project surroundings and public outreach

In addition to the impacts construction has on the natural and human environment, it can also have an effect on human social and economic systems. The corporate social responsibility (CSR) of a company is becoming increasingly important as these impacts become more widely acknowledged. CSR is defined by how a business affects people. This includes employees and people in the market, as well as neighbouring communities who are affected by the company's business practices. Companies across many sectors are measuring CSR as part of a 'Triple Bottom Line' approach to business, including the impacts of the business on society and on the natural environment as well as profits. In the construction sector, the social impacts of construction can be improved by creating a healthy workplace for construction workers, by promoting diversity and equal opportunity, and by using local resources to develop and support the project. Schemes such as Considerate Constructor in the UK and the Green and Gracious Builder Scheme in Singapore (see *Socially responsible construction*) provide useful frameworks for these practices.

Socially responsible construction

The construction industry historically has had what some call an 'image problem', perceived as being unfriendly to women and minorities, disruptive to the environment and more concerned with profit than the well-being of workers involved in construction projects. To address this problem, multiple schemes have been developed worldwide to recognize and promote more socially responsible behaviour on the part of construction firms.

Started in 1997, the Considerate Constructors Scheme in the UK (see www.ccscheme.org.uk) was formed as a non-profit organization to improve the image of the construction industry. The Scheme focuses not only on considerate practice for constructors on their job sites, but also on the vendors and suppliers that provide products and services to those companies. Its Code of Considerate Practice requires registered companies to care about appearance, respect the community, protect the environment, secure everyone's safety and value their workforce. Types of practices encouraged within the five main categories include:

continued

Appearance

- External image, including branding, communication, debris and dust prevention, entrance, graffiti, litter, signage, social media and others.
- Facilities, materials and plant, including layout and tidiness, procedures, remote compounds, screening of facilities, supervision, supply chain and others.
- Organisation and maintenance, including site waste, tidiness, vandalism, viewing points, workforce involvement and others.
- Workforce/Operative, including ashtrays, discreet areas, offsite appearance, onsite dress code and others.

Community

- Courtesy and respect, including company contact information, advance notice of disruptive works, public surveys, feedback/questionnaires, noise, sensitivity to neighbours and others.
- Promoting the Scheme Code, including meeting agendas, toolbox talks, legacy activities, workforce, charities/organisations and others.
- Supporting local communities, including businesses, CSR action plans, schools, careers advice, residents, community liaison and others.

Environment

- Ecology and landscape, including landscaping, post-completion impacts, trees, birds, rivers, wildlife, hazardous substance storage, spill control and others.
- Impact of pollution and vibration, including air quality, lighting, dust, working methods and equipment, fumes, noise and others.
- Management and communication, including environmental policy statement displays, notice boards, site-specific induction, incident procedures, promoting achievements and others.
- Waste and resources, including carbon footprint reporting, green purchasing, sustainable sourcing, harvesting rainwater, travel plans, offsite construction, water saving measures and others.

Safety

- Attitudes and behaviours, including alcohol and drugs testing, supplier engagement, attitudes, incentives, site-specific induction, drugs and alcohol policy and others.
- Minimise security risks, including boundary security, site security, child safety, traffic management, scaffold protection and others.
- Safety initiatives, including campaigns, recording accidents, communicating initiatives, training, identification of near misses and others.
- Safety systems, including drills, falling debris, reporting, signing out, visitor medical details, daily briefings, hazard board, emergency procedures, language differences and others.

Workforce

- Equality, diversity and respect, including counselling services, E-learning, financial advice, bullying, harassment, diversity, feedback, religious considerations and others.
- Health and well-being, including mental health, stress, healthy lifestyle advice, weather protection, health screening, worker fatigue and others.
- Training and development, including apprenticeships, illegal workers, disadvantaged groups, placements, training, careers advice and others.
- Welfare, including canteens, recreation, showers, changing rooms, toilets, lockers, secure storage, cleaning regime and others.

continued

Considerate Constructors has a best practices hub on its website, where case studies and other information are available to help constructors identify ways to meet the criteria. It also provides a mechanism for reporting violations that can be used by neighbours, workers or others affected by the company's practices.

The Green and Gracious Builder Scheme (GGBS) is a second example of socially responsible construction guidelines. Launched in February 2009, the GGBS was developed by the Singapore Building and Construction Authority (BCA) to promote both environmentally friendly and socially responsible practices during the construction phase of projects. Versions exist for both large and small/medium size firms, meant to target both general and speciality subcontractors. The GGBS works together with the BCA's Green Mark rating system and exists as a credit within that system. Becoming certified under the Scheme is also required for firms wishing to do public sector work in Singapore.

The GGBS covers three primary categories of considerations: Green Practices that focus on environmental impacts; Gracious practices that seek to reduce negative impacts to the public and workforce during construction; and Innovation, which is invoked for green and/or gracious performance not covered by other categories, or exemplary performance in existing categories. Bonus points are also available if the builder has received recognition and/or awards to recognize outstanding green and gracious performance or achievements.

Elements and best practices in the GGBS include:

Green practices

- Company Policy – procedures to inculcate green practices and raise awareness of all levels of staff; specific green goals; green supplier and subcontractor selection.
- Materials (Reduce/Reuse/Recycle) – construction waste management and minimization, including temporary structures used for construction, job site recycling and materials selection.
- Energy – monitoring energy use on site, including electricity and fuel consumption during construction; energy conservation in site office; alternative or renewable energy or fuels for equipment and office; and minimizing use of diesel generators.
- Environmental/Water – monitoring water consumption and pollution on site; water recycling for non-potable and construction use; and non-toxic pesticides and cleaning products.
- Housekeeping and Air Quality – proper site housekeeping and maintenance; proper material storage; dust mitigation; proper maintenance of equipment; designated personnel.

Gracious practices

- Company Policy – procedures for raising awareness/education; program promotion; and gathering feedback.
- Accessibility – maintenance of site entrance; access control; barrier-free public pathways; minimizing traffic obstruction; and signage.
- Public Safety – covered, protected walkways for public pathways; safety netting to catch debris; sufficient capacity and alternate routes for public pathways; site safety information; and monitoring of surrounding buildings.
- Noise and Vibration – activity scheduling and noise prevention/mitigation; equipment maintenance and training; alternative construction methods; and noise/vibration monitoring.
- Communications – communicating with neighbours in the community; informational signage; formal feedback system; enhanced site and neighbourhood security.
- Workforce Management – worker education and training; rest, recreation and welfare areas for workers; appropriate disciplinary procedures; enhancing on-site conditions for workers.

continued

Innovation – extraordinary measures may include:

- Energy-efficient Site Office – extra measures for energy efficiency in the site office.
- Green and Conducive Site Environment – conservation of site ecosystems; vegetative temporary and permanent structures; and vegetative erosion controls.
- Other Innovative Gracious Practices – extra measures for managing job site impact on surrounding communities.
- Demolition Protocol – protocol for waste recovery/recycling; pre-demolition audit; sequential demolition; site waste management plan.

Additional details about the GGBS are available in its Assessment Criteria Guidelines (www.bca.gov.sg/Awards/GGBA/others/Criteria.pdf) or Builder Guide (www.bca.gov.sg/Awards/GGBA/others/GGB_book.pdf).

Stakeholder involvement and community outreach

Involving stakeholders in the process of designing and managing construction operations is also important for project sustainability. While not all wishes of stakeholders can be accommodated in every situation, having a mechanism to inventory and address specific concerns can have a significant positive effect on stakeholder complaints and enforcement of operating constraints. Specific measures employed for this purpose include:

- **Community meetings** – these centralized meetings (also sometimes called public hearings) provide an opportunity for construction stakeholders to explain the project to interested members of the local community and its likely impacts on that community, as well as to explain any control measures that will be put in place to mitigate impacts.
- **Charrettes** – these centralized meetings involve local stakeholders more actively in the initial design of the project, which can lead to greater buy-in as construction occurs.
- **Tours** – providing physical tours of the project site after construction begins, or virtual tours using n-dimensional models or fly-throughs, can help to assuage concerns of stakeholders about the approach to construction being taken for the project.
- **Kiosks, web cams or displays** – these measures can be used to keep interested stakeholders apprised of the project's progress as well as alert them to upcoming activities of interest or concern.
- **Door-to-door meetings** – typically used with immediate neighbours, these meetings help to identify specific concerns from the stakeholders with the greatest potential to be impacted by the project. After concerns have been identified, measures can be designed to either address the concerns or provide some other compensating benefit to offset the concern.
- **Newsletters** – using printed or electronic newsletters to inform neighbours of upcoming project milestones or activities that may

Figure 7.25 The yPOD Kiosk, which is located in a construction classroom building overlooking a construction site, includes a Building Information Model and a 5-D schedule of the project that can be used by students to monitor the progress of the job

Source: Courtesy of Reneé Ryan, Department of Building Construction, Virginia Tech

influence their day-to-day activities can help residents plan ahead to accommodate the project.

Some measures such as community meetings may be required as a condition of permitting. Others such as kiosks, virtual tours or web sites/web cams can be used not only for public relations but also for other progress tracking or educational purposes (Figure 7.25). Meetings and charrettes can benefit from experienced facilitation to ensure that conflicts are resolved amicably. Benefits include preservation of community goodwill and reduction in delays due to complaint resolution.

Diversity and equal opportunity

A key part of social sustainability for construction projects is providing opportunities for workers to grow and develop in their profession. Diversity in the workforce, be it in terms of age, gender, ethnicity or other factors, benefits construction companies by bringing together people with a variety of backgrounds and skills to complete a job. It can also provide competitive advantage for companies by retaining employees and being able to attract different kinds of work in the marketplace. Companies can also enhance their level of diversity by partnering with other firms to pursue projects with diversity requirements.

Providing equal opportunity for employment, training and advancement can be advantageous for construction companies. The costs of employee turnover are significant. Recruiting and training new employees can be expensive. The need to fill positions on a crew can mean that work is delayed or does not get done correctly. Providing equal opportunities for workers to advance helps retain existing employees. These workers can then contribute to a more sustainable project.

Specific measures employed to increase diversity and provide equal opportunity include:

- **Enhanced recruitment** – this tactic includes measures targeted to attract underrepresented groups for employment with the firm, including recruiting at institutions with diverse student populations.
- **Targeted internship programs** – these programs focus on providing a synergistic experience for interns who are members of under-represented groups to encourage them to seek a career with the company.
- **Partnering with other firms** – this tactic involves developing temporary alliances with firms to pursue projects that have specific procurement requirements. Ongoing relationships should be maintained with these firms to ensure that working efficiency is sustained.

These practices help to develop capacity of underrepresented firms and individuals in local markets. Companies with active diversity programs may have increased employee retention and job satisfaction and may also qualify for additional work in the public sector. Different perspectives may enable better problem solving and creativity within the firm. Significant cost savings may occur with enhanced retention, including avoiding the need to recruit and train new employees. Overall, benefits may be difficult to calculate, but these measures can significantly contribute to a firm's reputation and public relations.

Community capacity building

As the activity that builds the infrastructure of society, construction can create powerful local impacts. Not only can construction firms contribute to sustainability by building green projects, but also the resources they employ to build those projects can help to build the economic sustainability of the local community or region as well.

Hiring local businesses and using local products and materials helps to support the local economy and build capacity for future work. Just as local farmers' markets provide a way for citizens to obtain fresher and healthier food, construction projects that support local businesses improve the health of the local economy. They provide a way for local citizens to contribute through their work to the buildings and facilities being built in their communities.

Using local products and services also improves the environment by reducing the need to transport goods and service providers from other places. The skills and capabilities built at the local level will contribute to more vibrant, resilient communities that can sustain themselves economically in the long term.

Specific measures employed on some projects include the use of local materials and services, investment in local recruiting and job training, and other community investment such as support of local non-profits and charities or encouraging employees to volunteer in the local community.

While patronizing local businesses may sometimes increase cost, enhanced relationships and increased economic sustainability at the local level can benefit the company in the long term. Volunteering may seem to be a distraction but can increase employee satisfaction and retention. Benefits include reduced transportation costs and improve responsiveness of local vendors, enhanced economic sustainability of the local community and enhanced local relationships for future projects.

To illustrate sustainable construction opportunities and strategies used in the construction phase of the facility lifecycle, the next sections of the chapter present sustainable construction best practices using two in-depth case studies: the Trees Atlanta building located in Atlanta, Georgia, USA, and the Proximity Hotel and Print Works Bistro in Greensboro, North Carolina, USA.

Case study: Trees Atlanta Kendeda Center construction phase best practices

To achieve the goals of sustainability in the construction project, it is necessary to implement sustainable strategies not just during design, but also during the construction phase of the building's lifecycle. The Trees Atlanta Building project incorporated a number of sustainability strategies at the construction phase, including:

- Erosion and sedimentation control through construction activity pollution prevention.
- Construction waste management.
- Construction indoor air quality management during construction and before occupancy.
- Building commissioning.

The following subsections describe these sustainability strategies in more detail.

Construction activity pollution prevention

During the development of the construction site, the removal of existing vegetation can increase erosion, which causes a variety of environmental problems. Erosion of soil from the construction site greatly reduces the soil's ability to support plant life, regulate water flow and maintain the biodiversity of soil microbes and insects that control disease and pest outbreak. In addition, the off-site consequences of erosion from developed sites include a variety of water quality issues. Water run-off from the construction site carries pollutants, sediments and excess nutrients that disrupt aquatic habitats in the receiving water. Sedimentation also contributes to the degradation of water bodies and aquatic habitats by lessening water flow capacity as well as increasing flooding and turbidity levels. Airborne dust from construction activities can also cause environmental and human health impacts including asthma.

In the Trees Atlanta building, several strategies for controlling erosion, sedimentation and airborne dust generation were implemented as part of an erosion, sedimentation and pollution control plan (Figures 7.26a, b and c) to reduce pollution from construction activities. The erosion, sedimentation and pollution control plan for the Trees Atlanta site was prepared using the same Georgia Department of Natural Resources and City of Atlanta design standards that apply to all construction sites of between 1 to 5 acres. The plans were designed with a three-phase approach

continued

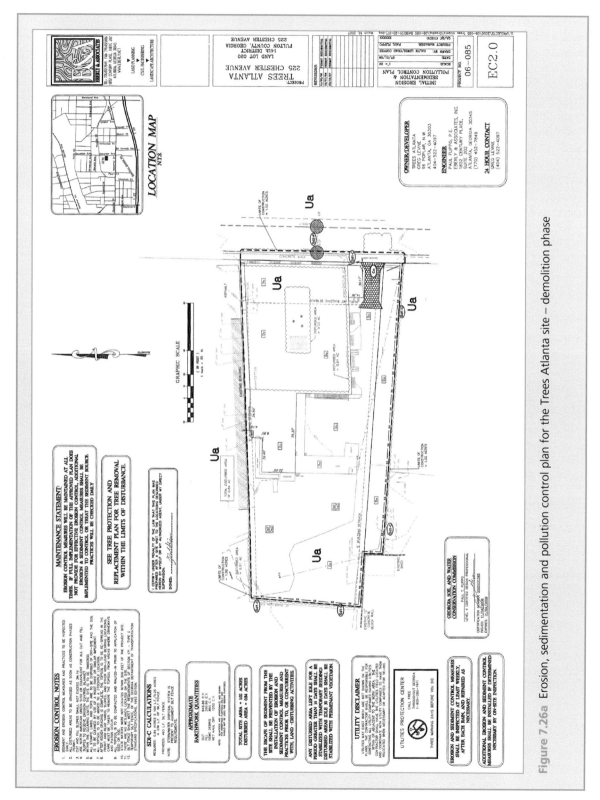

Figure 7.26a Erosion, sedimentation and pollution control plan for the Trees Atlanta site – demolition phase

continued

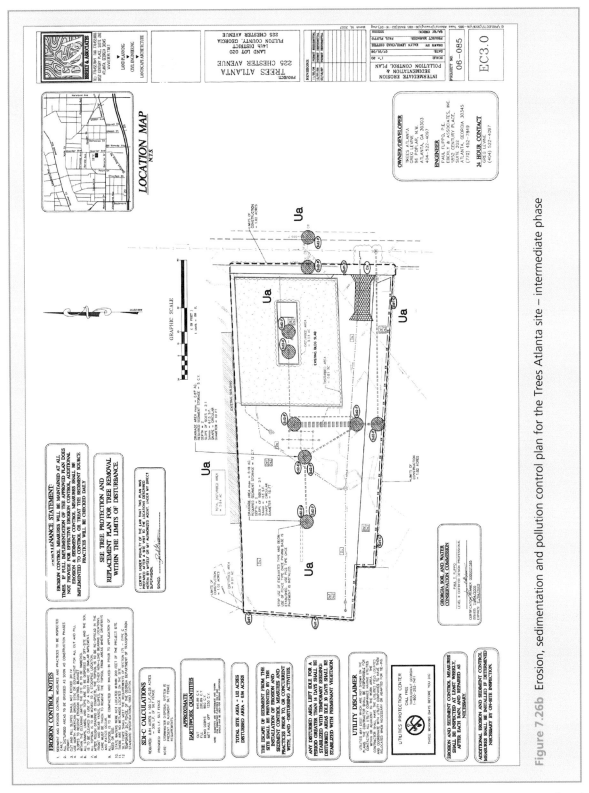

Figure 7.26b Erosion, sedimentation and pollution control plan for the Trees Atlanta site – intermediate phase

continued

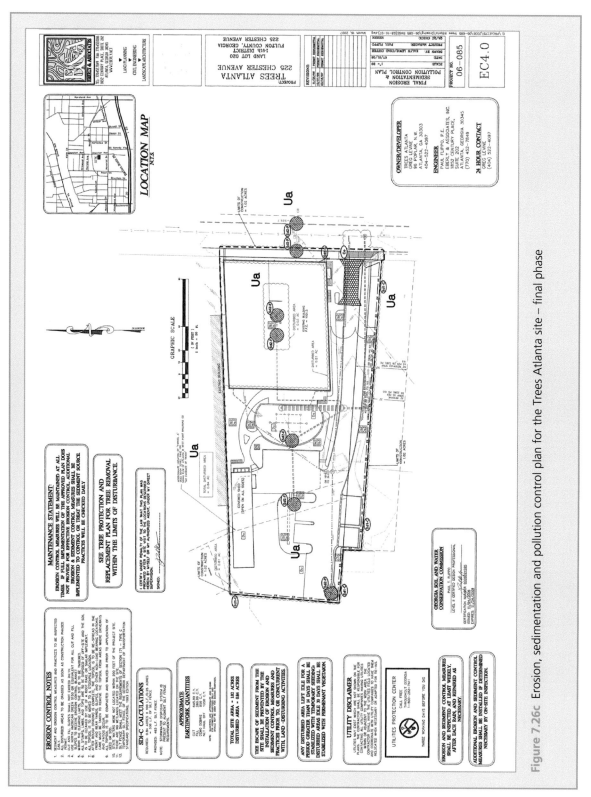

Figure 7.26c Erosion, sedimentation and pollution control plan for the Trees Atlanta site – final phase

continued

to show the demolition, intermediate and final phase best management practices (BMPs). Each phase of the plan demonstrated the use of BMPs such as silt fences, inlet sediment traps, seeding, grassing, dust control and construction entrances designed to provide at least 67 cu yd per acre of sediment storage within each phase. In addition to the measures proposed in the erosion, sedimentation and pollution control plan, while the Trees Atlanta parcel is 1.08 acres (net lot size), it was proposed to disturb only 0.84 acres, thus further limiting the need for damage control in the first place.

Construction waste management

Construction activities including both construction and demolition generate significant amounts of solid waste. According to the US Environmental Protection Agency (USEPA), building-related construction and demolition (C&D) debris totals approximately 160 million tons per year, accounting for nearly 16 per cent of total non-industrial waste generation in the United States (USEPA 2008). Because of the significant impacts of construction waste, source control and prevention of construction wastes can achieve the greatest environmental benefits. In addition, recycling of construction and demolition debris reduces demand for virgin resources and reduces the environmental impacts associated with resource extraction, processing and in many cases transportation. For the Trees Atlanta Building project, a construction waste management plan was implemented to divert the maximum amount of on-site construction waste from landfills through recycling and salvage efforts (see *construction waste management plan for Trees Atlanta*).

To achieve the goals of recycling and waste prevention, the general contractor and subcontractors diverted the following major types of wastes:

- concrete/masonry (679 tons)
- metal (13 tons)
- asphalt (161 tons)
- green waste (5.77 tons).

Due to its active recycling and waste prevention approach (Figure 7.27), the Trees Atlanta project recycled or salvaged just over 84 per cent of solid waste generated by the project. This equates to a total of 959 tons of construction waste diverted out of a total construction waste generated of 1139 tons.

Figure 7.27 Recycling of construction waste

continued

Construction waste management plan for Trees Atlanta

Purpose

To establish a procedure for the handling and disposal of construction debris. The plan is a component of the LEED Green Building Rating System, based on requirements of LEED-NC version 2.2, Credits MR 2.1 and 2.2.

Goal

To divert the maximum amount of on-site construction waste from landfills through recycling or salvage efforts.

Implementation

There will be two phases: demolition and construction.

Demolition

Gay Construction is responsible for ensuring that on-site demolition waste will be recycled or salvaged as much as possible. The materials to be recycled, but not be limited to, are: metals, concrete, concrete masonry block, ceiling tiles and unpainted wood. Individual containers will be clearly labelled (in English and Spanish) for each material. The demolition contractor will appoint an on-site member of his team to monitor the containers on a daily basis to be sure that the containers are being filled with the correct material. The demolition contractor is responsible for keeping an organized record of waste hauling and disposal receipts and reporting the percentages and totals diverted construction waste to Gay Construction while demolition is in progress. These reports/receipts shall include the quantity in volume and weight, hauler and destination.

Construction

Gay Construction is responsible for all recycling of construction waste and quality control during construction. Individual hauling containers will be provided for concrete, masonry, metals, wood, gypsum board, cardboard and general waste. Recycling containers will be provided for glass, aluminium, paper and plastic. All containers on site will be clearly labelled (in English and Spanish) and will be checked daily to ensure compliance with these terms. All subcontractors will be aware of and participate in this construction recycling effort. Waste haul receipts will be kept in an organized manner by material and month and must relate to the progress spreadsheets. Gay Construction shall submit waste management calculations and documentation on a monthly basis to confirm that materials are being diverted by recycling or salvage.

Orientation

Each subcontractor will be required to participate in recycling construction debris. As contractors begin work on-site, their foreman will be given a copy of the subcontractor guidelines and given a tour of the site recycling containers. The foreman will be required to convey this information to their labourers.

Reporting

Gay Construction will be responsible for the keeping a waste management report throughout the demolition and construction process. This report will be organized in the attached format and submitted on a monthly basis. Reports will include the type of material recycled, the recycling hauler and destination and the quantity of waste being diverted, by weight, as well as waste taken to the landfill.

continued

Daily activities

The Gay Construction site superintendent is responsible for the day-to-day activities on the jobsite. This is expanded to meet the demands of the CWMP. The site superintendent keeps a log of all containers removed from the site to compare to receipts from the container subcontractor. The superintendent also keeps track of waste created by the subcontractors and determines the appropriate means of disposal.

Construction indoor air quality management

Demolition and construction practices often lead to increased exposure to indoor air pollutants through the introduction of synthetic building materials, power equipment and vehicles and new furnishings and finishes. To reduce air quality problems resulting from construction, the Trees Atlanta building implemented an IAQ Management Plan based on the recommended design approach of the SMACNA *IAQ Guideline for Occupied Buildings under Construction*. The IAQ Management Plan (see *Indoor air quality management plan for Trees Atlanta*) describes the practices undertaken in the building to protect indoor air quality. Figure 7.28 shows some of practices employed to reduce indoor air problems. Although the plan discourages the use of HVAC equipment during construction, it provides for temporary operation if filters with a minimum efficiency reporting value (MERV) of 8 are used to protect the equipment from damage. In this project, the air handling units in the building were operated at various points during construction, but MERV 11 filters were installed and then replaced prior to building flush-out, thereby exceeding the minimum requirements set forth in the IAQ management plan.

In addition, a building flush-out was conducted that supplied a minimum of 3500 cu ft of outdoor air per sq ft of floor area (at a minimum rate of 0.30 cfm/sq ft of outside air) a minimum of 3 hours

Figure 7.28 IAQ protection measures undertaken during construction, including housekeeping, protection of materials from damage and plastic protection for ductwork and equipment in storage

continued

prior to occupancy and during occupancy, until a total of 14,000 cu ft/sq ft of outside air had been delivered to the space. By conducting these practices to improve indoor air quality, it was possible to prevent future indoor air quality problems resulting from construction and to promote the comfort and well-being of both construction workers and building occupants.

Indoor air quality management plan for Trees Atlanta

Purpose

The purpose of the IAQ plan is to reduce IAQ problems resulting from demolition and construction phases of this project. While work is being performed on the site, each subcontractor shall adhere to the following responsibilities, requirements and measures of this document to minimize construction related pollutants throughout the building system and contamination of materials. This plan is a component of the LEED Green Building Rating System, based on the requirements of LEED-NC version 2.2, Credit EQ 3.1 – Indoor Environmental Quality and the control measures of the SMACNA *IAQ Guidelines for Occupied Buildings under Construction*, 1995, Chapter 3.

Goal

The goal of the IAQ plan is to maintain a safe and healthy environment for construction workers while the project is underway and to ensure a healthful and comfortable environment for the building's future occupants.

Plan

The existing building will be partially demolished and the interior build-up will be completely demolished, leaving only a shell and structural members. During demolition and construction prior to dry-in, the building shall be kept open and well ventilated during working hours. Gay Construction will be responsible for making sure the building is swept out at the end of each workday. The existing roof will remain until the exterior walls are built and all added structural members are in place.

Moisture protection

Moisture-absorbent materials will be stored in a dry area inside the building and covered with plastic to keep dry. If these materials shall be installed before dry-in then the materials will be protected from the weather as needed. The general contractor will conduct a walk-through on a daily basis to ensure there is no moisture damage to installed and uninstalled materials. Any problems will be dealt with on a situational basis to decide whether the material can be salvaged or needs to be replaced.

Volatile organic compounds (VOC)

No VOCs that exceed LEED restrictions will be installed within the building envelope. All low-VOC emitting materials used within the envelope shall be kept in a closed container unless in immediate use. Any material used on the exterior of the building (i.e. exterior paint, roofing) that exceeds LEED restrictions or VOC content shall never enter the building. These materials shall be stored in ventilated storage outside and away from the building. The general contractor will continuously oversee this throughout construction and make a daily walk-through to ensure materials are stored in the appropriate places.

continued

HVAC protection

The HVAC system will not be used during the construction process. Once the system is installed all openings shall be sealed. The mechanical rooms shall be confined and used only in necessary situations and shall not be used to store construction materials.

Housekeeping

The building shall be kept clean and orderly on a daily basis throughout the construction process. Gay Construction will be responsible for the site in general and each subcontractor will be responsible for their particular work areas. This includes sweeping, mopping or vacuuming of dust, accumulated water and other potential contaminants. All spills shall be cleaned up immediately and reported to the IAQ supervisor. The subcontractors will be responsible for keeping their work areas clean and organized. All subcontractors will be responsible for their on-site equipment, tools and cleaning supplies and required to keep them contained within their appropriate areas. The IAQ supervisor shall ensure that the subcontractors conform to this act.

Implementation responsible parties

The general contractor will appoint a responsible person that will be responsible for implementing and monitoring the IAQ plan. This person will be responsible for but not limited to:

- **Coordinating inspections** – the IAQ supervisor will walk over the site at the end of each work day to ensure compliance with the IAQ plan. This will include making sure all materials on site are stored in the correct locations and protected from the elements, making sure the building and site are kept in a clean and orderly manner, and checking that the HVAC system is sealed.
- **Maintaining an inspection log** – the IAQ supervisor will keep a log of any non-compliant conditions and on corrective actions taken.
- **Coordinating meetings** – if the IAQ supervisor witnesses recurring problems then a meeting with the responsible parties may be scheduled. The general contractor and LEED coordinator may be present.

Orientation

Gay Construction will be responsible for ensuring that each subcontractor participates in the IAQ plan. As contractors begin work on site, their foreman will be given a copy of the subcontractor guidelines. The foreman will be required to convey this information to their labourers. All employees on the site shall understand and abide by the IAQ requirements set out in this plan.

Reporting

Gay Construction will provide documentation as required by LEED Version 2.2 in weekly IAQ reports with photo documentation that emphasize the following points (see box).

HVAC
- The installed HVAC system is not being used during construction.
- The system is sealed, and all openings are covered with plastic.

Materials and storage
- No materials exceeding the VOC limits are present in the building.
- All materials that contain VOCs are in a well-ventilated area.
- Stored materials are covered to protect from dust and moisture.
- All moisture sensitive materials are kept covered and dry.
- Materials are stored orderly and out of the way of walking traffic.

Housekeeping
- The floor is swept regularly to minimize dust particles.
- No liquid is standing on the slab.
- There are at least two clear means of egress.

continued

HVAC duct protection policy

If the HVAC system is operated during the construction process, temporary filters with MERV 8 will be used to protect all central filtration points and all return air inlets. Once the temporary filters are in place, photographs will be taken at each installed location. These photographs will be included in a filter replacement verification log, listing the date of initial installation, filter locations and filter MERV ratings. The filters will be inspected by the IAQ supervisor on a weekly basis and changed as needed. Each filter change out will be recorded in the filter replacement verification log, listing date, filter change location, filter MERV rating and a photograph of the newly installed filter.

SITE OBSERVATION REPORT

WORKINGBUILDINGS

Trees Atlanta Office
Trees Atlanta
Project #06090

Site Observation Report 1
Wednesday, September 26, 2007
Report by John McFarland

Present at Site:

Mechanical

Reference No.: SOR-1-1	Equipment: GSHP-2	Responsibility: Mechanical Contractor
Room No.: 128	Room Name: Future Office	Drawing:

Mechanical

Description:
A few plenum boxes were not covered with plastic at the time of the survey.

Recommended Action:
Cover plenum boxes with plastic to protect from dust.

Figure 7.29 Example site observation report

continued

Building commissioning

Facilities that do not perform as intended can consume significantly more resources over their lifetimes than they should. Thus, the building commissioning process has to be conducted to verify that the project's energy-related systems are installed, calibrated and performing according to the owner's project requirements, basis of design and construction documents. Major benefits of building commissioning include reduced energy use, lower operating costs, reduced contractor callbacks, better building documentation, improved occupant productivity after construction, and verification that the systems perform in accordance with the owner's project requirements. In the Trees Atlanta Building, the owner hired a commissioning authority to conduct detailed commissioning for the systems listed in Table 7.12.

As a result of the commissioning process, approximately 55 issues were identified, and all of them were fixed during the project. In addition, a detailed operation and maintenance manual was developed to help users to properly control all systems as designed, calibrated and installed. Figure 7.29 shows an example of one of the site observation reports generated during building commissioning of the Trees Atlanta Building. These forms were used to identify specific problems with the building where the constructed product did not meet the owner's design intent. The forms also identify recommended actions that should be taken to remediate all identified problems. In this way, commissioning was used to ensure that the owner received the facility for which it has paid.

Table 7.12 Building systems commissioned in the Trees Atlanta building

Mechanical systems:
- HVAC systems (ground-source heat pumps and energy recovery ventilation)
- HVAC control systems (standalone programmable thermostats)
- mechanical equipment.

Plumbing system:
- water heaters (electric and solar)
- rainwater harvesting system.

Electrical systems:
- lighting control systems (time clock)
- occupancy sensors
- photovoltaic systems
- electrical power distribution equipment.

All images courtesy of Smith Dalia Architects

The Trees Atlanta Kendeda Center employed multiple strategies to improve the sustainability of the construction process that were appropriate for adaptive reuse of an existing structure. In contrast, the Proximity Hotel and Print Works Bistro was a completely new construction project for a luxury hotel application, with very different design goals. Accordingly, the practices employed were different due to the different challenges posed by the project environment, although the overall considerations such as site protection, concern for indoor air quality, and waste/materials management were the same.

Case study: Proximity Hotel and Print Works Bistro, Greensboro, GA

The Proximity Hotel and Print Works Bistro represents one of the world's greenest, most energy-efficient and high-performance developments (Figure 7.30). The sustainably designed Proximity Hotel features 147 guest rooms and suites, a full service restaurant (Print Works Bistro), and 5,000 sq ft (465 sq m) of meeting and event space. The Proximity Hotel has incorporated many sustainable design and construction features, and the hotel has been recognized as the first hotel in the hospitality industry to obtain the US Green Building Council's top level of certification (LEED Platinum).

continued

This environmentally friendly hotel was developed to be a high-performance green building by Quaintance-Waver Restaurant and Hotels because the development team wanted to adopt sustainable construction practices that would not only achieve the benefits of sustainable development but also make sense to the bottom line in the long term. Since the developer had a passion for sustainable practices in its hotel development, the Proximity Hotel incorporated many such sustainability strategies.

Proximity Tower

- Project type: Hotel and restaurant.
- Project size: 102,000 sq ft with 147 rooms.
- Project cost: $26 million.
- Green rating: First LEED Platinum hotel.
- Developer: Quaintance-Waver Restaurant & Hotels.
- Architect: Centerpoint Architecture, LLP.
- Contractor: Weaver Cooke Construction, LLC.

The Proximity Hotel project began by assembling a collaborative development team including the developer, architect, contractor, landscape architect, engineer and other consultants. All team members worked in concert to not only maximize water and energy efficiencies in the design and achieve resulting sustainability benefits, but also avoid or surmount the barrier of increased first cost. This multidisciplinary and collaborative integrated design approach resulted in a hotel that is expected to use 39.2 per cent less energy and 33.5 per cent less water than a conventional hotel without reducing comfort or luxury and with minimal additional construction costs.

Sustainability features included in the design

The facility uses the sun's energy to heat hot water, with 100 solar panels covering the 4000 sq ft of rooftop (enough hot water for 100 homes). This system (Figure 7.31) heats around 60 per cent of the water for both the hotel and restaurant. On the building's site, 700 linear feet of stream were restored by reducing erosion, planting local, adaptable plant species, and rebuilding the buffers and banks. Approximately 700 cu yd of soil were removed to create a floodplain bench, and 376 tons of boulders and 18 logs were used to maintain grade control, dissipate energy and assist in the creation and maintenance of riffles and pools.

Inside the Print Works Bistro, the bar is made of salvaged, solid walnut trees that were felled because of sickness or storms (Figure 7.32). Room service trays are made of rapidly renewable Plyboo (bamboo plywood). Newly engineered variable speed hoods in the restaurant use a series of sensors to set the power according to the kitchen's needs and adjust to a lower level of operation, typically 25 per cent of their full capacity, when possible. The sensors also detect heat, smoke and other effluents, and increase the fan speed to keep the air fresh. Geothermal energy is used for the restaurant's refrigeration equipment instead of a standard water-cooled system, saving significant amounts of water.

In the hotel itself, North America's first regenerative drive model of the Otis Gen2 elevator reduces net energy usage by capturing the system's energy and feeding it back into the building's internal electrical grid. Abundant natural lighting, including large energy-efficient 7 ft x 4 ft square operable windows in guest rooms (Figure 7.33), connects guests to the outdoors by achieving a direct line of sight to the outdoor environment for more than 97 per cent of all regularly occupied spaces.

continued

In operations, water usage has been reduced by 33 per cent by installing high-efficiency Kohler plumbing fixtures, saving 2 million gallons of water the first year. Air quality is improved by circulating large amounts of outside air into guest rooms (60 cu ft/minute) and doing so in an energy-efficient way by employing energy recovery ventilation (ERV) technology, where the outside air is tempered by the air being exhausted. Regional vendors and artists were used for materials to reduce transportation and packaging while supporting the local economy (Figure 7.34).

From the standpoint of indoor air quality, low-VOC emitting paints, adhesives, carpets (Figure 7.35) and other products were used to reduce indoor air contamination. Guest room shelving and the bistro's tabletops are made of walnut veneer over a substrate of SkyBlend, a particle-board made from 100 per cent post-industrial recycled wood pulp with no added formaldehyde resins. A green, vegetated rooftop is planted on the restaurant to reduce the urban heat island effect. This roof reflects heat, thus reducing the amount of energy needed for refrigeration and air conditioning. It also slows stormwater runoff and insulates the rooftop, keeping the building cooler overall. Various types of plants were tried out on the roof in a test area before final planting. Including the green roof as part of the design helps to improve outdoor air quality.

Figure 7.30 Proximity Hotel and Print Works Bistro

Figure 7.31 A solar hot water heating system provides 60 per cent of the building's hot water

The hotel includes several features to support increased sustainability during operations. Among these is an education centre for sustainable practices that includes tours of the green hotel for visitors and guests, symposia of sustainable practice for local construction professionals, and outreach programmes for students of all ages. In addition, bicycles are available for use by guests to ride on the nearby 5-mile greenway, thus providing a means to improve health and fitness while travelling and reduce transportation impacts. All of these design features contribute significantly to the facility's sustainability. In addition, featured in this project was the use of a variety of sustainable construction practices, described next.

Sustainable construction practices

The Proximity Hotel implemented a variety of sustainable construction practices to achieve its goals of sustainability for the project. These practices, including sustainable site management with erosion and sedimentation control, waste management, materials management, indoor air quality management and building commissioning, are described in greater detail in the following subsections.

continued

Sustainable site management

The collaborative project team of the Proximity Hotel developed a sustainable site management plan to minimize disturbance to the project site during construction. One of the strategies in the project was to follow local erosion and sedimentation control standards and codes to minimize erosion and sedimentation at the job site. Since the project site was located next to a local stream, the project team led by the contractor properly implemented several erosion and sedimentation control strategies including silt fences, sediment traps and temporary seeding (Figure 7.36).

Figure 7.32 The Print Works Bistro bar, made from salvaged walnut

Figure 7.33 Operable windows in hotel rooms provide daylight, views and connection with nature through natural ventilation

Figure 7.34 Regional vendors and artists were used to reduce transportation impacts and support the local economy

Figure 7.35 Low-VOC finishes improve indoor air quality

continued

The project team led by the architect selected architectural precast concrete wall panels as the best cladding system to not only achieve the wall performance of the building but also facilitate sustainability at the construction phase. The hotel's insulated wall panels were made from recycled and recyclable materials only 90 miles (145 km) from the construction site. In addition, the contractor and the precast concrete supplier implemented just-in-time delivery to reduce the need for space on site and to improve productivity (Figure 7.37).

Construction waste management

The contractor of the Proximity Hotel developed a construction waste management plan to divert construction debris from disposal in landfills and incineration facilities. Due to the active implementation of the construction waste management plan, the project diverted 1535 tons (86.9 per cent) of on-site generated construction waste from landfill disposal. In addition, the project made significant use of precast wall panels, which also considerably reduced the generation of construction waste during the project.

Materials management

The Proximity Hotel used 22.4 per cent by cost of the total building materials with recycled content including reinforced steel with 90 per cent post-consumer recycled content, drywall with 100 per cent recycled content, asphalt

Figure 7.36 Silt fences and other strategies were used for erosion and sedimentation control

Figure 7.37 Precast concrete wall panel with just-in-time delivery

with 25 per cent recycled content, and staircase steel with 50 per cent recycled content. In addition, the concrete used on the project contained 4 per cent fly ash (224,000 pounds), the mineral residue left after the combustion of coal, which was thereby diverted from landfill disposal. By using recycled content materials, it was possible to avoid the impacts resulting from extraction and processing of virgin materials. In addition, 45.95 per cent of the total building material by cost was comprised of building materials and/or products that had been extracted, processed and manufactured within 500 miles of the project site.

continued

Figure 7.38a, b Indoor air quality management via covering of equipment openings

Indoor air quality management

The Proximity Hotel project team used a variety of strategies to manage indoor air quality during the project (Figure 7.38). The team developed and implemented a construction IAQ management plan that followed the referenced SMACNA Guidelines. To implement the IAQ management plan, the contractor protected all HVAC equipment, both temporary and permanent, from dust and odours during construction. After completing the construction, the contractor completed a whole building flush-out by supplying a total air volume of 14,000 cu ft of outdoor air per sq ft of floor area while maintaining an internal temperature of at least 60°F and relative humidity no higher than 60 per cent. By completing the flush-out prior to occupancy, the project team was able to reduce contaminants inside the building to result in greater occupant comfort and well-being.

Building commissioning

Given that the Proximity Hotel was a luxury hotel and restaurant project, many key systems required commissioning to ensure proper function after construction was complete, including HVAC systems, kitchen equipment and the solar hot water heating and geothermal systems. To meet this requirement, the project team along with a consultant conducted an enhanced commissioning process to ensure that building systems as installed could perform according to the design intent and owner's requirement. The building commissioning process of the Proximity Hotel was also similar to the process in the Trees Atlanta project. Requirements for fundamental and enhanced building commissioning are defined by the LEED rating system as shown in the following box.

Requirements for commissioning under LEED

Fundamental commissioning requirements

- Designating an individual as the commissioning authority to lead, review and oversee the completion of commissioning process activities.
- Reviewing owner project requirements (developed by the owner) and basis of design (developed by the design team) to ensure clarity and completeness.
- Developing commissioning requirements and incorporating them into construction documents.
- Developing and implementing a commissioning plan.

---

continued

- Verifying the installation and performance of systems within the scope of the plan.
- Completing a summary commissioning report to the owner.

Enhanced commissioning requirements

- Conducting a commissioning design review prior to the development of construction documents.
- Reviewing contractor submittals applicable to systems under the scope of commissioning.
- Developing systems manuals for commissioned systems.
- Verifying that requirements for training on commissioned systems are met.
- Reviewing building operation within 10 months after substantial completion.

Source: All images courtesy of Centerpoint Architecture

Sustainable construction in the future

Given the sustainable construction practices highlighted in this chapter, what might be the future of sustainable construction in the coming years? While a construction site might look the same in ten years to an uninitiated observer, advances in technologies, materials and processes are likely to significantly change the resource efficiency of construction practice, leading to reduced environmental impacts, improved project economics and increased health and well-being for workers involved in constructing the work.

While practices to improve health and safety of construction workers are already quite advanced in some parts of the world, best practices such as Prevention through Design (PtD) will become more widely used to ensure occupational health and safety on green projects, resulting in innovations across the spectrum of construction means and methods to reduce or eliminate risk to workers. Instead of more extensive personal protective equipment, construction projects in emerging economies will leapfrog directly to designed-in and engineering controls that eliminate health and safety threats.

Some technologies currently under development will be widely employed in the next 10 years to improve process efficiency. Automated equipment controls, augmented reality, project information models and improved information technology will make understanding the status of the job site instantaneous for project managers. On leading job sites today, four- and five-dimensional augmented reality (nDAR) enables project managers to see the current status of a project and compare that with its planned status. In the future, these technologies will enable project stakeholders to visualize other indicators of project sustainability at various points in the project delivery process, such as distance from the job site that a particular component was imported, carbon footprint or other key properties. These technologies will support more effective project planning and status monitoring which will, in turn, reduce waste and other inefficiencies that are presently difficult to identify and manage in a complex project environment.

Evolving equipment technologies are making better use of research and development in the automotive sector, such as hybrid gasoline-electric engines to eliminate idling or equipment fuelled by biodiesel. The energy recovery technologies in use today, such as regenerative drives on equipment such as elevators and energy recovery ventilation, will also be more commonly employed as permanent features of buildings, and may also be used to power construction processes. Technologies such as 3-D printing are now being tested at the building scale, and in future, printing whole buildings on site may offer more efficient and less impactful ways to develop building projects.

New approaches to reduced ecosystem impact on the job site will also be in use. Technologies are presently under development that will allow heavy traffic or weight-bearing storage on the job site without disrupting the growth of turf or vegetation on the original soil surface. With the growing awareness of the negative impact construction practices have on soil structure and viability, considerable research is being devoted to finding ways to reduce the literal footprint of a construction project on site. In the coming years, a detailed site soil and vegetation inventory will be conducted for every new project site, not just for the purposes of foundation design or stormwater management, but also to carefully design and manage the impacts of the construction process on the site and its biota.

Increased building reuse and adaptation in existing urban areas will also lead to a need for new technologies and approaches for high-resolution condition assessment and building inventorying. Increased use of deconstruction instead of demolition will be coupled with market systems for tagging viable building components before they even come out of a building. Developers may someday base their decisions for demolition on market demand for building components as much as the future potential a site might have for development.

The construction workforce of the future will also look considerably different than it does today. Increasing diversity and social equity, coupled with increased access to education worldwide, means that the workforce of tomorrow will be more heterogeneous. Enhanced technologies for construction mean that physical size and strength of construction workers will no longer be a requirement for a job in the construction industry. Higher levels of education will be required to understand, install, commission and operate the complex systems that comprise our built environment, leading to greater levels of workforce competency and achievement. These better-educated workers will be directed toward ever greater levels of specialization, leading to a greater need than ever before for an understanding of how their roles fit into the larger picture of the built and natural environments.

Along with the changing workforce, the corporation of the future will exist in a business environment requiring both greater agility and greater transparency of operations, as discussed in earlier chapters. Construction companies will be expected to understand and meet environmental performance regulations and standards, and to disclose and be accountable for their performance. Environmental and social factors will be key in evaluating corporate performance and will be inextricably linked to business success. Forward-looking firms will start now to evaluate where they stand with regard to this future business environment, and design a strategy for moving ahead in a world where doing more with less is the norm.

Discussion questions and exercises

7.1   Contact a local construction firm that offers preconstruction services. Interview or shadow a preconstruction services manager to learn more about the role of this stakeholder in project sustainability.

7.2   Locate a project in the construction phase in your local area. Visit the site during the process of site set-up and interview the construction manager or site superintendent. Document strategies used to manage the site sustainably, including pollution prevention and mitigation measures. How well is the project addressing opportunities to minimize disturbance and preserve important site assets? How might the site management be improved to increase sustainability?

7.3   Document the heavy equipment being employed on the site. Obtain a site plan and draw in the limits of site disturbance, material storage and staging areas, waste-tipping areas and equipment paths. How might the site plan be improved to increase project sustainability? What might be the resulting trade-offs for project productivity and safety?

7.4   Visit the site office and document measures used to increase the sustainability of business operations. How well is the project team addressing opportunities in site business operations? How could the sustainability be improved?

7.5   Visit the areas of the site where materials are stored and staged, talk with the project manager and document the procurement, storage, packaging and delivery practices being used on the project. Review the project's material management plan if one exists. How well is the project addressing opportunities to manage materials sustainably? Is product information being properly documented to prepare for project certification? How might materials management be improved to increase sustainability?

7.6   Visit the areas of the site where waste is separated, recycled and/or disposed, talk with the project manager and document the waste management practices being used on the project. What are the major waste streams that will be generated during the project? Review the project's waste management plan if one exists. How well is the project addressing opportunities to manage waste sustainably? Is waste recycling and/or disposal being properly documented to prepare for project certification? How might waste management be improved to increase sustainability?

7.7   After the project has been 'dried in', visit the enclosed areas of the site, talk with the project manager and document the indoor air quality management practices being used on the project. What are the major threats to indoor air quality that will be generated during the project? Review the project's indoor air quality management plan if one exists. How well is the project addressing opportunities to manage indoor air quality sustainably? Are activities and controls being properly documented to prepare for project certification? How might indoor air quality management be improved to increase sustainability?

7.8   Contact a local firm that offers commissioning services. Interview or shadow a commissioning authority to learn more about the role of this stakeholder in project sustainability. Review a commissioning report for a project. What types of issues are typically identified and resolved by commissioning a construction project?

Recommended resources

ASHRAE – American Society for Heating, Refrigeration, and Air-conditioning Engineers. (2009). *Indoor Air Quality Guide: Best Practices for Design, Construction, and Commissioning.* www.ashrae.org/resources–publications/bookstore/indoor-air-quality-guide (accessed 3 September 2016). Free guidebook contains best practices, tools and links for designing and building to promote IAQ.

BuildSmart – Greater Vancouver Regional District. (2004). *Green Construction: Introducing green buildings and LEED to contractors.* www.infrastructure.alberta.ca/Content/docType486/Production/LEED_PD_Appendix_10.pdf (accessed 3 September 2016). Contains examples of completed LEED worksheets for construction-related credits.

City of New York Parks Department. (2009). *Parks Tree Preservation Protocols.* www.nycgovparks.org/pagefiles/52/Preservation-of-Trees-During-Construction-FINAL-4–09.pdf (accessed 3 September 2016). Guidelines for protecting trees during construction.

Construction Industry Institute. (2010). *IR 257–3: Materials Management Planning Guide.* www.construction-institute.org/scriptcontent/more/ir257_3_more.cfm (accessed 3 September 2016). Guidelines for preparing a materials management plan.

Greater London Authority. (2014). *Control of Dust and Emissions During Construction and Demolition.* www.london.gov.uk/file/18750/download?token = zV3ZKTpP (accessed 3 September 2016). Contains guidelines for air quality risk assessment for construction projects.

King County, Washington. (2010). *King County Green Tools: Construction and demolition recycling.* http://your.kingcounty.gov/solidwaste/greenbuilding/construction-demolition.asp (accessed 3 September 2016). Includes examples of specifications and waste management plans as well as resources to estimate cost effectiveness.

National Institute of Building Science. (2016). *Whole Building Design Guide.* www.wbdg.org (accessed 3 September 2016). Contains guidance on a variety of construction best practices as well as model specifications.

SMACNA – Sheet Metal and Air Conditioning Contractors National Association. (2007). *IAQ Guidelines for Occupied Buildings under Construction,* 2nd edn. Available through www.smacna.org.

UK Department for Environment, Food and Rural Affairs. (2009). *Construction Code of Practice for the Sustainable Use of Soils on Construction Sites.* www.defra.gov.uk/publications/files/pb13298-code-of-practice-090910.pdf (accessed 3 September 2016). Includes guidelines and best practice case studies.

USEPA – U.S. Environmental Protection Agency. (2016). *National Menu of Best Management Practices (BMPs) for Stormwater.* www.epa.gov/npdes/national-menu-best-management-practices-bmps-stormwater (accessed 3 September 2016). Includes links to information on best practices for construction sites as well as guidelines for Stormwater Pollution Prevention Plans (SWPPPs).

US Department of Veterans Affairs. (2013). *Whole Building Commissioning Process Manual.* www.cfm.va.gov/til/Cx-RCx/CxManual.PDF (accessed 3 September 2016). Comprehensive overview and guidelines for whole building commissioning.

Virginia Department of Transportation. (2015). *Policy Manual for Public Participation in Transportation Projects.* www.virginiadot.org/business/resources/locdes/Public_Involvement_Manual.pdf (accessed 3 September 2016). Guidelines and criteria for involving the public in construction projects.

Washington State Department of Health. (2003). *School Indoor Air Quality Best Management Practices Manual.* www.doh.wa.gov/Portals/1/Documents/Pubs/333–044.pdf (accessed 3 September 2016). Strategies and recommended practices for design and construction to promote good IAQ.

References

Cole, R.J. (2000). "cost and value in building green." *Building Research & Information,* 28(5/6), 304–309.

Dewlaney, K.S., Hallowell, M.R., and Fortunato, III, B.R. (2012). "Safety risk quantification for high performance sustainable building construction." *Journal of Construction Engineering and Management,* 138(8), 964–971.

EASHW – European Agency for Safety and Health at Work. (2013). "OSH and small-scale solar energy applications." E-FACTS #68, EASHW, https://osha.europa.eu/en/publications/e-facts/e-fact-68-osh-and-small-scale-solar-energy-applications (accessed 14 January 2017).

Gil, N., Tommelein, I.D., Kirkendall, B., and Ballard, G. (2000b). "Lean product-process development process to support contractor involvement during design." *Proceedings, ASCE 8th International Conference on Computing in Civil and Building Engineering*, August 14–17, Stanford, CA, 1086–1093.

Grosskopf, K.R., and Kibert, C.J. (2006). "Developing market-based incentives for green building alternatives." *Journal of Green Building*, 1(1), 141–147.

Klotz, L., Horman, M., and Bodenschatz, M. (2007). "A lean modeling protocol for evaluating green project delivery." *Lean Construction Journal*, 3(1), 1–18, April.

Lapinski, A., Horman, M., and Riley, D. (2005). "Delivering sustainability: Lean principles for green projects." Proceedings, *2005 ASCE Construction Research Congress*.

Lapinski, A.R., Horman, M.J., and Riley, D.R. (2006). "Lean Processes for Sustainable Project Delivery." *Journal of Construction Engineering and Management*, 132(10), 1083–1091.

Matthews, J., Tyler, A., and Thorpe, A. (1996). "Pre-construction project partnering: Developing the process." *Engineering, Construction and Architectural Management*, 3(1/2), 117 – 131.

NFPA – National Fire Protection Association. (2010). *Fire Fighter Safety and Emergency Response for Solar Power Systems*. www.nfpa.org/~/media/files/news-and-research/resources/research-foundation/research-foundation-reports/for-emergency-responders/rffirefightertacticssolarpowerrevised.pdf (accessed 14 January 2017).

Oregon Solar Energy Industries Association. (2006). *Solar construction safety*. www.coshnetwork.org/sites/default/files/OSEIA_Solar_Safety_12–06.pdf (accessed 14 January 2017).

Pearce, A.R. (2008). "Sustainable Capital projects: Leapfrogging the first cost barrier," *Civil Engineering and Environmental Systems*, 25(4), 291–301.

Pulaski, M.H. and Horman, M.J. (2005). "Organizing constructability knowledge for design." *Journal of Construction Engineering and Management*, 131(8), 911–919.

Pulaski, M.H., Horman, M.J., and Riley, D.R. (2006). "Constructability practices to manage sustainable building knowledge." *Journal of Architectural Engineering*, 12(2), 83–92.

Pulaski, M., Pohlman, T., Horman, M., and Riley, D. (2003). "Synergies between sustainable design and construction at the Pentagon." *Proceedings, 2003 ASCE Construction Research Congress*.

Reed, W.G., and Gordon, E.B. (2000). "Integrated design and building process: What research and methodologies are needed?" *Building Research & Information*, 28(5/6), 325–337.

Riley, D., Pexton, K., and Drilling, J. (2003). "Procurement of sustainable construction services in the United States: The contractor's role in green building." *UNEP Industry and Environment*, 26(2/3), 66–69.

Rohracher, H. (2001). "Managing the Technical Transition to Sustainable Construction of Buildings: A Socio-Technical Perspective." *Technology Analysis & Strategic Management*, 13(1), 137–150.

Syphers, G., Baum, M., Bouton, D., and Sullens, W. (2003). *Managing the cost of green buildings*. State of California's Sustainable Building Task Force, Sacramento, CA.

Chapter 8

Post-occupancy sustainability opportunities and best practices

After a building has been delivered to the owner at the conclusion of construction, its life is only just beginning. The majority of a facility's impact from an energy and water standpoint occurs during its operational life. Similarly, the majority of its solid waste impact occurs at the conclusion of its service life. Although these phases have not historically been the focus of the AEC industry, they are critical in today's project environment where extended responsibility is becoming more common in the form of design–build–operate–maintain (DBOM) contracts. Even in the absence of such an agreement, the contractor's liability and reputation are contingent upon the continuing performance of a facility. The following sections describe best practices for sustainably operating and maintaining a facility after occupancy and for disposing of the facility at the end of its lifecycle.

Post-occupancy opportunities

During the use phase of a facility's lifecycle, significant opportunities exist to improve sustainability by virtue of managing the flows of matter and energy across site boundaries and choosing better sources and sinks for those flows. Other measures can also be taken to improve the sustainability of affiliated and unrelated systems outside the facility itself as a means to offset negative impacts. Within the facility boundaries, improvements also can be made to the user environment to improve sustainability.

Optimizing energy and water performance

Optimizing the performance of an existing facility in terms of energy and water consumption can yield significant benefits, both in operating costs and in user comfort, satisfaction and productivity. Multiple tactics are useful in optimizing energy and water performance, including operational adjustments, retrofit of energy and water consuming equipment and working with users to adjust their behaviour.

Opportunities are often overlooked for optimizing efficiency based on proper operations and maintenance. No matter how well it may be designed, equipment will function only as well as it is operated and maintained. Frequently, effective performance is based on not only environmental conditions, but also interactions with other facility systems and occupants. Sometimes, minor adjustments can yield significant results.

Processes such as continuous commissioning can be used for existing buildings to identify ways to improve efficiency with existing equipment. Continuous commissioning is a comprehensive and ongoing process to resolve operating problems, improve comfort, optimize energy use and identify retrofits for existing commercial and institutional buildings and central plant facilities. In contrast with building commissioning that occurs during the delivery process and focuses on design intent, continuous commissioning focuses on meeting changing use requirements during operations. Continuous commissioning typically pays for itself in less than three years (US FEMP 2002). It involves field metering and performance verification, adjustment of operational control schedules and set points, adjustments to building automation systems, and identification and recommendation of capital upgrades and retrofits that will improve building performance.

Operator training is also essential to ensure that building equipment is used most effectively, particularly in situations with frequent turnover of employees. This training ensures that maintenance schedules and procedures are understood and properly implemented. For example, even simple maintenance practices such as cleaning the transmissive surfaces of lighting fixtures can greatly increase their light output. This reduces the need for supplemental lighting in workspaces, thereby saving energy.

Often, energy audits identify opportunities to retrofit existing systems with more efficient equipment. A common retrofit with rapid payback in many buildings is a lighting retrofit (Figure 8.1). Replacing older magnetic ballasts with electronic ballasts in fluorescent lights, or replacing fluorescent lamps with LED lamps that do not require ballasts, are ways to begin. More sophisticated lighting retrofits may include replacing existing fixtures with solid-state LED lighting fixtures or T-5 fluorescent lighting. Replacing fixtures offers an opportunity to revisit lighting needs in the space and choose newer technologies that can more effectively meet those needs.

Other types of retrofits, including both energy- and water-consuming equipment and devices, may also be appropriate. Advances in the efficiency of building technologies may mean that it makes economic sense to replace building systems or equipment before the end of their service life. In some markets, external incentives such as tax credits, utility rebates and manufacturer rebates are available to encourage such retrofits. In addition to using technical audits to identify retrofit opportunities, external incentives should also be investigated to maximize the return on investment of energy upgrades.

Figure 8.1 Lighting retrofits can offer considerable energy savings

Landscaping can also offer performance improvement opportunities. In buildings deliberately designed to meet sustainability goals, landscaping is often selected to minimize water requirements. However, temporary irrigation measures may be included during the plant establishment period to ensure healthy initial growth. Discontinuing these measures at the end of the establishment period will help to ensure that water conservation goals are met. For buildings built before sustainability goals were established, replacing existing landscaping with native plants and water-conserving species will also offer significant water conservation potential, and may also help conserve energy if used to shade the building or reduce the urban heat island effect (Figure 8.2).

Clues to building performance problems can be identified by looking for occupant adjustments or compensating technologies. For instance, portable heaters, desk fans and occupant-supplied task lighting can be signals that a building is uncomfortable for its occupants. These and other adjustments (see Figure 8.3) should be noted and explored further as part of energy audits or continuous commissioning.

Information can also be a useful way to balance loads and shave peak demand. Large institutions such as universities often pay commercial rates for power based on time-of-day usage. During the hottest parts of the year, rates can easily triple or more when demand for power is highest. Some institutions send out email reminders (Figure 8.4) or use social media to alert users to shut down unnecessary equipment and lighting during these periods. Not only does this save electrical energy, it also saves considerable cost for the institution through avoided use during high-price periods.

For buildings designed with energy sustainability in mind, ongoing implementation of efficiency and operations plans is important to ensure

Figure 8.2 Using plantings to shade parking areas and buildings can reduce the urban heat island effect and offer energy savings potential

Figure 8.3 Occupant adjustments or compensating technologies are clues that a building may not be performing properly

that energy goals for the building are met in practice. Some of the efficiency and operations plans that may be included for a green building are:

- **General building operating plan**, including an occupancy schedule, equipment run-time schedules, design set points for heating, ventilation and air conditioning (HVAC) equipment, design lighting levels throughout the building, and seasonal or periodic schedule/set point changes.
- **Sustainable maintenance plan** that describes maintenance strategies and specific requirements for facility systems, including equipment maintenance schedules and procedures, spare parts stock requirements and other details.
- **Continuous commissioning plan** to adjust the facility for evolving operational requirements.
- **Measurement, monitoring and verification plan** to monitor ongoing building performance and identify performance problems.

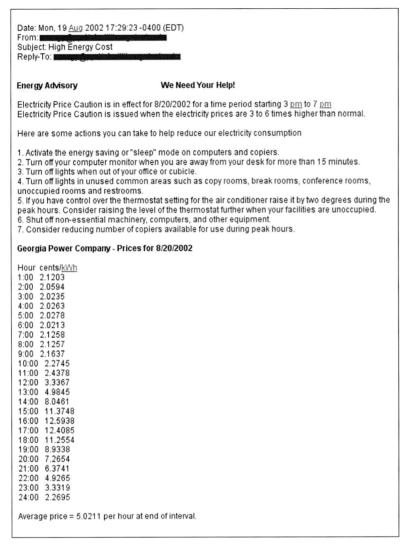

Date: Mon, 19 Aug 2002 17:29:23 -0400 (EDT)
From: ▮▮▮▮▮▮▮▮▮▮▮▮▮▮▮▮▮▮▮▮
Subject: High Energy Cost
Reply-To: ▮▮▮▮▮▮▮▮▮▮▮▮▮▮▮▮▮▮▮▮

Energy Advisory   **We Need Your Help!**

Electricity Price Caution is in effect for 8/20/2002 for a time period starting 3 pm to 7 pm
Electricity Price Caution is issued when the electricity prices are 3 to 6 times higher than normal.

Here are some actions you can take to help reduce our electricity consumption

1. Activate the energy saving or "sleep" mode on computers and copiers.
2. Turn off your computer monitor when you are away from your desk for more than 15 minutes.
3. Turn off lights when out of your office or cubicle.
4. Turn off lights in unused common areas such as copy rooms, break rooms, conference rooms, unoccupied rooms and restrooms.
5. If you have control over the thermostat setting for the air conditioner raise it by two degrees during the peak hours. Consider raising the level of the thermostat further when your facilities are unoccupied.
6. Shut off non-essential machinery, computers, and other equipment.
7. Consider reducing number of copiers available for use during peak hours.

Georgia Power Company - Prices for 8/20/2002

Hour cents/kWh
1:00   2.1203
2:00   2.0594
3:00   2.0235
4:00   2.0263
5:00   2.0278
6:00   2.0213
7:00   2.1258
8:00   2.1257
9:00   2.1637
10:00 2.2745
11:00 2.4378
12:00 3.3367
13:00 4.9845
14:00 8.0461
15:00 11.3748
16:00 12.5938
17:00 12.4085
18:00 11.2554
19:00 8.9338
20:00 7.2654
21:00 6.3741
22:00 4.9265
23:00 3.3319
24:00 2.2695

Average price = 5.0211 per hour at end of interval.

Figure 8.4 Email messages can be used to cue users to change their behaviour to save energy

In making operational adjustments to existing buildings, operators must be aware of the systems-level impacts that can result from adjustments to operational schedules. For instance, while it might seem to be a good financial decision to reduce or shut down ventilation and air-conditioning systems during unoccupied hours of a building, in humid climates this can result in disastrous impacts on indoor air quality as mould develops in warm, high-humidity interior conditions. Every proposed operational change or system retrofit should be carefully considered in terms of its effects on related systems and likely occupant responses before proceeding. This will help to ensure that high performing buildings continue to perform well over time.

Green product procurement

A second major area of sustainable facility operations is ensuring that all materials, resources and products procured for ongoing facility operations and maintenance meet sustainability goals. Depending on the specific type of facility being considered, there may be significant material or energy flows not directly associated with the facility itself, such as in a manufacturing facility. In these cases, environmental audit procedures such as those covered under the ISO 14000 Environmental Management standard can be used to identify improvement opportunities for material substitution and procurement from more sustainable sources.

For material and energy flows associated with facility operations, it is essential to ensure that products are being procured from the most sustainable source available and delivered to the facility in the most sustainable fashion possible. One key resource required by nearly all facilities is energy. The sustainability of power from outside sources can be improved in two major ways: by purchasing power from green-certified power providers who generate that power from renewable sources; and by purchasing so-called 'green tags' or carbon offsets to offset the negative impacts of power produced from conventional sources. One well-known green power certifying program, Green-e (www.green-e.org), is a programme of the Center for Resource Solutions (Figure 8.5). To be certified under this programme, power producers can apply for renewable energy certification if their power portfolio contains a sufficient proportion of power from certain renewable sources. They can also qualify for green pricing programme certification or for competitive electricity product certification, based on their market structure.

Figure 8.5
The Green-e certified power logo

For other types of products associated with building operations, green procurement practices can help to ensure that products are obtained from the most sustainable source and are as environmentally friendly as possible. As with construction materials, there are multiple considerations to take into account, including recycled content, recyclability, rapid renewability and non-toxicity. Products may be specified using the same approaches as in earlier phases, such as avoiding materials on the Red List or requiring third-party certification. These factors apply primarily to the products themselves. Product packaging and delivery pose another set of considerations, including using products that are local or regional to strengthen local economies and reduce transportation impacts. For instance, key factors in the US Department of Defense's Green Procurement Requirements include:

- products manufactured from recovered/recycled materials
- environmentally preferable products
- energy- and water-efficient products
- biobased products
- alternative fuels and fuel-efficient vehicles
- non-ozone-depleting substances.

Figure 8.6 Fluorescent lamps are one common building waste stream covered under Universal Waste Rules

Figure 8.7 Recycling receptacles can be designed to help users correctly segregate waste

Third-party standards such as Energy Star, WaterWise and GreenSeal are available to serve as a basis for green purchasing of products for facility operations. These standards and others are described in more detail in Chapter 4.

Green waste management

At the end of the product lifecycle, proper recovery and disposal of products at the end of their service life is also critical. Products such as analogue thermostats, fluorescent lamps and ballasts often contain hazardous components that could have significant negative environmental, health or safety impacts if handled improperly, especially when produced in large quantities. The Universal Waste Rule in the United States governs the proper disposal of these types of products, since they are generally produced in quantities below which laws associated with hazardous waste handling and disposal apply (Figure 8.6).

Recycling or otherwise diverting other types of waste is also a priority. Providing appropriate facilities for recycling can help direct user behaviour and result in high-quality recycled material waste streams. For instance, the containers shown in Figure 8.7 help users place materials in the proper receptacle to prevent cross-contamination. Along with user education programmes, these measures can help to meet operational recycling goals.

In some areas, user separation of recyclables has been replaced by off-site sorting, where either manual or automated processes are used to quickly and consistently extract materials for recycling from a commingled waste stream. Commingled recycling reduces confusion and the knowledge burden on the public. It is thought to significantly increase recycling rates even though it can also increase the amount of material collected that does not meet recycling requirements. Commingling also allows waste handlers to respond more quickly to market changes for new materials, since these materials may already be coming in as part of the commingled waste stream and public program changes do not need to be made to obtain them.

Green housekeeping, comfort and environmental quality

After a facility is occupied, green housekeeping practices can help to maintain a healthy indoor environment. The manufacturer's instructions should be followed for maintenance of all finishes. For instance, carpets should be vacuumed or cleaned at the recommended intervals, and can be kept in proper shape by hiring a maintenance contractor to

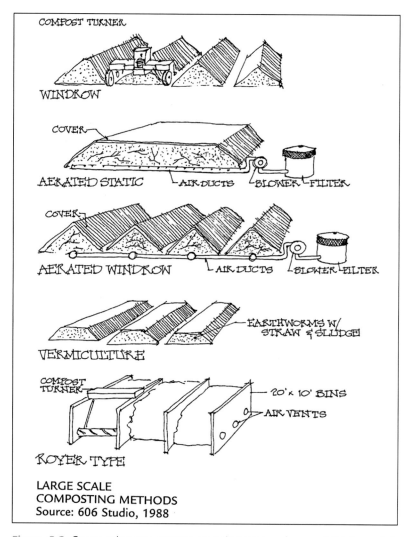

LARGE SCALE
COMPOSTING METHODS
Source: 606 Studio, 1988

Figure 8.8 Composting can capture organic waste and convert it into useful material for landscaping

perform this function. It is critical to read labels on cleaning chemicals and choose non-toxic, water-soluble cleaners wherever possible. GreenSeal (www.greenseal.org) is an organization that has numerous standards for non-toxic cleaning products. Trade associations such as the Carpet and Rug Institute (www.carpet-rug.org) also maintain standards and recommendations associated with green cleaning procedures for related products.

Building humidity levels should be carefully monitored and maintained, with relative humidity levels between 30 per cent and 50 per cent in all occupied spaces of the building. During summer, cooling coils may require maintenance to ensure that they are properly de-humidifying incoming air. All HVAC equipment should be regularly inspected and properly maintained to identify and fix problems before they create unhealthy conditions.

In addition to measures taken inside the facility, low-impact maintenance practices can also be employed for the site and exterior of the building. These practices may include using low-VOC paints and sealers; using non-toxic cleaners; irrigating landscape features only as necessary; minimizing the use of fertilizers, herbicides and pesticides (see *Pest control challenges in sustainable buildings*); composting landscape waste (Figure 8.8); creating wildlife habitat in the landscape; using mulching mowers to return nutrients to the landscape; and avoiding or removing invasive plant species.

Pest control challenges in sustainable building

Benjamin D. Chambers

In any building, there exists the possibility of invasion by insect pests. As long as a building provides food, shelter or warmth, it is a target. Pest prevention and control are achieved through a variety of methods, in both the design and management phases. Many of these remain applicable to sustainable building practice, and the popularity of building guidelines and certification programs mean that these can be increasingly adopted at the design phase. However, changing the ways in which building systems function makes new challenges appear. Current research on a species of insect that frequently spends winters in buildings has suggested some such challenges. This case study discusses some of the background and motivations for that research, and the implications for pests in green buildings.

The brown marmorated stink bug

The brown marmorated stink bug (BMSB) (*Halyomorpha halys*) (Figure 8.9) is an insect that is well known for invading homes in the fall, and spending winter and spring annoying occupants. As fall begins, BMSB begin to move from feeding grounds towards overwintering sites. When they find buildings, they can be observed in high numbers on the exteriors, on foundations, walls, eaves, windows and doors (Hoebeke and Carter 2003). Over the course of several weeks, they begin to find their way inside. They seek dark, close spaces in which to hide. Some of the literature suggests that warmer spaces are preferable (Kobayashi and Kimura 1969). Throughout the colder months,

continued

Figure 8.9
BMSB on an interior wall

BMSB will leave their hiding places and wander around building interiors. They leave their droppings and their dead all over, and their attraction to light bulbs (Aigner and Kuhar 2014) inspires flights around rooms.

This species originated in East Asia, where it has long been a nuisance. In the late 1990s, it was identified as an invasive species in the eastern United States. Since then, its range there has grown, and it has been found in several other parts of the world, possibly because of BMSB electing to overwinter in shipping containers. BMSB attack a wide variety of crops and have caused significant economic damage in the US since their arrival (Leskey et al. 2012). Therefore, the survivability of the species is of economic interest. It is still unknown how much the habit of invading buildings increases the winter survival of the species. They are known to overwinter in leaf litter, fallen trees, and dry mountain terrain (Lee et al. 2013, 2014), but their cold and chill intolerance (Cira et al. 2016; Kiritani 2007) suggest that buildings could play a major role in maintaining population size during colder winters.

Design and management for pest control

Sustainable building eliminates some conventional pest control challenges while also introducing new ones, and guidelines and reviews have been written with this in mind. The examination of a building's impacts over its lifetime means that designers and managers are seeking ways to reduce application of chemical control methods, through more advanced design and monitoring approaches. This has begun to appear in green building certification systems; for instance, the Leadership in Energy and Environmental Design (LEED) v4 for Existing Buildings certification now offers credit to buildings which have integrated pest management plans for both design and control (USGBC 2016). While credit guidelines are by necessity broad, they are based at least in part upon a detailed set of green building pest management guidelines produced by the city of San Francisco, based upon peer reviewed scientific literature (Geiger and Cox 2012).

A large portion of the recommendations deal with below-grade exclusion of termites, which are very small. Above grade, in traditional as well as green building guidelines, exclusion generally focuses on a rule of thumb of 1/4 in (6 mm), which is small enough to keep out mice. Any holes larger than that are to be treated in some way, with 1/4 in hardware cloth used for large openings. Unfortunately, a gap of that height is quite large enough for BMSB to fit through, provided that the gap is also wide. In the case of a smaller overwinter invader, ladybird beetles (ladybugs), guidelines call for 1/16 in (1.6 mm) mesh (Layton 2014), which is supported by the scientific literature (Nalepa 2009). Other exclusion guidelines suggest screen or mesh over all ventilation holes high on the building, such as in attics, in order to keep out wasps (Around the Clock Pest Control 2015; Beyond Pesticides 2015). Not all guidelines mention the very important fact that adding screening to a vent that was not designed for it can significantly reduce the effective area of that opening, and thus its functionality.

Gaps and cavities

The opportunistic nature of BMSB in seeking overwintering sites, combined with anecdotal evidence, suggests that wall cavities may be regularly used for winter shelter. In a poorly insulated or leaky home, these cavities might remain warm enough that BMSB inside enjoy high winter survival rates.

continued

However, in tightly sealed and well insulated homes, this may not be the case. Simple models of the thermal profiles of walls suggest that relatively minor improvements to insulation could drop cavity temperatures enough to significantly increase winter mortality rates.

Since BMSB tend to enter spaces in large numbers, increases in mortality have the potential to significantly affect the spaces in which they die. In a cavity behind a rain screen, a heap of dead bugs might collect and hold moisture, eventually leading to potential mold problems. If BMSB find their way into a cavity with weep holes or other small apertures, even a few fallen dead bugs could obstruct those holes, slowing ventilation and potentially creating moisture problems.

Thermal contrast

In green buildings, the goal of reduced energy usage often means reducing heat and air exchange between the building and the surrounding environment. The result is tighter building envelopes with fewer and smaller gaps, and more effectively insulated walls and windows. Tightening and eliminating gaps means fewer ways in for pests, and those that remain are more difficult for larger insects such as BMSB to access. However, tighter and better insulated envelopes also mean that the remaining gaps and thermal bridges have much higher thermal contrast with the building than would be present in a leakier envelope. Therefore, it is possible that invading species may actually have an easier time locating points of ingress.

References

Aigner, J.D., and Kuhar, T.P. (2014). "Using citizen scientists to evaluate light traps for catching brown marmorated stink bugs in homes in Virginia." *Journal of Extension*, (Awaiting Publication).

Around the Clock Pest Control. (2015). "Bee & Wasp Control – Around The Clock Pest Control." http://around theclockpest.com/bee-and-wasp-control/ (20 Nov 2015).

Beyond Pesticides. (2015). "Wasp and Yellowjacket Control Factsheet." www.beyondpesticides.org/assets/media/ documents/alternatives/factsheets/Wasp Control2.pdf (20 Nov 2015).

Cira, T.M., Venette, R.C., Aigner, J., Kuhar, T.P., Mullins, D.E., Gabbert, S.E., and Hutchinson, W.D. (2016). "Cold tolerance of *Halyomorpha halys* (Hemiptera: Pentatomidae) across geographic and temporal Scales." *Environmental Entomology*, 1(8).

Geiger, C.A., and Cox, C. (2012). *Pest Prevention by Design: Authoritative Guidelines for Designing Pests out of Structures*. San Francisco.

Hoebeke, E.R., and Carter, M.E. (2003). "*Halyomorpha halys* (Stal) (Heteroptera: Pentatomidae): A polyphagous plant pest from Asia newly detected in North America." *Proceedings of the Entomological Society of Washington*, Entomological Society of Washington, Washington, 105(1), 225–237.

Kiritani, K. (2007). "The impact of global warming and land-use change on the pest status of rice and fruit bugs (Heteroptera) in Japan." *Global Change Biology*, Blackwell Publishing Ltd., Oxford, UK.

Kobayashi, T., and Kimura, S. (1969). "Studies on the biology and control of house-entering stink bugs." *Bulletin of the Tohoku Agricultural Experiment Station*, 37, 123–138.

Layton, B. (2014). "Physical Exclusion: the Best Treatment for Home-Invading Insect Pests." *Mississippi State University Bug-Wise Newsletter*.

Lee, D.-H., Cullum, J.P., Anderson, J.L., Daugherty, J.L., Beckett, L.M., and Leskey, T.C. (2014). "Characterization of Overwintering Sites of the Invasive Brown Marmorated Stink Bug in Natural Landscapes Using Human Surveyors and Detector Canines: e91575." *PLoS ONE*, Public Library of Science, San Francisco, 9(4).

Lee, D.-H., Short, B.D., Joseph, S.V, Bergh, J.C., and Leskey, T.C. (2013). "Review of the Biology, Ecology, and Management of *Halyomorpha halys* (Hemiptera: Pentatomidae) in China, Japan, and the Republic of Korea." *Environmental Entomology*, Entomological Society of America, Lanham, 42(4), 627–641.

Leskey, T.C., Dively, G.P., Hooks, C.R.R., Raupp, M.J., Shrewsbury, P.M., Krawczyk, G., Shearer, P.W., Whalen, J., Koplinka, C., Myers, E., Inkley, D., Hamilton, G.C., Hoelmer, K.A., Lee, H., Wright, S.E., Nielsen, A.L., Polk, D.F.,

continued

Rodriguez, C., Bergh, J.C., Herbert, D.A., Kuhar, T.P., and Pfeiffer, D. (2012). "Pest Status of the Brown Marmorated Stink Bug, Halyomorpha Halys in the USA." *Outlooks on Pest Management*, Research Information Ltd., Saffron Walden, 23(5), 218.

Nalepa, C. A. (2009). "*Harmonia axyridis* (Coleoptera: Coccinellidae) in buildings: Relationship between body height and crevice size allowing entry." *Journal of Economic Entomology*, Entomological Society of America, 100(5), 1633–1636.

USGBC (2016). "Integrated Pest Management." *LEED O+M: Existing Buildings LEED v4*, www.usgbc.org/credits/existing-buildings-schools-existing-buildings-retail-existing-buildings-hospitality-exist-26 (27 Jun 2016).

Green renovations and indoor air quality

Any type of construction or renovation activity in an existing building has the potential to cause indoor air quality problems if not properly managed. Many construction activities such as cutting or sanding produce dust or airborne particulates. Other activities involve installing materials that release VOCs or moisture while they cure – like paint, sealants, adhesives, concrete and new lumber. Even with dust collectors installed on equipment, there is a need to protect building systems from absorbing these pollutants during construction. Proper protective measures can also relieve construction workers from having to constantly worry about dust and pollutants, speeding up construction and saving time and money (Figure 8.10).

At a minimum, construction isolation measures should include protection of all HVAC intakes, grills and registers. Use plastic sheeting and duct tape or ties to ensure that all possible points of entry to building ductwork or plenums are sealed off after first deactivating the HVAC system (Figure 8.11). Also isolate entryways to other parts of the building and stairwells or corridors outside the work areas.

Figure 8.10 Proper protection of work areas, unlike the example shown in this photo, can ultimately speed up construction and save both time and money

Figure 8.11 Plastic sheeting can be used to isolate ductwork and return air plenums during renovations

When possible, try to sequence the installation of absorptive materials to occur after dust-producing activities have taken place. Absorptive materials include but are not limited to carpet, wall coverings, fabrics, ceiling tiles and many types of insulation. If this is not possible, be sure to use plastic or paper sheeting securely taped in place to protect all absorptive surfaces.

Proper ventilation during construction should be maintained to ensure worker health and safety and to exhaust pollutants from the workspace. If possible, continuous negative pressure exhausted through a filtration system to the outdoors should be maintained within the workspace to prevent dust and contaminants from migrating to other parts of the building. Ventilation should be provided by a temporary ventilation system to avoid contaminating building ductwork with pollutants.

Regular housekeeping during construction will help to keep contaminants under control and prevent them from becoming airborne. Specialty contractors can be retained for this purpose, or work practices can be put in place to require workers to keep their areas clean and free of debris. All construction isolation measures should be regularly inspected for leaks and tears and should be replaced as needed.

At the completion of construction, all masking and sheeting used to isolate ventilation systems should be removed before restarting HVAC systems within the isolated area. Temporary high-efficiency filters may be used in the HVAC system during some construction activities and before building occupancy to trap remaining contaminants and protect them from entering the distribution system. These filters should be replaced prior to occupancy and properly disposed. The US Environmental Protection Agency (USEPA) recommends a two-week flush-out period for all areas in which construction has occurred prior to occupancy. During the flush-out period, ventilation systems should be run continuously with full outdoor air within a specified humidity and temperature range. While flush-out helps to accelerate the removal of offgassing from building materials, it can also introduce large amounts of humidity into the building depending on the climate and season. This humidity will supplement the already high humidity rates from the curing of materials inside the building itself. If a building flush-out is to be undertaken, operational guidelines for all HVAC systems should be consulted to ensure that humidity controls are able to handle the extra humidity load that may occur.

Material and equipment refurbishment and reuse

Another key strategy for green operations and maintenance is to avoid the consumption of new raw materials and energy by refurbishing or reusing existing elements in projects. Not all types of technologies are suitable for reuse – for instance, fixtures, systems and appliances that consume water and energy are often better to replace than reuse since newer technologies are generally far more efficient. Likewise, key building envelope elements such as windows may be better off replaced than reused, although in some situations window units can be retrofitted for enhanced performance (see *Case study: Empire State Building*, p. 461) or are even required to be maintained to meet historic preservation requirements.

Whether building envelope elements should be reused or not is dependent upon many factors including the climate in which the building is located, the conditioning requirements of interior spaces, the historic value of the building's appearance and other factors. Some facility and infrastructure systems are quite suitable for refurbishment or even direct reuse, and this can offer one of the more significant cost advantages of renovating existing buildings and infrastructure instead of constructing new facilities. For example, the roll-up doors on equipment bays in Homestead ARB's Fire Station project (see *Case study: Homestead Air Reserve Base*) were well-suited for reuse instead of replacement due to their unusual size and the fact that they were being used to enclose an area where thermal comfort expectations were low.

Case study: Homestead Air Reserve Base Fire Station, FL

This renovation/expansion project completed in 2002 is a classic example of turning lemons into lemonade. The original plan involved constructing an entirely new building, but budget constraints during the funding process made this option infeasible. To fit within the lower budget, a decision was made to adapt and expand the existing fire station to better serve the needs of the installation's fire fighters. A general contractor was hired based on willingness to commit to learning about and using sustainability principles, and a major design firm was selected as the architect/engineer due to its prior experience in sustainable design. The overall sustainability goals for the project were communicated during a charrette involving not only installation personnel, but also headquarters personnel including the command civil engineer, who took an active interest in the project.

The goal of lowering utility costs was achieved by using light shelves, low-e glass, clerestory windows, a high-efficiency HVAC system, photoelectric site lighting and shading the building with landscape plants. Although some of these measures increased the first cost of the project, increases were offset with savings in other areas, including a smaller HVAC system that could be substituted for the original capacity due to reduced heat loads from lighting. The photovoltaic site lighting was no more expensive overall due to the savings in energy supply lines.

Indigenous plants were used for landscaping to reduce water use, and stormwater is captured and retained through a percolation trench. The existing storm sewer in this area of the installation was already overtaxed, so use of the percolation trench also avoided the expense of having to upgrade the sewer system. Since a major use of water in fire stations is for washing vehicles, a vehicle wash water recirculation system was also included to minimize water demands.

Two other goals were to use recycled materials and to recycle as much waste material from demolition and construction as possible. The contractor was able to recycle nearly 100 per cent of the waste materials from demolition, saving substantial money over landfill tipping fees. As a result of experience on this project, the contractor began recycling waste on other jobs as well. On-site waste separation led to better housekeeping practices on the project, resulting in zero accidents due to a cleaner job site. Savings resulting from avoided waste disposal costs were used to offset the higher costs of using recycled materials in construction.

continued

Figure 8.12 The existing fire station was renovated and expanded to meet new requirements at Homestead Air Reserve Base

Figure 8.13 Demolition involved careful preservation of key structural elements

Figure 8.14 Nearly 100 per cent of demolition waste was recycled through on-site separation, and good housekeeping practices resulted in a project with zero injuries

Figure 8.15 Mature landscape was carefully preserved to provide passive shading

Figure 8.16 A high-albedo roof membrane was used to reduce heat gain and lower cooling costs

Figure 8.17 A percolation trench captures stormwater runoff and avoids overloading a treatment system already nearing capacity

Source: All photos courtesy of Bill Cadle, U.S. Air Force Reserve Command

At the infrastructure scale, system components may be custom-made to fit one-of-a-kind projects, and refurbishment may be a much more attractive option from both an environmental and economic standpoint than replacement. For example, the Chicago Red Line rehabilitation project (see *Case study: Chicago's Red Line rehabilitation*) required refurbishment of signalling and switchgear components for the project as part of the scope of work. This project also repurposed the existing third rail used to supply power for propulsion to trains as an auxiliary rail, replacing it with a new rail. This required half the materials compared to replacing both rails, although it also required very creative construction methods including the invention of a new kind of transport trolley to move long rail sections for storage during the project.

Case study: Chicago's Red Line rehabilitation, IL

Rehabilitation of the Red Line South Branch (Figure 8.18), a major section of the Chicago Transit System, was undertaken in 2012 to eliminate a series of slow zones that limited trains in this part of the network to speeds of no more than 15 miles per hour. Part of city government's 'Building a New Chicago' program, the project was made possible through state and local funding. The original south Red Line opened on 28 September 1969 and was more than 40 years old at the time of the project. Along with the faster travel times resulting from removal of slow zones, trains are now able to serve this area with higher reliability and lower operating and maintenance costs.

The entire scope of work, including complete system testing, commissioning and return to service after shut-down, was required to be completed within an accelerated 5-month time-span. During this time, the track was shut down completely, inconveniencing the passengers who complete roughly 80,000 journeys per day on this line. Ten hour work days were used, with contractors working six or seven days per week. The project was completed between May and October, with the majority of tasks completed in the middle of summer at a point when heat stress/heat exhaustion was a serious concern.

The logistics of the project were very challenging, with the 35-inch-wide project site located between north and southbound lanes of the Dan Ryan Expressway (Figure 8.18). Very limited access to the site was available, requiring pre-sorted material deliveries at night so that work could continue the next day. Multiple contractors were required to work in the same areas at the same time to complete the necessary scope of work, which further complicated the project.

Red Line rehabilitation

- Project Type: Transit reconstruction
- Location: Chicago, IL
- Contracting Method: Design-Bid-Build
- Owner: Chicago Transit Authority
- Duration: 5 months

- Overall Cost: $200 million
- Stakeholders Involved: Owner, General Contractor, numerous Subcontractors, City of Chicago, Traveling public

The scope of work included signals, traction power and communication systems along with track work including tracks, ties, ballast and drainage systems. Over 20 miles of new duct banks, new

continued

Figure 8.18 Project site

Figure 8.19 Trolley used to move long sections of rail

signals, traction power cable and auxiliary negative rail were installed, including a new third rail. The contract required removal, refurbishing and reinstallation of several key existing components of the project instead of replacement with new equipment, including signal and traction power equipment and third rails that were to be repurposed as auxiliary negative rails. The reuse of existing materials and systems was a major sustainability feature of the project. New third rails were installed to replace the repurposed rails and provide power to trains using the line. New elevators were also installed at all stations, improving the accessibility for passengers and bringing existing stations into compliance with the Americans with Disabilities Act (ADA).

Project contractors faced the challenge of finding the best way to meet the refurbishment requirements while working safely and efficiently in a very constrained project site. Particularly challenging were questions of how to cut and move sections of third rail and where to store this material while work was underway to refurbish the line's foundations. Coordination among project participants was essential given that the project work area was surrounded on both sides by active traffic. Communication was particularly important to ensure that no workers were trapped during lifting and transport of rail sections. After considering the pros and cons of cutting the rails into smaller pieces for easier handling vs. keeping larger sections intact to improve quality, a special trolley was designed that could move long rail sections without damage to storage areas at various points along the line (Figure 8.19).

From a social sustainability standpoint, the public works contract for this project required that at least 15 per cent of the labour force for the project be hired from local neighbourhoods, and 30 per cent of the total contract value had to be completed by Disadvantaged Business Enterprises (DBEs) as required by the Workforce Investment Act. Each major contractor on this project had to individually meet the requirements to ensure that the project as a whole remained in compliance (Figure 8.20). Special hiring practices and training programs were developed to ensure that work could be completed safely using workers inexperienced with the specialized equipment and skills required.

Although sometimes perceived as a burden, these affirmative labour requirements were considered by the project team to be beneficial from the standpoint of the project's safety culture. Local workers not only have personal relationships with each other, but can also bring additional

continued

Figure 8.20 Red Line Project team

insight through their relationships with surrounding neighbourhoods and cultural context. These relationships can facilitate effective communication among workers and encourage them to take personal responsibility for themselves and others. The Incident- and Injury-Free (IIF) principles used in this project aligned well with the goals of social sustainability by supporting local economic development while ensuring worker health, safety and well-being.

References

Besenger, E., Bradley, K., Khatri, S., Mills, O., Pearce, A., Seahorn, R., and Walker, R. (2015). "Case Study: Aldridge Electric's Red Line Rehabilitation." *Proceedings, Safe SWAG: Developing Safer Sites, Workers, Attitudes, and Goals Workshop*, Arlington, VA.

Freeman, Y. (2012). "Chicago plans to shut Red Line South to perform quick rehab." *The Transport Politic*, 4 June. http://thetransportpolitic.com/2012/06/04/chicago-plans-to-shut-red-line-south-to-perform-quick-rehab/ (accessed 2 January 2017).

Source: All photos courtesy of Aldridge Electric.

Other post-occupancy tactics

Ensuring that a green project remains green throughout its lifecycle requires attention not only to the project itself, but also to its occupants and people responsible for its operation and maintenance. Incentive programmes, education and training, and signage and information programmes can all achieve this end.

Owners can provide incentives and amenities that encourage facility users and occupants to use alternative means of transportation for getting to work, such as walking, car-pooling, taking a bus or train, or bicycling. They can also encourage practices such as telecommuting if appropriate for the type of work being done. Signage and information programmes, ranging from email or social media updates discussed earlier in this chapter to the educational signage shown in Figure 8.21, to real-time monitoring dashboards showing building performance, can also alert users not only to what has been done and how they should respond, but also why those measures were taken and why they should care. When coupled with a participatory design process, such measures can help facility operators and users take ownership for the systems in their facilities and keep them running smoothly. Some facilities provide occupant manuals or websites to ensure that occupants can find answers to their building-related questions over time.

Regular training for operations and maintenance personnel is also critical to ensure that the facility remains functioning in its optimal state. Especially for facilities where new technologies are employed, proper training and guidance in operations is essential. In some cases, it may make sense to outsource the maintenance function for complex systems such as building automation and controls. Ongoing training, in conjunction with proper reference material, as-built drawings, or building information models, is also essential, especially in situations where there is high turnover of building maintenance personnel.

Figure 8.21
Building signage can be used to explain innovative building features and guide occupant behaviour

Case study: Trees Atlanta Kendeda Center – operations

As a building with significant sustainability goals during design and construction, the Trees Atlanta building continues to set an example of sustainability best practice during the operations and maintenance phase of its lifecycle. The following subsections describe practices currently in use in the Trees Atlanta building to optimize energy and water performance, procure green materials and ensure comfort and a high-quality indoor environment for its occupants.

Optimizing energy and water performance

Optimizing energy and water performance of an existing facility is a critical strategy to achieve the goals of sustainability, not only to reduce annual utility costs but also to reduce the facility's carbon and water footprints. Lower energy and water consumption within the facility reduces greenhouse gas emissions from electricity generation, which accounts for the largest portion of US greenhouse gas emissions.

Three primary tactics – operational adjustments, retrofit of energy- and water-consuming equipment, and working with users to adjust their behaviour – are used by Trees Atlanta to achieve their goals. First, Trees Atlanta has a sustainable building operations and maintenance meeting every six months to review its monthly utility bills, including water and electricity bills. By comparing the building's monthly usage with building energy simulation and water-saving calculations, as well as with previous levels of consumption, it is possible to measure and verify the building's sustainable performance and to monitor the changing effects of occupants' behaviours.

continued

Second, installed systems and the building are continuously monitored to ensure that performance requirements are being maintained. Continuous commissioning involving a commissioning agent is also periodically undertaken to verify proper operation of all equipment and ensure that it is functioning correctly and as efficiently as possible compared to the as-built functional test procedures contained in each system manual (see *Continuous commissioning: commissioning agent responsibilities in the post-occupancy phase*).

Continuous commissioning: commissioning agent responsibilities in the post-occupancy phase

- Coordinate deferred and seasonal functional tests; verify correction of deficiencies.
- Periodically conduct on-site review with owner's staff:
  - Review the current facility operation and condition of outstanding issues related to the original and seasonal commissioning.
  - Interview staff to identify problems or concerns they have operating the facility as originally intended.
  - Make suggestions for improvements and for recording these changes in the O&M manuals.
  - Identify areas of concern that are still under warranty or are the responsibility of the original construction contractor.
  - Assist facility staff in developing reports, documents and requests for services to remedy outstanding problems.

Third, the Trees Atlanta building has regular operator training sessions, primarily because the nature of the organization is that it consists of many volunteers. Given that the building has green features that require regular inspection and maintenance, including a geothermal heat pump system with programmable thermostats, lighting occupancy controls, and a solar hot water heater and rainwater collection system for toilet flushing, there is a need to ensure that many people know how to properly operate the building's features. Therefore, the contractor of the Trees Atlanta building has developed an operations and maintenance manual that was verified by both an architect and a commissioning agent. In addition, the contractor was required to provide operations and maintenance training for all employees and several volunteers with the fully developed operations and maintenance manual. The commissioning authority also verifies with the owner that the operations and maintenance training is adequate and addresses their needs to properly operate and maintain the building. Finally, the commissioning officer of the project developed an annual operation and maintenance education programme for each member of staff and several volunteers. The education programme provided information on building and building systems operation, maintenance, and achieving sustainable building performance.

Green materials procurement

One of the major areas of sustainable facility operations is ensuring that all materials, resources and products procured for ongoing facility operations and maintenance meet sustainability goals. One key opportunity in green procurement is to purchase green certified power from outside sources. Green power improves sustainability because the production of certified electricity reduces air pollution impacts of electricity generation by relying on renewable energy sources such as solar,

continued

**renewable choice
E N E R G Y**

Be Part of the Solution, Not the Pollution
4041 Hanover Ave, Second Floor
Boulder, CO 80305
Phone 303.468.0405   Fax 303.648.4324

Renewable Energy Certificate
Purchase Agreement

Client Name:	Trees Atlanta
Contact Name:	Marcia Bansley
email address:	marcia@treesatlanta.org
Shipping Address:	225 Chester Avenue
	Atlanta, GA 30316
Phone:	404-522-4097
Billing Contact:	Regina Clifton
email address:	regina@treesatlanta.org
Billing Address:	225 Chester Ave.
	Atlanta, GA 30316
Phone:	404-522-4097

Summary of Annual Electric Use (kWh)

Project Name: Trees Atlanta	
Location: Atlanta, GA	
Estimated Annual kWh:	39,790 kWh

SALESPERSON	PO NUMBER	ORDER DATE
Jason Wykoff		
PAYMENT TERMS		
Net 30		

Annual Environmental Impact of 100% offset
Avoided Emissions:
        55,388 Pounds of CO2

Has the same environmental impact as:
        Not Driving: 60,467 miles in the average car
        Planting: 7 acres of fully mature trees

Purchase Details:

Option	Total kWh	Description	PRICE ($/kWh)	TOTAL AMOUNT	Initials
70% offset for 2 years to meet LEED E&A Credit 6 and ID point	79,580	Green-e certified American Wind™	$0.0080	$ 636.64	

The undersigned agrees that he/she is authorized to engage in an agreement on behalf of the company below.

By signing, the company agrees to the purchase amount and terms outlined on this page.

Confidential and Proprietary Information

American Wind™ is Green-e certified.

MARCIA D. BANSLEY
        Printed Name

Marcia D. Bansly   June 18, 07
        Signature / Date

TREES ATLANTA   Executive Director
        Company Name / Title

Order Acceptance: the undersigned agrees that he/she is authorized to accept this order and the terms outlined on this page on behalf of Renewable Choice Energy.

                          1/26/07

Aran Rice Director of Operations, Renewable Choice Energy

PRODUCT CONTENT LABEL

This is a renewable certificate product. For every unit of renewable electricity generated, an equivalent amount of renewable certificates is produced. The purchase of renewable certificates supports renewable electricity generation, which can help offset conventional electricity generation in the region where the renewable generator is located. You will continue to receive a separate electricity bill from your utility.

The product will be made up of the following new renewable resources averaged annually.

New[1] Renewable Resources in *American Wind*™		Generation Location
Wind	100%	Nationwide

[1]Includes renewable generators that first started operating after January 1, 1997 or as regionally defined.

For comparison, the current average mix of energy sources supplying the U.S. includes the following: Coal (52%), Nuclear (20%), Oil (3%), Large Hydroelectric (7%), Natural Gas (16%), and Renewables (2%). Source: from USEPA E-Grid. For specific information about this product contact Renewable Choice Energy at 303-468-0405 or on the web at www.renewablechoice.com.

This product is certified by the Green-e Program. For more information, call 888-63-GREEN or visit www.green-e.org.

Questions? Contact us
        303.468.0405
        www.renewablechoice.com

**AMERICAN
WIND**

100% Renewable

Figure 8.22 Purchasing power from green-certified power providers can reduce environmental problems in the built environment. *Source: Smith Dalia Architects*

continued

water, wind, biomass and geothermal sources instead of fossil fuels. Conventional energy production from traditional sources such as coal, natural gas and other fossil fuels, on the other hand, is a significant contributor to air pollution and negative environmental impacts, especially in the United States and other countries heavily relying on coal for energy purposes. The Trees Atlanta building purchased its green power credit of 79,580 kWh from Renewable Choice Energy in Boulder, Colorado (Figure 8.22).

To reduce the environmental and air quality impacts of the materials acquired for use in the operations and maintenance of the Trees Atlanta building, the Trees Atlanta staff purchases green materials, supplies and equipment for ongoing operations, including:

- recycled paper products
- sustainable cleaning products
- remanufactured toner cartridges
- rechargeable batteries
- recycled binders.

Green waste management including a recycling programme has been a major focus of Trees Atlanta. The director of Trees Atlanta believes that recycling can change Americans' attitudes and behaviour toward sustainability. Thus, Trees Atlanta has a strong recycling programme under the supervision of the director.

Comfort and indoor environmental quality

Regular housekeeping is the key to improving indoor environmental air quality within the Trees Atlanta building. Thus, staff and the janitor of the building diligently maintain the building using stringent cleaning guidelines and requirements. The copy room has a separate ventilation system to maintain the area under negative pressure and exhaust dusts and airborne contaminants generated by printing and copying activities. This separate ventilation system contributes to a high-quality environment that helps occupants enjoy working in the building. Figure 8.23 shows the return air vent for the separate ventilation system in the copy room to maintain negative air pressure that can sustain high indoor air quality.

Figure 8.23 This ceiling vent is part of a separate ventilation system used to maintain negative pressure in the copy room to prevent contaminants from entering other parts of the building

While in operation, the Trees Atlanta Kendeda Center demonstrates a variety of tactics in keeping with its organizational mission and building use. The second case study of the Empire State Building energy-efficiency retrofit illustrates a very different set of considerations that are relevant for an urban high-rise building in one of the most populous cities of the world.

Case study: Empire State Building energy-efficiency retrofit, New York, NY

A second example of sustainability activities post-occupancy is the Empire State Building (EBS). The EBS is no ordinary office tower. It is one of the world's most famous office buildings, located in the heart of New York City (Figure 8.24). Each year, it attracts between 3.5 million to 4 million visitors to the Observatory on the 86th floor. As the third tallest building in New York City, with a height of 1454 ft (443.2 metres), the ESB's spire is used for broadcasting by most of the region's major television and radio stations. The ESB has 12.8 million sq ft of leasable office space that houses a variety of large and small tenants including Skanska, the Federal Deposit Insurance Corporation, Taylor Communications and Funaro. The building first opened in 1931 and has undergone upgrades of lobbies, hallways and other common areas including renovation of the observation deck, restoring the building to its original grandeur.

Since the ESB is one of the largest buildings in the world, it has a significant impact on the environment as a result of the process of its operations and maintenance. According to the building's website www.esbnyc.com/esb-sustainability prior to 2008, the performance of the ESB was as follows:

- Annual utility costs: US$11 million (US$4/sq ft)
- Annual carbon dioxide emissions: 25,000 metric tons (22 lbs/sq ft)
- Annual energy use: 88 KBtu/sq ft
- Peak electric demand: 9.5 mW (3.8 W/sq ft, including HVAC).

To reduce negative environmental impacts, efforts to make the ESB and similar buildings more environmentally sustainable have resulted in significantly greener office spaces. However, most sustainability-related efforts worldwide have focused on new building construction, while the

Empire state building

- Location: 350 Fifth Avenue, New York, NY 10118.
- World's tallest building from 1931 to 1973.
- Construction period: 1929 to 1931 (excavation of the site on 22 January 1930).
- Building cost: US$40,948,900 (1931 dollars).
- Project duration: 410 days after construction commenced.
- Opening day: 1 May 1931.
- Antenna or spire: 1454 ft (443.2 m).
- Roof: 1250 ft (381.0 m).
- Top floor: 1224 ft (373.2 m).
- Floors: 102 storeys (85 storeys of commercial and office space and 16 storeys of Art Deco Tower, which is capped by a 102nd floor observatory.

- 6500 windows, 73 elevators and 1860 steps from street level to the 103rd floor.
- Floor area: 2,768,951 sq ft (257,211 sq m).
- 1000 businesses with 21,000 employees work in the building each day.
- Architect(s): William F. Lam at the architectural firm of Shreve, Lamb and Harmon.
- The building design was based on the Reynolds Building in Winston-Salem, North Carolina.
- Contractor: Starrett Brothers and Eken.
- Project was financed by John J. Raskob and Pierre S. du Pont.
- Management: W&H Properties.

continued

majority of existing office buildings have made little or no progress in the areas of energy efficiency and sustainability. Because of this opportunity, Anthony E. Malkin of the Empire State Building Company committed to establish the ESB as one of the most energy-efficient and sustainable existing buildings in the world, and used the project to explore the cost-effectiveness of energy-efficient and sustainable retrofits for existing office buildings. Through retrofit of the ESB, it has been possible to reduce greenhouse gases and utility costs to demonstrate how to retrofit a large commercial building cost-effectively, and to demonstrate that such work makes good business sense in the commercial office market. Proving the benefits of energy-efficient and sustainable retrofits in the ESB has resulted in a replicable model for whole-building retrofits for existing offices around the world as well as significant reduction greenhouse gas emissions and operating costs, all while pro- viding state-of-the-art indoor environments to office tenants. As a result, sustainable retrofits of the ESB have enhanced its long-term value based on the opportunity for higher occupancy and rents over time.

Retrofit sustainability team and approach

The desire to sustainably retrofit the ESB was mot- ivated by the owner's goals to reduce greenhouse gas emissions, to demonstrate a way for sustainable retrofitting of such a large commercial building cost- effectively, and to demonstrate that such work makes

Figure 8.24 The Empire State Building

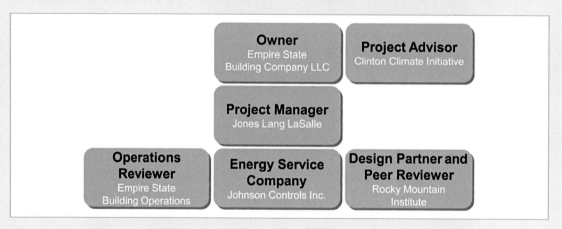

Figure 8.25 Project team members

continued

good business sense. To achieve these sustainability goals, five dedicated project partners were recruited from several key areas of expertise, including environmental consulting, engineering, project management and commercial real estate, to develop a proposal for retrofit that would achieve project goals while making good business sense. The five key partners in the ESB project development process were the Clinton Climate Initiative (CCI), Johnson Controls Inc. (JCI), Jones Lang LaSalle (JLL), the Rocky Mountain Institute (RMI) and the Empire State Building Company (ESB), representing both the owner and operations perspectives (Figure 8.25).

These core project team members brought their expertise to achieve the goal of analysing what actions would make the most sense in sustainably retrofitting the ESB. They also closely collaborated in many different activities including goal setting, brainstorming charrettes, tool development, energy modelling and financial/lifecycle cost analysis modelling (Figure 8.26).

After forming a collaborative project team, stakeholders established action goals to achieve the overall sustainability goals for the project, including:

- developing a replicable model for retrofitting the pre-World War II building in a cost-effective way.
- developing practices to lower energy consumption costs by as much as 20 per cent.
- increasing overall environmental benefits of the building retrofit through an integrated sustainability approach to maximize opportunities and market advantage.
- encouraging the team to be objective, creative and provocative in its approach.
- developing a model that is marketable to existing and prospective clients in its approach.
- coordinating with the ongoing capital projects within the building.
- developing a financial structure that is efficient and achievable.

Motivation for sustainable retrofit of ESB

The goal with ESB has been to define intelligent choices which will either save money, spend the same money more efficiently, or spend an additional sum for which there is reasonable payback through savings. Addressing these investments correctly will create a competitive advantage for ownership through lower costs and better work environments for tenants. Succeeding in these efforts will make a replicable model real for others to follow.

Anthony E. Malkin,
Malkin Holdings

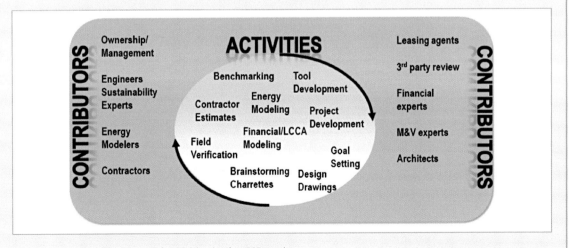

Figure 8.26 Contributors and activities in the ESB project

continued

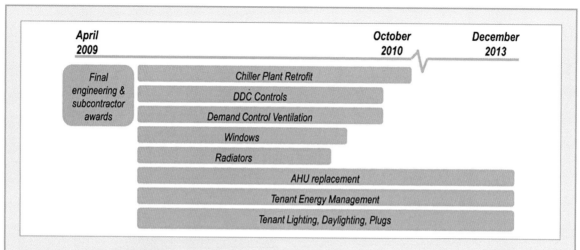

Figure 8.27 Project completion schedule

To achieve these actionable goals, the collaborative project team developed an optimal proposed solution for sustainable retrofit of the ESB through a rigorous and iterative process that involved experience, energy and financial modelling, rating systems, technical advice and robust debate. The ideas proposed as a result of this project were implemented in a retrofit process begun in 2010 and concluded by the end of 2013 (Figure 8.27), with measurement and verification of impacts continuing through 2025.

Solution development process

The solution development process started with assembling a collaborative team of world-class sustainability and energy specialists as core collaborative team members. The team developed a four-phase iterative process and rigorous cost/benefit analysis to design an optimal solution that would lead to significant reductions in greenhouse gas emissions and promote sustainable design and operations in the existing building. To support the project development process, the project team used various industry standards and state-of-the-art design tools (eQUEST, AGI32 and Google Sketch Up), decision-making tools and green building rating tools (Energy Star, LEED and Green Globes) to evaluate and benchmark existing and future performance. Figure 8.28 shows the whole four-phase process used to create a replicable model for the ESB.

At the 'identify opportunities' phase, the project team first fully investigated resource use at the existing ESB by examining energy usage between April 2007 and May 2008. They also reviewed tenants' surveys. Second, the project team brainstormed over 60 energy-efficiency ideas and strategies that might be implementable for the project. They also estimated theoretical minimum energy use to address occupant comfort requirements, passive measures and other system impacts, system design characteristics, technology, controls and changed operating schedules. The identified ideas and strategies were then narrowed to 17 implementable projects (Table 8.1). Next, the project team developed an eQUEST energy model to use for cost/benefit analysis of future improvements, modifications and operational changes (Figure 8.29).

In the second phase of design development, the project team calculated the net present value (NPV) of selected energy-efficiency strategies and also estimated greenhouse gas emission savings for each strategy (Figure 8.30). In addition, the project team estimated dollars per metric ton of carbon reduced by installing each energy efficiency strategy.

continued

Determining the optimal package of retrofit projects involved identifying opportunities, modeling individual measures, and modeling packages of measures.

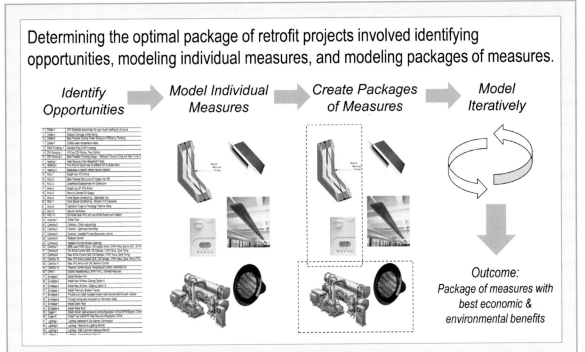

Identify Opportunities ➤ *Model Individual Measures* ➤ *Create Packages of Measures* ➤ *Model Iteratively*

Outcome: Package of measures with best economic & environmental benefits

Figure 8.28 Project development process for the ESB

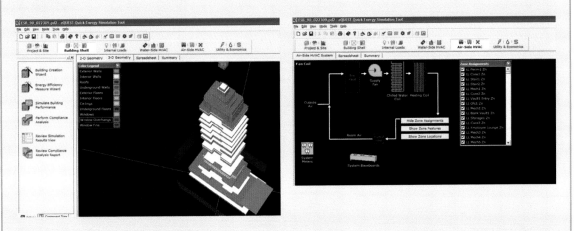

Figure 8.29 Energy modelling tool for energy-efficiency strategies

continued

Table 8.1 Implementable energy-efficiency strategies for the ESB retrofit

Energy efficiency strategies	Description
Building windows (plus t-operative temperature transducer and infiltration); SC75/TC88 heat mirror film	Refurbish all existing double-pane windows (6500 windows). Insert suspended heat mirror film (SC75 or TC88) and krypton/argon gas mix.
102nd floor windows	Add film to existing single-pane windows on 102nd floor event space.
Radiative barrier	Install reflective barrier behind 6500 steam radiators; seal all gaps with foam and insulate the enclosure pockets in which radiators sit.
Green roof/cool roof	Add green roof on 5th floor (20,000 sq ft) and 21st floor (10,000 sq ft); add white roof on 25th floor (6000 sq ft) and 30th floor (6000 sq ft).
Balance of direct digital controls	Install mechanical system controls throughout the building (2nd floor, 2-way valves, outdoor air (OA) control, exhaust).
Tenant demand control ventilation on existing units (OA)	Add carbon dioxide demand control ventilation for all existing constant volume air handling units (485 total) to reduce outside air conditioning needs.
Additional direct digital controls	Add controls for all radiator valves and main steam valves.
LED tower lighting	Replace existing metal halide tower lighting with LEDs.
New chiller plant	Install four 1400-ton chillers with primary variable secondary flow using all new pumps.
Retrofit chiller plant	Retrofit existing chillers (keep existing industrial grade shells). Replace refrigerant, new variable speed drives (VSDs), compressors and motors.
Corridor lighting	New fixtures, lamps, ballas, and digitally addressable lighting interface (DALI) controls for all corridors.
Restroom lighting	New fixtures, lamps and ballasts for all restrooms.
Variable air volume air handling units	Replace all existing constant volume air handlers with new variable air volume (VAV) units. Install two larger units per floor instead of four units (~240 new units).
Tenant daylighting/lighting and plugs	Install 0.8 W/sq ft lighting with dimmable ballasts and photocells (for daylight dimming on perimeter). Add plug load occupancy controls.
Tenant wall insulation	Add 1 inch of R-5 rigid insulation to the inside of all exterior walls in tenant spaces.
Tenant energy management	Provide for end-use sub-metering plus one full time employee (FTE) to manage a tenant energy management programme via the Gridlogix website interface.
Aggressive tenant energy management	Provide for an additional FTE to allow additional programming and tenant education/reporting within tenant energy management programme.

continued

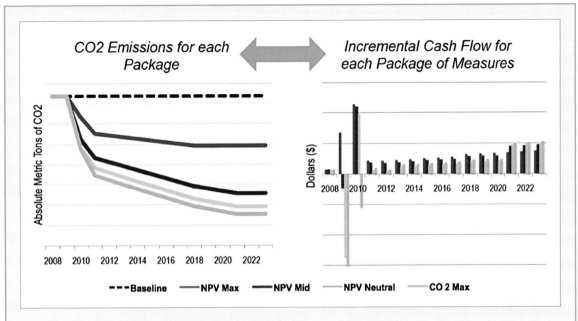

Figure 8.30 Carbon dioxide emissions for each package and incremental cash flow for each package

In the third phase of development, the team created different packages or combinations of energy-efficiency strategies with four different goals:

- to maximize Net Present Value (NPV)
- to balance NPV and carbon dioxide savings
- to maximize carbon dioxide savings for a zero NPV
- to maximize carbon dioxide savings overall.

Figure 8.30 demonstrates the potential carbon dioxide emission reduction for different packages and the incremental cash flow for each package. At this stage, the project team spent significant time to refine energy and financial model inputs to ensure the reliability and accuracy of outputs and to understand the critical relationship between economic investments and carbon dioxide reductions.

In the final phase of design development, the project team iteratively modelled the package measures with the best economic and environmental benefits. By conducting iterative analysis, the project team was able to choose the optimal package of eight energy-efficiency strategies, not only to save energy and reduce carbon dioxide emissions but also to enhance work environments within the building. In addition, these eight sustainable retrofit strategies were also determined to be cost-effective, with an appropriate return on investment (ROI) for the owner.

Optimal energy efficiency strategies

The final eight strategies selected for implementation were:

- building window retrofit
- radiative barrier installation

continued

- tenant demand control ventilation
- tenant daylighting, lighting and plugs
- direct digital controls (DDC)
- chiller plant retrofit
- variable air volume (VAV) air handling units
- tenant energy management.

These eight strategies have not only reduced energy consumption and carbon dioxide emissions, but also have delivered an enhanced environment for tenants, including improved air quality resulting from tenant demand-controlled ventilation; better lighting conditions that coordinated ambient and task lighting; and improved thermal comfort resulting from better windows, the radiative barrier and improved controls. Furthermore, these eight strategies are also cost-effective because the additional investment for them is being repaid by a combination of reduced operating expenses, higher rental rates and greater occupancy levels. This payback is occurring even though implementation required investing an incremental cost for retrofitting the building to achieve energy performance. The following subsections describe each of these strategies in greater detail.

Building windows

One of the first energy-efficiency strategies recommended was to refurbish all existing insulated glass units (IGU) consisting of 6500 double-hung windows to include suspended coated heat mirror film (SC75 or TC88) and krypton/argon gas mix (Figure 8.31). By improving the thermal performance of windows, it was possible not only to reduce energy consumption by about 5 per cent and reduce greenhouse gas emissions correspondingly, but also to improve thermal performance and comfort for occupants in the building.

Figure 8.31 Building windows refurbishment

continued

Radiative barriers

The second recommended strategy for the project involved the installation of more than 6500 insulated reflective barriers behind radiator units located around the perimeter of the building (Figure 8.32). In addition, all gaps between the radiators and wall were sealed with foam to insulate the enclosure of the building. Finally, each radiator was cleaned and its thermostat repositioned to the front side of the radiator. The installed radiative barriers reduced summer heat gain and winter heat loss, and hence reduced building heating and cooling energy usage.

Figure 8.32 Radiative barrier installation between radiators and walls

Chiller plant retrofit

The recommended chiller plant retrofit project (Figure 8.33) included the retrofit of four industrial electric chillers (one low-zone unit, two mid-zone units and one high-zone unit) in addition to upgrades in controls, variable speed drives (VSD) and primary loop bypasses. All existing pumps and steam chillers were allowed to remain. For the low-zone chiller, the retrofit involved the installation of a new chiller-mounted VSD, a new VSD-rated compressor motor, a new IEEE filter in the variable frequency drive (VFD) to reduce harmonic distortion, and a new Optiview Graphic Control Panel with the latest software revisions. For all other chillers, retrofit involved the installation of new drivelines, new evaporator

Figure 8.33 Chiller plant retrofit

and condenser water tubes, new Optiview Graphic Control Panels with the latest software revisions, chiller water bypasses with two-way disk type valves, new piping in place of back-wash reversing valves, new automatic isolation valves on the chilled and hot water supplies to each electric chiller, and temperature and pressure gauges on all supply and return lines. In addition, existing R-500 refrigerant was removed per EPA guidelines and replaced with R134A refrigerant. The existing steam chillers were retained. All electric chillers, condenser water (CW) and chilled water (CHW) pumps, pump VFDs, and zone bypass valves are controlled by the Metasys control system.

Variable air volume (VAV) air handling units

The sustainable retrofit of the ESB also recommended replacing existing constant volume air handlers with a new air handling layout (two floor-mounted units per floor instead of four ceiling-hung units) as well as the use of VAV units instead of existing constant volume units (Figure 8.34). This sustainable retrofit required little additional capital cost while reducing maintenance costs and

continued

Figure 8.34 VAV air-handling units and controls

significantly improving comfort conditions for tenants by reducing noise and increasing thermal accuracy and control.

Direct digital controls (DDC)

The control system retrofit involved upgrading existing control systems (Figure 8.35) at the ESB to Johnson Controls Metasys Extended Architecture BACnet controllers that include Ethernet and BACnet risers with all of the necessary devices and equipment, along with an ADX server/ workstation, printer, software and web access capability. The retrofit provided an upgrade in controls for the following building systems:

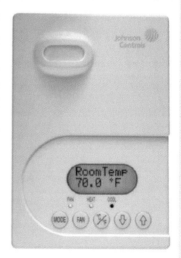

Figure 8.35 System controls

continued

- refrigeration plant building management system
- condenser water system upgrades
- chiller water air handling
- direct exchange (DX) air handling units
- exhaust fans
- standalone chiller monitoring
- miscellaneous room temperature sensors
- electrical service monitoring.

By upgrading to DDCs, the HVAC system provides the environmental conditions needed to improve productivity and indoor environmental quality, lower total energy consumption, improve the reliability of all installed systems, provide access to every aspect of the HVAC system through a secure web browser interface, and monitor and diagnose all HVAC systems installed in the ESB. The DDC upgrade also reduced maintenance time and costs.

Tenant demand control ventilation

Another energy-efficiency strategy recommended for the ESB was the installation of carbon dioxide sensors for control of outside air introduction to chilled water air handling and direct exchange (DX) air handling units (Figure 8.36). One return air carbon dioxide sensor was installed per unit in addition to removing the existing outside air damper and replacing it with a new control damper. Since a tenant demand control ventilation (DCV) system provides just the right amount of outside air for the occupants, it can save energy by not heating or cooling unnecessary quantities of outside air and by providing assur-

Figure 8.36 Tenant demand control ventilation

ances that sufficient outside air is being supplied to the occupants. In addition, active control of the ventilation system also provides the opportunity to control indoor air quality, which has been correlated with increased occupant satisfaction.

Tenant daylighting, lighting and plugs

A key recommendation of the sustainable retrofit plan was a strategy for tenant daylighting, lighting and plug-ins that involved reducing lighting power density in tenant spaces using ambient, direct/indirect and task lighting (Figure 8.37). This strategy also involved installing dimmable ballasts and photosensors for perimeter spaces that can operate with electric lights off or dimmed, depending on daylight availability. The strategy also provides occupants with a plug load occupancy sensor for their personal workstations.

The final analysis conducted by the design team assumed that these measures would not be part of the base upgrades conducted by the building owner but rather encouraged to be completed by tenants as they build out their spaces in the building. Examples (via pre-builts), data (the team analysis) and tools (the eQUEST model) have been provided to tenants to help them understand the cost savings of these measures over the term of their lease as a means of encouraging them to

continued

Figure 8.37 Tenant daylighting features

undertake these projects of their own volition. Tenant compliance with these recommendations has resulted in lower overall cooling demand and higher sustainability ratings for the building, and tenants have benefitted from reduced utility costs and higher-quality, more productive spaces as a result.

Tenant energy management

The sustainable retrofit plan built into each space a web-based tenant energy management system (Figure 8.38) that affords tenants instant feedback to measure, control and lower their energy costs, with actionable recommendations for improving efficiency and sustainability tips. An installed

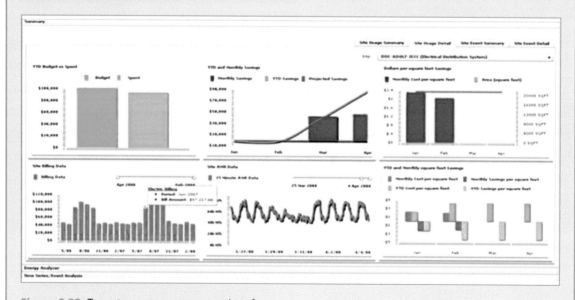

Figure 8.38 Tenant energy management interface

continued

EnNET/Active Energy Management (AEM) platform is provided for collecting 15-minute meter data and for creating a normalized database that can be used to support time series profiling, reporting and integration in the future with property management software for creating a bill based upon a current meter read. In addition, the AEM application can be commissioned and web pages can be created to properly display metering data, time series analysis, real-time metering information and notifications based on usage parameters. Installing this tenant energy management system has made it possible to reduce energy bills and greenhouse gas emissions, encourage occupants' participation and environmental responsibility, help occupants dispel the misconception that energy-saving measures would result in loss of amenity, boost staff morale and pride as people are enabled to feel part of the solution, and make involvement of key personnel in energy efficiency activities a priority.

Outcomes from sustainable retrofit

Modelling to pull the project together via iterations between the energy model (eQUEST) and financial models resulted in an analysis (Figure 8.39) that calculated the NPV of the selected packages of measures as opposed to cumulative carbon dioxide savings for the four different scenarios considered. In selecting an approach for the retrofit, the project team chose the NPV midpoint because it included strategies based on a balance of NPV with the amount of carbon dioxide avoided. Based on the analysis, the sustainable retrofit was expected to reduce energy use and greenhouse gas emissions by 38 per cent, saving 105,000 metric tons of carbon dioxide over the next 15 years.

The sustainable retrofit of the ESB has achieved many benefits, but it also required additional cost premiums for integrating the eight recommended energy-efficiency strategies. In 2008, the planned capital budget for baseline energy-related projects was about US$93 million. Based on the eight energy-efficiency strategies carefully integrated to maximize the benefits of energy and operational savings, the capital project budget for renovations required an additional initial net cost premium of US$13 million (US$20 million in new expenditure for energy-efficiency strategies minus

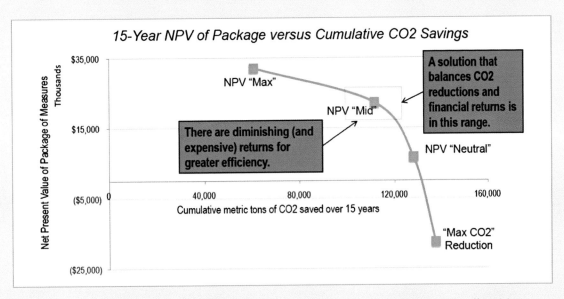

Figure 8.39 The ESB can achieve a high level of energy and carbon dioxide reduction cost-effectively

continued

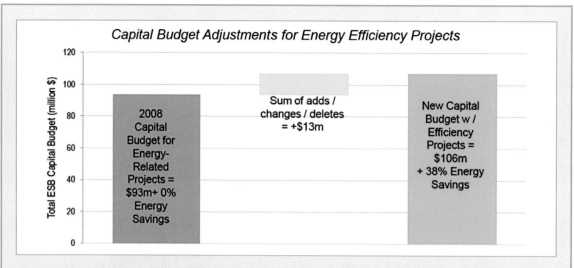

Figure 8.40 Capital budget adjustments for energy-efficiency projects

US$7 million of initial cost savings from downsized systems). The estimated return includes 38 per cent energy savings as a result, or around US$4.4 million annually (Figure 8.40).

Table 8.2 shows the more detailed incremental cost estimate for each strategy compared with the original budget for the retrofit of the ESB in 2008. This analysis shows that the incremental cost of the eight energy-efficiency strategies would be easily offset by annual energy savings resulting from the investment after three years of operation with all improvements completed. Since some retrofits involving tenant spaces are being completed over time, savings in initial years was expected to be less than the $4.4 million per year total (see ESB year 3 findings).

Table 8.2 Incremental cost for eight energy-efficiency strategies

Project description	Projected capital costs	2008 capital budget	Incremental cost	Estimated annual energy savings
Windows	$4.5 million	$455,000	$4 million	$410,000
Radiative barrier	$2.7 million	$0	$2.7 million	$190,000
DDC controls	$7.6 million	$2 million	$5.6 million	$117,000
Demand control ventilation	Included above	$0	Included above	$675,000
Chiller plant retrofit	$5.1 million	$22.4 million	-$17.3 million*	$702,000
VAV AHUs	$47.2 million	$44.8 million	$2.4 million	$941,000
Tenant daylighting/plugs	$24.5 million	$16.1 million	$8.4 million	$396,000
Tenant energy management	$365,000	$0	$365,000	$320,000
Total	$106.9 million	$93.7 million	$13.2 million	$4.4 million

*Negative capital costs result from downsizing equipment compared to baseline

continued

ESB year 3 findings

Three years of energy performance findings have been developed and archived to document the performance outcomes for the ESB energy efficiency retrofit. While the core elements of the retrofit were completed by 2011, additional tenant space improvements and automated controls are being added incrementally in new tenant build-outs over time. As of the first three years of measurement, the building has exceeded its performance targets consistently each year:

- Year 1 (2011): Exceeded energy target by 5 per cent ($2.4 million savings)
- Year 2 (2012): Exceeded energy target by 4 per cent ($2.3 million savings)
- Year 3 (2013): Exceeded energy target by 15.9 per cent ($2.8 million savings)

By the time all tenant spaces have been fully upgraded, the building is expected to save on the order of $4.4 million per year resulting from energy savings of 38 per cent or greater each year over the pre-retrofit building performance. These savings will be achieved based on an estimated incremental investment of $13.2 million beyond the costs of infrastructure upgrades already planned. The retrofit is also credited with creating over 250 new jobs.

(http://esbnyc.com)

The greatest reduction in annual energy consumption and carbon dioxide from the baseline was calculated to result from completing the task of installing DDC control systems (Figure 8.41). The DDC strategy alone would reduce energy use by approximately 9 per cent from the baseline. The tenant daylighting system, along with working with tenants to ensure that layouts maximize the use of natural lighting, would save 6 per cent from the baseline. Three other strategies including replacing constant volume air handling units with VAV units, retrofitting the chiller plant and

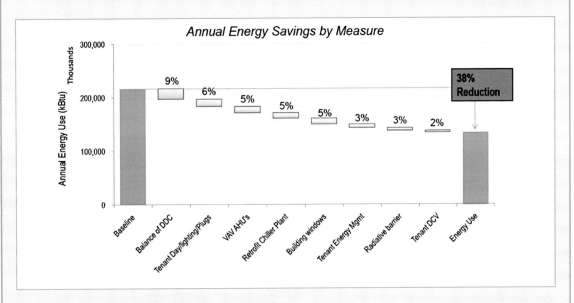

Figure 8.41 Annual energy savings by energy-efficiency strategies

continued

addressing window glazing would reduce energy consumption and carbon emissions by another 5 per cent. The tenant energy management system and the installation of radiative barriers would save 3 per cent each. The tenant demand control system is also estimated to reduce the energy consumption and carbon emissions by about 2 per cent. After all eight strategies were completed; the ESB was expected to be able to achieve an Energy Star score of 90, performing better than 90 per cent of buildings in the United States regardless of age.

The optimal package of eight strategies was also expected to reduce peak electricity demand by 3.5 mW, from 9.6 mW to just over 6 mW shown as in Figure 8.42. The reduction of peak demand can be beneficial to both the ESB in terms of energy costs, and to the utility company that provides electricity to the ESB in releasing capacity to serve other clients.

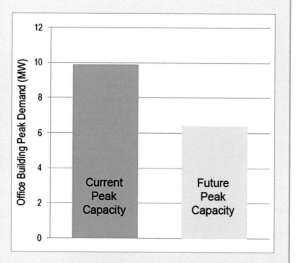

Figure 8.42 Peak demand change

In addition to reducing energy and carbon dioxide emissions, the sustainable retrofit has delivered an enhanced environment for tenants, including improved air quality resulting from tenant demand-controlled ventilation; better lighting conditions that coordinate ambient and task lighting; and improved thermal comfort resulting from better windows, the radiative barriers and better controls. Overall, these strategies contributed to achieving green certifications including an Energy Star rating of 90 out of 100 and a Gold level of certification under LEED for Existing Buildings.

Based on this case study of the ESB's sustainable retrofit project, sustainable retrofits for other existing buildings may not only achieve an attractive return on investment but also reduce energy consumption and carbon emissions. In addition, sustainable retrofits such as the ESB upgrades also enhance indoor environments for building occupants, which will ultimately enhance the desirability of projects like this in the future.

The case study of the Empire State Building in this chapter was adapted by the authors based on the content in www.esbsustainability.com/. The images in the case study were received from Empire State Building Company, LLC and used with permission under a licensing agreement.

End-of-lifecycle opportunities

At the end of a facility's useful service life, additional opportunities exist to provide an end to the building's life in a green way. Deconstruction and ecosystem restoration are two major strategies that can be used to ensure that a building's life ends as green as it began.

Deconstruction best practices

There are many different fates for built facilities at the end of their service lives. An alternative to the traditional demolition process is gaining popularity as the costs of disposal and raw materials continue to increase. Deconstruction is a process of taking a building apart piece by piece to recover materials in a usable state for use in other buildings. The process of salvaging components in old buildings is not new and has been in practice since construction first began. However, deconstruction takes salvage to a higher level by attempting to recover nearly all the components of a building, not just the most obvious or high-value materials. The level of deconstruction appropriate for a particular project depends on many factors including the condition of building elements, type of construction used and possible uses for recovered products and systems.

The benefits of deconstruction are numerous, including the generation of high-quality, reusable materials at lower cost than virgin materials and reduction of environmental impacts both locally and in landfills. Deconstruction can provide access to building materials that are no longer available in today's market, such as timber and wood products from old-growth forests. It can also provide jobs in urban areas and preserve architectural features that would otherwise be lost (Falk and Guy 2007).

Not all facilities are appropriate candidates for deconstruction. Key to the success of such an approach is finding a good balance between the additional labour required and the reduced costs of disposal vs. the value of materials recovered. Assessing the local market for recovered materials is essential to making the business case for deconstruction. Depending on disposal costs, it may be cost-effective to donate materials even though no revenue results. Often, donations can be used for tax deductions by the donating company. Organizations such as Habitat for Humanity or local salvage stores (Figure 8.49) can be good places to donate material (see *Case Study: Habitat for Humanity UnBuild*).

Case study: Habitat for Humanity UnBuild Project: Oakbridge Apartment Complex, Blacksburg, VA

Occupying a prime location immediately adjacent to the Virginia Tech campus in Blacksburg, VA, the 197-unit low rise Oakbridge Apartment Complex provided affordable and convenient housing for students in its 354 bedrooms since its initial construction in the 1950's. By 2013, however, the complex had seen better days, and anticipated growth of the student population at Virginia Tech promised significant new demand for off-campus housing. Blacksburg-based owner Campus Management Group, Inc. and partner CampusWorks of Charlotte, NC elected to demolish the existing apartments to make way for a higher density infill development of 254 furnished apartments with 911 bedrooms, set to open in 2014.

In May 2013, volunteers from the local Habitat for Humanity organization undertook a week-long deconstruction and salvage operation at the site to recover usable and recyclable materials before demolition (Figure 8.43). From refrigerators, stoves and dishwashers to doors, windows, cabinets, ceiling fans and even outdoor landscape plants, volunteers from campus and the local community removed saleable or reusable items that were donated to be resold at the local Habitat ReStore or incorporated into other local Habitat for Humanity projects (Figure 8.44). Items such as

continued

Figure 8.43 Low-rise apartment buildings at Oakbridge after removal of windows for reuse

Figure 8.44 A volunteer carefully removes a replacement window unit for resale

Figure 8.45 Students working to identify items to be removed from the apartments

Figure 8.46 Prototype greenhouse designed and built by students to demonstrate uses for recovered windows

aluminium gutters, steel radiators and copper piping were removed and sold for their scrap value. Over 400 appliances were recovered for resale, and the total revenue to Habitat from the effort exceeded $25,000. These materials would otherwise have been demolished and disposed in a landfill as part of the conventional demolition process.

Prior to the UnBuild event, sustainable construction students from Virginia Tech worked in teams with Habitat and the facility owner to identify items that could be safely removed from the apartments (Figure 8.45). Each team in the class developed safe work guidelines for the removal of one type of material or building system. Their work helped Habitat for Humanity plan and execute the deconstruction efforts, and they learned more about sustainable deconstruction and safe work planning in the process.

continued

Figure 8.47 Oakbridge window units were also incorporated as part of the Pantherhouse, an expansion of a local animal shelter that was designed and built by students

Figure 8.48 Oakbridge cabinets were a popular item for resale and have been incorporated in applications such as this garage storage unit

Materials from the Oakbridge UnBuild have found new life in a variety of different projects around the New River Valley. Virginia Tech students designed and built a prototype greenhouse using recovered windows that was on display at the local Habitat for Humanity ReStore (Figure 8.46). Oakbridge windows were also incorporated into an expansion of a local animal shelter that was designed and built by students as part of a sustainable building class (Figure 8.47). Recovered cabinets were a popular item for sale at the Habitat ReStore and have been used in applications such as this garage storage unit (Figure 8.48).

References

Bisnow. (2013). "Blacksburg comes around." *Multifamily Bisnow Newsletter,* 19 July. https://bisnow.com/archives/newsletter/multifamily-bisnow/how-to-get-fat (accessed 02 January 2017).

Gangloff, M. (2013). "Habitat shifts gears to unbuild housing." *Roanoke Times,* 23 May. http://roanoke.com/news/local/christiansburg/habitat-shifts-gears-to-unbuild-housing/article_ca86cf41-5525-52e7-8068-326183e2efc8.html (accessed 2 January 2017).

Jackson, T. (2013). "Construction from destruction." *Roanoke Times,* 19 May. http://roanoke.com/community/the_burgs/construction-from-destruction/article_9fd941d4-e87d-52b0-a396-3b743a6ca216.html (accessed 2 January 2017).

Figure 8.49
This local store sells salvaged or
deconstructed materials for reuse
in other projects

During deconstruction, the facility is typically 'soft stripped', where easy-to-remove components such as fixtures, appliances, doors, plumbing and electrical components, and flooring are removed and stored for future use. Candidate facilities must also be carefully assessed for environmental hazards, including asbestos, lead-based paint and other types of contamination. Hazardous components such as thermostats, ballasts and fluorescent lamps should also be removed, and elements such as asbestos or lead-based paint are also mitigated or abated according to applicable laws. Care must be taken during this time to properly capture and store materials and prevent them from being damaged, especially hazardous components (Figure 8.50). After this process, a more intensive approach is used to take apart the building's primary structure, including roof, walls and other components. This process requires careful planning and attention to safety. Not only are deconstruction personnel exposed to ordinary hazards present on a construction site, but also they may be subject to:

- pathogens resulting from corrosion or bacterial/fungal growth on recovered systems.
- possible shock hazards from live electrical circuits.
- cuts from broken glass and sharp edges.
- exposure to hazardous animal detritus such as bird or mouse droppings.
- strains from lifting heavy elements such as window assemblies or appliances.
- striking or crushing hazards when working in congested areas with heavy equipment.
- falls through openings or collapse of unstable structural elements as building elements are removed.

Figure 8.50
Careful storage is required to prevent problems with deconstructed building components, especially those containing hazardous substances such as the mercury in these fluorescent tubes

After the building has been deconstructed, additional processing is necessary to render building components suitable for reuse. Wood-based products such as flooring or timbers may need to be remilled to restore their aesthetic surface. Dimensional lumber will need to be de-nailed before future use, and may also need to be regraded to ensure that it is structurally sound. Some components such as windows and plumbing fixtures should not be reused in climate controlled buildings, since they are likely to be substantially less efficient than contemporary models. These elements may find alternative use in buildings such as garden greenhouses or other buildings where human comfort is not a key driver. After all reusable components have been extracted, the remainder of the building's components are then recycled if possible, or disposed properly in a landfill. Demolition waste recycling is similar to construction waste recycling, and can be managed using the procedures and resources discussed in Chapter 7.

Ecosystem restoration best practices

In addition to measures taken with built facilities at the end of their lifecycle, the site on which those facilities are located can also be dealt with in a sustainable fashion. This may at a minimum involve remediation of contamination on site, or more extensively, restoring the natural ecosystems that were located on the site before it was originally developed. Ecosystem restoration is a specialized undertaking that requires knowledge of the types of flora and fauna present on the site before development and well adapted to local conditions.

The degree to which a site is restored is often dependent on planned future uses. At a minimum, all hazards and environmental contaminants should be removed from the site and/or properly remediated. All

Figure 8.51 Ecosystem restoration using reseeding with native plants

Figure 8.52 This constructed wetland is used to treat acid mine drainage from coal mining before it enters the local river. Credit: *Smith Dalia Architects*

exposed soil should be properly graded, stabilized and replanted with proper vegetation. The topography and permeability of the site should also be carefully restored to ensure that erosion and sedimentation do not cause subsequent problems. Beyond these very basic measures, the possibilities for ecosystem restoration are limited only by the level of investment the owner wishes to make.

In some cases, sites may be deliberately rehabilitated to offset damage to other sites. For instance, development of wetland areas in the United States may be permitted if an equivalent wetland is restored or created elsewhere to offset the wetland that was removed. Wetlands can be created only on sites where conditions exist that can sustain the wetland over time. The same is true for other types of restored ecosystems; they must be carefully sited and designed to ensure that environmental conditions support their ongoing sustainability and evolution.

Ecosystem restoration can range from simple reseeding with native plants (Figure 8.51) to more complex development of synergistic natural systems to remediate and enhance a contaminated site or stabilize an unstable area (Figure 8.52). The possibilities for ecosystem restoration are dependent on the local conditions of the site and the types of plants and animals that are native to the area.

Case study: Trees Atlanta Kendeda Center – end-of-lifecycle

While the Trees Atlanta Kendeda Center has not yet reached the end of its lifecycle in its present use, as an adaptive reuse project it experienced end-of-lifecycle activity that transformed it for its new purpose. The following case study illustrates deconstruction best practices used in the Trees Atlanta project as part of its transformation to its new life as the Kendeda Center.

continued

As an example of adaptive reuse of an existing building, the Trees Atlanta headquarters in Atlanta offers some good examples of strategies for minimizing the negative impacts of the end of the building lifecycle and for recovering useful products and materials that can be put to beneficial use. The following subsections describe practices for deconstruction, recycling and waste management employed in the Trees Atlanta building project.

Deconstruction best practices

At the end of a facility's useful service life, it may be necessary to adapt or demolish the facility, which provides additional opportunities to achieve the goals of sustainability in the built environment. Since the Trees Atlanta project included the partial deconstruction and adaptive reuse of parts of an old warehouse, it had the potential to produce much construction waste. To minimize construction waste in the project and eventually reduce the amount of construction waste and debris in landfills, waste management and recycling strategies were used to divert the maximum amount of on-site construction waste away from landfills through recycling and salvage efforts (Figure 8.53).

Figure 8.53 Recycling reduced construction waste and debris in the deconstruction phase
Credit: *Smith Dalia Architects*

To achieve the goals of recycling and waste prevention during the end-of-lifecycle process for the old warehouse, the general contractor and subcontractors diverted construction and demolition wastes including:

- concrete/masonry (679 tons)
- metal (13 tons)
- asphalt (161 tons)
- green waste/land clearing debris (5.77 tons).

As discussed in Chapter 7, due to its active recycling and waste prevention approach, Trees Atlanta recycled or salvaged nearly 85 per cent (equal to the total construction waste diverted (959 tons)/Total construction waste generated (1139 tons)) of its total construction and demolition waste stream.

Post-occupancy sustainability in the future

How will our post-occupancy facility practices be different in the coming years? Some of the same drivers that influence design and construction – increasing resource scarcity and better knowledge and information about the built environment – will shape our facility practices over the coming years. In particular, as the extraction and harvesting of resources from finite stocks becomes more challenging, we will have to greatly improve our ability to deal with existing buildings and adapt them to meet changing human needs and environmental requirements. Better technologies for retrofit as well as new approaches and techniques for facility condition assessment and characterization will be common practice in the future. Augmented reality (AR) will be used to fuse building performance data streams and archival information with current building information models, providing the data necessary to design effective retrofits that optimize facility performance. Nondestructive testing methods and technologies will also improve, and new types of data-mining algorithms will enable better predictive modelling based on historical data.

Resource scarcity will lead to an increase in adaptive reuse of existing facilities as well as safer and more efficient techniques for deconstructing existing facilities in ways that permit recovery of valuable materials with minimal expenditure of energy. New facilities will have been designed for adaptivity or disassembly, allowing them to be rapidly reconfigured or adapted to changing requirements over time (see *Case Study: One Island East*). Modular materials and systems will be interoperative and upgradable over time as new technologies emerge, without extensive disturbance of related systems or construction time and effort.

Case study: One Island East, Hong Kong

Located in the heart of Island East in Hong Kong and completed in March 2008, One Island East is a premium 1.5 million sq ft, 70-storey commercial office tower (Figure 8.54). From design and construction through operations and maintenance to the planning for its ultimate demolition, One Island East incorporates multiple technologies and strategies to meet its sustainability goals (Figure 8.55). During demolition of the original building on site, concrete crushers were used instead of typical percussive equipment to not only reduce noise, dust and accidents but also minimize the use of water for suppressing dust. This approach also allowed for easier recycling.

Swire Properties and Gammon Construction worked together to develop a rigorous waste management plan that reused or recycled 99 per cent of demolition wastes. One Island East also reserved 75 per cent of the project site to provide a green area for leisure and recreational activities. Swire Properties incorporated multiple sustainable design strategies to reduce the consumption of energy, water and raw materials throughout the lifecycle of the building. One major sustainability strategy in One Island East was to choose grade 100 high-performance concrete for the columns and core walls, which has the benefits of offering the same reinforcing ratio as grade 60 concrete but with a 26 per cent reduction in volume (Figure 8.56). The project team incorporated sophisticated energy modelling tools including E20-II, Energy Plus, and computational fluid dynamics to improve energy performance. The building also includes a rainwater recycling system that can collect rainwater and reuse for cleaning and irrigation to reduce water consumption.

Modular and standardized designs such as movable external pedestal paving, wall panels, light troughs and air handling units were adopted for One Island East to maximize production and material efficiency as well as to minimize waste generation. One Island East incorporated an innovative

continued

Figure 8.54 One Island East incorporated holistic sustainable design strategies to achieve the goals of sustainability in the built environment
Source: Courtesy of One Island East

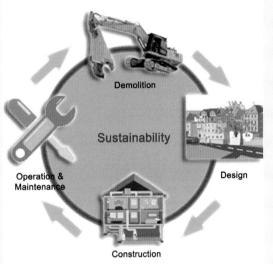

Figure 8.55 Sustainability strategies in One Island East's lifecycle
Source: Courtesy of One Island East

indirect lighting system to allow for dual lighting levels. Carbon dioxide and carbon monoxide sensors have been installed on all the office floors and car parks to control ventilation for optimal indoor air quality. The result is a building that is both comfortable and enhances occupant productivity.

(Swire Island East)

Figure 8.56 High-performance concrete reduces the weight of concrete along with associated carbon dioxide CO_2 emissions
Source: Courtesy of One Island East

Deconstruction of existing facilities will be better coupled with market demand for recovered materials via social network-based technologies for sharing information about the supply and demand of those systems. New occupations and specialities will be created to handle specialized tasks of facility deconstruction and recovery in a way that minimizes waste and maximizes output.

New generations of self-maintaining materials and smart materials that can self-adapt to environmental conditions will reduce the need for human intervention in building maintenance. Artificial intelligence-based control systems are already in use in many commercial facilities, and their functionality will continue to evolve in conjunction with smart infrastructure systems. Infrastructure systems themselves will change, with intelligent controls that better balance supply and demand more closely in ways that optimize the performance of the whole system, not just the individual facility. With net-zero energy/water buildings and distributed energy and water infrastructure at the building and district scale, our infrastructure networks of tomorrow will be a complex, hybrid grid of production, recycling and use of resources, requiring new approaches to investment and management that are only being envisioned today.

Summary of sustainable facility practices over the lifecycle

Both now and in the future, many opportunities exist to introduce sustainability throughout the facilities realization process. Some of these opportunities exist at the project level, while others require changing the organizational context and processes in which individual projects occur. Across all lifecycle phases, sustainable buildings and infrastructure systems must:

- Expand consideration in decision making to include the *whole lifecycle* of the project, not just what it takes to get the project built. This includes building strategic partnerships with internal and external stakeholders to ensure that funding and stewardship exists to sustain the project in the long term.
- Expand consideration in decision making to take into account the *context* of the project, and how the project will interact with that context over its lifecycle. This includes coordination with larger strategic or master plans, developing business cases and funding plans that consider long-term future trends such as resource scarcity and environmental constraints, and maximizing opportunities to tie the project into multiple organizational sub-units.
- Maximize *communication* among project stakeholders, and include feedback loops and education to ensure people are aware of and in alignment with project sustainability goals.

These themes are essential for successfully achieving sustainability goals in building and infrastructure projects, and are necessary to develop facilities that meet stakeholder needs in the present while preserving an organization's long-term capabilities to sustain itself in both the community and global context.

Discussion questions and exercises

8.1   Consider the building in which you are presently located, and review historical utility bills for the facility. You can obtain these bills from the facility's owner or possibly also from the local utility offices. How much energy and water are used by the building, and what systems within the building need those resources to function? What are the biggest users of energy and water within the facility? What upgrades would provide the most significant return on investment? What combinations of upgrades would offer synergistic effects?

8.2   Talk with stakeholders responsible for building operation and maintenance. What measures are in place to ensure proper functioning of the building? How could the building be optimized to improve energy and water performance?

8.3   What evidence can you find in the buildings you occupy to indicate suboptimal performance of the building with respect to meeting its occupants' needs? What compensating actions and technologies have been implemented by occupants to adjust the building to their needs?

8.4   Identify any efficiency and operations plans developed for your building, including general operating plans, sustainable maintenance plans, continuous commissioning plans, and measurement, monitoring and verification plans. How are problems identified and resolved by these plans?

8.5   Does your organization have any provisions for green product procurement? What opportunities are available for green power in your area, both for your organization and your own home? Conduct an inventory of the products used in maintaining your facilities. For which of these are there more sustainable alternatives that could be employed?

8.6   How does your organization presently manage solid waste in its facilities? What opportunities exist to improve its waste management practices? Inventory the waste management options available in your area to determine what can be done to improve solid waste management in your facility.

8.7   Conduct an inventory of the housekeeping practices in your current facility. Review your facility's green housekeeping plan if one exists. What measures are being taken to ensure that housekeeping practices contribute to building sustainability? How could current housekeeping practices be improved?

8.8   Locate a building renovation project in your local area and visit the site to inventory job site practices. What measures are being taken to protect occupied areas during construction? What controls are in place to ensure future indoor air quality? What special measures are used to ensure worker health and safety? What could be improved to increase the sustainability of the renovation project?

8.9   Identify a green building technology or practice in your facility that is observable by users of the facility. Design a sign to describe how the technology works and educate users on how to interact with it.

8.10   Determine whether your building has a tenant manual or green occupant manual. If yes, review the manual and identify what user measures are recommended to keep the building functioning properly. If not, what would you include in a green occupant manual to help building occupants interact more sustainably with the building's features?

8.11 Visit a local salvage yard or building material reseller if one exists in your community. What types of products are for sale? What cost or quality advantages do they have over comparable new products?

8.12 Consider the building in which you are located and develop an inventory of materials and components that could be recovered through deconstruction. What hazardous materials and components are part of the building that would require special management if the building were to be deconstructed?

8.13 Contact the nearest office of government responsible for ecosystem management and restoration in your region or country. What efforts are underway to restore ecosystems in your area? If possible, visit a restoration site and talk with the people who are managing that site. What are the critical measures being taken to restore environmental quality on the site?

References and resources

AFCEE – Air Force Center for Environmental Excellence. (2000). *C&D Waste Management Guide*. Brooks AFB, TX. http://infohouse.p2ric.org/ref/24/23088.pdf (accessed 7 January 2017).

Falk, R.H., and Guy, B. (2007). *Unbuilding: Salvaging the architectural treasures of unwanted houses*. Taunton Press, Newtown, CN.

Fishbein, B.K. (1998). *Building for the future: Strategies to reduce construction and demolition waste in municipal projects*. http://informinc.org/buildforfuture.php (accessed 7 January 2017).

Lund, E., and Yost, P. (1997). *Deconstruction–building disassembly and material salvage: The Riverdale case study*. NAHB Research Center, Upper Marlboro, MD. http://lifecyclebuilding.org/docs/Riverdale%20Case%20Study.pdf (accessed 7 January 2017).

Lund, H.F. (1993). *The McGraw-Hill recycling handbook*. McGraw-Hill, New York.

NAHBRC – National Association of Home Builders Research Center. (2000). *A Guide to Deconstruction*. US Department of Housing and Urban Development, Office of Policy Development and Research, Washington, DC. https://huduser.gov/publications/pdf/decon.pdf (accessed 7 January 2017).

Tchobanoglous, G., Theisen, H., and Vigil, S. (1993). *Integrated solid waste management: Engineering principles and management issues*. McGraw-Hill, New York.

US FEMP – US Federal Energy Management Program. (2002). *Continuous Commissioning Guidebook for Federal Energy Managers*. Office of Energy Efficiency and Renewable Energy, US Department of Energy, Gaithersburg, MD. http://eber.ed.ornl.gov/CommercialProducts/ContCx.htm (accessed 7 January 2017).

Wilson, A. (2001). *Greening federal facilities*. 2nd ed. Office of Energy Efficiency and Renewable Energy, US Department of Energy, Gaithersburg, MD. http://nrel.gov/docs/fy01osti/29267.pdf (accessed 7 January 2017).

Chapter 9
The business case for sustainability

The implementation of high performance building technologies and practices in built facilities has significantly increased over the past 25 years. Evidence is growing that considering sustainable strategies in the development of built facilities can provide financial rewards for building owners along with social and environmental benefits for both occupants and society at large. This chapter emphasizes the business case for sustainable design and construction, including a detailed example of the business case with regard to the energy and water optimization of a building. This is one of the key considerations in implementing sustainability strategies in buildings, because significant opportunities exist to lower annual costs for energy and water. However, the reduction of annual operating utility costs sometimes requires the additional expense of higher first costs. Therefore, it is necessary to identify the relationship between first cost premiums for sustainable strategies versus their lifecycle costs within their payback period. The case study covered in this chapter does just that.

In addition to direct cost saving opportunities, the implementation of sustainability strategies can also provide indirect economic benefits to facility owners, occupants, society and the environment. For example, reducing energy consumption in built facilities can significantly lower greenhouse gas emissions, which reduces the vulnerability of society to climate change at a global level. In addition, energy and water-efficient facilities offer society a range of economic benefits, including cost reduction from lower civil infrastructure costs (including avoided waste-water treatment plant expansions, power plant development and additional transmission/distribution lines) and preserving non-renewable resources, including fossil fuels and water aquifers, for future generations. Furthermore, a number of green features such as daylighting in buildings can also promote better health, comfort, safety and well-being of facility occupants, which can reduce levels of absenteeism, increase productivity in the workplace and provide a greater quality of life for facility users in general.

To illustrate the process of developing a business case, the chapter begins with a detailed case study of the planning, design, construction and operation of Reedy Fork Elementary School located in Greensboro,

North Carolina, United States. This case illustrates multiple ways to improve energy and water efficiency in buildings including an overview of potential strategies to lower or equal the first cost, provide annual energy and water cost saving opportunities, and deliver other economic, social and environmental benefits. The chapter then discusses the range of costs and considerations that should be taken into account when preparing a business case.

Case study: Reedy Fork Elementary School, Greensboro, NC

Reedy Fork Elementary School (RFES) is an 87,000 sq ft school facility that includes classroom space for 725 students and 70 staff plus dining, gymnasium, auditorium, science, art, music, computer, library and administration facilities (Figure 9.1). Construction of the project began in 2006 and was successfully completed in 2007. Since this school incorporated many high performance building strategies, especially related to energy savings, it was among the top 10 per cent of schools in the United States designed to earn the Global Energy Star label. To achieve the project, RFES involved a collaborative project team comprising:

- Owner: Guilford County Schools
- Architect: Innovative Design, Inc.
- Civil engineering consultants: B & F Consulting, Inc.
- Landscape architects: Landis, Inc.
- Structural engineering consultants: Lysaght and Associates, P.A.
- General contractor: Barnhill Contracting, Inc.
- Construction management: HiCAPS, Inc.

Figure 9.1 An overview of Reedy Fork Elementary School

continued

Reedy Fork Elementary School

Location: 4571 Reedy Fork Parkway, Greensboro, NC, USA
Size: 87,000 sq ft
Started: 2006
Completed: 2007
Use: school for 725 students in grades pre-kindergarten through five
Cost: $13.6 million
Distinction: Among the top 10 per cent of schools in the nation designed to earn the Energy
Star label

Reedy Fork sustainable design features

The RFES project started its pre-design and design based on the G3-Guilford Green Guide that
was the School Board's commitment to promoting sustainable design and construction in the school
district. As a result, the RFES project incorporated holistic, innovative, sustainable design and
construction solutions that were strongly tied to the students' curriculum. The school featured a
comprehensive and well-integrated set of sustainability strategies that included:

- appropriate building orientation.
- design for natural daylighting.
- an energy-efficient building shell, with radiant barriers and solar reflective roofs.
- underfloor air distribution system.
- indirect lighting with photocells and occupancy sensors.
- solar water heating and photovoltaic (PV) systems.
- a holistic water cycle approach (rainwater for toilet flushing coupled with bioswales and wetlands
  for stormwater management).
- recycled materials and use of local products.
- indoor environmental quality management.
- computer based real-time monitoring of sustainable systems.

Facility orientation

RFES was oriented on an east–west axis to maximize the southern solar potential for daylighting,
passive solar, solar domestic hot water and PV applications. East and west glazing was minimized
to reduce heat gain in summer. Small courtyards were also provided on the east side of the school
to achieve the same effect through buffering. Figure 9.2 shows the east–west orientation of the three
main wings on the west side of the school.

Innovative daylighting design

The use of natural daylighting is a key strategy to reduce energy consumption as well as a significant
factor in improving students' performance, health, comfort and well-being. RFES adopted several
daylighting design strategies to achieve potentially significant benefits. Specific daylighting design
features include two south-facing clerestories that use curved interior translucent light shelves. These
light shelves can filter sunlight down to occupied areas and bounce the remaining light deep into
the classroom (Figure 9.3). In addition, highly reflective ceiling tiles also enhance the light being
reflected deep into the space. These two strategies resulted in requiring 40 per cent less glass than

continued

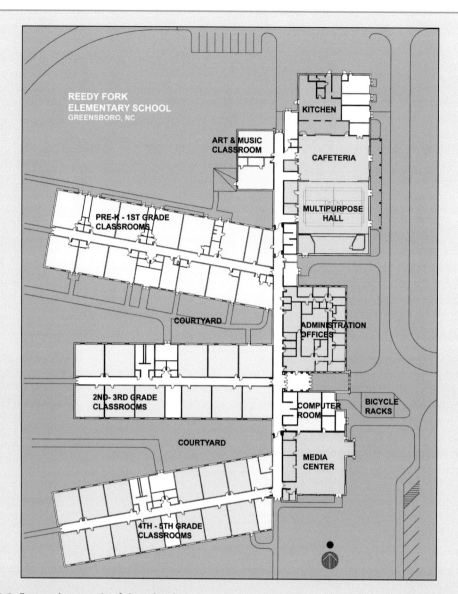

Figure 9.2 East and west axis of the school

typical side-lit glazing solutions (Figure 9.4), both reducing initial wall system and HVAC cost and reducing cooling loads in this warm, humid climate. Furthermore, due to the 20 per cent visible light transmittance of the translucent panel used in the light shelves, glare was also reduced and a soft light was well distributed within the space. The curved translucent light shelves could provide light immediately under each shelf, bounce light back into the space and diffuse the light. This allowed for glazing at the wall clerestory aperture with maximum visible light transmission. In addition, because the white single ply roofing and the curved translucent light shelves provided adequate daylight by themselves, the glass-to-floor ratio was reduced significantly over side-lit solutions. To avoid glare, interior dropped soffits were situated to intentionally shade the projection

continued

screen area and the television monitors in classroom spaces without blocking views and without the need for operable clerestory window shading devices. These innovative daylighting strategies were the primary lighting source for all education and administrative spaces during two-thirds of the daylight hours, which reduced the need for fluorescent lighting while also diminishing the school's need for air conditioning.

South-facing roof monitors with translucent fabric baffles in the light wells also provided daylighting in the gymnasium, multi-purpose and dining areas of the school (Figure 9.5). These features

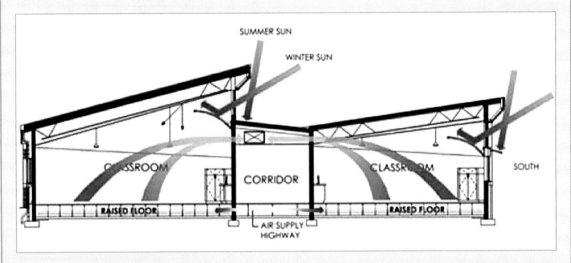

Figure 9.3 Schematic section of classroom showing daylighting

Figure 9.4 Daylighting strategies in RFES

continued

Figure 9.5a, b Daylighting monitors in the gymnasium and multi-purpose room

Figure 9.6 Indirect fluorescent lighting with a dimmable lighting system

eliminate direct glare and effectively diffuse light throughout the spaces. Clear double-glazing was also used to maximize visible light transmittance and minimize the glass-to-floor ratio. In addition, adequate overhangs over the monitor windows protect the spaces from direct light during peak cooling periods in the summer. This monitor approach was used because of the large room dimensions and because the ceiling cavity was limited to the thickness of the roofing system, which was shallower than conventional systems using dropped ceilings. Therefore, the reflective losses associated with deep ceiling cavities were eliminated. In addition, indirect fluorescent lighting was installed through the building to further improve energy savings (Figure 9.6) and provide the minimal additional lighting needed to supplement daylighting. The lighting is dimmable and can be controlled by an occupancy sensor and a photocell sensor that work in conjunction with natural daylight to minimize the need for artificial light in the building when there is sufficient daylighting.

Underfloor air distribution

An underfloor air distribution system has been incorporated in classrooms, the media centre and administration offices (Figure 9.7). This raised floor system improves thermal comfort, indoor air quality, flexibility in space usage and energy consumption. In addition, the system also saved on initial construction costs by reducing the need for expensive steel ductwork at ceiling level. Several courses of masonry were also eliminated because of reducing the ceiling cavity height by 2–3 ft throughout, which also improved the performance of the daylighting strategies. The underfloor strategy also eliminated scaffolding costs and associated safety risks during installation of HVAC, while simultaneously easing the installation and coordination problems associated with overhead ductwork, plumbing, electrical and control wiring.

continued

Figure 9.7 Underfloor air distribution system

Solar energy

A PV system (see Figure 9.8) was incorporated into the entry canopy to feed 1.75kW of electricity into the computer lab. Stand-alone PV systems were also used to power the building's entrance sign and pond aerator. These on-site renewable systems reduce the negative environmental and economic impacts associated with fossil fuel energy use while providing needed power and supporting necessary functions both inside and outside the building.

Figure 9.8 PV panels used for electricity generation

A solar thermal system (Figure 9.9) has also been installed to provide approximately 75 per cent of the hot water for the school, the majority of which is used for the kitchen. These solar water heating systems can also be used as an educational tool to teach students about renewable energy sources.

continued

Figure 9.9
Solar collectors
provide hot
water for the
kitchen

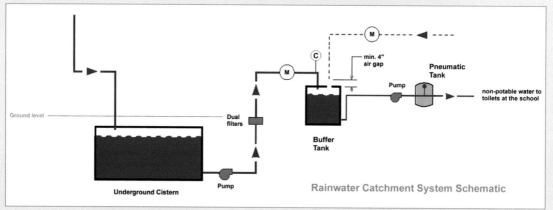

Figure 9.10 Rainwater harvesting system

Rainwater harvesting system

The RFES project incorporated a rainwater harvesting system to reduce consumption of potable water from municipal systems throughout the school. Reedy Fork's rainwater harvesting system collects rain from half the roof area of the school and sends it to a 45,000 gallon underground storage tank (Figure 9.10). The collected rainwater is then pumped from the tank to the school, filtered, chlorinated, dyed light blue and used for flushing each toilet in the school. By using rainwater for toilet flushing, the school can save over 767,000 gallons of water annually that would otherwise be purchased from the City of Greensboro. The use of rainwater replaces 94 per cent of the total water required for toilet flushing. By diverting and treating stormwater that would otherwise be directed from the site's roof into the storm sewer collection system, the rainwater harvesting system also reduces the generation of wastewater. This system also protects the natural water cycle and saves water resources for future generations.

Bioswales and constructed wetlands

A series of bio-retention swales and constructed wetlands on site capture all the rainfall not falling onto the roof areas, minimizing nitrogen runoff in the stormwater before it is absorbed into the soil. No stormwater is discharged into local storm sewers, which significantly reduced infrastructure costs because the existing central sewer line was located miles from the site. Avoiding the cost of

continued

Figure 9.11 Bioswales and constructed wetland

stormwater infrastructure was also a major driver for the decision to adopt the rainwater harvesting system in the building itself. Special soils and a variety of aquatic plants such as pickerel weed, soft rush and spike rush reduce pollutants from the stormwater, again returning clean water to the aquifer. PV-driven aerators are also used in the constructed wetland to move surface water and minimize mosquito problems.

Materials recycling and local products

Materials with recycled content used on the project included carpeting, metal roofing and acoustical ceiling tiles. A construction waste management plan was required by the G3-Guilford Green Guide (http://gcsnc.com/construction/pdfs/G3.pdf) during construction to minimize waste going to landfills. Almost 60 per cent of the total construction waste from the project was diverted for recycling during construction. The school also implemented a programme for daily recycling which continues on an ongoing basis. Locally manufactured masonry products were the predominant structural and finish materials for the school. The specifications for these products were developed to encourage the use of local products and manufacturers, and preference was given to local manufacturers during the bidding process.

Indoor environmental quality management

Poor indoor environmental conditions are well known to affect the health, safety, performance and comfort of students. The quality of indoor environment is linked to satisfaction, productivity and educational learning in school facilities (Heerwagen 2000; Paul and Taylor 2008; Preiser and Vischer 2005; Riley et al. 2010). To improve the occupant experience, RFES implemented the following strategies as part of the project to improve indoor environmental quality:

- low VOC adhesives used for carpet tiles
- no VOC paints and low-VOC adhesives
- high MERV filters used throughout
- xeriscaping to minimize use of pesticides and irrigation
- indoor air quality management plan required during construction
- air quality testing prior to occupancy
- increased ventilation using outdoor air
- carbon dioxide sensors to determine need for outside air
- 100 per cent daylighting in all classrooms.

continued

Through implementing these strategies to improve indoor environmental quality, RFES was able to pursue goals of increasing comfort levels, reducing absenteeism and increasing student productivity and performance.

Experiential learning

Implementing sustainability features at the school was an important opportunity to enhance experiential learning for students. The entry area at the school features a sundial, which allows students to connect seasonal changes with the different positions of the Earth and sun (Figure 9.12a). Integrative signage installed throughout the school buildings and site also helps to educate students, staff, the community and visitors about sustainable features of the building and their benefits (Figure 9.12b). RFES incorporates real-time monitoring of the sustainable design features in the building so that students can use their computer monitoring systems to compare the performance of their system with similar sustainability features in other schools across the country and around the world (Figure 9.13). With these assets, the school can be used as a case to teach students about the concept of sustainability and influence their attitudes and behaviour toward sustainability.

Figure 9.12a, b Experiential learning opportunities for students

Breaking through the first cost barrier

Pursuing a higher degree of sustainable design and construction is often challenged by higher first costs, even though some sustainability features can lower operational costs compared to conventional approaches. To achieve the highest level of sustainability within a constrained budget, a project has to use an integrative design approach as described in Chapter 6. The first step of the integrative design approach is to form a collaborative team including the owners, architects and engineers; sustainable design consultants; landscape architects; operations and maintenance (O&M) staff; health, safety and security experts; the general contractor and key subcontractors; cost consultants; value engineers; and occupant representatives. This collaborative team must work together from the start to develop innovative sustainability solutions that meet energy, environmental and social goals while maintaining first costs within budget.

continued

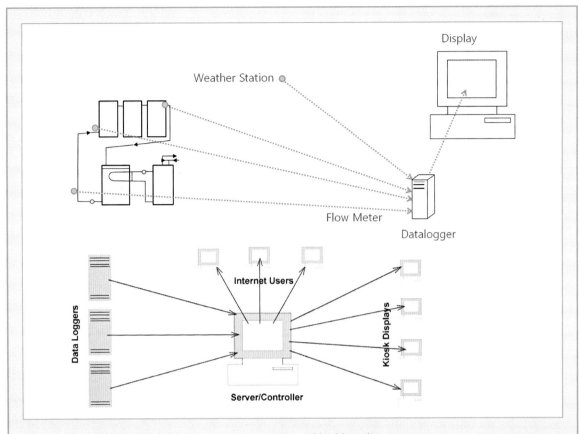

Figure 9.13 Web-based sustainability education using real building data

Since adding sustainability strategies into a project in isolation from one another potentially increases first costs compared to conventional approaches, the integrative design team has to combine sustainability strategies together and optimize them simultaneously to reduce first costs while achieving the goals of sustainability. For example, by improving the building envelope, the design team can often downsize or eliminate the heating, ventilation and air conditioning (HVAC) system around the perimeter of the building and also downsize the primary HVAC system. Downsizing the HVAC system can also reduce the size of ducts and overall building height, which may help pay for the envelope improvements. The RFES project is a good example of how to pursue sustainability on a budget. The next section describes specific sustainable design strategies that have been implemented in the RFES project expressly to reduce or eliminate first cost premiums while saving lifecycle energy and water costs as well.

Site design

One obvious strategy to reduce first costs is to take advantage of the site by properly orienting the building to maximize the southern exposure, if the building is located in the northern hemisphere, and minimize east and west glazing. In addition, it is possible to use existing trees, landscaping and natural berms to protect against winter winds and reduce unwanted solar gain to reduce cooling loads. These site design strategies can save operating costs by reducing energy consumption, and may also reduce the first costs for building envelope and HVAC requirements.

continued

At Reedy Fork, the design team chose to retain indigenous vegetation on site to minimize water needs, and adopted xeriscape planting strategies to reduce the first cost from that of non-native plants. In addition, the design team created two functionalities from one system by using the pavement that was part of secondary fire lanes as hard-surface play areas for uses such as basketball courts. This strategy saved the first cost of building two systems instead of one. Finally, the school structure was located on the higher part of the site in order to take advantage of natural slopes for drainage, reducing the need for drainage infrastructure at no additional cost.

Daylighting and underfloor air distribution system

Daylighting has many benefits including saving energy, increasing productivity and improving health. Integrated daylighting strategies along with adjusting the size of the HVAC system achieved simple dollar paybacks ranging from two or three years in the most expensive scenario, to having both first cost and lifecycle cost advantages in the least expensive scenario. On the RFES project, the design team actively used daylighting strategies with south-facing clerestories. This strategy was coupled with underfloor air distribution to improve air quality and to control the first costs of the HVAC system. While most buildings incorporating daylighting and underfloor system strategies generally add substantially to the first cost of a project (at least $15/sq ft in this context), the design team at Reedy Fork adopted a holistic approach to lower the cost premium to $2.73/sq ft. The main mediators of first costs in this case were as follows:

- Daylighting performance was improved, causing glass-to-floor ratio to fall below a typical side-lit solution, thus saving on envelope costs since glazing is a significant cost factor.
- The white single-ply roofing further reduced daylighting glass areas by increasing reflected lighting into the clerestory areas of the classrooms on the north side.
- Installing clear double-glazing in the glass areas helped to maximize the visible light transmission and lower the first cost from that of low-E glazing.
- The single sloped ceiling improved reflectance back into the space and eliminated the need for a roof monitor well cavity.
- The elimination of east- and west-facing glass except view glass not only lowered the first costs but also reduced peak cooling loads and, in turn, reduced the size of installed cooling equipment.
- Through the incorporation of an optimally sized daylighting strategy, peak cooling loads were significantly lower and chiller capacity was able to be reduced.
- By implementing indirect lighting, the ceiling cavity was reduced from 14 in to 8 in.
- The underfloor air distribution system reduced the ceiling cavity by 2 ft to 3 ft. This reduced high-end finishes on the exterior of the facility due to shorter floor-to-floor height and also eliminated several masonry courses.
- The underfloor system strategy also eliminated scaffolding costs and associated safety concerns during construction while easing the installation and coordination problems associated with overhead ductwork, plumbing, electrical and control wiring.
- The simplified construction framing of the roof assembly resulted in material and installation cost savings.

Figures 9.14a to 9.14e illustrate the various classroom design options for daylighting and underfloor systems, along with their first cost premiums.

Electrical systems for lighting

RFES used PV-powered exterior lighting for remote locations where conduit and trenching cost exceeded the marginal cost of the PV system. Figure 9.8, p. 495 shows an example of the light for

continued

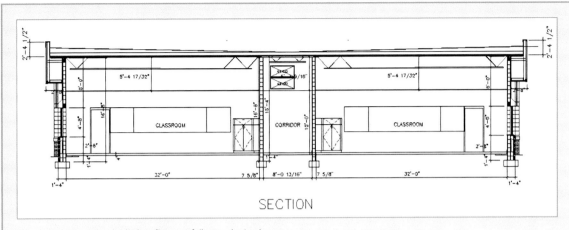

Figure 9.14a No-daylight, flat roof (base design)

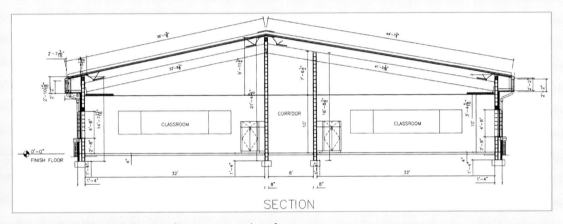

Figure 9.14b No-daylight, standing seam metal roof

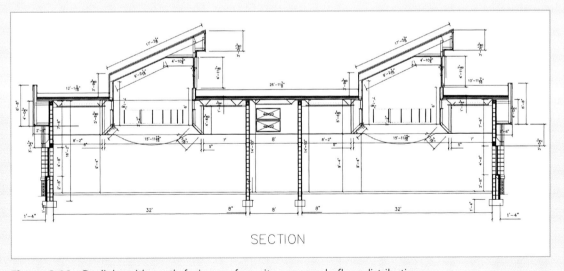

Figure 9.14c Daylight with south-facing roof monitors, no underfloor distribution

continued

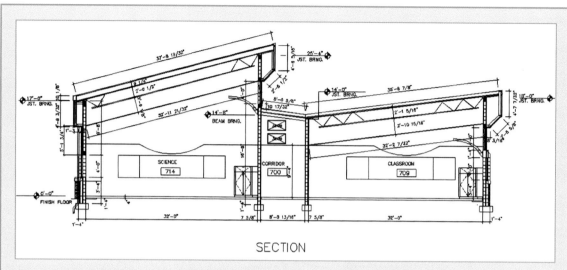

SECTION

Figure 9.14d Daylight with south-facing clerestories, no underfloor air distribution

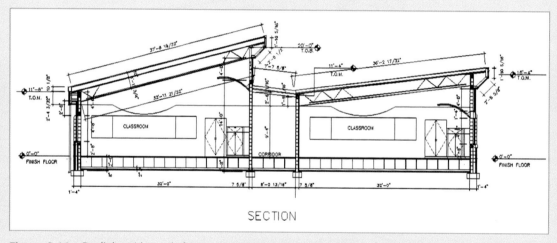

SECTION

Figure 9.14e Daylight with south-facing clerestories and underfloor air distribution

the school sign at the front of the school with its top mounted PV system, installed at a reduced first cost. In addition, the school used ganged fluorescent light fixtures in the gymnasium, which cost less and provide an additional advantage by being dimmable. The design team also specified the minimal number of lighting fixtures in hallways to minimize overlighting, which lowered first costs as well as operating costs.

Building shell

RFES used white, single-ply roofing materials and radiant barriers as shown in Figure 9.15. Since 90 per cent of heat gain from the roof was the result of radiant gains, this roof had several advantages: it stays reflective for a long time; it can be used to bounce light into daylighting apertures, thereby reducing the size of glazing needed; and it is also good for rainwater catchment system collection areas. RFES also properly placed radiant barriers that could reduce installed cooling equipment enough to offset the cost of the barrier material. To improve lighting inside rooms, all

continued

interior walls were painted light colours, and the rooms were provided with highly reflective ceiling materials and light floor finishes. Lighter-coloured finishes reduced the number of lights that had to be installed in the building. This consideration not only reduced the initial cost of installing lighting fixtures but also reduced operating costs over the building's life.

Mechanical systems

In RFES, seasonal and hourly space conditioning loads were carefully analysed to determine full-load conditions and fully account for the benefits of daylighting in terms of cooling load reduction. In addition, the

Figure 9.15 Radiant barrier installed beneath a white single-ply roof membrane

integrative design team optimized the mechanical system as a complete entity to allow for the interaction of various building system components, avoiding oversizing of equipment. Finally, the team carefully investigated the unit sizes of mechanical systems because they wanted to create an opportunity to reduce the overall cooling load to the next smaller chiller unit size by investing in other design elements (Table 9.1). These approaches further reduced the first cost premium and simultaneously reduced future annual operating costs.

First cost comparison

Although RFES included many sustainability features, the actual construction cost was similar to the average bid at the time for an elementary school in the state of North Carolina. This low first cost premium was primarily the result of whole-building, whole-site approaches to the project during the early design process. The final bid for RFES was $151/sq ft in 2006. This bid price was subsequently increased to $157/sq ft because of the need to have an accelerated schedule to 13 months. The average bid for schools in North Carolina during this time period was $147/sq ft in 2006. Since RFES incorporated many sustainability features, the school was expected to be among the top 10 per cent of schools in the nation designed to earn the Energy Star® label. Therefore, RFES would enjoy not only lowering annual energy consumption in the building, but also reduced greenhouse gas emissions that would have resulted from energy production.

Integrative design to achieve energy cost savings

Many sustainability features can help effectively minimize a building's energy consumption during its service life. As previously discussed, integrative design (whole building design) considering all architectural and mechanical features at very early stages helps to minimize a building's annual energy use and reduce energy costs while maintaining comfort and quality. For example, the daylighting strategies coupled with the underfloor system installed in RFES resulted in a slightly higher first cost, but the resulting annual cost savings resulted in a lower lifecycle cost. To illustrate the whole building design approach, energy-efficient strategies that can lower the annual energy costs are discussed in this section.

To explain this integrative design concept for energy efficiency strategies, the annual energy costs of RFES were analysed. This approach illustrates how an integrative design approach to alter architectural elements and mechanical systems can lower annual energy costs. The base case

continued

Table 9.1 Parameters of energy simulation

Parameter	Value
Weather and climatic data	Greensboro, NC, USA (TMY2)
Run period	31 December – 1 January
Cooling design day conditions	Indoor 75°F, 50% RH Outdoor 90°F Dry Bulb 70°F Wet Bulb
Heating design day conditions	Indoor 70°F Outdoor 10°F
Building area	Building area = 85,000 ft² (approx)
Construction	Ext. Wall (16") U Value = 0.027 Btu/hr·ft²·F° Ext. Wall (20") U Value = 0.026 Btu/hr·ft²·F° Monitor U Value = 0.042 Btu/hr·ft²·F° Raised Floor U Value = 0.415 Btu/hr·ft²·F° Slab Floor U Value = 0.709 Btu/hr·ft²·F° Metal Roof U Value = 0.032 Btu/hr·ft²·F° Membrane Roof U Value = 0.031 Btu/hr·ft²·F° View Glass U Value = 0.34 Btu/hr·ft²·F° Daylight Glass U Value = 0.55 Btu/hr·ft²·F° Ext. door U Value = 0.067 Btu/hr·ft²·F°
Occupant schedule	Students 8 am to 4 pm (year round) Admin 8 am to 4 pm (year round) Media 8 am to 4 pm (year round) Gym 8 am to 4 pm (year round) Cafe 8 am to 4 pm (year round) Kitchen 8 am to 4 pm (year round)
Ventiltion (outdoor air)	15 cfm per person (gymnasium) 15 cfm per person (classroom use) 20 cfm per person (office use) 55°F DB Economizer Setting
Natural gas	PSNC Energy
Electricity	Duke Power
Interest rate	6% interest rate

building, to which the more sustainable building with integrated energy-efficient features was compared, was assumed to meet the levels of energy efficiency in the American Society of Heating, Refrigerating and Air Conditioning Engineers (ASHRAE) 90.1–2001 standard. The total construction cost of this base case building was estimated and the incremental first costs for sustainable features were also estimated. The annual energy consumption was simulated using DOE 2.1 based on energy simulation models, eQUEST 3.63 and other daylighting simulation tools. After the annual energy uses were simulated using energy modelling, the annual energy requirement was compared with the energy-use estimates of the base case building, and estimated energy savings and the associated

continued

incremental costs were calculated. After these calculations, a lifecycle cost analysis was conducted to calculate the net present value (NPV) and payback periods to compare the base case school building with the actual design that incorporated various energy-efficiency features.

Energy simulation for the RFES building

Annual energy consumption was calculated using DOE 2.1-based energy simulations. To support this analysis, it is necessary to identify important parameters used in energy simulations including envelope thermal properties, internal loads and schedules, and HVAC system operation schedules (see Table 9.1). Several resources were used to assist in the energy modelling including ASHRAE Standard 90.1–2001 and ASHRAE Standard 62–2001. The base case school design is based on these ASHRAE standards.

Selection of sustainable features

Many sustainability features affecting energy performance were available for use as part of the RFES project. These energy-efficient strategies and technologies included basic building design, use of passive systems, and use of high-performance mechanical systems (Figure 9.16). However, given the large number of sustainability features related to energy, it was necessary to narrow down to a number of design options to maximize energy efficiency in the building.

Figure 9.17 shows the general procedure to select sustainable features for energy optimization. The initial step was to collect and identify all sustainability design strategies that could improve energy efficiency in the building or that could generate energy using renewable energy sources. The second step was to have an energy charrette and integrative design workshop involving project stakeholders including architects, engineers, users and others in the process of narrowing down the set into implementable sustainability features for this specific building project. The third step was to finalize a number of design options for which energy models were developed and detailed cost studies conducted.

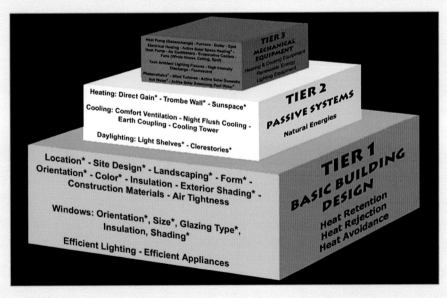

Figure 9.16 Three-tier approach to the design of heating, cooling and lighting

continued

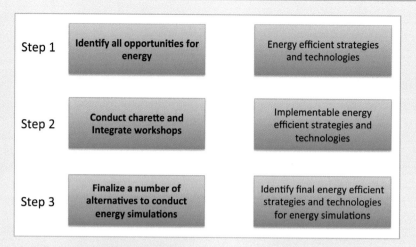

Step 1	Identify all opportunities for energy		Energy efficient strategies and technologies
Step 2	Conduct charette and Integrate workshops		Implementable energy efficient strategies and technologies
Step 3	Finalize a number of alternatives to conduct energy simulations		Identify final energy efficient strategies and technologies for energy simulations

Figure 9.17 Procedure to choose appropriate energy efficiency strategies

Using these three steps, a number of alternatives were identified to describe how sustainability features for energy could lower the annual energy consumption in the RFES project compared with the base case school building that complied with ASHRAE Standard 90.1–2001. The alternatives used in this analysis for the school building are described in the following subsections.

Base case school building

The base case building was constructed of systems based on the minimum require-ments set forth by ASHRAE Standard 90.1. No daylighting glass was included. In this case the building was simulated with the front facing east (Figure 9.18). The central plant in the base case building consisted of one standard efficiency air-cooled screw chiller using a $10°\Delta T$ and two 80 per cent efficient condensing-type boilers. Classroom wings, the media centre and administration areas were served by variable air volume (VAV) systems. The VAV air handlers were equipped with constant volume fans, chilled water coils for cooling and hot water coils for preheating. Air was distributed overhead (mixing ventil-ation). Each classroom and major spaces were

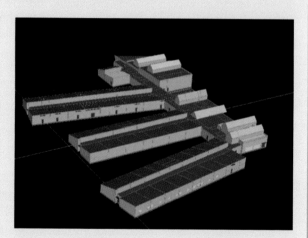

Figure 9.18 Rendering of the base case school building

served by VAV terminal boxes equipped with hot water heating. Multipurpose, dining, kitchen and music/art areas would be served by overhead constant volume air systems. Constant volume air handlers were equipped with constant volume fans, chilled water coils for cooling and hot water coils for preheating and reheating. Ventilation (outside) air was provided through the central air-handling units. An air-side economizer cycle was used to take advantage of free cooling when outside air temperature was at 55°F or lower.

continued

Alternative 1: Daylighting

The daylighting alternative was one of the most beneficial in that it was very cost effective while at the same time having the potential to improve both the health and well-being of the students and teachers. As noted earlier, daylighting strategies included improved glazing, south-facing clerestories, white single-ply roofing, curved translucent light shelves, highly reflective ceiling tiles and shading devices.

Alternative 2: Improving insulation

Increasing the levels of insulation in wall and roof surfaces would reduce energy consumption in the building and improve occupant thermal comfort. Accordingly, the second alternative was to improve wall, roof and floor insulation values beyond the ASHRAE Standard 90.1 to a wall insulation level of R-38 and a roof insulation level of R-32.

Alternative 3: Underfloor air distribution systems

In the third alternative, underfloor air distribution systems were added. The central plant would consist of one standard-efficiency air-cooled screw chiller using a 10°ΔT and two 80 per cent efficient condensing-type boilers, the same as in the base case building. Classroom wings, the media centre and administration areas would be served by standard underfloor air distribution (UFAD) systems. The UFAD air handlers were equipped with VAV fans, chilled water coils for cooling, and hot water coils for preheating. Air was distributed underfloor (mixing ventilation). Each classroom and major space would be served by underfloor air terminal boxes equipped with a control damper and hot water heating. Multipurpose, dining, kitchen and music/art areas were served by overhead constant volume air systems. The constant volume air handlers were equipped with constant volume fans, chilled water coils for cooling, and hot water coils for preheating and reheating. Ventilation (outside) air was provided through the central air-handling units. An air-side economizer cycle was used to take advantage of free cooling when outside air temperature was at 60°F or lower.

Alternative 4: Premium underfloor air distribution system

In the fourth alternative, the building would be served by an upgraded underfloor air distribution system. The central plant would consist of one standard-efficiency air-cooled screw chiller using a 10°ΔT and two 80 per cent efficient condensing-type boilers, the same as in the base case building. Classroom wings, media centre and administration areas would be served by premium UFAD systems. UFAD air handlers were equipped with VAV fans, chilled water coils for cooling and hot water coils for preheating. Air was distributed underfloor (mixing ventilation). Each classroom and major space would be served by premium underfloor air terminal boxes equipped with a constant volume fan and hot water heating. Air would be supplied through the floor by varying the geometry of the floor outlets. Multi-purpose, dining, kitchen and music/art areas were served by overhead constant volume air systems. Constant volume air handlers were equipped with constant volume fans, chilled water coils for cooling, and hot water coils for preheating and reheating. Ventilation (outside) air was provided through the central air-handling units. An air-side economizer cycle was used to take advantage of free cooling when outside air temperature is at 60°F or lower.

Alternative 5: Premium efficiency air cooled screw chiller

In the fifth scenario considered, the central plant would consist of one premium efficiency air-cooled screw chiller using a 14°ΔT and two 94 per cent efficient condensing-type boilers. Classroom wings,

continued

Table 9.2 Input parameters and energy consumption

		Base case	Alt 1: Day- lighting	Alt 2: Improved insulation	Alt 3: UFAD	Alt 4: Premium UFAD	Alt 5: Premium chiller
SHELL	Wall insulation	R-6.6	R-6.6	R-38	R-38	R-38	R-38
	Roof insulation	R-15.4	R-15.4	R-32	R-32	R-32	R-32
	Glazing	View glass: ASHRAE 90.1 No daylighting	View glass: low-e insulated Daylight: clear insulated	View glass: low-e insulated Daylight: clear insulated	View glass: low-e insulated Daylight: clear insulated	View glass: low-e insulated Daylight: clear insulated	View glass: low-e insulated Daylight: clear insulated
BUILDING SYSTEM	Cooling plant	Air cooled screw chiller (10° ΔT)	Air cooled screw chiller (10° ΔT)	Air cooled screw chiller (10° ΔT)	Air cooled screw chiller (10° ΔT)	Air cooled screw chiller (10° ΔT)	Air cooled screw chiller (14° ΔT)
	Heating plant	80% efficient hot water boiler	80% efficient hot water boiler	80% efficient hot water boiler	80% efficient hot water boiler	80% efficient hot water boiler	94% efficient hot water boiler
	Air handling units	Variable air volume Air side economizer cycle and constant volume systems	Variable air volume Air side economizer cycle and constant volume systems	Variable air volume Air side economizer cycle and constant volume systems	Standard underfloor air distribution Air side economizer cycle and constant volume systems	Premium underfloor air distribution Air side economizer cycle and constant volume systems	Premium underfloor air distribution Air side economizer cycle and constant volume systems
	Air distribution	Overhead air distribution	Overhead air distribution	Overhead air distribution	Underfloor and overhead air distribution	Underfloor and overhead air distribution	Underfloor and overhead air distribution
	Cooling load peak	257 tons 3081 kBtu/hr	209 tons 2510 kBtu/hr	201 tons 2415 kBtu/hr	200 tons 2403 kBtu/hr	197 tons 2364 kBtu/hr	197 tons 2364 kBtu/hr
ENERGY	Heating load peak	(2536) kBtu/hr	(2509) kBtu/hr	(2362) kBtu/hr	(2398) kBtu/hr	(2398) kBtu/hr	(2398) kBtu/hr
	Energy performance	46.3 kBtu/sqft/yr	38.9 kBtu/sqft/yr	36.8 kBtu/sqft/yr	36.8 kBtu/sqft/yr	36.5 kBtu/sqft/yr	32.7 kBtu/sqft/yr
	Electric end use	673,835 kWhr	508,371 kWhr	491,206 kWhr	459,487 kWhr	454,562 kWhr	439,961 kWhr
	Natural gas end use	17,420 Therm	16,601 Therm	15,415 Therm	16,413 Therm	16,382 Therm	13,563 Therm
	Lighting end use	258,977 kWhr	131,611 kWhr	131,611 kWhr	131,611 kWhr	131,611 kWhr	131,611 kWhr

continued

media centre and administration areas would be served by premium UFAD systems. UFAD air handlers were equipped with VAV fans, chilled water coils for cooling and hot water coils for preheating. Air was distributed underfloor (mixing ventilation). Each classroom and major space was served by premium underfloor air terminal boxes equipped with a constant volume fan and hot water heating. Air was supplied through the floor by varying the geometry of the floor outlets. Multipurpose, dining, kitchen and music/art areas would be served by overhead constant volume air systems. Constant volume air handlers were equipped with constant volume fans, chilled water coils for cooling and hot water coils for preheating and reheating. Ventilation (outside) air was provided through the central air-handling units. An air-side economizer cycle was used to take advantage of free cooling when outside air temperature is at 60°F or lower.

Table 9.2 summarizes the detailed input parameters of five alternatives plus the base case along with. The next step was to simulate energy consumption using the DOE 2.1E-based energy simulation engine to calculate cooling peak load, heating peak load, energy performance and

Table 9.3 First cost of HVAC systems for alternatives and base case (in US$)

Economic component	Base case	Alt 1: Daylighting	Alt 2: Improved Insulation	Alt 3: UFAD	Alt 4: Premium UFAD	Alt 5: Premium Chiller
Cooling equipment	90,000	80,000	80,000	78,000	78,000	85,000
Heating equipment	20,000	20,000	20,000	15,000	15,000	30,000
Hydronic pumps	28,000	28,000	28,000	25,000	25,000	20,000
Hydronic piping and accessories	220,000	215,000	215,000	215,000	215,000	200,000
Air handling units	200,000	200,000	200,000	225,000	225,000	225,000
VAV boxes w/HW coil	85,000	85,000	85,000	100,000	100,000	100,000
Ductwork and accessories (air hwy. incl.)	350,000	350,000	350,000	350,000	300,000	300,000
Air distribution equipment	55,000	55,000	55,000	55,000	90,000	90,000
Exhaust fans	5000	5000	5000	5000	5000	5000
Unit heaters	5000	5000	5000	5000	5000	5000
Breechings and vents	5000	5000	5000	5000	5000	5000
Controls and instrumentation	175,000	175,000	175,000	175,000	200,000	200,000
Test and balancing	25,000	25,000	25,000	25,000	30,000	30,000
Miscellaneous	50,000	50,000	50,000	50,000	50,000	50,000
Total first cost for HVAC	**$1,313,000**	**$1,298,000**	**$1,298,000**	**$1,328,000**	**$1,343,000**	**$1,345,000**
Incremental first cost	**$0**	**$15,000**	**$15,000**	**$15,000**	**$30,000**	**$32,000**

continued

end-use consumption of electricity, natural gas and light. All simulated data is also summarized in Table 9.2.

Different alternatives that can reduce annual energy consumption have different design and construction costs. Thus, it was necessary to estimate incremental costs of energy efficient features that could enhance energy performance. The cost of each feature was estimated based on drawings and construction documents under the support of the architecture firm and the general contractor. To estimate the incremental first costs of energy efficient features, it was necessary to estimate the first cost adjustments based on the base case because energy efficient features could potentially reduce the size of HVAC systems. Thus, the final incremental first cost for energy efficient features was estimated by combining the change of construction costs and HVAC cost change. The incremental first costs related to energy saving features are summarized in Tables 9.3 and 9.4.

Table 9.4 Summary of incremental first costs (US$)

	Incremental first cost ($/sq ft)	Incremental first cost ($)	HVAC cost adjustment ($)	Total accumulated incremental cost ($)
Base case	0	0	0	0
Alternative 1: Daylighting	2.77	240,990	15,000	225,990
Alternative 2: Improved insulation	1.28	111,765	15,000	337,755
Alternative 3: UFAD	1.37	119,190	15,000	486,945
Alternative 4: Premium UFAD	1.37	119,190	30,000	501,945
Alternative 5: Premium chiller	1.37	119,190	32,000	503,945

Annual energy cost savings by energy-efficient features

In order to determine the best, most cost-effective energy-saving strategies for the RFES project, the annual energy consumption of the five alternatives as well as the base case building were simulated using the DOE 2.1E-based simulation tool. The unit price of electricity and gas were based on data from electricity and gas providers including Duke Energy and Piedmont Natural Gas. The electricity unit price was $0.09725/kWh and the natural gas unit price was $1.080130/therm. Total energy cost was equal to the cost of electricity plus the cost of gas.

Table 9.5 shows the total annual energy savings of the five alternatives compared with the base case design (Figure 9.19 and Table 9.5). The annual energy cost of the base case was $84,346, combining the electricity cost of $65,530 and the gas cost of $18,816. The following subsections describe each alternative compared with the base case and the previous alternative. In other words, the alternatives are considered to have a cumulative effect on energy consumption, with each alternative including the features of all previous alternatives.

continued

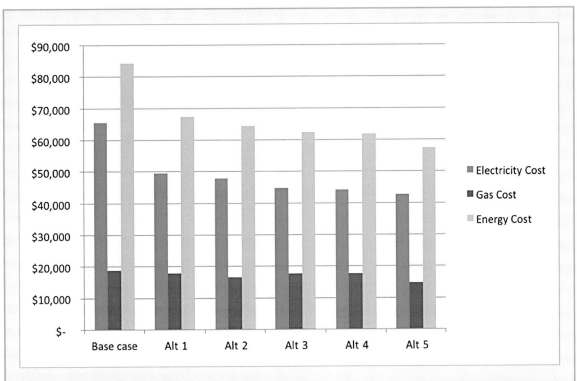

Figure 9.19 Annual energy costs for alternatives

Alternative 1: Daylighting

The implemented daylighting system would reduce annual energy consumption within the building, which ultimately reduced annual energy costs. Since the proposed daylighting strategy reduced electrical lighting use from 258,977 kWh to 131,611 kWh based on the simulation result, it would be possible to reduce annual electricity cost by $12,386 for lighting, from $65,530 to $49,439. Daylighting strategies also reduced the gas cost by $885, from $18,816 to $17,931, by reducing the building's heating demand in winter. Thus, total reduction of energy costs was about $16,976.

Alternative 2: Improving insulation

Improving the level of insulation would reduce both heating and cooling loads of the building. Based on the annual energy consumption simulated by energy modelling, improving insulation would further reduce annual electricity cost by $1670 compared to Alternative 1. In addition, better insulation would also further reduce the annual gas cost by $1281 compared to Alternative 1. Therefore, improving insulation could reduce the annual energy cost by $19,926 compared to the base case building, or by $2950 compared to Alternative 1.

Alternative 3: Underfloor air distribution systems

The underfloor air distribution system was estimated to additionally reduce the annual electricity cost by $3084 compared to Alternative 2, although the annual gas cost would increase by $1078. Since the annual electricity cost saving exceeded the increase of annual gas cost, the underfloor air distribution cost could still save annual energy costs of $21,933 compared to the base case, or $2007 compared to Alternative 2.

continued

Table 9.5 Annual energy cost savings by energy-efficient features

	Base case	Alt. 1: Daylighting	Alt. 2: Improved Insulation	Alt. 3: UFAD	Alt. 4: Premium UFAD	Alt. 5: Premium Chiller
Electricity (kWhr)	673,835	508,371	491,206	459,487	454,562	439,961
Total electricity cost ($)	65,530	49,439	47,770	44,685	44,206	42,786
Electricity savings ($)	—	16,091	17,761	20,845	21,324	22,744
Gas (therms)	17,420	16,601	15,415	16,413	16,382	13,563
Total gas cost ($)	18,816	17,931	16,650	17,728	17,695	14,650
Gas savings ($)	—	885	2166	1088	1121	4166
Total energy cost ($)	84,346	67,370	64,420	62,413	61,901	57,436
Savings compared to base case	—	16,976	19,926	21,933	22,445	26,910
Savings compared to previous alternative	—	16,976	2950	2007	512	4465

Alternative 4: Premium underfloor air distribution system

The premium underfloor air distribution system would further reduce the annual electricity cost by $479 and the gas cost by $33 compared to Alternative 3. Overall, this option would reduce the annual energy cost by $22,445 compared to the base case, and by $512 compared to Alternative 3.

Alternative 5: Premium efficiency air-cooled screw chiller

This option would additionally reduce the annual electricity cost by $1420 and the annual gas cost by $3045 compared to Alternative 4. It would lower the annual energy cost to $57,436, resulting in a potential annual energy savings over the base case of $26,910.

Lifecycle cost and payback period

Lifecycle cost is a very important decision-making criterion because it considers all the costs associated with a facility, from construction costs to operation, maintenance, repair and replacement costs throughout the facility's life span. Thus, calculating a net present value (NPV) by lifecycle cost analysis (LCCA) for all alternatives and the base case could help decision makers evaluate the financial effectiveness of energy-efficiency strategies in the RFES project because it could identify the relationship between first cost premiums of energy-efficient strategies and their potential O&M savings during the operation phase. To conduct LCCA, the following assumptions were applied:

- real discount rate for the analysis: 3.0 per cent (OMB Circular No. A-94)
- length of analysis: 25 years
- energy price escalation: 3.0 per cent.

Table 9.6 shows both the first cost premium over the base case and the total lifecycle cost for each alternative as a result of this analysis.

continued

Table 9.6 First cost premium and lifecycle costs

	First cost premium ($)	Total lifecycle cost ($)	Annual energy cost ($)
Base case	—	1,659,670	84,346
Alternative 1: Daylighting	225,990	1,557,746	67,370
Alternative 2: Improved insulation	337,755	1,611,390	64,420
Alternative 3: UFAD	486,945	1,721,105	62,413
Alternative 4: Premium UFAD	501,945	1,726,005	61,901
Alternative 5: Premium chiller	503,945	1,581,147	57,436

Based on the LCCA for all alternatives and the base case, the daylighting strategy (Alternative 1) resulted in the lowest LCC of $1,577,746, which was $101,924 lower than the base case. However, Alternative 5 (premium chiller plus insulation, daylighting and premium UFAD) had the second lowest LCC of $1,581,147 and resulted in the lowest annual energy consumption in the building, even though it required first cost premiums.

After estimating all first cost premiums and LCC of energy-efficient alternatives, the project team, mainly the architecture firm and the owner, made a decision to implement Alternative 5 because it would not only minimize annual energy consumption but also save operating cost over the building's life. Therefore, RFES would consume less than half the energy of typical schools and was chosen as one of those in the top 10 per cent of school facilities in the United States designed to earn the Energy Star label. This business case for implementing energy-efficiency strategies illustrates how holistic, innovative and sustainable energy-efficiency strategies can contribute to the goal of sustainability in the building while reducing or eliminating first cost premium barriers.

Integrative design to achieve water cost savings

Water use is the other cost considered for this business case analysis. Water efficiency in a capital project can be achieved by a number of technologies available on the market that reduce indoor water consumption compared to standard technologies. These include ultra-low-flow fixtures and tap aerators, no-water urinals and dual-flush toilets. A building can also reduce potable water consumption by substituting non potable water from harvested rainwater or treated wastewater, using recirculation water systems, undertaking leak detection and repair, employing sustainable landscaping, and other strategies.

Additionally, reduction of wastewater exported to municipal wastewater treatment systems can also result in significant cost savings from avoided infrastructure costs. Some reductions can be achieved through upstream water conservation efforts for potable water use in the building. Significant reductions are also possible through managing stormwater on site, which can result in the reduction or elimination of infrastructure to collect and convey stormwater to off-site treatment systems.

The reduction of water consumption and wastewater generation by a building reduces annual water utility bills for the owner, minimizes the need for municipal infrastructure investments, and reduces energy used for transporting and treating water. In cases where rainwater is captured to displace potable utility-supplied water, these strategies can also reduce pollutants resulting from

continued

stormwater runoff and reduce local erosion and flooding. To achieve annual water cost savings within its building, the RFES design team considered two alternatives: including water-efficient fixtures, and installing a rainwater harvesting system, as well as both together. The choice of water saving strategy was based on the preliminary cost-benefit study that compared different alternatives.

Design information for water saving

Occupants' estimated water consumption must be determined by calculating full-time equivalent (FTE) and transient occupants and applying appropriate fixture use rates to each. Before calculating FTE and transient occupants, it was necessary to identify the number of students and full-time staff, including administrators, support staff, teachers and custodians. Therefore, the design team assumed and estimated that the school operated during a normal school year (175 days) in a co-education session with 750 students, 75 administrators and support staff, and four custodians with an approximately even gender split (50 per cent male and 50 per cent female). This assumption helped the design team to estimate annual water consumption in the building.

Water-efficient fixtures

The water use from toilet and urinal fixtures within the building was estimated by using the Raincheck water use program developed by Innovative Design Inc. located in Raleigh, North Carolina. The program estimated the volume of annual water use within the school facility in North Carolina. The flow rates of water efficient fixtures used in the simulation are summarized in Table 9.7.

Estimated water savings for advanced fixtures

Based on these estimates, RFES would annually need 818,730 gallons of potable water from the city water authority or a well under the base case scenario (baseline fixtures) without any on-site recovery of rainwater for reuse. If advanced water-efficient fixtures were chosen, the annual water consumption was decreased to 409,365 gallons, which saved about 50 per cent of potable water required. In addition, 50 per cent of wastewater generation within the building would also be eliminated, which could also reduce the need for municipal sewer facilities and costs associated with larger sized collection and conveyance piping.

Table 9.7 Water efficiency fixtures – flow rate (gallons/flush)

Fixture type	Advanced fixtures	Baseline fixtures*
Water closet (female)	0.8	1.6
Water closet (male)	0.8	1.6
Urinal (male)	0.5	1.0

* Set by the Energy Policy Act of 1992.

Incremental costs for advanced fixtures

Under this scenario, using advanced fixtures that exceed the minimum standards set by the Energy Policy Act would increase the cost of fixtures and installation. Table 9.8 shows the incremental costs of advanced fixtures. Based on a discussion with the architecture firm and construction company, the incremental cost of advanced fixtures was $0.75/sq ft in the RFES project. The total incremental

continued

costs were $65,250 (87,000 sf * $0.750/sf) in 2006. Since that time, the cost differential for fixtures has diminished, and some low flow fixtures and fittings are available for the same cost as conventional alternatives.

Rainwater collection system

Using an integrated rainwater harvesting system was a second strategy considered in this project to reduce potable water consumption and minimize the need to collect and treat stormwater off-site at the municipal treatment plant. A rainwater harvesting system collects rain from half the roof area of the school and diverts it to an underground storage tank. The rainwater is then pumped from the tank to the school, filtered, chlorinated, dyed light blue, and used for flushing each toilet in the school. Considering the annual rainfall and volume of water used in the building, a 45,000 gallon water storage tank was selected to store rainwater from 37,000 sq ft of collection area on the roof. The primary driver of tank sizing was based on the amount of rainfall expected to be received by the roof area, since the design team wished to eliminate the need to treat stormwater off-site. Capture of rainwater by the rainwater harvesting system along with other on-site stormwater management features would significantly reduce the need for infrastructure to capture and convey stormwater to the municipal treatment plant, which was located at a significant distance from the project site. The design details of the system are described in Figure 9.20.

Estimated water savings for the rainwater harvesting system

Using rainwater for toilet flushing could significantly reduce potable water use. Based on the rainwater calculation algorithm developed by Innovative Design, Inc., the rainwater storage system would collect 859,063 gallons of rainwater annually that could be used for toilet flushing, saving an estimated 767,565 gallons annually (Table 9.8). This would not only reduce 94 per cent of potable water used, but also eliminate 90 per cent of stormwater runoff from the site to the local wastewater treatment system (Figure 9.21). This assumption was based on the notion that all stormwater collected from the roof and other impervious surfaces would be captured and sent for treatment off-site in the base case scenario.

Incremental first cost for the rainwater harvesting system

Installing the rainwater harvesting system also required additional costs for excavation, components of the rainwater harvesting system including storage tanks, conveyance piping, filtration/treatment systems, and other infrastructure. As shown in Figure 9.22, the first cost premium for the rainwater harvesting system was $122,962, including $59,653 for the underground storage cistern, $46,309 for equipment, and $17,000 of additional plumbing components.

Comparing two water-saving alternatives

For the RFES project, the design team and the school board made a decision to only install the rainwater harvesting system instead of water-efficient fixtures after considering, among other things, first cost premiums and potential water cost savings (Table 9.8). On the surface, this decision may not appear to make economic sense on the basis of an LCCA. However, in this case, decision makers had a perception that high-efficiency fixtures were extraordinarily expensive in 2006 when the decision was being made. Given projected future water shortages in the Atlantic region of the United States, external environmental conditions may also have contributed to the decision to invest in a rainwater harvesting system as a high priority for the RFES project. Upon being interviewed several years after project completion, members of the design team believe that they would definitely make

continued

Guilford Reedy Fork Elementary School		August 12, 2011

STATE: NORTH CAROLINA

REGION: 3103 Northern Piedmont   Without Irrigation

Building Data		**Return on Investment (ROI)**	
Total Roof Area:	86,400 sf	Estimated Net First Cost	$119,636
Target rain water supply:	100%	Net ROI after 15 years:	($49,922) COST

Estimated Water Use Data:		**Cistern Size and Optimized Collection Area**	
Fixture Type:	Conventional	Cistern Size	45,000 gal
Number of Students:	750	Optimized Collection Area:	37,000 sf
Number of Staff:	75	Override Collection Area:	10,000 sf
Number of WC groupings:	4	Projected Rainwater Supply:	93.75%

Estimated Water Use Volume		**Estimated Volume of Makeup and Overflow**	
School Calendar Format	Conventional	**With Cistern**	**Average Year**
Spring term begins:	6-Jan	Yearly Volume from City/Well:	51,165 gal
Summer break begins:	6-Jun	Yearly Volume to Stormwater System:	89,765 gal
Fall term begins:	11-Aug	**Without Cistern**	
Winter break begins:	18-Dec	Yearly Volume from City/Well:	818,730 gal
Special Events per month:	1 /mon.	Yearly Volume to Stormwater System:	859,063 gal
Average attendance per event:	750		

Irrigation Data		**Cost Savings from Water Use**	
Irrigation area:	0.000 acre	Water cost (make-up volume):	$243
Irrigation Season Begins:	1-May	Cost with conventional water supply:	$3,889
Irrigation Season Ends:	30-Sep	First Year Water Cost Savings:	$3,646

Data for Yearly Life-Cycle Cost Analysis	
Local Water Cost (per 1,000 gal)	$4.75
Inflation Rate	2.0%
Water Rate Escalation	5.0%
Number of Years for ROI	15 yrs.

Volume of Water Use		**Links on the Web**
Typical School Day:	3,510 gal	North Carolina State Energy Office
Typical Weekend:	750 gal	American Rainwater Collection Systems Association
Typical Holiday:	162 gal	American Water Works Association
Holiday Transition:	1,230 gal	USGS Water Resources
Weekly Irrigation Water Load:	0 gal	Innovative Design

Environmental Benefits		**High Performance Ratings Systems**
Nitrogen reduction from stormwater	18.007 lbs/yr	NC High Performance Building Guidelines ver 2.0 - Section 3.3: 18-104 points
Less energy used to transport and treat water		USGBC LEED ver 2.0 - Section 3.1/3.2: 1-2 Points
Lessens local erosion and flooding		
Soft water - less impact on pipes, water heater, etc.		
Lessens impact on stormwater infrastructure		

© 2004 Innovative Design, Inc.   NC ver 100

Figure 9.20 Rainwater harvesting system model

continued

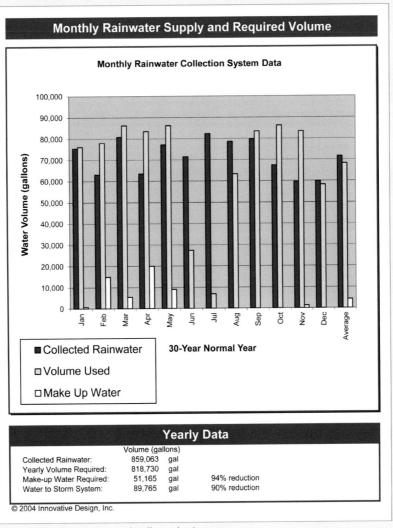

Figure 9.21 Refigured water volume and collected rainwater

Table 9.8 Comparison between two water-saving alternatives

Alternatives	Annual water saving (gallons)	Annual cost saving ($4.75/1000 gallons) ($)	First cost premiums ($)
Advanced fixtures	409,365	1944	65,250
Rainwater harvesting system	767,565	3646	122,962

continued

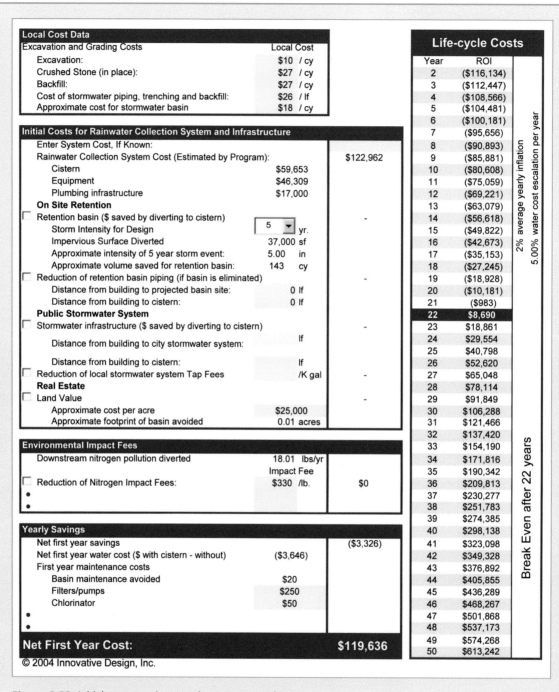

Local Cost Data		
Excavation and Grading Costs	Local Cost	
Excavation:	$10 / cy	
Crushed Stone (in place):	$27 / cy	
Backfill:	$27 / cy	
Cost of stormwater piping, trenching and backfill:	$26 / lf	
Approximate cost for stormwater basin	$18 / cy	

Initial Costs for Rainwater Collection System and Infrastructure		
Enter System Cost, If Known:		
Rainwater Collection System Cost (Estimated by Program):		$122,962
Cistern	$59,653	
Equipment	$46,309	
Plumbing infrastructure	$17,000	
On Site Retention		
☐ Retention basin ($ saved by diverting to cistern)		-
Storm Intensity for Design	5 ▾ yr.	
Impervious Surface Diverted	37,000 sf	
Approximate intensity of 5 year storm event:	5.00 in	
Approximate volume saved for retention basin:	143 cy	
☐ Reduction of retention basin piping (if basin is eliminated)		-
Distance from building to projected basin site:	0 lf	
Distance from building to cistern:	0 lf	
Public Stormwater System		
☐ Stormwater infrastructure ($ saved by diverting to cistern)		-
Distance from building to city stormwater system:	lf	
Distance from building to cistern:	lf	
☐ Reduction of local stormwater system Tap Fees	/K gal	-
Real Estate		
☐ Land Value		-
Approximate cost per acre	$25,000	
Approximate footprint of basin avoided	0.01 acres	

Environmental Impact Fees		
Downstream nitrogen pollution diverted	18.01 lbs/yr	
	Impact Fee	
☐ Reduction of Nitrogen Impact Fees:	$330 /lb.	$0
•		
•		

Yearly Savings		
Net first year savings		($3,326)
Net first year water cost ($ with cistern - without)	($3,646)	
First year maintenance costs		
Basin maintenance avoided	$20	
Filters/pumps	$250	
Chlorinator	$50	
•		
•		

Net First Year Cost:		**$119,636**

© 2004 Innovative Design, Inc.

Life-cycle Costs	
Year	ROI
2	($116,134)
3	($112,447)
4	($108,566)
5	($104,481)
6	($100,181)
7	($95,656)
8	($90,893)
9	($85,881)
10	($80,608)
11	($75,059)
12	($69,221)
13	($63,079)
14	($56,618)
15	($49,822)
16	($42,673)
17	($35,153)
18	($27,245)
19	($18,928)
20	($10,181)
21	($983)
22	$8,690
23	$18,861
24	$29,554
25	$40,798
26	$52,620
27	$65,048
28	$78,114
29	$91,849
30	$106,288
31	$121,466
32	$137,420
33	$154,190
34	$171,816
35	$190,342
36	$209,813
37	$230,277
38	$251,783
39	$274,385
40	$298,138
41	$323,098
42	$349,328
43	$376,892
44	$405,855
45	$436,289
46	$468,267
47	$501,868
48	$537,173
49	$574,268
50	$613,242

2% average yearly inflation
5.00% water cost escalation per year

Break Even after 22 years

Figure 9.22 Initial cost premiums and net present values

continued

a different choice and also incorporate advanced fixtures as part of the project if they had the opportunity to make the choice today.

In the RFES project, the sizing of the rainwater harvesting system was driven primarily by its use as a stormwater management strategy rather than as an alternative water source. Although employing water efficient fixtures would significantly reduce the demand for water provided by the rainwater system and might therefore allow the capacity and cost of the system's storage cistern to be reduced, in this case, the storage capacity had to be maintained at the same size to handle surges from rainfall on the building's roof area. Thus, cost savings were not able to be realized from tightly coupling the two alternatives and reducing costs of one system through investment in another.

This example highlights the fact that sometimes, sustainable building decisions take into account factors beyond simple economic analysis. Water-efficient fixtures in particular must consider operating and maintenance requirements and costs, and the ability of facility maintenance staff to keep them in proper operating condition during their service life. For instance, costs associated with rainwater harvesting systems to displace municipal water may include filter media, disinfection chemicals, dye to distinguish rainwater from municipal water in toilet fixtures, replacement UV lamps for treatment fixtures, and the energy to operate UV disinfection and circulation pumps. Regular water testing is also required by regulations. They should also consider the relative importance of water as an environmental constraint in the specific context of use.

Annual water cost savings and lifecycle costs

In the chosen scenario, the rainwater harvesting system would reduce the estimated annual potable water demand by 767,565 gallons (94 per cent of required water volume for toilet flushing in the building). It also was estimated to reduce the annual stormwater discharge from the site by 89,765 gallons (90 per cent reduction). Even though the break-even point of this investment was about 22 years with assumptions of 2 per cent inflation and 5 per cent water rate escalation, this water saving investment could decrease energy consumption required for pumping and treating potable water, mitigate local erosion and flooding, and significantly reduce the impact of the project on stormwater infrastructure. It could also reduce the risk of future water shortages expected to occur with greater frequency over the coming years, and reduce nitrogen pollution from stormwater runoff, which results from air pollution deposition on roof surfaces as well as fertilizers used in landscaping that run off from ground-level surfaces. Since the school was projected to operate over a 50-year service life, the calculated NPV for a 50-year-service life was estimated to have a financial benefit of $613,242 compared to the base case. This would result primarily from avoided costs of utility-supplied potable water for building operations. Although not included in the calculations, savings would also accrue from reduced environmental impact fees for nitrogen runoff and avoidance of costs to construct and maintain a stormwater detention or retention basin. These factors were considered negligible in this case, but could be very significant in projects where the cost of land is high, since stormwater basins can require significant land area to construct.

Post occupancy evaluation of Reedy Fork Elementary School[1]

The performance of a building during operations is based on a large number of multi-level factors, including thermal comfort, which relies on heating, ventilation and air-conditioning; illumination and visual comfort; occupants' satisfaction; and physiological and psychological comfort. Post occupancy evaluation (POE) is a method used to verify and measure occupant satisfaction as well as the performance of a green building in terms of water efficiency, energy efficiency and/or other key design goals after the building is in operation. Physical measurement of environmental conditions

continued

in the occupied building, including indoor air quality, thermal comfort, lighting quality, design quality, space location and satisfaction, are often part of a POE. POE can also be used to identify relationships between occupant satisfaction and perceptions of green building (Abbaszadeh et al. 2006; Bonde and Ramirez 2015; Gou et al. 2012; Wilkinson et al. 2014). The key aims of a POE are to (Aranda-Mena et al. 2009; Clara and Stefano 2011; Cooper 2001; Ku et al. 2008; Riley et al. 2010; Sobek et al. 1999):

- learn how a building actually performs in practice
- generate valuable feedback regarding specific green building strategies and technologies during the operation phase
- provide lessons learned for the future design of similar buildings
- provide building operations and maintenance staff with recommended operational adjustments that can help to improve indoor quality (usually measured in terms of occupant comfort).

A POE of Reedy Fork Elementary School was conducted to compare actual energy and water consumption to predicted consumption. In addition, the POE included a post occupancy survey and interviews with the principal, the energy manager and the facility managers to identify any issues they had with the building and their perceptions regarding its influence on students. Monthly energy and water consumption data were obtained from the maintenance department of Guilford County Schools and the city of Greensboro, North Carolina.

Post-occupancy survey results

Respondents to the occupant survey included teachers, teacher assistants, facility and operation managers, the principal and other staff. Respondents specifically identified classrooms and the library as the most satisfying spaces in the building, with an average Likert rating of 6.2 and 6.1 on a seven point satisfaction scale. They also indicated a high degree of satisfaction with the overall design quality of the building (6.12 out of 7) and green building strategies (5.8 out of 7).

With regard to their level of satisfaction with the building's indoor environment, the study asked participants to rate their perceptions of the quality of air, temperature, lighting and daylighting, acoustic quality and aesthetic quality (Figure 9.23). On a 7-point Likert scale, with 1 being low and 7 high, 62% of the respondents indicated a 6 or 7 for air quality, for an overall mean of 5.46, although

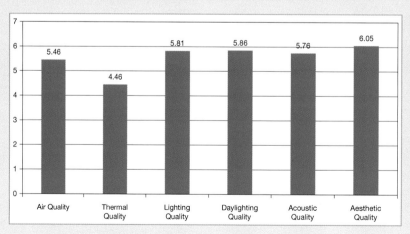

Figure 9.23 Satisfaction with indoor environmental quality

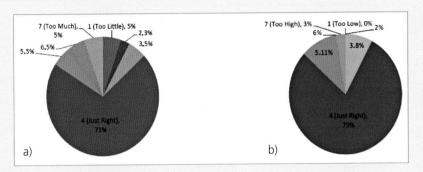

Figure 9.24 Daylighting a) and artificial lighting b) satisfaction

five out of 38 respondents scored this factor below 3. The respondents indicated their lowest satisfaction to be the thermal quality, with a score of 4.46. An interview with two facility managers and the school principal revealed this to be largely due to issues related to the control of the HVAC system and thermostat settings. This was confirmed by the survey respondents, who indicated a relatively low satisfaction of 3.49 out of 7 for temperature controllability. The facility managers indicated that thermostat settings and the HVAC system of the school are controlled by the central facility office of the school district. With regard to lighting and daylighting, the respondents indicated high levels of satisfaction, with scores of 5.81 and 5.86, respectively. As Figure 9.24 shows, 71% of the respondents indicated that they had just the right amount of daylight in their work spaces and 79% believed that they had the right level of artificial lighting. However, the librarian and the principal indicated that there was insufficient daylight in the library after changing the wall colours to a darker colour scheme.

In addition to strategies to improve the occupant experience, Reedy Fork Elementary School also incorporated green building technologies as part of their instructional mission, using technologies such as a real time monitoring system for energy consumption, a sundial and integrative signage in order to serve as a experiential learning tool for sustainability. To assess the effectiveness of these strategies, respondents were asked to assign a rating to each of the green building strategies. The top five green building strategies chosen by the participants were the innovative daylighting design, photovoltaics, rainwater for toilet flushing, recycled materials and the sundial. In addition, 27 out of 38 respondents indicated a score of 5 or above from a 1–7 Likert scale for the questions 'Do you think green building strategies and technologies in Reedy Fork Elementary school building will help students change their attitude and behaviour toward sustainability?' and 'Do you think green building will help students' learning and academic achievement?' This indicates that the participants were confident that green school building can help students change their attitude toward sustainability and that green building can also contribute to learning and academic achievement.

Energy performance

The school's integrative energy saving strategies and technologies included an innovative daylighting design, high efficiency envelope, efficient HVAC systems and onsite renewable energy sources in order to reduce annual energy consumption and carbon footprint of the building. During design, the project team developed energy simulations to evaluate the building's energy performance using many different energy saving strategies and technologies. Their simulation predicted an annual gas consumption of 13,563 Therms and an annual electricity consumption of 439,961 KWh. However, the annual gas consumption in 2014 was actually 25,382 Therms, 1.87 times higher than predicted, and the annual electricity consumption was 880,690 KWh, two times more than the energy simulation prediction. Despite these discrepancies, the school building was still identified as the most efficient elementary school building of the 53 elementary schools in the Guilford County School District (Graydon 2009).

While the discrepancy between simulated and actual energy consumption may seem large, such discrepancies are actually quite common. The primary purpose of simulation in design is to compare different possible building configurations to *each other* in choosing the best design solution, not to accurately predict the eventual performance of the final building. However, such discrepancies can ultimately pose challenges to building owners and developers who make investment decisions based on assumptions about likely energy costs. These results suggest that all the stakeholders need to revisit energy simulation to more accurately predict likely annual energy consumption. Measurement and verification of actual energy consumption in the building is also a key part of diagnosing possible problems. The principal and administrative staff at the school are not responsible for measuring the school's energy consumption and their bills are automatically paid by the school district, so suboptimal functioning of the building may go unnoticed without deliberately measuring and verifying the building's actual performance.

Water conservation

Reedy Fork Elementary sought to achieve water conservation goals primarily through the use of a rainwater harvesting system, which was predicted to save an estimated 767,565 gallons of water annually as well as prevent up to 90 per cent of the stormwater runoff that would otherwise be deposited into the municipal sewer system for treatment. Design simulations predicted an annual water demand of 50,000 gallons from municipal sources after taking into account water supplied by the rainwater harvesting system. This demand would include all potable water uses besides toilet flushing in the building, such as lavatory and kitchen uses, plus any additional water needed for toilet flushing in the event that the rainwater harvesting system did not provide an adequate supply.

The POE revealed an actual demand of 72,556 gallons from the municipal supply, using 2011 data. In this case, the net effect was to consume approximately 45 per cent more water than anticipated from the municipal system. The reasons for this discrepancy may be the result of multiple factors, including:

1 Greater demand than anticipated, including greater demand for non-toilet flushing uses of potable water or toilet flushing needs beyond what the rainwater system could provide, or
2 Lower yield than expected of the rainwater harvesting system, including a drought year in which rainfall is less than average or where the variability of rainfall exceeds the design capacity of storage tanks to provide a continuous water supply, poor filter maintenance practices that result in lower capture rates or water quality problems that require greater-than-anticipated system flushing using rainwater reserves.

In the case of Reedy Fork Elementary, 2011 was a year in which North Carolina was affected by greater than usual heat as well as relative drought conditions for rainfall (Messick 2011), so it is possible that the yield of the rainwater harvesting system was lower than expected. Since the system is not separately metered, the details of its performance are not known. However, the design of rainwater harvesting systems varies greatly, and standards for system sizing and design are only now being codified in the United States.

As discussed in Chapter 6 (see *Unanticipated consequences of water conservation*, p. 284), designing rainwater harvesting systems to balance water supply and public health concerns can be challenging. As with energy, this poses an additional challenge for building owners considering investing in decentralized infrastructure, since accurate prediction of performance requires information not only about average rainfall and expected use, but also the *timing* of rainfall to ensure adequate storage sizing while taking into account water age and associated water quality problems that may require flushing. Similar issues may also affect on-site renewable energy systems, particularly as weather patterns change.

Ultimately, while the business case for decentralized infrastructure such as on-site power production and rainwater harvesting is increasingly compelling, our understanding of the potential unanticipated consequences of these systems remains meagre. Owners considering the business case for green technologies and features must remember that all simulations are predictions, with many variables that could lead to performance discrepancies. Their primary utility is in comparing possible design scenarios, not necessarily providing pinpoint forecasts of performance under field conditions. Appropriate measures of risk and uncertainty should be included as part of the analysis to ensure that the results of energy and water modelling are not used out of context as the basis for operational decisions.

Sources: Images 9.3–9.15 and 9.17–9.24 courtesy of Innovative Design; Figure 9.16 courtesy of Lechner (2008)

Project planning with the business case in mind

Both immediate and long-term cost savings can be realized by taking into account environmental factors from a performance standpoint when planning and designing a project. Providing performance-based standards, in which the desired performance of the building is specified instead of specific materials or systems, can help to ensure that the desired quality of a school or other facility is obtained while allowing designers the flexibility to take into account the interaction between and among the features. The G3-Guilford Green Guide played such a role in the RFES case study. For example, a common environmental goal is to reduce energy consumption of the building. If this goal is articulated as a performance standard (for instance, 'the facility should exceed Model Energy Code requirements by 30 per cent as demonstrated by whole building simulation'), designers can choose from a whole palette of options that could meet the goal, some of which may be less expensive than a generic design applied to many situations. If the designer is encouraged to optimize the design from a whole-building perspective, the first cost of the building can actually be reduced while also reducing lifecycle operational costs (Pearce et al. 2000; Weizsäcker et al. 1999).

Design integration across building systems saves money by acknowledging the interdependencies among these systems. The key to achieving synergies in design is to provide design standards that emphasize the desired level of performance for the building, then suggest a variety of high-performing building system types and technologies that can be used to achieve them. Even in spaces with special functional requirements, integrative design allows the combination of strategies that can achieve required levels of performance at costs that are competitive with or even lower than average. The choice of materials and construction techniques used in today's school construction has been a result of several factors.

- First, policies, rules or standards. For example, in some states, the Department of Education pre-qualifies certain materials for construction. If a school system wants to use materials that are not pre-qualified, they must get approval. If the materials cost more than the standard material, the school system bears the additional cost.
- Second, financial constraints and resources of the local school system.
- Third, competitive market pressures on the design professionals to maintain the owner's construction budget. Design professionals must choose materials and systems with lower first costs to maintain the owner's budget.
- Fourth, local climate and environmental constraints. Local schools are more comfortable with products and building systems that perform best in their local school environment based on past performance.

Balancing the initial cost against serviceability and continuing operational costs for maintenance and energy is necessary. However, the significance of improvements in energy conservation should not be underestimated. For example, the City of Philadelphia school district found that the use of performance-based standards for energy-efficient operation of its 260 facilities resulted in impressive savings of $3.3 million in the first year in energy costs alone (Sender 2000). The cost of energy upgrades can also be supported by grant or incentive programmes. One example is the US Environmental Protection Agency's (USEPA's) Green Lights program, which has provided partners with connections to financing opportunities and extensive technical assistance to identify 'profitable' lighting upgrade opportunities, i.e. those with a rate of return of 20 per cent or more. A school designed to maximize the conservation of energy, minimize environmental impacts, be resource efficient and be aesthetically compatible with the site is desirable if first costs can be managed. Grant programs or energy services contracts in which third party companies provide energy at a reduced fixed rate through a leasing arrangement for renewable energy systems can provide a variety of options for ensuring that first costs premiums are minimized or eliminated.

Consideration of other environmental quality factors can provide additional positive benefits as well, which is of particular importance in educational facilities as well as other types. It is estimated that nearly 56 million students, teachers and staff spend a significant part of their day in schools in the United States (Pérez-Lombard et al. 2008), and the physical quality of the environment has been linked to their performance and health (e.g. Ahn et al. 2011; Ku and Taiebet 2011). By using environmentally sensitive building materials that improve indoor air quality and improving the performance of heating and lighting systems, some schools have reported lower rates of absenteeism and vandalism by 'creating an atmosphere in which students can take pride in their school' (Energy Smart Schools 1997). The USEPA further concluded that by improving indoor air quality in school facilities, it may be possible to reduce the following issues and problems:

- An average of one out of every 13 school-age children has asthma.
- Asthma is a leading cause of school absenteeism.
- 14.7 million school days are missed each year because of asthma.

Researchers at Georgetown University found that achievement scores in school buildings with 'poor' environmental conditions were over five percentage points below scores of students in buildings with 'fair' ratings, and 11 percentage points below those in schools with 'excellent' conditions (Edwards 1991). Another study in North Carolina found that children in daylit (rather than artificially lit) schools score higher on standard performance exams (up to 14 per cent increase over three years) and have better attitudes and attendance rates than their peers in non-daylit facilities (Nicklas and Bailey 1996). Similarly, a Canadian study of the effects of natural light in elementary schools found that students in classrooms with full spectrum light were absent less, grew taller, and had increased concentration levels and more positive moods (Alberta Dept. of Education 1992). Energy savings from daylit schools can also be significant: estimates are that $500,000 on average can be saved over a ten-year period in the average U.S. middle school that incorporates daylighting features (English 1997).

Establishing performance-based design standards to achieve environmental quality goals for schools can have significant impacts not only on the first and lifecycle costs of the building, but also on the basic health, achievement and learning of the students. By providing designers with the flexibility to seek innovative and synergistic design solutions for their projects, a variety of high-performance features can be built into the school that will benefit not only the budget of the school district and the state, but also the environment and the students themselves. Schools in the United States and elsewhere have successfully integrated advanced building technologies such as PV arrays, computerized energy management systems and multi-fuel boiler systems. These technologies not only help the school's maintenance staff do a better job of operating the building, but also they can double as a teaching tool, where students can monitor fuel, energy generation and use, and other variables as part of learning. The school then becomes a living laboratory to demonstrate the physical and scientific principles of sustainability and building science for the students and community (Augenbroe and Pearce 1998, 2009; Greven 1997). A variety of organizations in the United States have developed online educational tools including classroom activities, lab experiments and other exercises that use building-related problems and projects to teach math, science and economics.[2]

Capturing indirect benefits through holistic cost management

Despite commitment to developing sustainable facilities, many organizations still experience difficulty in implementing sustainability technologies and practices because of how funding is allocated to projects. Especially in the public sector, personnel responsible for developing project estimates have few resources for accurately estimating the first costs of a project that involves unfamiliar or innovative technologies, let alone potential lifecycle cost impacts of those technologies. Historically on public sector projects, a common approach for estimating sustainable project costs was to add a contingency factor to the estimate for a conventional project to cover anticipated cost increases for design, materials and systems, and other project costs, particularly for innovative projects in the early planning stages (Pearce 2008; Pearce et al. 2010). This approach inhibits the implementation of sustainability for two reasons, particularly in public agencies. First, projects are often funded based on efficiency of first cost, meaning that projects with a higher parametric cost estimate are less likely to be funded when competing with other projects for limited funding. Second, adding a contingency to the project estimate means that even if the project does get funded, there is often no incentive to seek cost savings since the money will be lost if it is not spent, creating a self-fulfilling prophecy of increased costs for projects with sustainability goals.

Figure 9.25 provides a means for examining expectations about capital project costs. The grid represents a way to plot the relative cost of a sustainable project against its traditional counterpart in terms of both lifecycle and first costs. Figure 9.25(a) shows what many decision makers expect about a sustainable project: it will cost more in the beginning, but will likely save money over the whole lifecycle due to waste reduction, increased durability, reduced operations and maintenance requirements, and other factors. This expectation is illustrated by the red circle in the lower left quadrant of the diagram.

In typical funding scenarios, particularly in the public sector, the projects that will be funded without special intervention lie within the region indicated by green in Figure 9.25(b): those that cost the same or less from a first cost perspective. In many cases, funding constraints mean that minimum first cost is the goal, even though overall lifecycle

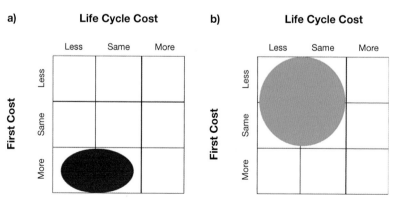

Figure 9.25 Expectations about sustainable project costs: (a) sustainable projects; (b) funded projects

Source: Pearce (2005)

costs may be greater (the third column in the diagram). This suboptimal result is possible because the sources of funding for first cost
are different and disconnected from funding sources for operations and
maintenance, especially in public-sector organizations, and may be
managed or controlled by entirely different people.

The concept of holistic cost management considers a larger set of
questions than traditional project costing from the very beginning of a
project. For instance, what will be the impacts of design/construction
decisions on lifecycle costs? What opportunities exist to offset increases
in first cost for design improvements (as in integrative design)? What
external costs and benefits should be considered that could result in
better decisions about costs?

Instead of the two-dimensional representation of cost shown in
Figure 9.25, holistic cost management expands the figure along a third
dimension to include additional cost/benefit considerations that are
associated with the project. Figure 9.26 illustrates the revised cost model.

The bottom layer of the figure represents the two-dimensional cost
comparison shown in Figure 9.25. This layer represents traditionally
considered, quantifiable costs such as the costs of material, labour,
equipment and cost of money (Table 9.9). Decisions about individual
products are sometimes made on a unit cost basis without necessarily
considering cost from a systems standpoint. This practice means that
some products offering sustainability advantages seem more expensive
than they really are. For instance, when asked why they don't use integrated building systems such as structural insulated panels or insulating
concrete forms, many owners reply that these systems are more expensive than traditional methods such as concrete masonry construction.

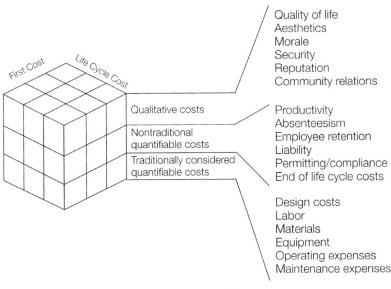

Figure 9.26 Revised whole project cost model
Source: Pearce (2005)

Table 9.9 Traditionally considered quantifiable costs

First costs	Lifecycle costs

Site acquisition
Design costs
Project management costs
Construction costs:
  • labour
  • materials
  • equipment
  • contingencies
  • financing and other costs of money
  • commissioning/turnover costs
	Operation/maintenance costs:
	  • labour
	  • materials
	  • equipment
	  • e n e r g y
	  • water
	Repair/remodel/rehabilitation costs:
	  • design costs
	  • labour
	  • materials
	  • equipment
	  • contingencies
	  • financing
	  • turnover

Yet if savings in labour cost and construction schedule were taken into account, these materials could show an immediate cost advantage to owners. For example, one federal project manager estimated a savings of $3,000 per day due to shortening the construction schedule on his project by using pre-engineered autoclaved aerated concrete systems – all from avoiding the cost of housing displaced personnel in hotels instead of the facility being built as their residence. Similar or even greater savings in opportunity costs can result on many other projects as well.

Other considerations, if taken into account, could more clearly indicate the benefits of sustainability in the built environment. Captured in the second level of the cost model shown in Figure 9.26, these considerations include benefits such as:

- Reduced costs of consumption, waste disposal and noncompliance.
- Reduced liability and environmental risk.
- Improved use of assets, particularly human assets (including increased productivity, reduced absenteeism and building-related health problems, improved morale and better employee retention).
- Reduced operational and disposal costs.
- Reuse of facilities that otherwise would be disposed.
- Preparedness for future regulations and requirements.

Each of these benefits reflects a potential cost savings for owners, although many of these costs are not typically associated with specific projects and the associated decision processes behind their funding. If these potential benefits can be realized, then more sustainable projects will truly have an economic advantage over their conventional counterparts. Table 9.10 shows examples of these types of costs that should be considered when making decisions about sustainable projects. The first category, definite costs, includes all costs that will happen as part of the project but typically are not considered as part of project costs, instead covered as part of overhead or administrative costs. The second category, contingent costs, consists of costs that may or may not occur – their probability of occurrence is less than 1.0. In other words, there is some likelihood that they will occur, and their total costs can be estimated using probabilistic methods such as decision trees. Although the listing is not comprehensive, it illustrates the kinds of costs for which dollar values could be calculated but are typically not included in project decision making.

An additional set of considerations in holistic cost management consists of qualitative costs: costs that have some real impact but are difficult to quantify because of societal values and other measurement challenges. It is at this top level of the holistic cost model that sustainable projects truly dominate conventional projects in their impact reduction. However, the difficulty of assigning actual costs and benefits to specific projects is significant, and therefore these are not typically considered as part of project decision making. Table 9.11 lists some of the cost items in this category.

Costs can be broken into two types at this level: internal costs and external costs. Internal costs are those difficult to quantify costs experienced directly by project stakeholders. Externalities, on the other

Table 9.10 Non-traditional quantifiable costs

Definite costs	Contingent costs
• Qualification of suppliers/contractors • Reporting and record-keeping • Monitoring and testing • Spill response readiness • Recycling/waste management • Facility decommissioning costs • Disposal costs	• Future compliance costs • Future liability/damage costs • Remediation costs • Responses to future releases or presently unknown hazards • Impacted productivity and/or absenteeism • Impacted staff retention

Table 9.11 Qualitative costs

Internal costs	Externalities
Impacts on quality of life Value of relationships with surrounding community Value of environmental image	Costs borne by society as a whole, such as: • global warming • ozone depletion • deforestation • resource degradation • ecosystem degradation • species/biodiversity loss • air pollution • water pollution

hand, are generally borne by society as a whole. While projects have some individual contribution to these costs, the net cost is a result of all human activities, and allocating specific responsibilities is difficult. Methods such as hedonic price models (see *Green building certification and property value of commercial buildings*) can be used to assign costs to these kinds of impacts, and are used frequently in risk analysis and policy development.

Green Building certification and property value of commercial buildings

Min Jae Suh

Property values are generally determined by the principle of supply and demand, and the property value of a specific building is affected by the actual property commodity itself as well as features of the external environment such as social factors, physical factors, economic factors, and government and political factors (Bloom et al. 1982; Boyce 1984; Epley and Rabianski 1981; Epley et al. 2002; Friedman et al. 2013). As a feature of an actual property, third party green building certification is a market-based way of signaling reduced energy consumption and lower energy cost as a result of superior energy performance (McGraw-Hill Construction 2010). Third party systems such as Leadership in Energy and Environmental Design (LEED) and Energy Star not only promote the reduction of negative environmental impacts to society as a whole but also provide directly measurable positive economic benefits to building stakeholders. McGraw-Hill Construction (2010) has argued that four factors, namely building value, return on investment, occupancy rate and rent ratio, increase after certification for both green retrofit and renovation projects and for new green buildings.

Researchers interested in the economic benefits of green buildings have studied the role of certification in creating a value premium by comparing green certified buildings and comparable non-certified buildings under similar conditions over time to determine the economic impact of achieving certification. For instance, multiple studies in the literature have examined the effect of LEED certification by comparing LEED certified vs. non-certified buildings in terms of rental cost, cost per unit area (price per sq ft), occupancy rate, and resale value rate (Table 9.12).

When comparing relative benefits of LEED vs. Energy Star certified buildings, the findings are less clear. For instance, although Eichholtz et al. (2010) concluded that Energy Star certified buildings enjoy a rent and sale price premium of about 3.3 per cent and 19 per cent, respectively, they found that LEED certified buildings do not carry as high a premium. Miller et al. (2008) and Wiley et al. (2010) disagreed, instead finding the sale price premium of LEED certified office buildings to be about 5 per cent higher than that of Energy Star certified office buildings. Fuerst and McAllister (2011) found little difference between the two certifications, reporting a rental premium of approximately 5 per cent for LEED certification vs. 4 per cent for Energy Star certification. In terms of sales prices in their study, LEED certified buildings showed a 25 per cent price premium, while Energy Star commanded 26 per cent. Despite these variations, certification seems to consistently offer economic benefits to property owners not only in its promotion of lifecycle operational savings, but also in other metrics of intrinsic property value.

Multiple studies have also explored the effects of environmental determinants external to a property on the value of that property, with measurable effects based on changes to external environmental features (Table 9.13).

continued

Table 9.12 Comparisons of the real estate value of LEED certified buildings and non-certified buildings

	Rental cost ($)		Cost per unit area (price per sq ft)		Occupancy rate (%)		Resale value rate	
Certified?	Yes	No	Yes	No	Yes	No	Yes	No
Miller et al. (2008)	42.15	28.00	—	—	92	88	1.1	1
Eichholtz et al. (2010)	29.84	28.14	289.22	248.89	89	81	1.16	1
Dermisi and McDonald (2011)	—	—	239.73	180.49	89	86	—	—

Source: Suh et al. 2013

Table 9.13 Effects of external environmental features on property values

	Property type	Measurement	Effects
Nelson (1980)	Residential	Airport	Negative impact on property values due to the airport noise levels
Weinberger (2001)	Residential or commercial	Light rail	Positive impact on property values and a rent benefit for the residential or commercial area due to the improved accessibility by a public transportation
Kim et al. (2003)	Residential	Air quality improvement	Positive impact on the willingness to pay by buyers due to a better quality of life
Mansfield et al. (2005)	Residential	Urban forests	Positive impact on the property value for residential property due to better quality of life
Hui et al. (2007)	Residential	Sea view Green belt area Air quality Noise level Accessibility	Positive impact on the willingness to pay by buyers due to a better quality of life
Lin et al. (2009)	Residential	Foreclosure	Negative impact on property values of neighbourhood area
Liu et al. (2010)	Commercial or residential	Land use	Positive impact on the property values for both property areas, residential or commercial, due to a better quality of life
Chang and Chou (2010)	Neighborhood	A building's green design	Positive impact on the property values of neighbourhood area due to the environmental improvement
Chun et al. (2011)	Commercial	Visible space	Positive impact on the property values due to a better quality of life
Saphores and Li (2012)	Residential	Urban green area	Positive impact on the property values due to a better quality of life

Source: Adapted from Suh 2015

continued

As a feature of the economic environment within a neighborhood, what effect might green building certification have on the buildings surrounding a property for which certification has been achieved? In a study of commercial buildings in New York City, Suh (2015) concluded that the improved market value of LEED and Energy Star certified buildings can also "spill over" as an external environmental factor to have a beneficial impact on neighborhood market values. Suh studied both LEED and Energy Star certified office buildings and the neighborhoods surrounding them. He concluded that not only do certified buildings offer direct economic benefit to their owners in terms of improved market value, but they also offer social benefits to the neighborhoods in which they are located due to spillover effects of that enhanced market value on the value of surrounding properties. This spillover benefit of green certification bridges the gap between direct benefits of green building investments to developers and overall societal benefits. The measurable economic benefits to surrounding buildings mean that efforts of one property owner to improve also helps neighbours, even if neighbouring property owners do not invest in certification of thier own buildings.

References

Bloom, G.F., Weimer, A.M., Fisher, J.D. (1982). *Real estate, 8th Ed.*, John Wiley & Sons, New York, NY.

Boyce, B.N. (1984). *Real estate appraisal terminology*, Ballinger Publishing Company, Cambridge, MA.

Chang, K., and Chou, P. (2010). "Measuring the influence of the greening design of the building environment on the urban real estate market in Taiwan," *Journal of Building and Environment*, 4S, 2057-2067.

Chun, B., Guldmann, J.M., and Seo, W. (2011). "impact of multi-dimensional isovist on commercial real estate values in CBD area using GIS," *Journal of Seoul Studies*, 12(3), 17-32.

Dermisi, S. and McDonald, J. (2011). "Effect of green designation on prices/sf and transaction frequency: the Chicago office market." *Journal of Real Estate Portfolio Management* , 17(1), 39-52.

Eichholtz, P., Kok, N., and Quigley, J.M. (2010). Doing well by doing good? Green office buildings, *American Economic Review*, 100(5), 2492-2509.

Epley, D.R., and Rabianski, J. (1981). *Principles of real estate decisions*, Addison-Wesley Publishing Company, Philippines.

Epley, D.R., Rabianski, J.S., and Haney, R.L.JR. (2002). Real estate decisions, Thomson Learning, Cincinnati, Ohio.

Friedman, J.P., Harris, J.C., and Lindeman J.B. (2013). *Dictionary of real estate terms*, Barron's Educational Series, Hauppauge, NY.

Fuerst, F. and McAllister, P. (2011). "Green noise or green value? Measuring the effects of environmental certification on office values." *Journal of Real Estate Economics*, 39(1), 45-69.

Hui, E.C.M., Chau, C.K., and Law, M.Y. (2007). "Measuring the neighboring and environmental effects on residential property value: Using spatial weighting matrix," *Journal of Building and Environment*, 42, 2333-2343.

Kim, K. and Son, J. (2011). *Real estate economics*, KU Smart Press, Seoul, South Korea.

Lin, Z., Rosenblatt. E. and Yao, V.W. (2009). "Spillover effects of foreclosures on neighborhood property values," *Journal of Real Estate Finance and Economics*, 38(4), 387-407.

Liu, Y., Zheng, B., Turkstra, J., and Huang, L. (2010). "A hedonic model comparison for residential land value analysis," *International Journal of Applied Earth Observation and Geoinformation*, 12(2), 181-193.

Mansfield, S., Pattanayak, S.K., McDow, W., McDonald, R., and Halpin, P. (2005). "Shades of Green: Measuring the value of urban forests in the housing market," *Journal of Forest Economics*, 11, 177-199.

McGraw-Hill Construction. (2010). *Green Outlook 2011 – Green Trends Driving Growth*, McGraw-Hill, New York, NY.

Miller, N., Spivey, J., and Florance, A. (2008). "Does green pay off?" *Journal of Real Estate Portfolio Management*, 14(4), 385-398.

Nelson, J.P. (1980). "Airports and property values: a survey of recent evidence," *Journal of Transport Economics and Policy*, 14(1), 37-52.

Saphores, J., and Li, W. (2012). "Estimating the value of urban green areas: A hedonic pricing analysis of the single family housing market in Los Angeles, CA," *Journal of Landscape and Urban Planning*, 104, 373-387.

continued

Suh, M. (2015). *The Spillover Effect of Proximity to LEED-Energy Star Certified Office Buildings on Neighborhood Market Values.* Dissertation, Virginia Tech, Blacksburg, VA.

Suh, M., Pearce, A.R., and Kwak, Y. (2013). "The effect of LEED certified building on the neighborhood market value in New York City," *Proceedings, The 5th International Conference on Construction Engineering and Project Management,* 9-11 Ja , Anaheim, CA, USA.

Weinberger, R.R. (2001). *Commercial property value and proximity to light rail: A Hedonic Price Application.* Dissertation, University of California Berkeley, CA.

Wiley, J.A., Benefield J.A., and Johnson K.H. (2010). "Green design and the market for commercial office space." *Journal of Real Estate Finance & Economics* , 41(2), 228–243.

With indirect methods for estimating qualitative costs, a price is inferred from actual choices to which monetary values can be assigned, such as consumers' choices about where to live. Market prices are not the same as the cost to the market of producing goods and services, and different methods are used to evaluate the price or worth of a product or service in the market (see *Economic sustainability: housing price vs. Cost in Turkey*). These methods may examine:

- Averting behaviours, such as how much people will pay to fix environmental damage; the cost of cleanup.
- Weak complementarity/travel cost, such as where the value of cleaner water is assumed to be connected to visits to a lake.
- Hedonic market methods where the price of a house or a job can be decomposed into attributes, one or more of which are environmental attributes.

Direct methods for estimating qualitative costs are also known as contingency valuation. These methods involve direct questioning about willingness to pay or willingness to accept compensation in exchange for damage of some sort.

Economic sustainability: housing price vs. cost in Turkey

A. Tolga Ozbakan

Housing is not only a shelter that satisfies a basic human need. It is also a commodity produced for exchange in markets and an asset for storing and enhancing wealth. In Turkey as well as in other countries, housing consumes a large share of people's income and often represents a significant portion of their net worth. To be specific, for Turkish households, average housing costs amount to 25 per cent of disposable income and primary housing units appear as the most important component of individuals' net worth (TURKSTAT, 2015). In addition, housing represents a majority of economic activity in the construction sector in many markets. In Turkey, for instance, seventy-five per cent of all new building construction is residential (KONUTDER, 2013).

Mispricing of housing can have serious repercussions for individuals, firms, industries, nationwide economies, and for the global economy. For example, beginning with the turn of the millennium,

continued

the real estate market in the United States 'took off'. Within a few years, the average real home price increased around 50 per cent until 2004 and reached its peak in 2006, at which point it came to an abrupt end, known as the infamous 'housing bubble' of 2006 (Shiller 2009). This instability in market valuation of housing drove the US economy into its most severe economic crisis since the Great Depression. As a result, by the end of 2008, over two million homes entered foreclosure proceedings and about 2.5 million people in the building industry lost their jobs (Magdoff and Yates, 2009). Between 2007 and 2013, house prices in the United States declined around 50 per cent (Shiller 2014). Similarly, home prices in the Netherlands, Spain, Iceland and Ireland suffered significant falls (Ambrose et al. 2013; Levitin and Wachter 2013).

While global housing prices have been declining over the recent years, prices in Turkey have been rising steadily. Between 2010 and late 2015, nationwide real house prices increased almost 33 per cent, while Istanbul's housing market nearly doubled this rate at 65 per cent (CBRT, 2016). Interestingly, during this upsurge, the supply of housing units kept increasing as well, amounting to a change in stock equal to almost 15 per cent and 13 per cent in Turkey and in Istanbul respectively (TURKSTAT, 2016). There are differing opinions about whether or not this trend is sustainable. On the one hand, some analysts argue that the dynamics of the market are consistent with economic fundamentals (Cushman & Wakefield, 2014; PricewaterhouseCoopers LLP, 2013), while others voice serious concerns about a potential mispricing in the Turkish residential market (Acemoğlu 2012; IMF 2013; Roubini 2013).

In parallel to the rising house prices in Turkey, the interest in 'green materials', 'sustainable' and 'environmentally friendly' housing projects has been growing over the recent years (Turkish Green Building Council 2016). Although the proponents of these practices seek to improve the

Figure 9.27 A high-income residential building with green features in Izmir, Turkey. *Source: A. Tolga Özbakan*

continued

environmental sensibility of the housing units and the health of their inhabitants, it is also important to recognize the social and economic risks that may stem from a price increase in the housing markets. While some researchers reflect on the far-reaching consequences of mobilizing sustainability as a marketing tool (Baldassarre and Campo 2016), others highlight the need to make green home features accessible to all communities (Mehdizadeh et al. 2013). More specifically, in examining the housing markets in China, Hu (2014) argues that sustainability practices lead 'to a significant market premium'.

The market price of something is not necessarily the same as its cost. Housing prices are a function of the market's demand, supply and *perceived* desirability of that supply, not necessarily the true cost of the labour and the materials that went into producing that housing. Some green building technologies and practices do increase the cost of producing a housing unit, but they can also fictitiously inflate its price as well. This may not only render environmentally sustainable housing unreachable by the majority of low and middle-income households but may also exacerbate the overall risks associated with the sustainability of housing market prices. While the environmental sustainability of housing is a global challenge that requires utmost attention, it is also important to consider the social and economic risks associated with unsustainable price trends in all housing markets.

References

Acemoğlu, D. (2012). *Renowned economist warns against Turkish Real Estate Bubble.* http://thefreelibrary.com/Renowned+economist+warns+against+Turkish+real+estate+bubble.-a0298862202 (accessed 21 June 2014).

Ambrose, B.W., Eichholtz, P., and Lindenthal, T. (2013). "House prices and fundamentals: 355 years of evidence." *Journal of Money, Credit, and Banking.* 45(2–3), 478–491.

Baldassarre, F., and Campo, R. (2016). "Sustainability as a marketing tool: To be or to appear to be?" *Business Horizons* 59(4), 421–429.

Central Bank of the Republic of Turkey (CBRT). (2016). *The New House Price Index,* The Central Bank of the Republic of Turkey, Ankara, Turkey.

Cushman and Wakefield (2014). *Turkish housing market: Price bubble.* Cushman & Wakefield, Istanbul, Turkey.

Hu, H. (2014). *Green home for whom? Estimating green housing opportunities of various socio-economic groups in Nanjing China.* XVIII ISA World Congress of Sociology, Utrecht, the Netherlands.

International Monetary Fund (IMF). (2013). *IMF Global Housing Watch.* http://imf.org (accessed 18 June 2014).

KONUTDER. (2013). *Housing sector assessment presentation,* Association of Housing Developers and Investors, Istanbul, Turkey.

Levitin, A.J., and Wachter, S.M. (2013). "Why housing?" *Housing Policy Debate* 23(1), 5–27.

Magdoff, F., and Yates, M.D. (2009). *The ABCs of the economic crisis: What working people need to know.* Monthly Review Press, New York, NY.

Mehdizadeh, R., Fischer, M., and Burr, J. (2013). "The green housing privilege? An analysis of the connections between socio-economic status of California communities and Leadership in Energy and Environmental Design (LEED) certification." *Journal of Sustainable Development* 6(5), 37–49.

PricewaterhouseCoopers LLP (2013). *Urban Land Institute.* http://uli.org/ (accessed 6 June 2014).

Roubini, N. (2013). *Back to Housing Bubbles.* http://project-syndicate.org/commentary/nouriel-roubini-warns-that-policymmakers-are-powerless-to-rein-in-frothy-housing-markets-around-the-world (accessed 8 January 2017).

Shiller, R.J. (2009). *Irrational exuberance.* 2nd ed. Broadway Books, New York, NY.

Shiller, R.J. (2014). Speculative Asset Prices [Nobel Prize Lecture]. *Cowles Foundation Discussion Papers,* Issue 1936, 1–45.

Turkish Green Building Council. (2016). *2016 Congress Documentation.* http://cedbik.org/zirve_p1c_tr_-8_.aspx (accessed 14 August 2016).

TURKSTAT. (2015). *Household consumption expenditures.* Turkish Statistical Institute, Ankara, Turkey.

TURKSTAT. (2016). *Building permits statistics.* Turkish Statistical Institute, Ankara, Turkey.

Figure 9.28
Sustainable vs. conventional
projects from a holistic standpoint

Source: Pearce (2005)

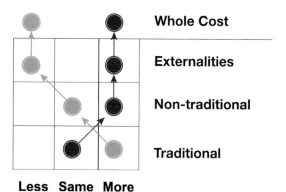

The ultimate outcome of considering the full spectrum of costs associated with a project is a true picture of what costs and benefits will stem from each alternative over the whole building lifecycle. If all costs are considered, then actions that might ordinarily cause problems later (such as endangered species habitat disturbance or use of hazardous materials) can be adequately considered for the risk they truly represent. Figure 9.28 shows how sustainable projects compare with conventional projects when all these factors are considered. The sustainable project, indicated by the green dots, has lower costs than the traditional project in terms of qualitative and non-traditional quantifiable costs, indicated by red. This two-dimensional representation corresponds to looking at the three-dimensional model from the side and reducing first cost and lifecycle costs to a single metric such as NPV for comparison purposes.

Making the business case for sustainability in the future

How will making the business case for sustainability be different in the coming years? With current trends, it is likely to be even easier than making the case today. Factors that may result in reduced quantifiable costs for sustainable projects over the next 10 years include:

- **Learning curve** – as sustainable products and strategies become more familiar to designers and builders, less effort will be required to use them correctly (arrow A in Figure 9.29).
- **Economies of scale** – as demand increases for sustainable technologies in the marketplace, production and distribution networks will grow and become more efficient (arrow B in Figure 9.29).
- **Resource scarcity** – as many of the materials presently used in building (such as fossil fuels, old growth timber and mineral-derived products) become more scarce, their costs will increase and the cost of using alternatives will become relatively less expensive (arrow C in Figure 9.29).
- **Stricter legislation** – as environmental, safety and occupational health (ESOH) regulations become more restrictive, projects that have lower risk of these threats will become relatively less expensive (arrow D in Figure 9.29).

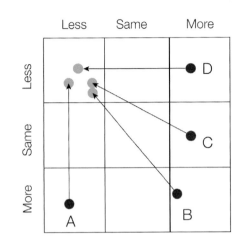

Life Cycle Cost

Figure 9.29
Influences on the cost of
sustainable projects
A – Learning curve
B – Economics of scale
C– Resource scarcity
D – Stricture legislation
Source: Pearce (2005)

Changes in legislation and policy in particular are poised to play a significant role in the near future, as evidence of the seriousness of climate change becomes more severe. As countries work together using treaties, agreements and policies such as carbon taxes to reduce their carbon footprints, reductions will come from real changes in the choices and behaviours of businesses and individuals in those countries. Choices with high carbon footprints are likely to become much more expensive as carbon trading and offsets become the norm in the marketplace, or become subject to phase-out as part of long-term regulatory strategies. Those projects that offer an energy advantage over the alternatives through greater efficiency, on-site production of power using renewables or a lower overall carbon footprint due to better design and construction choices will have a significant advantage over conventional projects. Market-based approaches to managing carbon emissions will make inefficient choices more expensive than green choices, and will drive innovation that makes more sustainable projects the norm.

All of these external influences will tend to give advantage to sustainable projects from an economic standpoint, at least in the long run. In addition, items presently classified as nontraditional quantifiable costs or externalities are also increasingly likely to be part of any business case evaluation. With the growth in popularity of metrics such as the Dow Jones Sustainability Index (see Chapter 4), companies will be increasingly evaluated based on a Triple Bottom Line of social, environmental and economic performance. Our ability to evaluate the true value of common resources and assets will only improve as well.

The ultimate goal of sustainability is our long-term survival as the human species. The negative impacts of our present methods of construction can no longer be denied. The time has come to seek better ways of building that will maintain our quality of life and ensure that our children and their children can continue to live as well as we do.

Thinking strategically about how we increase the sustainability our building and infrastructure projects can ensure that we are able to meet sustainability goals without breaking the project budget.

Discussion questions and exercises

9.1 List some of the most promising recommendations you have made in previous chapters for measures to improve the sustainability of the facility in which you are located. What are the first costs associated with implementing those measures? Develop a cost estimate for each of the most promising recommendations. How much more or less expensive is each than the status quo it would replace? Which options cost the least from a first cost standpoint? Do any of the recommendations offer a lower first cost than the status quo?

9.2 For each recommendation considered, how would its lifecycle costs compare with the status quo it would replace? Develop a cash flow profile to show major maintenance and replacement costs. Which recommendations are most preferable in terms of lifecycle costs?

9.3 Do any recommendations offer synergy from a cost standpoint? How might strategies be combined to help break through the first cost barrier? What complementary actions could be undertaken to further reduce costs, such as downsizing related systems?

9.4 Inventory the non-traditional quantifiable costs and benefits associated with each recommendation over its lifecycle, such as increased productivity or user satisfaction, reduced environmental liability, or diminished risk of price fluctuations associated with resource depletion. Which are definite costs, and which are contingent costs?

9.5 What qualitative costs and benefits are associated with each of your recommendations? Which are internal to your organization, and which are externalities that must be borne by society as a whole?

9.6 For the most promising recommendations you have identified, compare the costs and benefits of each option with the status quo. Given this analysis, which alternative(s) would you most highly recommend?

9.7 Which of the external factors influencing cost – learning curve, economies of scale, resource scarcity and stricter legislation – should be most carefully considered by your organization in determining how to manage its facility portfolio, given the context of your operations? Why? What measures can you take to manage that risk?

9.8 Consider the local construction project you visited in earlier chapters. Which of the synergistic design and construction strategies discussed in the chapter could be used to reduce the first cost of that project while improving overall project sustainability?

9.9 Search online for resources discussing the costs and benefits of green or sustainable buildings. Which references have been developed based on scientific studies? Which are presented by organizations or individuals who have a vested interest in green building? Which seem most credible?

Notes

1 This section is based on the paper "Post Occupancy Evaluation for Green School Facilities: Case Study of Reedy Fork Elementary School" by Yong Han Ahn and Annie R. Pearce, presented at the 2012 International Conference on Sustainable Design, Engineering, and Construction.

2 See www.engineeringpathways.org for examples. The business case study of the Reedy Fork Elementary School has been developed with the support of Innovative Design, Inc.

References and recommended reading

Abbaszadeh, S., Zagreus, L., Lehrer, D., and Huizenga, C. (2006). "Occupant satisfaction with indoor environmental quality in green buildings." *Proceedings, Healthy Buildings*, Vol. III, 365–370, Lisbon, Portugal, June.

Ahn, Y.H., Choi, Y.O., Koh, B.W., and Pearce, A.R. (2011). "Designing sustainable learning environments: Lowering energy Consumption in a K-12 Facility." *Journal of Green Building* 6(4), 112–137.

Alberta Department of Education. (1992). *Study into the effects of light on children of elementary school age: A case of daylight robbery.* Alberta, Canada.

Aranda-Mena, G., Crawfor, J, Chevez, A, and Froese, T. (2009). "Building information modelling demystified: Does it make business sense to adopt BIM?" *International Journal of Managing Projects in Business* 2(3), 419–434.

Augenbroe, G.L.M., and Pearce, A.R. (1998). "Sustainable construction in the USA: A perspective to the year 2010." In L. Bourdeau, P. Huovila, R. Lanting and A. Gilham (eds), *Sustainable Development and the Future of Construction: A comparison of visions from various countries.* CIB Publications, Rotterdam, Netherlands.

Augenbroe, G.L.M., and Pearce, A.R. (2009). "Sustainable construction in the USA: a perspective to the year 2010." In A.K. Pain (ed.), *Construction industry: Changing paradigm.* Icfai University Press, Hyderabad, India.

Bennett, M., and James, P. (eds). (2000). *The green bottom line: Environmental accounting for management – current practices and future trends.* 2nd ed. Greenleaf Publishing, Sheffield, UK.

Bonde, M., and Ramirez, J. (2015). "A post-occupancy evaluation of a green rated and conventional on-campus residence hall." *International Journal of Sustainable Built Environment* 4(2), 400–408.

Clara, P., and Stefano, S. (2011). "Indoor Environmental Quality Surveys." A Brief Literature Review. *Indoor Air,* Dallas, TX.

Cooper, I. (2001). "Post-occupancy evaluation – Where are you?" *Building Research and Information* 29(2), 158–163.

Edwards, M. (1991). *Building conditions, parental involvement, and student achievement in DC public schools.* Georgetown University, Washington, DC.

Energy Ideas. (1997) "Energy smart schools." *Energy Ideas* 4(3), Winter/Spring.

English, H. (1997). "Students shine in daylit classrooms." *Energy Ideas* 4(3), Winter/Spring.

Gou, Z., Lau, S.S.-Y., and Zhang, Z. (2012). "A comparison of indoor environmental satisfaction between two green buildings and a conventional building in China." *Journal of Green Building* 7(2), 89–104.

Graydon, C. (2009). *Guilford County Schools.* http://utilitydirect.schooldude.com (accessed 11 September 2009).

Greven, E. (1997). "Investments that last a lifetime." *Energy Ideas* 4(3), Winter/Spring.

Heerwagen, J. (2000). "Green buildings, organizational success, and occupant productivity." *Building Research and Information* 28(5), 353–367.

Halliday, C.O. Jr., Schmidheiny, S., and Watts, P. (2002). *Walking the talk: The business case for sustainability.* Greenleaf Publishing, Sheffield, UK.

Kats, G. (2003). *The costs and benefits of green buildings: A report to California's sustainable building task force.* California's Sustainable Building Task Force, Sacramento, CA.

Ku, K., Pollalis, S.N., Fischer, M.A, and Shelden, D.R. (2008). "3D Model-Based Collaboration in Design Development and Construction of Complex Shaped Building." *Journal of Information Technology in Construction* 13, 458–485.

Ku, K., and Taiebet, M. (2011). "BIM experiences and expectations: The constructors' perspective." *International Journal of Construction Education and Research* 7(3), 175–197.

Lechner, N. (2008). *Heating, cooling, lighting: Sustainable design methods for architects.* 2nd ed. Wiley & Sons, New York.

Matthiessen, L.F., and Morris, P. (2004). *Costing green: A comprehensive cost database and budgeting methodology.* Davis Langdon Adamson.

Messick, J.K. (2011). "North Carolina Drought Overview," Presentation to N.C. Drought Advisory Management Council, NC Department of Agriculture. http://ncdrought.org/presentations/20110721/Messick_NC%20Agricultural_Drought_Overview_07.21.2011.pdf (accessed 15 January 2017).

Morton, S. (2002). "Business case for green design." *Building Operating Management,* November. http://facilitiesnet.com/bom/article.asp?id = 1481 (accessed 8 January 2017).

Nattrass, B., and Altomare, M. (1999). *The natural step for business: Wealth, ecology, and the evolutionary corporation.* New Society Gabriola Island, BC.

Nicklas, M., and Bailey, G. (1996). "Energy performance of daylit schools in North Carolina." http://innovativedesign.net/files/Download/Energy%20Performance%20of%20Daylit%20Schools%20in%20North%20Carolina.pdf (accessed 8 January 2017).

Paul, W.L., and Taylor, P.A. (2008). "A comparison of occupant comfort and satisfaction between a green building and a conventional building." *Building and Environment* 43, 1858–1870.

Pearce, A.R. (2005). "Leapfrogging the first cost barrier in sustainable construction: A taxonomy and recommendations for the next generation of costing tools." *Proceedings of the 2005 Mascaro Sustainability Initiative Sustainable Engineering Conference*, Pittsburgh, PA, 10–12 April.

Pearce, A.R. (2008). "Sustainable capital projects: Leapfrogging the first cost barrier." *Civil Engineering and Environmental Systems* 25(4), 291–301.

Pearce, A.R. (2010). "Costing sustainable capital projects: The human factor." *Proceedings, 2010 Conference, New Zealand Society for Sustainable Engineering and Science*, Auckland, NZ.

Pearce, A.R., Fischer, C.L.F., and Jones, S.J. (2000). *A primer on sustainable facilities and infrastructure.* Georgia Tech Research Corporation, Atlanta, GA.

Pearce, A.R., Sanford-Bernhardt, K., and Garvin, M.J. (2010). "Sustainability and socio-enviro-technical systems: Modeling total cost of ownership in capital facilities." Invited paper, *Proceedings, 2010 Winter Simulation Conference*, Baltimore, MD.

Pérez-Lombard, L., Ortiz, J., and Pout, C. (2008). "A review on buildings energy consumption information." *Energy and Buildings* 40(3), 394–398.

Preiser, W., and Vischer, J. (2005). *Assessing building performance.* Elsevier, New York, NY.

Riley, M., Kokkarinen, N., and Pitt, M. (2010). "Assessing post occupancy evaluation in higher education facilities." *Journal of Facilities Management* 8(3), 202–213.

Romm, J.J., and Browning, W.D. (1995). *Greening the building and the bottom line: Increasing productivity through energy-efficient design.* Rocky Mountain Institute, Snowmass, CO.

Sender, M. (2000). *Energy conservation program coordinator.* School District of Philadelphia, Philadelphia, PA.

Singh, A., Syal, M., Grady, S.C., and Korkmas, S. (2010). "Effects of green buildings on employee health and productivity." *American Journal of Public Health* 100(9), 1665–1668.

Sobek, W., Ward, A.C., and Liker, J.K. (1999). "Toyota's principles of set-based concurrent engieering." *Sloan Management Review* 40(2), 67–83.

Thatcher, A., and Milner, K. (2014). "Changes in productivity, psychological wellbeing, and physical wellbeing from working in a 'green' building." *Work* 49, 381–393.

USDOE – US Department of Energy. (2003). *The Business Case for Sustainable Design in Federal Facilities.* Energy Efficiency and Renewable Energy Program, Federal Energy Management Program, USDOE, Washington, DC.

USGBC – US Green Building Council. (2015). *The Business Case for Green Building.* http://usgbc.org/articles/business-case-green-building (accessed 8 January 2017).

Weizsäcker, E., Lovins, A.B., and Lovins, L.H. (1999). *Factor four: Doubling wealth, halving resource use.* Earthscan, London.

Wilkinson, S., Kallen, P.V.D., and Kuan, L.P. (2014). "The relationship between the occupation of residential green buildings and pro-environmental behavior and beliefs." *Journal of Sustainable Real Estate* 5(1), 1–22.

Wilson, A. (2005). "Making the case for green building." *Environmental Building News* 14(4), 1 ff.

WGBC – World Green Building Council. (2013). *The Business Case for Green Building.* http://worldgbc.org/files/1513/6608/0674/Business_Case_For_Green_Building_Report_WEB_2013–04–11.pdf (accessed 8 January 2017).

WGBC – World Green Building Council. (2014). *Health, wellbeing, and productivity in offices: The next chapter for green building.* http://worldgbc.org/activities/health-wellbeing-productivity-offices/ (accessed 8 January 2017).

Chapter 10
The future of sustainable buildings and infrastructure

What seemed only recently like a fringe movement in green, sustainable and high-performance facilities has grown with surprising speed across the AEC industry. Once the province of only a few progressive designers and builders in the field, high performance green building has become mainstream as owners realize the benefits associated with creating environments that better suit their occupants while relying less on the finite resources and fragile ecmany ways as the construction industry evolves. Over the past 30 years, the way in which we approach the goal of sustainability in the built environment has also evolved to reflect our greater understanding of the connectedness of human and natural systems and the wisdom of local solutions customized to meet individual needs. Organizations that consider the potential impacts of this movement will be well prepared for multiple eventualities and can respond with agility to potential threats and opportunities in both the short and long term.

Where are we going, and what comes ahead? This final chapter begins by exploring 'over the horizon' possibilities for the context in which the construction industry will operate in the future, followed by a perspective on three major paradigm shifts developing to shape our path toward greater sustainability in the built environment:

- From 'less bad' to regenerative
- From prescriptive to performance-based
- From rigid to resilient

Together, these trends represent fundamental shifts in how we understand the world and our role in it as professionals in the AEC industry. In combination, they will shape the way in which we respond to the unknown challenges facing us in the next 100 years, both sociopolitical and ecological. Despite the sometimes dramatic social, cultural and political shifts that are increasingly part of our daily life, these larger trends guide us toward a hopeful future where we work together to find the best path forward toward sustainability.

Looking over the horizon: strategic sustainability

What challenges will we face in the future, and how should we adjust our path to accommodate them? Ultimately, the long-term strategic success of business enterprises depends on predicting, adapting to, and seeing as opportunity the social, resource and environmental forces asserted by a growing global population with expectations and aspirations for increased standards of living. While concern about the environment and energy resources is not new and has been influencing corporate and institutional policy since the 1970s or before, the need to actively integrate these types of contextual considerations as part of corporate strategy is now and will continue to be essential to the ultimate economic sustainability of these enterprises.

Coupled with the uncertain but threatening effects of climate change and the rapidly evolving human role in mitigating it, anticipating the future is a challenge that faces communities, businesses and individuals who must balance the opportunity costs of proactively planning for an uncertain future versus continuing with 'business as usual'. As the nature of our enterprises changes in response to contextual constraints, so too must the built environment change in how it supports and enables those activities. The following subsections highlight ways in which the built environment and industries responsible for delivering and maintaining it and managing its end of lifecycle are evolving over time.

Recent trends affecting the industry

Multiple trends have occurred in recent years that have affected the strategic operation of firms in our industry, including:

- Increasing price volatility for raw materials, particularly nonrenewable resources and renewables whose harvest or manufacture depends on heavy use of non-renewables (such as fuel costs). Prices of fossil fuels in particular have experienced severe volatility over the past several years due to development of new supplies of energy from tar sands and **fracking**, among others, and ongoing changes in public policy that affect the regulation and development of such resources.
- Increasing savvy among end consumers about the ecological and social impacts of purchasing decisions driven by improved availability of information and tools to support decision making, made widely available on the Web.
- Forays among companies worldwide into sustainability through making changes in their capital facilities as well as their industrial processes and supply chains. Many organizations elect to begin 'greening' their programmes by starting with changes to capital facilities since it is simultaneously less threatening and more visible to make changes to capital facilities than it is to reengineer industrial processes themselves.

- An increase in environmental threats from severe weather events and changes in ecological systems, including increased incidents of heat waves, droughts and flooding, saltwater intrusion in coastal fresh-water supplies resulting from sea level rise and aquifer drawdown and others.
- An increase in human threats against the operations and facilities of municipalities, corporations and individuals. This trend has led to increased needs to consider security issues in all capital construction projects, including new ways to harden facilities against attacks and new approaches for detecting and verifying threats.

These trends have led to changes in the expectations our customers have about the built environment which we provide to them, including new requirements and aspirations for the projects we deliver. Successful companies with strategic visions will look for ways to anticipate client expectations and position themselves to deliver projects that can meet future needs that clients may not yet even realize exist. Being aware of continuing trends that will influence the market is a starting point for this strategy.

Continuing challenges affecting project sustainability

Continuing challenges are of interest in establishing proactive strategies that can benefit AEC companies in the changing global context. These concerns include:

- Increasing volatility in the basic sources of supply for raw materials, due to both increasing scarcity of supply and growing sociopolitical and ecological volatility in regions where some raw materials are harvested. This will correspond with increasing volatility and overall costs for energy in all sectors of the economy.
- Reduced supplies/reliability of both quantity of supply and quality for essential commodities such as water, coupled with price volatility and overall price increases.
- In developed countries, continued consumer awareness and pressure on the retail and supply sectors for environmentally and socially responsible products and products which maximize resource effi-ciency over their lifecycles. This trend will be amplified by increased consumer ability to investigate and purchase products and services online without being constrained by what is available locally.
- Increased access worldwide to ever-greater amounts of informa-tion, without corresponding quality control on the content of such information or ethical constraints on its use. New techniques for micro-marketing will allow companies to target information at the individual level to achieve business aims, using the growing stream of data generated by people as they use smart devices to interact online. Ethical constraints on the use of such data are presently lagging far behind its widespread availability for a range of purposes.

- In developing countries, continued population growth and increased purchasing power due to rapid economic growth and spread of both economic opportunity as well as the information and infrastructure to exploit it.
- Increased vulnerability of populations in both developed and developing countries to natural disasters because of the greater density in settlements in vulnerable areas. Increased vying for future stocks of natural resources from developed countries, such as oil reserves.

Long-term possibilities

While increased economic growth is bringing new opportunities to people worldwide, the world's resources remain finite, as does its ability to absorb the impacts of human activities without degrading. Over the long term, human ingenuity must ultimately lead to new kinds of solutions in the built environment that enable all life to thrive and prosper (du Plessis 2012). The following are some of the solutions presently being considered by progressive thinkers in the AEC industry:

- Transformational process changes across multiple industries to address resource supply constraints. These process changes will result in a need to change the types of facilities that support industrial enterprises and dramatically increase both the efficiency of processes and the ability to recover resource streams formerly considered to be waste.
- Colocation of complementary industries to create industrial ecosystems whose inputs and outputs are useful to one another, thereby increasing efficiency and robustness while reducing transportation overhead to achieve useful aims. The construction industry must consider ways in which it can find niches in this industrial ecosystem in order to maximize value to industry.
- Distribution of production capabilities rather than centralization because of the high costs of transportation, and an increase in goods produced locally. The globalization of ideas and services will increase as a means of stimulating the development of emerging countries. Effective distribution of production capabilities will require process scalability, flexibility in required operating resources and new approaches to capital construction that increase cost-effectiveness for multiple small operations. The AEC industry must respond to this trend with new ways of building and managing projects.
- Increased reliance on distributed instead of centralized infrastructure, especially to meet energy and fuel needs.
- Increasing reliance on locally renewable resources (including energy sources); for instance, tapping of decommissioned landfills to obtain methane to power industries located near such facilities. Overall, industry may begin to view sites formerly considered to be nuisances

(such as contaminated brownfields or waste disposal sites) as potential assets as new technologies emerge to recover the resources that contaminate these sites.

- Increasing need to develop both facilities and industrial processes that are robust and versatile in terms of raw material supplies and environmental conditions for operation. The uncertain impacts of global climate change will potentially affect raw material sources, reliability of transport and supply chains, and even the physical operation of facilities that house industrial processes. For instance, a rise in sea level might necessitate the relocation of industrial operations currently located near coasts or other bodies of water.

Ultimately, the forces we are beginning to see today represent both threats and opportunities to the organizations that comprise the construction industry. Proactively considering future possibilities will afford those organizations the chance to take a leadership role among their peers and direct their courses of action to be successful and sustainable over time. Concurrent with these forces driving decisions in today's strategic business environment, three fundamental paradigm shifts are transforming the way in which we think about sustainable solutions to the challenges facing humanity. The next sections discuss each of these paradigm shifts in turn.

Moving from 'less bad' to regenerative

The first major paradigm shift has been gaining momentum over time and is finally reaching the forefront of sustainability thinking in the AEC industry. The original focus in the early days of green building was on increasing efficiency and reducing waste, resulting in tighter, leaner and more cost-effective projects. While these measures did indeed show return on investment for those who pursued them, they ultimately fell prey to the law of diminishing returns, with each increment of additional efficiency becoming harder and harder to obtain. Performance improvements were considered in terms of their cost tradeoffs, with each action taken to reduce impact adding another burden to the project budget.

Inspiration for a new approach emerged from taking a fresh look at ecological systems and considering ways in which human-designed solutions could contribute as part of the larger set of living systems of which we are a part (Cole 2012a, and 2012b; Conte and Monno 2016; Mang and Reed 2012). By considering how ecological systems in nature are able to sustain themselves on an ongoing basis, new ideas emerged not only for creating human systems to mimic these mechanisms, but also to integrate natural systems in synergistic ways with human systems. Today, the expectation is growing that sustainable solutions should be *regenerative*, not just low impact or 'net zero'.

Coupling eco-efficiency and eco-effectiveness

One of the earliest divergences in sustainability thinking, discussed further in Chapter 2, was the debate between eco-efficiency and eco-effectiveness. Although both perspectives have persisted over time and there is broad agreement that both have lessons to contribute in improving the performance of the built environment, they represent fundamentally different paradigms about what is possible for humans to achieve. Both paradigms are based on an underlying assumption that human activity by its very nature functions as a metabolic process which consumes resources and produces waste. The eco-efficiency paradigm seeks to maximize the ratio of benefit to humans per unit of resources consumed or waste produced. By becoming increasingly efficient in our industrial processes, we can meet the more needs of more humans without depleting finite resources or destroying the ecological basis for our own survival. High performance, energy- and water-efficient buildings are the result of this paradigm, and we have continued to evolve new technologies and processes over time that increase the amount of value per unit resource consumed or impact created, often with increased initial investment required.

In contrast, the eco-effectiveness paradigm considers the technological metabolism of our built environment not as separate from, but rather as an integral part of, the larger natural and human systems of which it is a part. Efficiency is a key factor in the overall fitness and evolutionary advantage of solutions within this ecosystem, but it is not the only factor. Complementarity with other systems becomes even more important, with built facilities filling open ecological niches in the natural-human ecological system. In other words, the built environment is constructed and operated to depend on and recycle flows of matter and energy within the larger system, where cyclical resource loops are refreshed via solar input and waste is metabolized and used by other elements in the system. In this paradigm, minimizing waste may actually be counterproductive if one's waste is food for another complementary species. Likewise, if one's resources required to meet human needs are derived from the waste stream of another species, then the role of waste recovery has value to the health of the larger system, and one need only balance consumption with available supply, not necessarily minimize consumption itself.

While both paradigms are essential for achieving sustainability, one without the other is unlikely to be successful. This progression of thought is reflected in some of the newest rating systems such as the Living Building Challenge and the Envision Infrastructure rating system, where projects are rewarded if they can function within their own ecosystems at a net zero or net positive, regenerative level. Indeed, exploiting opportunities to provide complementary services to neighbouring sites or projects at the local, district or regional level is at the foundation of cutting-edge solutions for sustainable buildings and infrastructure systems, helping to provide multiple paths for meeting the needs of system stakeholders and increasing resilience to possible threats.

Seeking bioinspired solutions

One key strategy to achieve the complementarity required by the eco-effectiveness paradigm while increasing the efficiency with which human needs are met is to look to nature for bioinspired solutions. The range of bioinspired solutions includes varying degrees of blending ecological with human-created systems at different scales, from individual materials and systems to whole projects, developments or communities.

Biomimicry

Biomimicry is one type of bioinspired solution, where technology is used to imitate nature and create new solutions modelled after natural systems. Nature produces robust solutions with optimum efficiency: plants and animals that cannot compete simply do not survive. There are nine basic attributes of nature that can serve as a core set of biomimicry principles for design and construction of human systems (see *Biomimicry principles for sustainable design and construction*).

Here follows some examples from nature that have been applied in the AEC industry:

- The lotus leaf, where air bubbles trapped under the microscopically rough surface of lotus leaves make water droplets roll off the surface, taking with them dust particles and contaminants. This results in a self-cleaning surface.
- African termite mounds, where the geometry of the mound creates natural convective pathways that encourage ventilation, and the thermal mass of the mound walls buffers temperature changes. Together, these properties maintain temperatures within the mound that are significantly more comfortable than ambient conditions.
- Spider silk, which is as strong as Kevlar but manufactured into extremely durable molecular protein structures on a diet of insects at ambient temperatures instead of using hazardous chemicals at extremely high artificial temperatures and pressures.
- Sheep's feet, which provide even and thorough compaction of clay soils and can be mimicked for soil compaction on the construction site.
- Pine cones, which have a structure that is naturally 'breathable'. This structure can be mimicked to create breathable fibre-reinforced plastic plates that can be used for structural retrofits.
- Self-assembling, self-healing biological organisms – all animals self-assemble their bodies during gestation and are able to self-heal many wounds. Materials such as concrete can be imbued with these qualities to self-repair cracking that would otherwise lead to erosion of reinforcing steel or other damage.
- Natural fibres, which are comparable in terms of stiffness and strength to glass fibres presently employed in building materials, and are more resilient and damage tolerant. They can also be recovered for fuel at the end of their service lives (Nguyen 2008a).

Examples of commercialized building products designed using the principles of biomimicry are:

> **Biomimicry principles for sustainable design and construction**
>
> - Nature runs on sunlight.
> - Nature uses only the energy it needs.
> - Nature fits form to function.
> - Nature recycles everything.
> - Nature rewards cooperation.
> - Nature banks on diversity.
> - Nature demands local expertise.
> - Nature curbs excesses from within.
> - Nature taps the power of limits.
>
> (Benyus 2002)

- I2 Entropy, a carpet tile product by Interface FLOR that contains patterns based on observation of natural landscapes (as discussed in Chapter 3). These tiles are designed in such a way that directional installation is not required. This saves installation time and results in a pattern that looks natural no matter how the product is installed.
- TX Active concrete products, made by Essroc Corporation, that contains titanium dioxide photocatalytic. In the presence of sunlight, it releases an electrical charge at the surface that oxidizes contaminants on the surface of the material and makes them release easily from the surface, resulting in a self-cleaning material. The Aria line of products is claimed to actually clean the surrounding air by attracting and depositing particulates from it.
- Self-deploying structures, currently under development, that react to changes in their environment such as loads and movement to remain structurally viable. These structures may also be constructed of biomimetic materials themselves in addition to functioning biomimetically.

Biointegration

Biointegration is a second strategy for creating regenerative systems. Biointegrated systems are human-developed systems that incorporate or merge with natural systems as part of their basic function. Biointegrated design is becoming more widely used in many applications to provide core functions required by humans. From constructed wetlands that treat human wastewater as a food source, to living walls and roofs that help mitigate temperature swings and absorb stormwater while providing habitat and absorbing carbon dioxide and pollutants, Green Infrastructure/Living Architecture (GILA) is an emerging design specialty in the bioengineering domain that creates integrated human and natural systems with the express purpose of meeting human needs. The **POSCO Green Demonstration Building in Songdo**, South Korea contains multiple examples of biointegration (see *Case study*), including a green roof on part of the building, a living wall in the main entry area, a bamboo-filled interior atrium, indoor hydroponic growing areas in the living units for raising food, and outdoor plantings of edible trees and vines to provide food for inhabitants.

Case study: POSCO Green Building, Songdo, South Korea

POSCO Green Building (Figure 10.1) is a future-oriented project that features environmentally-friendly and energy reduction strategies and technologies. The building is located on the Yonsei University International Campus located in Songdo, South Korea and consists of an office building (5 floors, B1-4F, total 59,965 sq ft), a joint housing building of five units (three floors), four modular home units, and an exhibition hall. The POSCO Green Building was a test bed for the collaborative research project 'Promoting Market-based Green Buildings', funded by POSCO Co., Ltd. and the Korea Agency for Infrastructure Technology Advancement. This green building research project led

continued

Figure 10.1 Rendering of POSCO green building concept

by the Center For Sustainable Building at Yonsei University with the architectural firm of POSCO A&C, the construction firm POSCO E&C, and the Research Institute of Industrial Science and Technology (RIST), is operated with funding from the POSCO Group.

The main goal for the POSCO Green Building was to develop an integrated team and process to implement green building strategies and technologies throughout the building's lifecycle from design through operation. In addition, the research team also wanted to promote a sustainable future in harmony with nature, based on new and innovative materials and technologies related to green building. To achieve the green building goals, the project team used an Integrative Design Process (IDP) (Figure 10.2).

The research and project team explored many different strategies to achieve green building with respect to water, energy and comfort, and structure and material using 10 main principles: site planning and landscape design; water conservation; daylighting and natural ventilation; energy efficient building shell; sustainable building products; heating, cooling and ventilation system; lighting and electrical system; renewable energy sources; building energy management; and eco-education. The design team also used integrative design across multiple systems to optimize overall energy and water performance as well as provide a good indoor environment to occupants (Figure 10.3).

continued

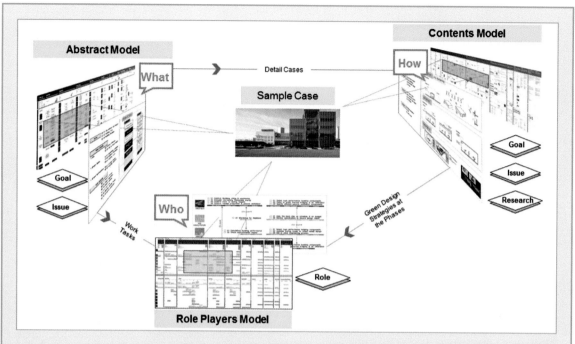

Figure 10.2 Integrative design process model for POSCO green building

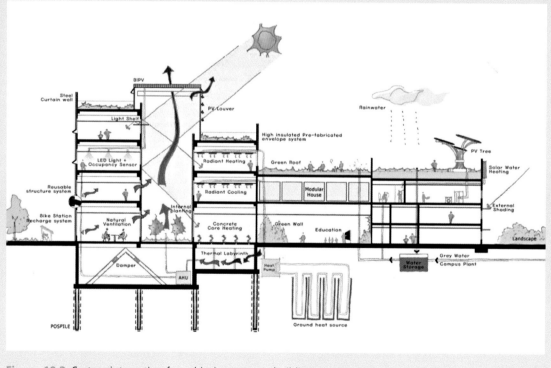

Figure 10.3 System integration for achieving a green building

continued

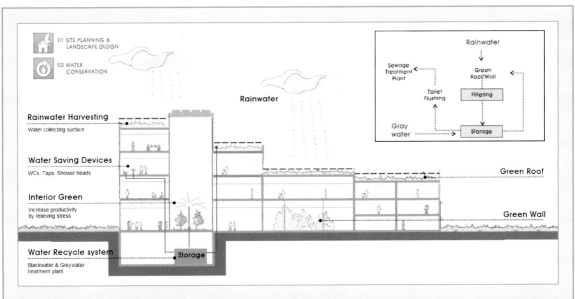

Figure 10.4 Sustainable site and water conservation strategies at the POSCO green building

The project site offered opportunities to incorporate green site planning, landscape design and water conservation strategies (Figure 10.4) in the POSCO Green Building.

Plantings both inside and outside the building improve the environment for occupants by improving indoor air quality (Figure 10.5) and also providing a productive outdoor landscape for producing food, including grapes, fruit trees and other edible landscaping (Figures 10.6a and 10.6b). One of the high performance prototype housing modules also includes an integrated hydroponic unit designed to allow families to grow their own food inside the home (Figure 10.7). Since these modules are designed to be incorporated as new or retrofitted homes in high rise construction, the food growing modules provide an option for families that would not otherwise be available.

Figure 10.5 Indoor atrium with bamboo improves air quality in the office area of the building

continued

Figure 10.6a, b The outdoor landscape includes multiple examples of food-producing species

As a major steel manufacturer, the parent company POSCO Ltd. saw this building as an opportunity to demonstrate cutting edge materials research. Accordingly, the POSCO Green Building included innovative structure and material strategies to achieve the goals of green building including a high insulation steel curtain wall system, a modular building unit system and salt tolerant concrete to increase its durability given its location near the coast of South Korea (Figures 10.8a and 10.8b).

The POSCO Green Building implemented many different green building technologies and strategies to improve energy efficiency and indoor comfort and environment. The primary green strategies related to energy and comfort included light shelves and shading devices to prevent glare and distribute daylight, organic LED (OLED) lighting and building-integrated photovoltaics (BIPV) (Figures 10.9a and 10.9b); a radiant floor system for heating and a thermal labyrinth for storing heat energy (Figure 10.10); a double skin façade, chilled beam system, and heat pump system (Figure 10.11) and a highly insulated pre-fabricated envelope system with concrete core cooling (Figure 10.12).

Figure 10.7 An indoor hydroponic unit allows families to grow their own food in high-rise homes

The POSCO Green Building also uses a Building Energy Management System for office spaces (Figure 10.13a) and a Home Energy Management System (HEMS) for residential spaces to control the building's energy use (Figure 10.13b).

continued

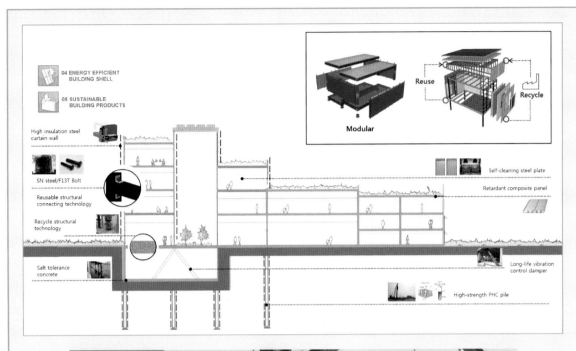

Figure 10.8a, b Green materials (a) and modular construction strategies (b) used in the POSCO Green Building

continued

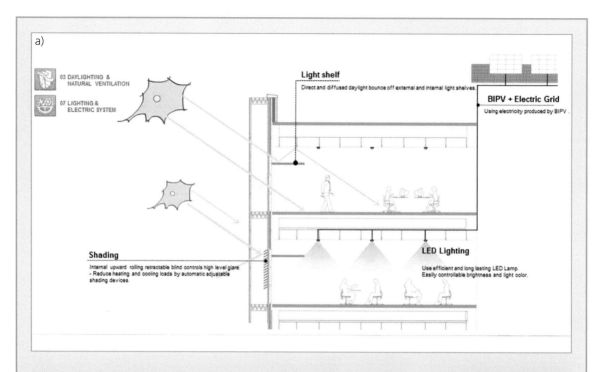

Figure 10.9a, b Shading, light shelves (a) and BIPVs (a) are key solar energy strategies

continued

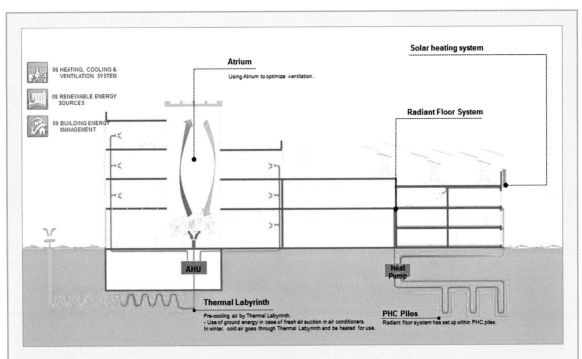

Figure 10.10 Passive heating and cooling strategies include an underground thermal labyrinth and pre-stressed, high-strength concrete (PHC) piles for heat exchange

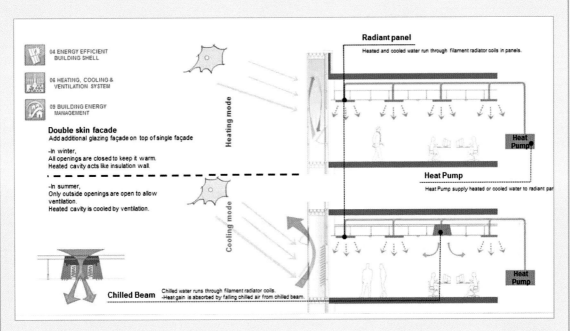

Figure 10.11 Thermal comfort is provided using radiant ceiling panels and chilled beams, and a double skin façade controls solar heat gain

continued

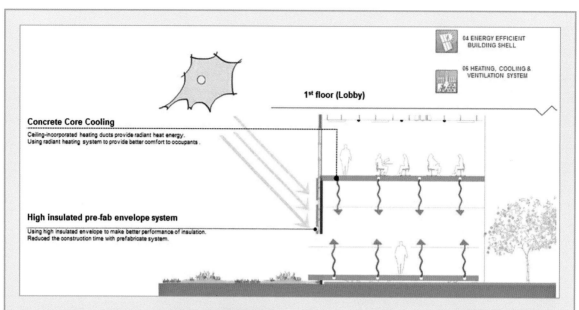

Figure 10.12 The prefabricated envelope system is highly insulated, and ducts integrated with the concrete ceilings provide conduits for heating and cooling

Figure 10.13a, b BEMS (a) and HEMS (b) provide controllable thermal comfort and indoor air quality to occupants

HEMS can provide the occupants with information about how they use energy in the home and provide feedback to occupants on how to modify behaviour to reduce energy consumption. HEMS can also provide the household with the ability to control energy consuming processes in the home remotely, via a smart phone or web service. The BEMS controls and monitors the building's mechanical and electrical equipment such as ventilation, lighting, power systems, fire systems and security systems. In addition, it also integrates with the indoor environment quality measurement sensors to provide a comfortable indoor environment to occupants in the building.

continued

The POSCO Green Building is a showcase of green technologies and strategies to educate building-related professionals, university students and school students including elementary students. The prototype modular apartments are examples of green building models for energy-efficient homes and office buildings in South Korea. The POSCO Green Building is a destination for visitors to South Korea to learn green building strategies and technologies and to explore new building materials developed and produced by POSCO.

Source: All images courtesy of POSCO E&C

Green infrastructure is defined as the deliberate and systematic use of plants and organisms in the context of biological processes to provide organized services to human communities. Most often deployed for stormwater and wastewater management and control, green infrastructure (along with its building-scale application, living architecture) has seen successful application in a variety of contexts and is growing in acceptability as a core solution in the toolset of urban and suburban infrastructure managers. The WaterHub at Emory University, presented as a case study in Chapter 6 (p. 290), is an example of green infrastructure at the campus or district level and includes both outdoor and indoor ecosystems to treat wastewater from the Emory University campus.

Bioinclusivity

A different philosophy is embodied in the bioinclusivity movement. Multiple trends reflect growing interest in our society to deliberately include plants and animals as key elements within in the built environment. From pets in the workplace, urban agriculture and rooftop farming to resident animals and gardens in senior living and healthcare facilities, humans are realizing measurable benefits from intentionally incorporating other species as part of our activities. However, in the majority of building and development projects, living species are not actively considered as integral stakeholders of the design. If they are included at all, it is as ornamentation (landscaping), as isolated systems providing a specific function such as stormwater control, or after the fact as an operational consideration. Instead, we design facilities such as animal shelters, plant nurseries or farms where other species are segregated from everyday human activities, thereby missing opportunities for synergy that can lead to positive impacts for humans and non-humans alike.

Real concerns inhibit the broader adoption of bioinclusive features and practices in our society, including potential auditory or olfactory nuisance, hygiene concerns, difficulty of maintenance and negative impacts on people with allergies or phobias. Such concerns drive the development of policies at the organizational or government level to manage and control (and sometimes even incentivize) how other species are included in human development (e.g., urban agriculture policies, workplace pet policies, green roof policies). Many of these concerns can be addressed through improved design and construction of the built environment in which humans and other species interact. However, while concrete evidence exists of the benefits to humans afforded by interaction with other species, we are still learning how the design of the built environment can act as an effective platform for successful interactions between humans and other species to achieve those benefits without concurrent negative impacts.

Urban ecology

Ecological systems can serve as a metaphor for community interactions with the aim of collaborative sustainability. Beyond the building scale, urban ecology involves mimicking the function of entire ecosystems in designing human communities and enterprises. Industrial ecology involves the deliberate design of industrial or urban systems such that the waste from one component can be used as the feedstock to another. Complementary enterprises may be colocated in industrial 'ecoparks' to minimize the transport requirements for sharing waste streams. Some ecoparks, such as the well-known park in Kalundborg, Denmark, also include non-industrial parts of the community as part of the symbiosis achieved. Figure 10.14 shows the major components included in the Kalundborg network.

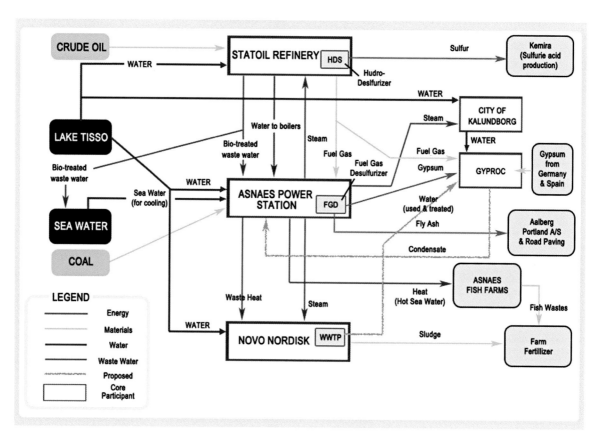

Figure 10.14 Components in the Kalundborg industrial ecology network

Source: Chris H. Lewis, American Studies professor (retired), University of Colorado Boulder

One common example of industrial ecology is the colocation of cement and concrete facilities near coal-fired power plants (Figure 10.15). The flyash produced during combustion of coal to make power can be used in place of Portland cement for making concrete, and waste heat from the power plant can also be used to support the enormous energy requirements for operating the cement kiln itself. Other examples of possible synergies include co-location of utility service lines required by multiple facilities in the same excavations, coordination of preventive maintenance across facilities, exchange of complementary inputs and outputs and cooperative procurement of common resources. New techniques for information modelling and sharing are also being developed to facilitate these inter-organizational relationships and collaborative management and decision-making. The exchange of resources between members of the larger system is known as the urban or industrial metabolism, which mimics the behaviour of elements within a living organism in converting food/fuel into necessary function. Such collaboration among a community of enterprises can result in greater efficiency, reduced transaction costs, exploitation of experience curves and creation of critical mass required for innovation that would otherwise be unavailable to small firms (Meneghetti and Chinese 2002).

Overall, the principles of bioinspiration allow designers to learn from natural systems that have been evolving since the dawn of life on Earth. They facilitate taking advantage of waste as feedstock and cycling resources in complementary ways, rather than using a linear process of taking resources from the Earth, using them and generating waste that must be distributed back to the natural environment.

Figure 10.15 A concrete plant is located near Power Plant #3 in Ulan Baatar, Mongolia

Regenerative business models

New kinds of business and operating models are also emerging that challenge our contemporary cultural assumptions about how to interact with one another as members of the larger economic and societal ecosystem. These approaches may seem new in the context of our current practice, but in many ways they reflect the wisdom of ecological systems that we are only now rediscovering. Alternative currencies, open source development and collaborative consumption all reflect new ways of thinking that expand the assets upon which we can draw to meet human needs and aspirations.

Alternative currencies

The notion of using items for barter and trade instead of currency is as old as commerce itself. While the world's economic engine runs on official currencies issued and backed by government organizations, the widespread emergence of new alternative currencies such as Bitcoin has provided a new basis for interactions among parties that can occur outside the normal constraints of trade. Proponents of alternative currencies argue that they are less susceptible to manipulation by governments and inflation, while their detractors argue that they can also more easily be used unethically or for criminal activities since they exist beyond the supervision capacity of government.

Microcurrencies are emerging in some markets as a way to facilitate localized transaction without the inherent overhead involved in normal currency transactions such as taxation. Some local currencies mimic official currency in having actual bills or coins that are the means of exchange, while others rely on online trading and documentation to track accumulation and exchanges. One alternative currency growing in popularity is time banking or time exchange, in which individuals trade services and goods based on the number of hours required to produce or deliver them. For example, Ithica Hours were introduced in the New York city of Ithica in the early 1990s as one of the first community currency based on time banking (Ward 2013), and the trend has recently gained momentum in the United States (see *Case Study: The Livability Initiative* for an example).

Alternative currencies allow parties that might not otherwise be able to participate in economic exchange to bring value to transactions, thereby increasing the possible levels of synergy and interaction within the human ecosystem. At the community scale, they add important redundancy and empower citizens to be beneficial participants of the larger system. They also recognize value not captured by conventional currency and can more accurately reflect costs and benefits that might otherwise be externalized.

Open source development

Open source development is another trend that has the potential to affect how business is done in the AEC industry. Like evidence-based design discussed in Chapter 6 (p. 338), open sourcing allows the sharing of information that might conventionally be considered proprietary in hopes of advancing the state of knowledge more rapidly. By establishing a common platform of information on which innovation can build, open sourcing facilitates more rapid development of new ideas and broader diffusion of benefits.

Open source products are characterized by universal access at no cost to the design or blueprint of the product, which can be used as-is or adapted and evolved as needed. Becoming widely known as a result of its use in the software industry, open source development has also been applied to appropriate technologies for building and community development, including basic modular equipment design that can be easily fabricated using simple materials and equipment and then used as a platform to fabricate more complex designs that build upon it.

Open Source Ecology (www.opensourceecology.org) is an example of a community toolkit of modular open source designs including the Global Village Construction Set (GVCS), a collection of the '50 most important machines that it takes for modern life to exist'. This set includes basic components that allow fabrication of other components (e.g., a circuit making machine, a laser cutter and an open source welder) and thus comprises self-replicating production tools that can be assembled in various ways to meet human needs. Designs for the machines are shared online for free. The first machine in the set was a compressed earth brick press used to develop modular masonry units for building construction. Together with a tractor machine and a soil pulverizer, the brick press was used to fabricate a prototype microhome, which meets another essential human need using the basic set elements.

Open source development promotes sustainable and regenerative development by allowing an evolutionary approach to local adaptation of solutions while minimizing the resources required for 'overhead' activities such as transportation and distribution of goods from central-ized manufacturing. As with evidence-based design, open source development allows everyone to start from the same basic set of knowledge and therefore contribute to more rapid innovation rather than starting from scratch.

Creative commons licensing

A variant on open source development is creative commons licensing, a legal mechanism for open source sharing and distribution of creative work. Creative Commons (CC) is a U.S.-based non-profit organization (see http://creativecommons.org) that manages different types of licenses for work available free of charge to the public. These licenses replace conventional copyrights and allow their holders to specify particular levels of rights that are reserved in the sharing and distribution of creative work. Examples of different levels of licensing include the ability to use with attribution, use for non-commercial purposes with attribu-tion, use or modification of work as long as the same terms are applied to the modified work, and others.

Several well-known platforms such as YouTube, Flickr and Wikipedia provide their users with the ability to license and share works using creative commons licensing. This mechanism is one way in which information models and design information could be openly shared and protected in the AEC industry as an alternative to conventional practice.

Collaborative consumption

Collaborative consumption is a set of alternative economic concepts in which consumers can both obtain and provide resources and services through direct interaction with each other or via a third-party organization serving as a mediator or broker. Popularized in 2010 by Botsman and Rogers, well-known examples of this idea have existed in the form of flea markets, garage sales and other consumer-to-consumer modes of economic exchange. New mediums of exchange are flourishing with the growth of new kinds of information platforms and social media, allowing new kinds of economic ecosystems to emerge in response to local needs and opportunities.

Two kinds of collaborative consumption have been identified in the literature: mutualization, which focuses on the temporary exchange of access to resources for a fee or at no cost (i.e., rental or borrowing systems), and redistribution, which involves permanent exchange of resources, again either for a fee or at no cost. The latter often involves resale or donation of pre-owned goods, thereby keeping them 'in circulation' and providing beneficial service rather than languishing unused. Both mutualization and redistribution displace the consumption of virgin materials and energy resources, resulting in reuse, recycling or downcycling of goods that can still provide serviceability.

The collaborative consumption model introduces new roles to existing players in conventional consumption economies. Whereas previously consumers were only passive receivers of goods and services, in collaborative consumption they now have agency to play an entrepreneurial role or collaborate with existing producers to create new synergistic relationships. As such, collaborative consumption introduces the potential for new kinds of synergies, mimicking the richness of natural ecosystems with organisms playing multiple roles to support the health and function of the whole. Examples of collaborative consumption in the current market include Air BnB in the hospitality industry and Lyft, Uber and Zipcar in the transportation field. All of these enterprises allow greater levels of benefit and value to be derived from existing resources that would otherwise be underutilized, thus increasing both the efficiency and effectiveness of those resources in meeting human needs and aspirations.

Moving from prescriptive to performance-based

A second major paradigm shift has occurred with respect to how we seek to implement sustainability objectives in creating the built environment: the shift from standardized, prescriptive solutions toward more customized, tailored approaches that focus on how a solution will respond in its particular context of use. As early as the 1990s, DuBose (1994) observed that sustainability is an inherently contextual concept that favours tailored solutions designed to address relevant situational challenges, rather than solutions sufficiently robust to handle any possible challenge that might come along.

Today, this principle is being applied to achieve better-performing buildings at lower capital cost. For example, the 6 New Street mixed-use building in East Boston, Massachusetts used site-specific wind analysis to justify the use of lower grade steel at considerable initial cost savings, compared to the structural design that would have been required under default code requirements (Pearson 2016). Not only were there capital cost savings as a result of this analysis, but also the building owner was able to make the case for reduced loss risk to the building's insurance provider, justifying an insurance premium 10 times lower than would have been quoted for a comparable conventional building (ibid.).

Tailoring solutions to situations is becoming a part of sustainable practice throughout the industry, from the ideation stage through capturing and using relevant building information, predicting and managing outcomes more effectively, making better choices and customizing implementation to context.

Coming up with better ideas

As more stakeholders become involved in the processes of planning, designing and delivering of sustainable projects, new ideas can be brought to the table that take into account considerations not ordinarily considered by specialized designers. Multiple players must be involved in the integrative project design and delivery process for new facilities, starting in the very earliest phases of consideration for potential projects. Understanding the organizational requirements for physical space along with the nuances of how well existing facilities meet current needs can help to identify design goals for new facilities that will result in a better fit between organization and physical environment. Stakeholders responsible for operating and maintaining facilities during the occupancy phase of the lifecycle, the period during which a building requires its largest share of energy and water resources, also have important knowledge to contribute. Representing a broad range of stakeholders as part of integrative design helps to ensure that the resulting project will fit well with both organizational needs and operational capabilities.

Even if they do not participate directly in the integrative delivery process, future building occupants and users are also key to understanding what are appropriate goals for a project and what conditions and constraints will govern its operation. New approaches are being developed to involve not only professional stakeholders but also laypersons as part of participatory planning and design. At a passive level, the perspective of markets and stakeholders is increasingly part of business strategy and marketing through data analytics. **Analytics** are the qualitative and quantitative techniques used to process large data sets to extract an understanding of specific segments of the population by whom the data were generated. With particular focus on behavioural data and patterns that characterize consumer decisions and choices, data analytics allows design to be undertaken with a greater awareness of what key issues will guide stakeholder choices.

At a more active level, social media and new tools for collaborative planning and decision making provide a basis to support the identification and alignment of goals across more heterogeneous communities of people. Long-term comprehensive planning is now possible at regional scales using software that engages members of the community to identify key issues and prioritize potential solutions. The Livability Initiative is one example of a multi-year planning process used in the United States to develop a regional strategy for sustainable development supported by new organizations and programs (see *Case Study: The Livability Initiative*).

Case study: The Livability Initiative: a sustainable future for the New River Valley

Begun in 2009, the Livability Initiative is a regional planning process for the New River Valley region in southwest Virginia, United States. Originally sponsored by a federal grant, this initiative brought together citizens, business, elected officials and staff from local municipalities to identify key challenges and goals for the region. Multiple participatory processes were used to collect citizen input (Figure 10.16), including a facilitated charrette held at a local park facility, a web-based survey, public input meetings held by the initiative's leaders at various locations throughout the region, and other summary meetings.

The initial charrette was held at a local park facility and involved stakeholders from four counties plus one incorporated city. Input was collected through multiple means throughout the day including team exercises, keypad clickers and open discussion, with participants allowed to regroup according to topics of personal interest. Topics included local food networks, buildings and energy, transportation, land use and others. Each group of stakeholders met around a table, and members of the facilitation team captured ideas in real time and entered them into a database. Electronic voting was used to prioritize ideas and issues in sessions, resulting in a prioritized set of needs based on input from all participants.

The online survey was used to engage people who could not attend the original meetings held during the business day, and involved a web-based interface that allowed people to allocate funding from a limited pool to various initiatives based on which they believed

Figure 10.16 Multiple methods were used to invite public input and civic dialogue

continued

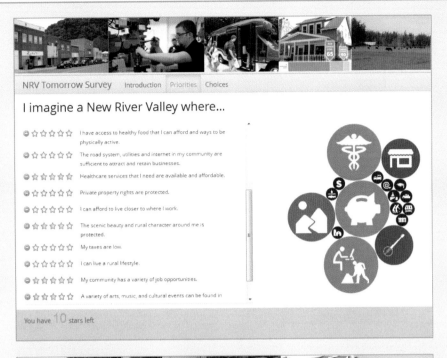

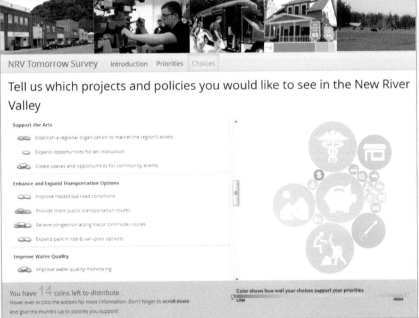

Figure 10.17a, b
A customized web-based survey with an engaging graphical interface involved members of the local community in identifying preferences for how their tax dollars should be spent

continued

Figure 10.18 The BUILT NRV board game

were most critical for the future of the region. The outcome of this process was a set of priorities for addressing needs identified in the first charrette (Figure 10.17).

Additional efforts to obtain public input to the process included town hall meetings, story circles, performance theatre events and a board game called BUILT NRV (Figure 10.18), in which players experiment with different adjacent land uses and investments to explore core values and priorities, key tensions and tradeoffs, obstacles, challenges and potential action strategies. In addition to obtaining public input in multiple ways, the facilitators of these events also conducted studies to develop new knowledge about innovative methods for civic dialogue.

The ultimate outcome of the funded portion of the effort was a three-part report (see http://nrvlivability.org) summarizing participant ideas and input. This report is presently being used by local governments within the region as part of comprehensive planning to better understand priorities of citizens both within and around their municipalities. Several other outcomes have also resulted from the initiative, including a Time Bank (http://nrvtimebank.org) that was identified as a high priority in response to concerns of residents wishing to age in place in their own homes within the community.

This level of regional planning is unusual in the United States, particularly in rural areas. Historically, each municipality, including small towns and cities as well as larger county areas, has had its own government which makes decisions and infrastructure investments on behalf of its constituencies, leading to many independent solutions for infrastructure challenges. Increasingly, the potential for synergies and efficiency at a larger scale is driving regional planning efforts, and funding trends support investments in regional infrastructure rather than solutions serving only one municipality. Whether

continued

greater efficiency also means greater sustainability remains to be seen in the long term. While regional-scale infrastructure may promise greater operating efficiencies, it may also reduce redundancy of function that contributes to resilience. Determining the best scale for investment is a complex question that depends on many factors. From an informational and planning standpoint, however, regional-level planning efforts provide a 'big picture' perspective that introduces new possibilities for cooperation among municipalities that may not otherwise be identified.

Source: All images courtesy of New River Valley Regional Commission

Overall, the future of sustainable building and infrastructure depends on earlier involvement in planning and design at both strategic and operational levels to enable more radical approaches to facility sustainability (Galamba and Nielsen 2016). Improved ways to recognize and incorporate the needs and perspectives of multiple stakeholders such as social media will allow future facilities to be more responsive to human needs and aspirations, and more agile processes for facilities management will lead to buildings and infrastructure that better support the needs for which we develop them. Further evolution of both process and technology can create opportunities for broader application of the knowledge base of facilities management to support human enterprise.

Capturing and using relevant building information

High quality sustainable design requires an understanding of how a building will perform after it is built, which in turn requires computer-based simulation software for rigorous building analysis. The advent of building information models (BIM) offers even greater opportunities for building analysis by pairing the analysis software and BIM for the seamless assessment of building performance. BIM refers to the creation and coordinated use of a collection of digital information about a building project. The information is used for design development and decision making, production of high-quality construction documents, predicting performance, cost estimating and construction planning, and eventually, for managing and operating the facility. BIM has the potential to support lifecycle management and decision making for built facilities, and represents considerable potential for eliminating process waste as part of the facility delivery process. As such, it is well suited to support efforts to increase project sustainability, although opportunities for additional development of BIM tools abound.

A well-populated building information model carries a wealth of information necessary for many aspects of sustainable design and green certification. For instance, schedules of building material quantities can be obtained directly from the model to determine percentages of material reuse, recycling or salvage. Various design options for sustainability can be pursued in parallel and automatically tracked in the model. Advanced visualization techniques can be used for solar studies and to produce 3D renderings and construction animations of a green project.

In order to achieve more comprehensive sustainable solutions using BIM, an expansion of traditional thinking is required while making decisions during the design process, including:

- understanding climate, culture and place
- understanding the function of the building
- understanding the needs of the community
- minimizing the consumption of resources
- using locally available resources and natural systems
- using efficient human-made systems
- applying renewable energy generation systems.

A fundamental tenet of sustainable design is the integration of all the building systems among themselves as well as with the external economic, social and environmental context of the project. Therefore, the combination of high-performance sustainable design strategies and BIM technology has the potential to change the profession dramatically and bring a higher-quality decision making to architectural design, construction management and facilities management. BIM technologies can be characterized and described in terms of the sustainable design strategies that can be investigated using these capabilities, including:

- **Parametric modelling**: building orientation, massing.
- **Simulation and analysis tools**: daylighting, water harvesting, energy modelling, renewable energy, materials.
- **Integration**: combining parametric automation capabilities with simulation tools for design optimization.
- **Agent-based modelling**: integrated software solutions that track a building's carbon footprint such as embodied energy of materials, emissions from construction and emissions from the fuel it takes to get work crews to the job site.

Emerging tools such as augmented and virtual reality offer new opportunities to overlay information from BIM or n-D models developed to support the construction process (see *IT-enabled design and construction for a sustainable future*). The use of BIM for sustainable facilities management is still in its infancy and has incredible potential as new technologies are developed to automate the arduous and intensive process of developing models for existing buildings and structures. The ever-increasing use of robot and drone devices as well as RFID tags and other sensors to track the movement of materials, labour and equipment within and around projects offers exciting opportunities to mine position data as a means of evaluating project parameters such as productivity and carbon footprint.

Interoperability issues remain a significant barrier to more effective use of BIM throughout a project's lifecycle. At present, critical questions remain to be addressed regarding who owns the intellectual property captured in a BIM model, who is responsible for translation errors that

IT-enabled design and construction for a sustainable future

Sandeep Langar

Given the nature of climate change and the hazards it represents, future designers and builders will be challenged with creating adaptive and resilient facilities that are dynamic and responsive to external natural hazards while providing safety and security to their occupants. Such attributes will be required in addition to established expectations of providing facilities that are ecologically responsive, economically feasible, aesthetically pleasing, functionally satisfying and socially acceptable. No single strategy will achieve adaptive and responsive buildings. Multiple approaches and a combination of methods will be necessary, including education of project stakeholders; use of an integrative design process; alternative materials, technologies and strategies for facility construction; alternative sources of energy and water; and new kinds of information technology (IT).

The use of IT in the design analysis is not novel, and the roots of its origin can be traced to the early 1970s (Eastman et al. 1974). In the last few decades, IT within design and construction has evolved considerably. The most recent development has been Building Information Model (BIM) adoption and implementation among project stakeholders (Ku and Taiebat 2011; McGraw-Hill 2010; Langar and Pearce 2014). Along with other benefits, BIM can analyze data, simulate impacts of technologies and strategies on a project, and derive results that can help reduce ecological impacts. BIM has also been used to help achieve green certification (Azhar et al. 2008; Mahdavinejad and Refalian 2011; Siddiqui et al. 2009) and enhance the sustainability of projects (Ku and Mills 2010; Strafaci 2008). Various software associated with BIM, such as eQUEST, Ecotect, Radiance and others, help designers incorporate measures to enhance the energy use, lighting and airflow around the facility, and improve the indoor air quality of the facility. Software can help identify the most optimum shape for a building so that the structure can maximize the use of freely available renewable resources such as the wind, rainwater, sun and others along with synergies generated from terrain. In addition to supporting design, IT also contributes to building operations by automating building elements such as window shading or other dynamic components of the exterior façade. These dynamic elements enable optimum efficiency within the building's surroundings and enhance value of the facility. Projects such as 'Sharifi-Ha House,' Tehran, Iran (Figure 10.19), have incorporated project components that are dynamic in nature. The rooms of the house on the exterior façade can rotate to maximize natural lighting penetration into the structure, provide dynamic views and respond to the dynamics of the exterior. This is the very first step towards creating structures that are dynamic and responsive to their environment. In the future, IT will enable structures to become more dynamic and responsive like ecosystems, while maintaining safety, functionality and usability.

Another recent breakthrough supported by BIM is the emergence of three-dimensional (3D) printing at a large scale. The world's first 3D printed office building '*Office of the Future*,' located in Dubai (Figure 10.20), is evidence that this technology has real commercial applications. In the future, with the commercialization and application of 3D printing technologies for construction projects, designers will no longer be constrained by standard geometrical forms. 3D printing can be used to build structures that are organic and complex with the aim of mimicking natural organisms. Even the cartridge used to print structures can be customized to use unconventional materials to create building components that react to the environment in a complementary manner. For example, the combination of lime with sequestered carbon dioxide has been used to create a product similar to concrete (CO2CONCRETE), using 3D printers (UCLA 2016). Researchers at UCLA have produced the material in lab conditions, although it remains to be commercialized.

continued

Figure 10.19 Office of the future, Dubai
Source: Golnar Ahmadi

With integrative project design and delivery, experts such as chemists, biologists and scientists may be involved in facility design to incorporate new interactive composite materials as part of designs that mimic natural organisms or ecosystems.

These recent technological developments represent possibilities for humanity to alter the paradigm under which facilities have been historically designed and constructed. The future for design and construction of facilities will involve professionals from multiple disciplines in an integrated environment who possess the ability to design, assess, analyse and test strategies and technologies. In the process, future facilities using BIM will be inspired from successful ecological systems, thereby paving the way for facilities that are not only intelligent and dynamic but also responsive and resilient to externalities.

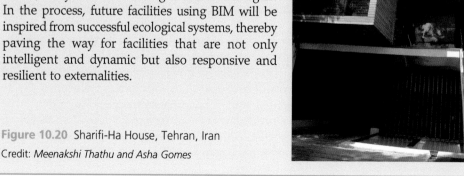

Figure 10.20 Sharifi-Ha House, Tehran, Iran
Credit: *Meenakshi Thathu and Asha Gomes*

continued

References

Azhar, S., Nadeem, A., Mok, J., and Leung, B. (2008). "Building Information Modeling (BIM): A New Paradigm for Visual Interactive Modeling and Simulation for Construction Projects." *First International Conference on Construction in Developing Countries*, Karachi, Pakistan, 4–5 August.

Eastman, C., Fisher, D., Lafue, G., Lividini, J., Stoker, D., and Yessios, C. (1974). *An outline of the building description system*. Institute of Physical Planning, Carnegie-Mellon Univ., Pittsburgh, PA.

Ku, K., and Mills, T. (2010). "Research needs for Building Information Modeling for Construction Safety." *Associated Schools of Construction 46th Annual International Conference*, Boston, MA, 7–10 April.

Ku, K., and Taiebat, M. (2011). "BIM experiences and expectations: The constructors' perspective." *Intl. J. of Const. Edu. and Res.* 7(3), 175–197.

Langar, S., and Pearce, A.R. (2014). "State of Adoption for Building Information Modeling (BIM) in the Southeastern United States." *Proc., 50th Annual Intl. Conf., Associated Schools of Construction*, Washington DC.

Mahdavinejad, M.J., and Refalian, G. (2011). "Architectural design process simulation and computer-based modeling." *J. Adv. Comp. Sci. Tech. Res.*, 1, 74–83.

McGraw-Hill. (2010). "Green BIM: How building information modeling is contributing to green design and construction." *McGraw-Hill Construction*.

Siddiqui, M.Z., Pearce, A.R., Ku, K., Langar, S., Ahn, Y.H., and Jacocks, K. (2009). "Green BIM Approaches to Architectural Design for Increased Sustainability," *Proceedings of the International Conference on Construction Engineering and Management/Project Management* (ICCEM-ICCPM). Jeju, Korea, 27–30 May.

Strafaci, A. (2008). "Building Modeling and Sustainable Design for Information Civil Engineers." http://cadapult. net/company/whitepapers/BIM_civil%20Engr.pdf (12 September 2009).

University of California, Los Angeles (UCLA). (2016). "UCLA Researchers Turn Carbon Dioxide into Sustainable Concrete." http://newsroom.ucla.edu/releases/ucla-researchers-turn-carbon-dioxide-into-sustainable-concrete (6 August 2016).

may occur during handoff from one lifecycle phase to another, and how data sources external to the BIM model may most effectively be integrated to support building-related workflows.

Predicting and managing outcomes more effectively

Coupled with the evolution of BIM is a broad array of modelling and simulation tools that enable comparison of design and construction alternatives long before a project becomes reality. From illumination, thermal comfort and energy performance to construction productivity, cost and schedule, current generations of modelling and simulation tools provide the basis for enhanced optimization such as the 6 New Street project in East Boston mentioned earlier in this chapter. Modelling during integrative design is critical for evaluating and comparing possible alternatives, and many of the leading rating schemes depend on models as the basis for awarding credits for building performance. As discussed in Chapter 9, quite often these models do not correlate well to actual performance for multiple reasons, sometimes causing misunderstandings in their use to support ongoing performance management. Additional work is needed to better reconcile modelling with the kinds of decision making that occurs as part of facility and portfolio management, to ensure both that models are used in appropriate ways, and that appropriate models are available to support different kinds of proactive decisions throughout the facility lifecycle.

Monitoring is at the heart of managing the performance of facility and infrastructure systems, including tracking the amount of energy, water and other resources consumed as well as the corresponding conditions created using those resources. Since energy and water use are typically cost items for a facility, they are conventionally monitored periodically at the whole building or development scale by external utilities that supply those resource needs. While this coarse resolution of information can be useful in identifying large deviations that require attention, the time lag between the problem's origin and being billed for resources on a monthly basis makes effective management difficult. Higher resolution monitoring and performance verification at the system or device level is now possible for energy and water use, and is often integrated as part of building management systems to allow for more timely identification and resolution of problems. Monitoring devices can be used in conjunction with opperationally callibrated design-phase performance models to identify discrepancies and track them to their likely source where they can be quickly resolved.

Providing feedback at the organizational level about facilities and operational practices is an important practice for the sustainable management of an organization. Facilities are a key element of the cost of doing business, and they contribute significantly to the negative environmental impacts of an organization. As such, their performance is monitored and positive changes in environmental impacts are often tracked as part of Triple Bottom Line or corporate social responsibility accounting. Since facilities are often the primary public 'face' of an organization, improvements to the environmental performance are more easily observable and may be easier to achieve than process improvements or changes to the core organizational mission. As such, planning and reporting on improvements to environmental and social performance through facility upgrades and management practices is a large part of many organizations' CSR/TBL reporting efforts. New approaches to mine and integrate data streams to monitor performance in real time are evolving that can allow adaptive adjustments based on immediate local conditions rather than rule-based control.

Making better choices

As our ability to use information in new ways continues to evolve, so also does our understanding of how to use that information to make better decisions in designing and constructing the built environment. Evidence-based design, discussed at length in Chapter 6 (p. 338), represents a growing movement to carry the lessons learned from both science and professional experience forward in a systematic way to inform future projects. Extensive efforts have been made in the United States to develop repositories of evidence to support the effective and sustainable design of healthcare and educational facilities. This shared knowledge base represents a new way of thinking about knowledge as a common rather than a proprietary advantage. By establishing a knowledge base upon which all designers can draw, evidence-based design efforts raise the performance standard for all design and allow designers to focus their creativity on innovation beyond the baseline.

At a larger scale, the lessons of choice architecture, community-based social marketing and decision science (discussed in Chapter 2 (p. 66)) are being applied to the development of policies and incentives to shape the choices of owners and the behaviour of building occupants. Rating schemes such as the Envision Sustainable Infrastructure rating system are being rethought as ways to adjust what people consider the norm vs. the exception, thereby encouraging them to make more aggressive choices that move us toward truly regenerative infrastructure.

The democratization of decision making in communities of multiple stakeholders is also part of a sustainable future, but it is not well-supported by existing tools and technologies in the sustainability realm. For example, multiple sustainability assessment tools have been developed to evaluate sustainability at the urban development scale, including LEED for Neighbourhood Development, BREEAM Communities, CASBEE for Urban Development and others. Many of these tools evolved from building-scale tools and have a similar structure and implementation process. Boyle and Michell (2016) point out that while these tools have some utility in evaluating the physical and material properties of the built environment, their focus is primarily on environmental impacts, and their application requires expertise of professionals rather than the perspectives of stakeholders in the community. While such tools may be useful for campuses or other neighbourhood developments with a single major stakeholder, they are less useful in communities of multiple independent stakeholders that evolve organically over time, since they lack explicit frameworks for collaboration among stakeholders and do not provide information relevant to manage individual actions (ibid.). Opportunities exist for new tools that focus not only on environmental but also on social and economic impacts, which can be customized to context and include objectives that are responsive to local stakeholder values.

Customizing implementation to context

New technologies for 3-D printing and rapid prototyping have begun to transform the ways in which we think about designing and producing products, and are only now beginning to take hold at the scale of buildings and infrastructure. These technologies are disrupting long-held assumptions about economies of scale associated with centralized production, and providing new ways to rapidly evolve custom solutions through trial and error at the local level.

Representing the ultimate in localization, innovations in distributed manufacturing will change the way we approach construction and maintenance of facilities and infrastructure, with products and materials being manufactured and adapted locally using shared information rather than produced centrally and shipped over long distances. Solutions will be customized to context at the local scale, allowing for increased efficiency in producing facility services under local conditions. Large scale urban retrofitting will be necessary to adapt existing facilities and infrastructure to function as part of these networks.

Moving from rigid to resilient

Awareness of the interdependency among infrastructure systems in particular has led to increased recognition of the lack of resilience these systems have under likely future conditions. New approaches to buildings and infrastructure, including decentralized infrastructure as part of net-zero or net-positive buildings and smart grids that connect these nodes together, can help create greater resilience in shifting loads across networks of sources rather than relying on single providers. This distributed capacity requires new approaches to planning, design and construction as well as facilities management, where the interactive behaviour of facility stakeholders and their buildings can be anticipated and accommodated as part of larger network stability. Together with improvements in the infrastructure elements themselves, improvements in ubiquitous data collection and analysis will enable real-time adaptive management of heterogeneous infrastructure networks. Intelligent controls will improve responsiveness of facilities to variations in supply and demand and will better coordinate the performance of physical facilities and infrastructure with the human activities they support.

From mechanical control to passive survivability

As the global climate changes, we can expect more severe weather in the future. Severe storms such as Hurricane Katrina in the United States in 2005 will continue to threaten human development, especially as population growth increases the number of people who live in vulnerable areas such as coastal regions or flood plains. For example, an estimated 53 per cent of the population of the United States lives in coastal zones, making the possibility of rising sea levels and increasingly severe storm events of critical interest there. Other areas are not immune. Tornados, floods, heatwaves, earthquakes and a variety of human-created disasters or errors can cause critical infrastructure and building services to go down at any time. The potential for cyber-terrorism that interrupts control systems for critical infrastructure, along with resource shortages and fluctuations, exacerbates this problem. How will buildings of the future fare when this happens?

Passive survivability is the ability of a building to continue to offer basic function and habitability when its supporting infrastructure, such as power, heating fuel or water supply, fails. There is growing interest in designing buildings around this idea and incorporating passive survivability requirements into the building code. Many contemporary buildings fail miserably when deprived of power. Modern high-rise construction relies on electrical power for everything from ventilation to elevators to pumping water to upper floors. Users may not even be able to open a window for ventilation because of the building design.

Passive survivability is a feature of natural systems that can be modelled in human design. For instance, termite mounds in Australia and Africa are able to control temperature, humidity and ventilation better than most mechanically conditioned buildings designed by humans. Historical vernacular architecture developed before complex analysis methods evolved can also offer key lessons. For instance, the high-mass adobe buildings common in the southwestern United States, Middle East and northern Africa have the ability to absorb and release heat slowly to buffer severe outside temperature swings that occur during a typical diurnal cycle.

The principles of passive survivability align well with sustainable design and construction. Designing buildings to function in the absence of external supplies of energy, water and materials makes them less vulnerable to external threats. It also means that they rely less on the environment for these resources. Passive solar design, natural ventilation, rainwater harvesting, durable materials and greywater reuse are all tactics that both increase passive survivability and increase building sustainability at the same time. Specific areas of consideration in passive survivability design include:

- natural lighting and incorporation of windows in areas that might be critical during power outages, such as stairwells.
- building orientation for minimizing cooling and heating requirements.
- liveable thermal conditions, the design of which depends greatly on climate requirements and local resources; use of local renewable resources for fuel.
- local renewable energy generation using micro-hydro, photovoltaics, wind and/or others.
- sanitation, especially in the absence of conventional water supply; waterless urinals and composting toilets are examples of technologies that can provide this function.
- water collection and reuse, including rainwater harvesting, greywater recycling systems and bio-based treatment systems.
- food production as part of the building structure or landscape and site plan.

Although these features may not be incorporated into every structure, they should be seriously considered as part of a community design strategy and incorporated into facilities that may be used as emergency shelters during crisis situations, such as schools, hospitals, emergency-service buildings and government buildings. Residential construction can also benefit from these strategies coupled with sustainable landscape designs, thus enabling people to remain in place during crisis situations instead of being displaced to centralized areas where resources may be scarce and conditions are difficult.

From centralized authority to distributed autonomy

A second trend in achieving resilient systems is moving away from centralized services and controls to distributed autonomy. This trend is happening at multiple scales from individual materials all the way to infrastructure systems on a municipal or even regional scale. At the material level, new types of high-tech smart materials that are self-managing, self-deploying and/or self-healing are enjoying new applications in the construction industry. These materials are able to automatically activate or initiate repair to themselves as needed, without human diagnostics or interventions. Self-healing materials with applications in construction include polymers and elastomers, metals, ceramics and cementitious materials. Self-healing concrete in particular shows considerable promise for reducing vulnerability to failure resulting from corrosion of reinforcement due to its ability to maintain continuous cover and prevent intrusion of environmental substances that accelerate decay.

Self-constructing structures are another concept under development, of particular interest for exploration and discovery in extreme environments such as space missions or underwater applications. New systems for smart cladding on buildings use the differential responses to environmental conditions of different types of materials to affect changes in shape, form or other physical properties. For instance, bi-layer materials with different coefficients of thermal expansion can be made to curl or change shape as the layers expand or contract at different rates due to temperature changes. These types of materials can be employed as part of building envelopes to control heat gain in buildings through shading, opening and closing ventilation pathways and other mechanisms.

The *Internet of Things* is an emerging concept describing the collection of networked devices throughout the human environment that exchange information as part of their regular function. From computers and vehicles to buildings and infrastructure systems, these products contain electronics, software, sensors, actuators and other components that are networked together and enable the products to respond to information received from the network as well as provide information to the network about their own performance. In some cases such as self-driving vehicles, products are able to use adaptive automation to operate without human intervention to achieve specified goals. The simplest example of adaptive automation is automated robotic vacuums or lawnmowers, which use simple on-board adaptive controls to modify their path in response to obstacles within a constrained area to cover all parts of that area over a period of time. Similar devices are possible at a larger scale to undertake construction or maintenance tasks without human intervention.

At the infrastructure scale, distributed autonomy is becoming a key part of effective management of the smart grid for both energy and water. Smart grid technologies are receiving significant investment as a means to use the power of information to optimize the provision of electricity and water. Combined with integrated sensing and information systems and smart control algorithms, smart grids allow industry and consumers alike to do more with less, an essential goal for a more sustainable world. In both cases, smart grids balance discrepancies in supply vs. demand by adjusting resource flows for nonessential uses. Likewise, smart grids can be used with new algorithms to optimize cost by shifting demand for time-flexible uses to non-peak periods.

The smart grid must be able to balance many distributed sources and sinks within the network in a sustainable and resilient fashion. A smart energy grid, for instance, can take advantage of such sources as small building-integrated PV arrays to displace centralized power generation when conditions are favourable. Water technologies are also likely to be closely integrated in a distributed network with vertical construction to provide local water capture and reuse for purposes such as urban or vertical agriculture or beneficial landscaping. In terms of the physical infrastructure that comprises today's smart grids, significant advances are also being made in transmission and distribution technology such as self-healing coatings on transmission lines and pipes, and automated condition assessment and management systems.

Underlying all of these functions is the information technology provided by telecommunication systems. Information technology (IT) is at the heart of many of the smart infrastructure systems of tomorrow. As the backbone of today's information-driven society, the telecommunications infrastructure that allows information to flow is dependent upon reliable sources of energy even as the systems that generate that energy depend upon it. While wireless telecommunication is becoming the standard that allows us to be constantly connected to information no matter where we are, it ultimately depends on hard-wired systems that transfer data at high speeds from point to point. Investment in new infrastructure to support this skeleton is growing, and its vulnerability from an energy standpoint is well understood. New approaches to greening data centres and operating them with lower electrical demand are now being pioneered and are already becoming commonplace. The movement of data from local storage to the cloud as part of the revolution in cloud computing will necessitate ever more attention to the core infrastructure of server farms, data centres and backbone transmission lines as data storage becomes dematerialized, a commodity service that can be outsourced instead of something that exists locally on power-consuming hard drives.

Ultimately, many people envision a world that has access to high-speed data communications no matter where one is on the planet, although a large part of the planet remains disconnected even today, especially in developing countries and rural areas where only low bandwidth is available. Integrated telecommunications are at the core of the built environment of the future. Information will allow humans to continue to reduce our carbon footprint and further dematerialize and reduce demands on other types of infrastructure. For example, better bandwidth will facilitate telecommunication, thereby reducing demands for transportation infrastructure. Online billing, banking and tax returns already save megatons of carbon through reduction in raw materials used and the physical transport of those materials throughout their life-cycle (Scheck 2008). Without a doubt, telecommunications will be an integral part of the sustainable world of the future.

From isolation and specialization to integration

In both product and process, the trend is toward increasing integration of function rather than isolation and specialization as a way of increasing resilience. Notions of function 'stacking' have long been a part of permaculture design principles (Mollison 1990), where each function within a designed ecosystem has redundant mechanisms for achieving the function, and each element itself provides multiple functions. Similarly, contemporary approaches to integrative design involve multiple stakeholders in multiple ways at multiple phases of the project lifecycle to ensure that all perspectives are represented in generating the broadest variety of potential ideas.

Likewise, building elements and components are being called upon to provide multiple functions such as insulation, structure, cladding, finishes, etc. as part of a single prefabricated building system, thus reducing construction time and cost that can offset additional costs of manufacturing the material itself. For example, structural insulated panels (SIPs) and insulating concrete forms (ICFs) can be prefabricated by manufacturers off-site using commonly available materials such as polystyrene and oriented strandboard or cold-rolled steel, and shipped to the job site where they are quickly installed to shorten the construction schedule (Figure 10.21). Both of these products have significantly improved thermal performance compared to site-built walls due to the air or vapour gaps in the wall and eliminating thermal bridging.

New combinations of conventional materials such as laminated photovoltaics applied to standing seam metal roof panels over solar water heating piping represent synergistic combinations of commercially available off-the-shelf technologies that can be used to achieve new levels of performance. In this 'solar sandwich' example (Figure 10.22), the heating array beneath the roof acts as a heat sink for the photovoltaic panels, reducing their operating temperature and subsequently increasing efficiency of energy production while simultaneously providing the useful benefit of heating the heat exchange fluid in the piping beneath the roof panels.

At the broadest level, designers are beginning to consider the coupled relationships between our food systems, water systems and energy systems at the network scale. By exploring ways in which urban development and infrastructure can exploit complementarities, designers can reduce inefficiencies in our current approaches that result from separating these functions, often by significant geographical distance that requires considerable investments of energy and resources to overcome. The Food-Energy-Water (FEW) Nexus is a term used to describe the interdependencies between systems developed by humans to provide for our most basic needs: food, water and energy (see *Food-Energy-Water Nexus*).

Figure 10.21
Structural insulated panels improve both construction delivery time and thermal performance

Figure 10.22
A 'solar sandwich' roof system uses waste heat from dark-coloured photovoltaic panels to heat domestic hot water while increasing the efficiency of power production

Source: Englert Metal Roofs, Inc.

Food-energy-water Nexus

With increases in drought due to changing weather patterns, the overharvesting of water in some parts of the world has fostered a growing appreciation of the interdependent relationship between water and energy. Many solutions to water shortage problems have been proposed, ranging from desalination plants to convert abundant seawater into potable water, construction of new pipelines to redistribute water from areas where it is abundant, or even towing freshwater icebergs from polar regions through the ocean to areas where water is needed. All of these solutions require extensive amounts of energy to sustain, in addition to the extraordinary costs their construction would require.

At the same time, many of our current methods for producing energy are extremely water-intensive. It has been estimated that 1 kW of energy on average is required to produce every gallon of potable water produced by U.S. water treatment plants, including both energy used for treatment and energy required to pump the water to its point of use. This means that every gallon of potable water used for non-potable purposes is also consuming energy unnecessarily, if the same needs could be met locally with non-potable sources of water such as rainwater or greywater.

Figure 10.23 Aquaculture combined with intensive plant arrangements allows production of multiple food sources in a very small footprint

continued

Since both water and energy are necessary for the production of food, all three of these critical resources for human survival are tightly coupled. In many cases, water and energy spent for agricultural food production is extremely inefficient using current methods, and modern agriculture has come to rely on ready supplies of fossil fuels for the production of fertilizer, fuels and energy needed to pump water to where it is needed. Even with our modern agricultural systems, the distribution of food to points where it is used is extremely energy-intensive, with many areas even in developed nations suffering from low access to healthy food. These so-called 'food deserts' further reinforce the structural inequality of the poor, and reduce their overall resilience to both environmental stress due to climate change and economic stressors such as the loss of a job.

Figure 10.24 Goats, chickens and other domestic animals can combine with fish and plants to provide protein as well as other essential services such as landscaping and sustainable brush-clearing

Building on the awareness of the fragility of our food supply system and the lack of capacity of current water and energy systems, new approaches (e.g. Figures 10.23 and 10.24) are being developed to localize and distribute these critical capacities as a way of increasing community resilience. Vertical farming, aquaculture, urban agriculture and other techniques are being explored on scales ranging from the individual household to production farming scale, often in neighbourhoods formerly used for industrial and manufacturing purposes.

Future-proofing our built environment

Ultimately, striving to improve the sustainability of the built environment is about 'future-proofing' our buildings and infrastructure systems. In light of current trends and future possibilities, we owe it to ourselves to be strategic and smart about how we invest our efforts. Just as electrical subcontractors think ahead to future information technology requirements and install chases and conduits for future network wiring, so too should we anticipate what lies ahead for all the systems in our buildings and infrastructure. Leading thinkers in the field have identified four core issues that should be considered in pursuing resilient and sustainable design (Pearson 2016; see box).

The design features and processes of a project should be selected with consideration to promote resilience to future threats, but there are other ways to future-proof projects as well. Function stacking and multi-function materials and systems, discussed earlier in this chapter, can help to ensure that innovative systems important for a project's sustainability are not removed as part of 'value engineering' or cost

Four core issues to tackle for resilient design

1   **Address the most likely hazards** in the context of each specific project – design to be context-specific. Don't invest in solutions to harden your project against threats that are not likely to ever occur in your context.

2   **Factor in climate change** – emerging tools such as the Projected CREAT Climate Scenarios tool by ArcGIS (available at http://arcgis.com) will apply available prediction data to your specific project location to estimate changes in future average temperature, precipitation and storm intensity. Other tools including the Climate Resilience Toolkit can help to identify relevant strategies.

3   **Look for ways to foster social cohesion** through community engagement during planning, design and operation of buildings, especially through the public use of building resources.

4   **Find ways to problem-solve across scales**, including collaborating and cooperating with neighbouring sites and owners to develop aligned and coordinated investments in key supporting infrastructure.

(Pearson 2016)

reduction exercises. It is critical to consider not just the external forces and threats that may affect a building or infrastructure system in the long term, but also the internal organizational forces that may affect how it is used, what is expected of it, and what resources will be available for its ongoing operation, maintenance and retrofit. Planning and management at the enterprise scale or even the community scale can help to provide the perspective needed to foresee what actions may become important in the long term.

Finally, being 'future-fit' requires thinking about the planetary context in which we all exist. As stated by the Natural Step as part of its Future-Fit Business Benchmarking program, 'A future-fit business is one that does nothing to undermine the potential for humans and all life to flourish on Earth forever' (www.thenaturalstep.org/future-fit/). If we are to continue as a species to increase our prosperity and well-being, we must consider the consequences of our actions at every step in the lifecycle of our projects and make thoughtful and careful choices. Working together toward this end, we can direct our efforts toward a sustainable built environment to meet not only our needs and aspirations, but also contribute to the ability of future generations to meet their own needs.

Discussion questions and exercises

10.1   Choose one of the emerging green materials or systems discussed in this chapter to investigate further. What product or products does the emerging product supersede? How does it compare in terms of first cost with its predecessor? Be sure to include all components necessary for functional equivalence in your analysis. What is the expected payback period for investing in the new technology, given expected savings over its lifecycle? Is the payback period shorter than the service life of the product or system?

10.2   Choose one of the regenerative business models described in the chapter to investigate further. What are some examples of this model that you have experienced in your life? What conventional business models compete with it? How does it compare?

10.3   Identify one aspect of the building in which you are located that could lead to its eventual obsolescence based on the types of change identified in the chapter. How might the building's design be adapted to accommodate that change and still remain functional?

10.4   Identify a bioinclusive strategy or biointegrated system in a building on your campus or in/around your local community. Talk to the stakeholders involved in operating and maintaining the system. What challenges do they face? Has their perception of the value of the system changed over time? If so, how?

10.5   Use an Internet search to determine whether any of the sustainable infrastructure technologies mentioned in the chapter exist in your area. Schedule a visit to the site and learn more about the project, or identify a project with case study information online. How is the project more sustainable than the conventional alternative? What challenges exist, both technical and social, in implementing the project? What lessons can be learned for similar projects in the future?

10.6   Visit the online AskNature database at www.AskNature.org. Explore the biomimicry taxonomy of functions to identify database entries with applications to the built environment. Choose one or more functions and describe how they could be used as a model for technologies in buildings.

10.7   Consider the building in which you are located. What would it be like if all external infrastructure services were suddenly to disappear? What would be the biggest threat to passive survivability in the facility? What retrofit strategies or technologies could be employed to reduce this threat?

10.8   Which of the sustainability trends discussed in the chapter is likely to be the most significant to you and your professional or personal life? How could you adjust your current practice to be more resilient to adapt to this trend? What resources would you need? What actions would you need to take? What trend do you believe is most significant for your organization or nation? How can you take action as an individual to increase resilience at a larger scale?

References and resources

Ausubel, J.H. (1996). "Can technology spare the earth?" *American Scientist Magazine* 84(2), 166–178.

Benyus, J. (2002). *Biomimicry: Innovation inspired by nature.* Harper Perennial, New York.

Botsman, R., and Rogers, R. (2010). *What's mine is yours: The rise of collaborative consumption.* Harper Business, New York, NY.

Boyle, L., and Michell, K. (2016). "Urban facilities management: A systemic process for achieving urban sustainability," Working paper, Department of Construction Economics and Management, University of Cape Town, South Africa.

Cole, R.J. (2012a). "Transitioning from green to regenerative design." *Building Research and Information* 40(1), 39–53. doi:10.1080/09613218.2011.610608.

Cole, R.J. (2012b). "Regenerative design and development: Current practice." *Building Research & Information* 40(1), 1–6.doi:10.1080/09613218.2012.617516.

Conte, E., and Monno, V. (2016). "The regenerative approach to model an integrated urban-building evaluation method." *International Journal of Sustainable Built Environment* 5, 12–22. http://dx.doi.org/10.1016/j.ijsbe.2016.03.005.

DuBose, J.R. (1994). *Sustainability as an inherently contextual concept: Some lessons from agricultural development.* M.S. Thesis, School of Public Policy, Georgia Institute of Technology, Atlanta, GA.

Du Plessis, C. (2012). "Towards a regenerative paradigm for the built environment." *Building Research & Information* 40(1), 7–22. doi:10.1080/09613218.2012.628548.

Galamba, K.R., and Nielsen, S.B. (2016). "Towards sustainable public FM: Collective building of capabilities." *Facilities* 34(3/4), 177–195.

Mang, P., and Reed, W. (2012). "Regenerative development and design." In R.A. Meyers, ed. *Encyclopedia of Sustainability Science & Technology,* Chapter 303. Springer-Verlag, New York, NY.

Meneghetti, A., and Chinese, D. (2002). "Perspectives on facilities management for industrial districts." *Facilities* 20(10), 337–348.

Mollison, B. (1990). *Permaculture: a designer's manual.* Island Press, Covelo, CA.

Nguyen. (2008a). "Biomimicry in Construction." BTSC, Construction Industry Institute, Austin, Ted. https:// construction-institute.org/scriptcontent/btsc-pubs/CII-BTSC-112.doc (accessed 19 September 2010).

Pearson, C. (2016). "The four core issues to tackle for resilient design (And the programs that can help)." *Environmental Building News* 25(3), 7 March. www-buildinggreen-com.ezproxy.lib.vt.edu/feature/four-core-issues-tackle-resilient-design-and-programs-can-help (accessed 6 Feburary 2017).

Scheck, H.O. (2008). "Power consumption and energy efficiency of fixed and mobile telecom networks." ITU-T Conference, Kyoto, Japan, April.

Ward, K. (2013). "A Fistful of Lindens." *New York Magazine, 7* April. http://nymag.com/news/intelligencer/topic/alternative-currencies-2013-4/ (accessed 6 February 2017).

Glossary

Absorptive finish – a surface finish that will absorb dust, particles, fumes and sound.

Acidification – a process that converts air pollution into acid substances. This leads to acid rain, best known for the damage it causes to forests and lakes. Also refers to the outflow of acidic water from metal and coal mines.

Agricultural land – property suitable for raising crops or livestock. Includes land cultivated for crops, orchards or vineyards, as well as meadows and pastures used for grazing.

Airborne dust – particles of soil or other materials that are small enough to be suspended in the air and breathed by organisms.

Albedo – the extent an object reflects light from the sun. It is a ratio with values from 0 to 1. A value of 0 is dark. A value of 1 is light.

Alternative fuel – any material or substance that can be used as a fuel other than conventional fossil fuels. They typically produce less pollution than fossil fuels and include biodiesel and ethanol.

ANSI – the American National Standards Institute, an organization that sets standards for product performance and testing.

Analytics - qualitative and quantitative techniques for processing large data sets to extract an understanding of the population that created the data.

Aquifer – an underground layer of water-bearing rock or soil from which groundwater can usefully be extracted using a water well.

Aquifer depletion – a situation when the withdrawal of water from the aquifer is greater than the rate of natural recharge.

ASTM – the American Society for Testing and Materials, an organization that sets standards for material performance and testing.

Best practice – an action or technology to achieve a desired end that is agreed upon by industry to represent the most effective or efficient way to achieve that end.

Bio-based – a material made from substances derived from living matter. It typically refers to modern materials that have undergone more extensive processing. Linoleum, cork and bamboo are examples.

Biodegradable – organic material such as plant and animal matter and other substances originating from living organisms and capable of being broken down into innocuous products by the action of microorganisms.

Biodiversity – a measure of the variety among organisms present in different ecosystems.

Bio-fuel – a solid, liquid or gas fuel consisting of or derived from recently dead biological material. The most common source is plants.

Bioinclusivity – a design philosophy that incorporates provisions for non-human species as part of the design of human environments.

Biophilia – a love or affinity for nature and the natural environment.

Bioretention cells – engineered areas designed to capture and treat stormwater runoff and remove impurities before they percolate into the soil.

Bioswales – constructed watercourses covered in appropriate vegetation to slow and retain stormwater (see also *Vegetated swales*).

Blackwater – water or sewage that contains faecal matter or food waste.

BMP (best management practice) – a way to accomplish something with the least amount of effort to achieve the best results. This is based on repeatable procedures that have proven themselves over time for large numbers of people (see also Best Practice).

Brownfield – property that contains the presence or potential presence of a hazardous substance, pollutant or contaminant. Cleaning up and reinvesting in these properties reduces development pressures on undeveloped open land and improves and protects the environment.

Building envelope – the elements and systems of a building that separate the interior of the building from the outside, including the roof, walls, windows and doors.

Building-integrated photovoltaics (BIPVs) – photovoltaic systems built into other types of building materials.

Building massing – the process of determining the overall shape and volume of the building envelope with respect to its location on site.

Business case – the set of pros and cons, often expressed in economic terms, associated with a proposed action that defines how that action will contribute to a business's enterprise.

Byproducts – secondary products derived from a manufacturing process or a chemical reaction. It is not the primary product or service being produced.

Carbon cycle – the movement of carbon between the biosphere, atmosphere, oceans and geosphere of the Earth. This biogeochemical cycle has sinks, or stores, of carbon and processes by which the various sinks exchange carbon.

Carbon footprint – a measure of the impact that human activities have on the environment. It is determined by the amount of greenhouse gases produced and is measured in units of pounds, kilograms or tons of carbon dioxide.

Carbon offsets – a donation or other act that aims to remove a certain amount of carbon dioxide from the atmosphere to compensate for the same amount added to the atmosphere by another activity.

Carpool – an arrangement where several people travel together in one vehicle. The people take turns driving and share in the cost.

Ceramic frit – vitreous compounds, not soluble in water, obtained by melting and then rapidly cooling carefully controlled blends of raw materials.

Certified sustainable harvest – the extraction or collection of a natural product in a way that can be continued indefinitely without damage, observed and verified by a third party.

Certified wood – wood or wood products that have been verified as coming from sustainably harvested sources by a third-party agency.

Charrette – a meeting of project participants that is an intense period of design activity. They are a way of quickly generating a design solution and also integrate the abilities and interests of a diverse group of people.

Chlorofluorocarbons (CFCs) – a set of chemical compounds that deplete ozone. They are widely used as solvents, coolants and propellants in aerosols, and are the main cause of ozone depletion in the stratosphere.

Commissioning – a process of quality assurance undertaken during project delivery to ensure that all building systems function according to their design intent.

Compact fluorescent lamps (CFLs) – a fluorescent lamp that is designed to fit in a normal light fixture. They use less energy and last longer than incandescent lamps.

Composting – purposeful biodegradation of organic matter, such as yard and food waste, into a soil amendment to improve soil quality.

Conservation – using natural resources wisely and at a slower rate than normal.

Constructed wetland – an artificially made wetland designed for the purposes of treating human-generated wastewater with plants and microorganisms before returning it to the natural environment.

Corporate social responsibility (CSR) – the practice of managing or operating a company to have an overall positive impact on society.

CRI Green Label – a rating system developed by the Carpet and Rug Institute to evaluate the indoor air quality performance of carpet and flooring products.

Daylighting – the use of windows and reflective surfaces to bring natural light into a space and reduce or eliminate the need for artificial light.

Deconstruction – the careful disassembly of a built facility in a way that enables components to be reused on future projects.

Deforestation – the removal of trees without sufficient replanting.

Desertification – the creation of deserts through degradation of productive land in dry climates by human activities.

Design review – a meeting of stakeholders in which the design solution for a project is evaluated to assess whether it meets the owner's project requirements and identify opportunities for improvement. Typically undertaken at 30, 60 and 90 per cent of design completion.

Downcycling – recycling one material into a material of lesser quality. An example is the recycling of plastics, which are turned into lower-grade plastics.

Downstream impacts – effects occurring as a result of the use of a product, or as a result of its disposal at the end of its service life.

Earthbag construction – a construction method using sandbags or other fabric containers filled with soil and stacked as modular construction units. Layers of barbed wire are often placed between horizontal layers to stabilize layers and prevent displacement. This method was developed specifically for war-torn areas where barbed wire and sandbags are commonly left behind after combat.

Ecolabel – a mark or certifying logo that indicates an environmentally friendly attribute of a product.

Ecological footprint – a measure of human demand on the Earth's ecosystems and natural environment. It compares human consumption of natural resources with the Earth's capacity to regenerate them, and is measured in units of area such as hectares or acres.

Ecological fragmentation – the division of ecosystems or undeveloped areas into smaller pieces as a result of human development of roads and buildings.

Ecological rucksack – the total mass of materials moved or used to create a product or service.

Ecosystem – a combination of all plants, animals and microorganisms in an area that complement each other. These function together with all of the non-living physical factors of the environment.

Embodied energy – the total energy that was used in all the activities associated with producing a product and delivering it to its point of use.

Emissions – a substance discharged into the air or water as a pollutant.

End of lifecycle – the point in time when a product is no longer suitable to be used for its intended purpose.

Endangered species – a population of a species at risk of becoming extinct. A threatened species is any species that is vulnerable to extinction in the near future.

Energy efficiency – the percentage of energy expended to do work that results in a useful work output.

Energy performance credits – a series of credits in the Energy and Atmosphere category of the LEED rating system that reward projects for exceeding the energy code by a certain amount.

Energy Star – a US government programme to promote energy efficient consumer products. It is a joint program of the US Environmental Protection Agency and the US Department of Energy.

Enhanced commissioning – a process of quality assurance that begins during design and continues throughout and following construction to ensure that all building systems function according to their design intent.

Environmental stewardship – responsibility for environmental quality that is shared by those whose actions affect the environment.

Equipment idling – the operation of equipment while it is not in motion or performing work. Limiting idle times reduces air pollution and greenhouse gas emissions.

Erosion – the displacement of solids by wind, water, ice or gravity or by living organisms. These solids include rocks and soil particles.

Erosion and sediment control plan – a plan developed for a project that identifies risks for erosion and sedimentation during the project and specifies measures for preventing or mitigating them.

Externalities – consequences of an action that are experienced by a third party not participating in the action.

Fenestration – pertaining to windows.

Fitness for purpose – the ability of a technology or product to fully meet the intent for which it was designed.

Flush-out – using fresh air in the building HVAC system to remove contaminants from the building.

Fly ash – fine solid particles of ash that are carried into the air when a fuel is burned.

Footprint – an outline or indention of a foot or shoe on a surface. Also refers to the effect of human activity on the natural environment, as in the phrases **ecological footprint** and **building footprint**.

Fossil fuels – hydrocarbon deposits such as petroleum, coal or natural gas derived from living matter of a previous geologic time and burned to produce useful work.

Fracking – hydraulic fracturing of subsurface rock to release deposits of natural gas, which is then harvested for fuel.

FSC Certified – wood or wood products that have met the Forest Stewardship Council's tracking process for sustainable harvest.

Fugitive emissions – pollutants released to the environment other than from stacks or vents. They are often caused by equipment leaks, evaporative processes and wind disturbances.

Furred out – extended away from a solid wall through the use of a false stud wall or furring strips to provide insulation and/or attachment points for a finished surface.

Geothermal – heat that comes from within the Earth.

Global climate change – changes in weather patterns and temperatures on a planetary scale. This may lead to a rise in sea levels, melting of polar ice caps, increased droughts and other weather effects.

Greyfield (grayfield) – a previously developed, unused site with no real or perceived environmental contamination. Often used to refer to sites with extensive hardscape including parking lots where little natural ecology remains.

Green – having some environmental benefit compared to a conventional alternative.

Green Seal Certified – a certification of a product to indicate its environmental friendliness. Green Seal is a group that works with manufacturers, industry sectors, purchasing groups and governments at all levels to 'green' the production and purchasing chain. Founded in 1989, Green Seal provides science-based environmental certification standards.

Green space – the undeveloped areas around built facilities that are covered in landscaping or vegetation.

Greenfield site – a project site that has not been previously developed except for agricultural use.

GreenGuard – a rating and certification system used to evaluate the indoor air quality performance of building products and systems.

Greenhouse gas – a substance in the atmosphere that acts to insulate the Earth's surface and retain heat. Examples include carbon dioxide (CO_2) and methane (CH_4).

Greenwash – dissemination of misleading information by an organization to conceal poor environmental practices and present a positive public image.

Greywater (graywater) – a non-industrial waste-water generated from domestic processes. These include dish washing, laundry and bathing. Greywater comprises 50–80 per cent of residential waste water.

Halons – an ozone-depleting compound consisting of bromine, fluorine and carbon. Halons were commonly used as fire extinguishing agents in both built-in systems and handheld portable fire extinguishers.

Heat island – an area with a higher temperature than its immediate surroundings caused by greater absorption of solar energy. Often has a different microclimate than its surroundings.

Hybrid vehicle – a vehicle that uses two or more distinct power sources to propel the vehicle.

Hydrochlorofluorocarbons (HCFCs) – a group of human-made compounds containing hydrogen, chlorine, fluorine and carbon that are used for refrigeration, aerosol propellants, foam manufacture and air conditioning. They are broken down in the lowest part of the atmosphere and pose a smaller risk to the ozone layer than CFCs.

Hydrologic cycle – the circulation and conservation of Earth's water. The process has five phases: condensation, infiltration, runoff, evaporation and precipitation.

IAQ (indoor air quality) – the level of contamination of interior air that could affect health and comfort of building occupants.

IAQ management plan – a plan developed for a construction project that inventories possible threats to future indoor air quality arising from the construction process or materials used in a project, then identifies strategies to mitigate or avoid them.

Infiltration trenches – excavated areas filled with plants and vegetation that are designed to allow stormwater to percolate back into the soil.

Innovation credit – a credit under the LEED rating system awarded for actions that greatly exceed the credit requirements defined under LEED.

Insulating concrete form (ICF) – rigid forms that hold concrete in place during curing and remain in place afterwards. The forms serve as thermal insulation for concrete walls.

Integrated design – a collaborative design methodology. It emphasizes knowledge integration in the development of a complete design.

Integrative design – a process involving stakeholders from all phases of a project's lifecycle and all disciplines involved in its design to develop a design solution that optimally meets the owner's project requirements.

ISO – the International Standards Organization, a collection of national standards organizations that coordinates international standards.

Just-in-time delivery – the practice of arranging for materials to be delivered to the project site just before they are installed, so that they are not at risk from being damaged during storage.

Leadership in Energy & Environmental Design (LEED) – a family of rating systems developed by the US Green Building Council to rate and certify the environmental performance of built facilities.

LEED Online – the website used by members of a project team to manage the process of applying for LEED certification.

Lifecycle – the useful life of a system, product or building.

Lifecycle assessment – an analytic technique to evaluate the environmental impact of a system, product or building throughout its lifecycle. This includes the extraction or harvesting of raw materials through processing, manufacture, installation, use and ultimate disposal or recycling.

Lifecycle cost – the cost of a system or a component over its entire life span.

Lifecycle costing – the process of identifying and tabulating all the costs associated with a product or process over its whole lifecycle, including first costs, operating and maintenance costs, and end-of-lifecycle costs.

Light pollution – excessive or unwanted light from artificial sources.

Light shelf/shelves – horizontal panels located outside or inside windows to reflect light further into interior spaces and provide shading for perimeter areas.

Light tube – a cylinder penetrating the plane of a roof and extending into an occupied area to direct natural daylight to occupied building spaces.

Liquid waste – materials discharged from a process or entity in liquid form as a pollutant.

Lithium ion (LiIon) battery – a type of rechargeable battery in which a lithium ion moves between the anode and cathode; commonly used in consumer electronics.

Local materials – materials that come from within a certain number of miles from the project. Materials produced locally use less energy during transportation to the site.

Long-cycle renewable materials – materials that, while bio-based and renewable, take many years to regrow. Examples include many types of hardwood species.

Low-emission vehicle – vehicles that produce fewer emissions than the average vehicle on the road.

Low-emitting substance – a material that emits fewer volatile organic compounds than conventional materials of the same type.

Materials management plan – a plan developed for a construction project that inventories materials involved in the project and identifies strategies to protect them until they can be installed, thereby preventing waste.

MERV (minimum efficiency reporting value) – a measurement scale designed in 1987 by the American Society of Heating, Refrigerating and Air-Conditioning Engineers (ASHRAE) to rate the effectiveness of air filters. The scale is designed to represent the worst case performance of a filter when dealing with particles in the range of 0.3 to 10 microns.

Multi-function material – a material that can be used to perform more than one function in a facility.

Multifunctionality – serving more than one functional purpose within a design, such as enclosure

and thermal insulation, or waterproofing and power production.

Nano-material – a material with features smaller than a micron in at least one dimension.

Nickel cadmium (NiCad) – a popular type of rechargeable battery using nickel oxide hydroxide and metallic cadmium as electrodes.

Noise pollution – environmental sound that is annoying, distracting or potentially harmful.

Non-renewable – a material or energy source that cannot be replenished within a reasonable period of time.

Non-toxic – substances that are not poisonous.

NSF – the National Sanitation Foundation, an organization that evaluates and certifies plumbing components and other products.

Occupancy sensor – a sensing and control device that switches on or off depending on the presence of humans in an area. Occupancy sensors rely on changes in heat or motion to detect human presence.

Offgassing – the evaporation of volatile chemicals at normal atmospheric pressure. Building materials offgas by releasing chemicals into the air through evaporation.

Overhang – a horizontal panel or roof protrusion designed to shade and protect the area immediately beneath it.

Overpopulation – excessive population of an area to the point of overcrowding, depletion of natural resources or environmental deterioration.

Ozone depletion – a slow, steady decline in the total amount of ozone in the Earth's stratosphere.

Ozone hole – a large, seasonal decrease in stratospheric ozone over Earth's polar regions. The ozone hole does not generally go all the way through the layer.

Papercrete – a construction material using pulped paper fibre, often from waste paper, mixed with water and Portland cement, clay or other binder to cure into a rigid but comparatively lightweight structure.

Particulate – a small mass of solid or liquid matter that remains suspended in a gas or liquid.

Passive survivability – the ability of a system to function at a basic or enhanced level without inputs and outputs from and to centralized infrastructure systems such as power, water and wastewater.

Payback period – the amount of time it takes to recover costs in savings resulting from an investment.

Pervious (permeable) concrete – a mixture of coarse aggregate, Portland cement, water and little to no sand. It has a 15–25 per cent void structure and allows 3–8 gallons of water per minute to pass through each square foot.

Phase change materials – materials that convert between solid and liquid at a desirable or convenient temperature that can store and release large amounts of energy as they change state. Often used for passive cooling or thermal control.

Photovoltaics (PVs) – a technology that converts light directly into electricity.

Pollution prevention – the prevention or reduction of pollution at the source.

Pollution prevention plan – a plan developed for a construction project that inventories potential sources of pollution or waste during construction and identifies strategies to keep them from happening.

Post-consumer – products made out of material that has been used by the end consumer.

Post-consumer recycled content – waste that comes from a product or material after it has served the useful purpose for which it was made.

Post-industrial/pre-consumer – material diverted from the waste stream during the manufacturing process.

Potable – of suitable quality for safe human consumption.

Pre-consumer recycled content – waste that comes from the production process for a product or material.

Preservation – the act and advocacy of the protection of the natural environment.

Rainwater harvesting – the gathering and storing of rainwater. Systems can range from a simple barrel at the bottom of a downspout to multiple tanks with pumps and controls.

Rammed earth – an ancient building technique similar to adobe, using soil that is mostly clay and sand. The difference is that the material is compressed or tamped into place, usually with forms that create very flat vertical surfaces.

Rapidly renewable – a material that is replenished by natural processes at a rate comparable to its rate of consumption by humans or other species. The LEED system of building certification offers points for rapidly renewable materials that regenerate in 10 years or less, such as bamboo, cork, wool and straw.

Raw material – a material that has been harvested or extracted directly from nature, and is in an unprocessed or minimally processed state.

Recyclable – material that still has useful physical or chemical properties after serving its original purpose. It can be reused or remanufactured into additional products. Plastic, paper, glass, used oil and aluminium cans are examples of recyclable materials.

Recycled content – a property of a material meaning that it contains components that have been recycled. Made from materials that would otherwise have been discarded.

Recycled materials – materials that have been taken from the waste stream and reprocessed for further use.

Recycled plastic lumber (RPL) – a wood-like product made from recovered plastic or recovered plastic mixed with other materials. It can be used as a substitute for concrete, wood and metals.

Recycling – the reprocessing of old materials into new products. A goal is to prevent the waste of potentially useful materials and reduce the consumption of fresh raw materials.

Regenerative – having the ability to recover or regrow following disruption or consumption.

Regionality – the degree to which a material or product is produced locally or within the local region.

Renewable – a resource that may be naturally replenished.

Renewable energy – electricity generated from renewable sources such as photovoltaics or solar thermal, wind, geothermal or small hydroelectric facilities.

Resilience – the ability of a system to recover function following a stressor that interrupts normal function.

Reusable – a material that can be used again without reprocessing. This can be for its original purpose, or for a new purpose.

Salvage – a discarded or damaged material that is saved from destruction or waste and put to further use.

SCS – Scientific Certification Systems, a third-party organization that certifies environmental claims.

Sediment – small particles that settle to the bottom of a liquid, such as soil particles in a body of water.

Sedimentation – a process of depositing a solid material from a state of suspension in a fluid. The fluid is usually air or water.

Service life – the length of time during which a technology or product can be used for its intended purpose.

Site disturbance plan – a plan that indicates the types and limits of disturbance to be allowed during construction on a project site, and identifies strategies for ensuring that disturbance remains within those limits.

Smart material – materials that have one or more properties that can be significantly changed. These changes are driven by external stimuli.

Social networks – groups of people or organizations connected by relationships and interactions between and among each other. Also refers to the online infrastructure through which these interactions occur.

Soil compaction – the compression of soil by heavy weight or pressure, which causes damage to plants and makes it difficult for vegetation to grow.

Solar energy – energy from the sun in the form of heat and light.

Solid waste – products and materials discarded after use in homes, businesses, restaurants, schools, industrial plants and elsewhere.

Solvent-based – a material that consists of particles suspended or dissolved in a solvent. A solvent is any substance which will dissolve another. Solvent-based building materials typically use chemicals other than water as their solvent, including toluene and turpentine, with hazardous health effects.

Source control – deliberate actions taken to prevent waste at the point in a process where pollution or waste is generated.

Sprawl – unplanned development of open land.

Stormwater runoff – the unfiltered water that reaches streams, lakes and oceans after a rainstorm by means of flowing across impervious surfaces.

Straw bale construction – a building method that uses straw bales as structural elements, insulation or both. It has advantages over some conventional building systems because of its cost and easy availability.

Structural insulated panel (SIP) – a composite building material used for exterior building envelopes. It consists of a sandwich of two layers of structural board with an insulating layer of foam in between. The board is usually oriented strand board (OSB) and the foam either expanded polystyrene foam (EPS), extruded polystyrene foam (XPS) or polyurethane foam.

Sustainable – able to continue or persist indefinitely without depleting resource bases or damaging natural ecosystems.

Sustainably harvested – a method of harvesting a material from a natural ecosystem without damaging the ability of the ecosystem to continue to produce the material indefinitely.

Takeback – a requirement that waste from packaging or products themselves be recovered by the manufacturer or provider at the end of their lifecycle.

Temporary utilities – services such as electrical power, water and communications during construction that are known in advance to be of limited duration.

Thermal bridge – a condition created when a thermally conductive material bypasses an insulation system, allowing the rapid flow of heat from one side of a building wall to the other. Metal components, including metal studs, nails and window frames, are common culprits.

Thermal mass – a property of a material related to density that allows it to absorb heat from a heat source, and then release it slowly. Common materials used for thermal mass include adobe, mud, stones and volumes of water.

Triple Bottom Line (TBL) – accounting for and reporting on not only the financial profits and losses for a company but also the impacts of that company on the environment and society.

Upstream impacts – the effects of human activity occurring before the activity happens that can be attributed to that activity.

Urban heat island effect – the net rise in temperature in an area resulting from greater absorption of heat by the buildings and paved surfaces in the area.

Urbanization – the removal of the rural characteristics of an area. A redistribution of populations from rural to urban settlements.

Urea formaldehyde – a transparent thermosetting resin or plastic. It is made from urea and formaldehyde heated in the presence of a mild base. Urea formaldehyde has negative effects on human health when allowed to offgas or burn.

Vapour resistance – the ability of a material to resist the flow of water vapour.

Vegetated swales – constructed watercourses covered in appropriate vegetation to slow and retain stormwater (see also *Bioswales*)

Virgin material – a material that has not been previously used or consumed. It also has not been subjected to processing. See also *Raw material*.

Volatile organic compounds (VOC) – gases that are emitted over time from certain solids or liquids. Concentrations of many VOCs are up to 10 times higher indoors than outdoors. Examples of products emitting VOC are paints and lacquers, paint strippers, cleaning supplies, pesticides, building materials and furnishings.

Walk-off mat – mats in entry areas that capture dirt and other particles.

Waste diversion – preventing solid waste from going to a landfill and directing it instead to recycling, salvage or other beneficial use.

Waste ratio – the proportion of extra material planned for during estimating or ordered during procurement that is expected to go to waste through process inefficiency.

Waste separation – separating waste into recyclable and non-recyclable materials.

Water efficiency – the planned management of potable water to prevent waste, overuse and exploitation of the resource. Includes using less water to achieve the same benefits.

Water resistant – a material that hinders the penetration of water.

Water-based – materials that use water as a solvent or vehicle of application.

Waterproof – a material that is impervious to or unaffected by water.

WaterSense – a rating and certification system developed by the US Environmental Protection Agency to evaluate the water efficiency of building products.

Wetlands – lands where saturation with water is the dominant factor. This determines the way soil develops and the types of plant and animal communities living in the soil and on its surface.

Wildlife habitat – an environment where an organism or community of organisms typically live or are found.

Index